**Practical Statistics for Geographers
and Earth Scientists**

Practical Statistics for Geographers and Earth Scientists

Second Edition

Nigel Walford

Registered Office(s)
John Wiley & Sons, Inc., 111 River Street, Hoboken, NJ 07030, USA
John Wiley & Sons Ltd, New Era House, 8 Oldlands Way, Bognor Regis, West Sussex, PO22 8NQ, UK

For details of our global editorial offices, customer services, and more information about Wiley products visit us at www.wiley.com.

Library of Congress Cataloging-in-Publication Data Applied for:
Paperback: 9781119526971

Cover Design: Wiley
Cover Image: © vectorfusionart/Adobe Stock

Set in 9.5/12.5pt STIXTwoText by Straive, Pondicherry, India

To Ann

Contents

Plate section: Statistical Analysis Planner and Checklist falls between pages 159 and 161

Preface to the Second Edition

Some 12 years on from publication of the first edition of *Practical Statistics for Geographers and Earth Scientists* provides the opportunity to reflect on what has changed and what has remained the same. The majority of undergraduate and postgraduate students on degree programmes in Geography, Earth and Environmental Sciences are still expected to carry out a piece of independent research in a similar way to architecture students design a new structure or engineering students fashion the prototype of a new product. The underlying aim of such projects persists, namely to enable students to produce new knowledge and understanding, but is now accompanied by an increased emphasis on how successfully completing a project develops practical skills and experience that are pertinent in the 21st century workplace.

More than 60 years ago the quantitative revolution in Geography and cognate disciplines established the basis of quantitative research that was subsequently re-enforced by the arrival of Geographical Information Systems and the 'spatialising' of statistical analysis, sometimes by researchers in other disciplines. Reaction against a perceived dominance by the quantitative paradigm contributed to the 'cultural turn' in human geography and the ascent of qualitative approaches. At the risk of inviting challenge and dispute, it could be argued that protagonists on both sides of the quantitative versus qualitative debate have now seen an opportunity for pragmatic compromise in the form of mixed methods research. Although this text is firmly grounded in statistical analysis of both spatial and non-spatial varieties, it also emphasises that a combination of approaches to addressing research questions will often produce a more holistic interpretation of the data and understanding of the topic.

Writing the second edition has provided the opportunity to update and revise the content of the first, although the fundamentals of statistical analysis have remained extant for decades, if not centuries in some cases. Three specific changes are worth highlighting. First, non-spatial and spatial statistical techniques are now separated into different sections of the book with their own chapters. This still allows readers to examine similarities and differences between the two groups of techniques, but also avoids confusion. Second, an entirely new concluding chapter focuses on four practical ways of helping students to carry out their independent research project and write their dissertation or report. The Chapter 12 sets out to:

- review the results obtained from previous chapters and identify unanswered questions arising from the analyses;
- examine how researchers have presented and discussed their findings in published journal articles;
- outline how to use information technology and software are used for quantitative analysis;
- describe three mini projects showing how to develop on previously published research.

The third change is that selected reading and references have now been added to the end of each chapter to make following up on points in other sources more direct and straightforward. The journal articles examined in Chapter 12 are all available on open access enabling readers to relate the discussion in the text to the articles themselves.

This book has been written in recognition that three aims or learning outcomes are common to many undergraduate and postgraduate degrees in Geography, Earth and Environmental Sciences:

1) development of a level of understanding and competence in statistical analysis.
2) completion of a carefully defined, independent research investigation involving qualitative and/or quantitative analysis.
3) interpretation of research findings and statistical analyses presented in journal articles and other published material.

Many features of the first edition have been retained to promote these aims including the use of self-assessment or reflection questions scattered through most of the chapters. Some of these could form the basis for discussion in class, others are intended for independent study. The *Statistical Analysis Planner and Checklist* between book sections II, *Exploring Geographical Data*, and III, *Testing Times*, has been kept to provide guidance on selecting techniques for different styles of project. The format of the chapters includes boxed sections where the spatial or non-spatial techniques are explained and applied to one of a series of datasets. These datasets are representative of teaching undergraduate and postgraduate students for over 35 years in practical computer laboratory workshop sessions or relate directly to group projects carried out during field work. The chapters in the book progress from consideration of issues related to formulating research questions, collecting data and summarising information, applying statistical techniques to test hypotheses, analysing associations and relationships between variables and investigating the inherently geospatial aspects of geographical data. Overall, the focus is on the practicalities of choosing statistical techniques for different styles of quantitative research, including those that could be incorporated into a mixed methods approach, but even the comprehensive introduction given here cannot hope to include everything.

April 2024

Nigel Walford
Wivenhoe

Acknowledgements

My initial point of contact at Wiley, Andrew Harrison, moved on to pastures new during the course of writing this book. I would like to thank him and his successors at Wiley for their patience and gentle reminders over the last few years and to others in the production and editorial offices for steering the manuscript through to publication.

Several of the original figures and diagrams, which have been included in this second edition, were produced by Claire Ivison formerly the cartographer in the Department of Geography, Geology and the Environment at Kingston University. I offer my sincere thanks to her.

I also acknowledge the support and assistance from colleagues at Kingston University and to the students whose questions prompted me to attempt this book as they attempted to understand the complexities of statistical analysis.

Finally, I extend a very sincere and heartfelt thank you to Ann Hockey, who has survived me writing a textbook four times over more than 30 years.

About the Companion Website

This book is accompanied by a companion website.

www.wiley.com/go/PracticalStatistics2e

This website contains:

- A Word file giving details of the variables in the dataset.
- An Excel file containing the columns of data.

Section I

First Principles

1

What's in a Number?

> *Chapter 1 provides a brief review of the development of quantitative analysis in Geography, Earth and Environmental Science and related disciplines. It also discusses the relative merits of using numerical data and how numbers can be used to represent qualitative characteristics. A brief introduction to mathematical notation and calculation is provided to a level that will help readers to understand subsequent chapters. Overall, this introductory chapter is intended to define terms and to provide a structure for the remainder of the book.*

Learning Outcomes

This chapter will enable readers to:

- Outline the difference between quantitative and qualitative approaches and their relationship to statistical techniques;
- Describe the characteristics of numerical data and scales of measurement;
- Recognise forms of mathematical notation and calculation that underlie analytical procedures covered in subsequent chapters;
- Plan their reading of this text in relation to undertaking an independent research investigation in Geography and related disciplines.

1.1 Introduction to Quantitative Analysis

Quantitative analysis is one of the two main approaches to researching and understanding the world around us. In simple terms, it is the processing and interpretation of data about things, sometimes called phenomena, which are held in a numerical form. In other words, from the perspective of Geography and other Earth Sciences, it is about investigating the differences and similarities between people and places that can be expressed as numerical quantities rather than words. In contrast, qualitative analysis recognises the uniqueness of all phenomena and the important contribution towards understanding that is provided by unusual, idiosyncratic cases as much as by those conforming to some numerical pattern. Using the two approaches together has become popular in recent years to develop a more robust understanding of how processes work that lead to variations in the distribution of phenomena over the Earth's surface than by employing either methodology on its own.

Practical Statistics for Geographers and Earth Scientists, Second Edition. Nigel Walford.
© 2025 John Wiley & Sons Ltd. Published 2025 by John Wiley & Sons Ltd.
Companion website: www.wiley.com/go/PracticalStatistics2e

If you are reading this book as a student on a university or college course, there will be differences and similarities between you and the other students taking the same course in terms of such things as your age, height, school level qualifications, home town, genetic make-up, parental annual income and so on. You will also be different because each human being, and for that matter each place on the Earth, is unique. There is no one else exactly like you, even if you have an identical twin, nor is there any place precisely the same as where you are reading this book. You are different from other people because your upbringing, cultural background and physical characteristics have moulded your own attitudes, values and feelings and how you respond to new situations and events. In some ways, it is the old argument of nature versus nurture, but we are unique combinations of both sets of factors. You may be reading this book on a laptop in your room in a university hall of residence, and there are many such places in the various countries of the world and those in the same institution often seem identical, but the one where you are now is unique. Just as the uniqueness of individuals does not prevent analysis of people as members of various groups, so the individuality of places does not inhibit investigation of their distinctive and shared characteristics.

Quantitative analysis concentrates on those influences that are important in producing differences and similarities between individual phenomena and to disregard those producing unusual outcomes. Confusion between quantitative and qualitative analysis may arise because sometimes numbers are used to represent the qualitative characteristics of people and places. For example, areas in a city may be assigned to a series of qualitative categories, such as downtown, suburbs, shopping mall, commercial centre and housing estate, and these may be given numerical codes, or a person may agree or disagree with the UK's exit from the European Union with these opinions assigned numerical values such as 1 or 2. Just as qualitative characteristics can be turned into numerical codes, numerical measurements can be converted into descriptive labels. This could be simple, such as grouping household income values into low, medium and high ranges. In contrast, a selection of different socio-economic and demographic numerical counts for various locations may have combined them in an 'analytical melting pot' to produce geodemographic descriptions or labels of what neighbourhoods are like, such as *Old people, detached houses; Larger families, prosperous suburbs; Older rented terraces;* and *Council flats, single elderly.*

The major focus of this book is on quantitative analysis in Geography and other Earth Sciences, although this emphasis does not imply that it is in some definitive sense 'better', or even more scientific, than qualitative analysis. Nor are they mutually exclusive since researchers from many disciplines now appreciate the advantages of combining both forms of analysis in a mixed methods approach. This book concentrates on quantitative analysis because many students, and perhaps researchers, find dealing with numbers and statistics difficult and are a barrier to understanding the 'real' Geography or Earth Science topics that interest them. Why should we bother with numbers and statistics, when what really interests us is why migrants are risking their lives crossing the English Channel to seek asylum in the UK, why we are experiencing a period of global temperature increase, or why people live in areas vulnerable to natural hazards?

We can justify working with numbers and statistics to answer such research questions in a few diverse ways. As with most research, when we try to explain things such as migration, global warming and exposure to natural hazards, the answers often seem like common sense and even rather obvious. Maybe this is a sign of 'good' research because it suggests the process of explaining such things is about building on what has become common knowledge and understanding. If this is true, then ongoing academic study both relies upon and questions the work of previous generations of researchers and puts across complex issues in ways that can be understood. Answers to research questions are commonly stated with a high degree of certainty, whereas they are often underlain by an analysis of numerical information that is anything but certain. The results of the research are likely to be correct, but they may be false. Using statistical techniques gives us a way of expressing this uncertainty and of hedging our bets against the possibility that our set of results has only arisen by chance. At some time in the future, another researcher might come along and contradict our findings.

But what do we really mean by the phrase 'the results of the research'. For the 'consumers' of research, whether the public at large or professional groups, the results or outcomes of research are often some pieces of crucial information. The role of such information is to either confirm facts that are already known or believed or to fulfil the unquenchable need for new information. For the academic, these detailed factual results may be of less direct interest than the implications of the research findings, about some overarching theory. The student undertaking a research investigation as part of their course sits somewhere, perhaps uncomfortably, between these two positions. Realistically many students recognise that their research endeavours are unlikely to contribute significantly to theoretical advance, policy decisions or commercial product development, although obviously there are exceptions. Yet they also recognise that their tutors are unlikely to be impressed simply by the presentation of new information. Such research investigations (often called independent projects) are typically included in undergraduate degree programmes to provide students with training that prepares them for a career where such skills as collecting and assimilating information will prove useful, whether this be in academia or more typically in other professional fields. Students face a dilemma to which there is no simple answer. They need to demonstrate that they have carried out their research in a rigorous scientific manner using appropriate quantitative and qualitative techniques, but they do not want to overburden the assessors with an excess of detail that obscures the implications of their results.

In the 1950s and 1960s, several academic disciplines 'discovered' quantitative analysis and few geography students of the last 50 years will not have heard of the 'quantitative revolution'. There was, and still is, a belief that the principles of rigour, replication and respectability enshrined in scientific endeavour sets it apart from, and superior to, other forms of more discursive academic enquiry. The adoption of the quantitative approach was seen implicitly, and in some cases explicitly, as providing the passport to recognition as a scientific discipline. Geography and other Earth Sciences were not alone, and perhaps were more sluggish than some disciplines, in seeking to establish their scientific credentials. The classical approach to geographical enquiry followed from the colonial and exploratory legacies of the 18th and 19th centuries that permeated regional geography into the early 20th century and concentrated on the gathering of information about places. Using this information to classify and categorise places seemed to correspond with the inductive scientific method that recognised pattern and regularity in the occurrence of phenomena with any preconception or theory about what would be found. However, the difficulty of establishing laws about intrinsically unique places and regions led other geographers to search for ways of applying the deductive scientific method. The deductive method involves devising hypotheses with reference to existing conditions and testing them using empirical evidence obtained through the measurement and observation of phenomena.

Geography and to a lesser extent the other Earth Sciences have emerged from a period of self-reflection on the scientific nature of their endeavour with an acceptance that various philosophies can coexist and further their collective enterprise. Thus, many university departments include physical geographers and Earth scientists, adhering to positivist scientific principles, working alongside human geographers following a range of traditions including humanism, Marxism and structuralism as well as more positivist social science. A further important development over the past 40 years has been the growing influence of information and communications technologies on the conduct of academic and other types of research. The most obvious illustration of this being the use of computers for collecting, storing, processing and visualising data. The development of Geographical Information Systems (GIS) and remote sensing has especially prompted the emergence of new directions in geographical enquiry and fostered an expanded range of quantitative techniques for the analysis of spatial data. Students in academic departments need to be equipped with the skills not only to undertake quantitative and qualitative research investigations but also to handle geographical and spatial data in a digital environment.

Johnston (1979) commented that statistical techniques provide a way of testing hypotheses and the validity of empirical measurements and observations. However, the term statistics has several meanings. In general usage, statistics typically refers to information obtained from censuses, surveys and administrative processes, such as recording waiting times in hospital accident and emergency departments, that are published in books, newspapers, on the Internet or other media. The associated term 'official statistics', usually reserved for information collected, analysed and published by national, regional or local government, are commonly regarded as having more authority and being more reliable than those statistics disseminated by commercial or other types of organisation that wish to put across a certain message. This belief may be founded upon a presumption of impartiality and rigour, although recent events may have increased people's scepticism about the neutrality of method or intent in all 'official statistics'. Statistics also refers to a branch of mathematics that may be used in quantitative investigations to substantiate or refute the results of scientific research. Used in this way the term statistics has a double meaning: first as a series of **techniques** ranging from simple summarising measures to complex models involving many variables and second as the numerical **quantities** produced by these techniques. All these meanings of the term statistics are relevant to this text, since published statistics may well contribute to research investigations, and statistical techniques and the measures associated with them are an essential part of quantitative analysis. Such techniques are applied to numerical data and serve two general purposes:

- to confirm or otherwise the significance of research results towards the accumulation of knowledge with respect to a particular area of study;
- to establish whether empirical connections between distinctive characteristics for a given set of phenomena are likely to be genuine or spurious.

Different areas of scientific study and research have over the years carved out their own particular niches. For example, in simplistic terms the Life Sciences are concerned with living organisms, the Chemical Sciences with organic and inorganic matter, Political Science with national and international government, Sociology with social groups and Psychology with individuals' mental condition. Subdivision has often occurred when these broad categories became too general, which led to the emergence of fields in the Life Sciences such as cell biology, biochemistry, and physiology. Geography and the other Earth Sciences do not seem to fit easily into this straightforward partitioning of scientific subject matter, since their broad perspective leads to an interest in all the things covered by other academic disciplines, even to the extent of Earth Scientists transferring their allegiance to examine **terrestrial** processes on other planetary and celestial bodies. No doubt if, or when, ambient intelligent life is found on other planets, geographers will be there investigating the interactions and associations between members of these non-human societies and their environment. It is commonly argued that the unifying focus of Geography is its concern for the spatial and earthly context in which those phenomena of interest to other disciplines make their home. Thus, for example human geographers are concerned with the same social groups as the sociologist but emphasise their spatial juxtaposition and interaction rather than the social ties that bind them. Similarly, geochemists focus on the chemical properties of minerals not only for their individual characteristics but also for how assemblages combine to form different rocks in distinctive locations. Other disciplines may wonder what Geography and Earth Science add to their own academic endeavour, but geographers and Earth scientists are equally certain that if their area of study did not exist, it would soon need to be invented.

The problem that all this raises for geographers and Earth scientists is how to define and specify the units of observation, the things that are of interest to them, and them alone. One solution that emerged during the era of quantification was that geography is pre-eminently the science of spatial analysis and its focus is to discover the laws that govern the processes giving rise to spatial patterns of diverse types of human and physical phenomena (Hartshorne, 1958). A classic example of this approach in human geography was the search for regions or countries where the collections of settlements conformed to the spatial hierarchy theorised by central place theory (e.g. Getis and Getis, 1966; Smith, 1986). Physical geographers and Earth scientists also became interested in spatial patterns. They had more success in associating the spatial occurrence of environmental phenomena with underlying explanatory processes, such as the lines of earthquakes and volcanoes associated with the movement of tectonic plates around the Earth. According to this spatial analytic approach, once the geographer and Earth scientist have identified some spatially distributed phenomena, such as settlements, hospitals, earthquakes or volcanoes, then their investigation can proceed by quantifying variables such as adjacency, connectivity and distance and from these describe the spatial pattern.

This foray into spatial theorising in the 1950s and 1960s became progressively undermined, when it became apparent that exception rather than conformity to proposed spatial laws was the norm. Other approaches, or possibly paradigms, emerged, particularly in human geography, that sought to escape from the straight jacket of positivist science. Advocates of Marxist, behavioural, politico-economic, structural and cultural geography have held sway at various times over the intervening 50 years. However, it is probably fair to say that none of these have entirely rejected using numerical quantities as a way of expressing geographical difference. Certainly, in physical geography and Earth Science where many would argue that positivism inevitably still forms the underlying methodology, quantitative analysis has never receded into the background. Despite the vagaries of all these different approaches, most geographers and Earth scientists still hold on to the notion that what interests them and what they feel other people should be reminded of is that the Earth, its physical phenomena, its environment and its inhabitants display differences and similarities and are unevenly distributed over space.

From the practical perspective of this text, we need to agree what constitutes legitimate data for the study of Geography and Earth Science. Let us simply state that a geographical or Earth sciences dataset is a collection of data items, facts if you prefer, that relate to a group of spatially distributed phenomena (things). Such a dataset should relate to at least one discrete, defined and delimited portion of the Earth, its surface or its atmosphere. This definition deliberately provides a wide scope for using many diverse types and sources of data to investigate and hopefully answer geographical and Earth science research questions.

The spatial focus of Geography and the Earth Sciences has two significant implications. First, the location of the phenomena or observation units, in other words where they are on or near the Earth's surface is regarded as important. For instance, investigations into landforms composed of calcareous rocks need to recognise whether these are in temperate or tropical environments. The nature of such landforms including their features and structures is in part dependent upon prevailing climatic conditions either now or in the past. Second, the spatial arrangement of phenomena may be important, which implies that a means of numerically quantifying the position of different occurrences of the same category of observation may be required. In this case, spatial variables such as area, proximity, slope angle, aspect and volume may form part of the data items captured for the phenomena under investigation.

1.2 Nature of Numerical Data

We have already seen that quantitative analysis uses numbers in two ways, as shorthand labels for qualitative characteristics to save dealing with long textual descriptions or as actual measurements denoting differences in magnitude. Underlying this distinction is a division of data items into

attributes and **variables**. Williams (1984: 4) defines an attribute as 'a quality ... whereby items, individuals, objects, locations, events, etc. differ from one another'. He contrasted these with variables that 'are *measured* ... assigned numerical values relative to some standard – the *unit of measurement*' (Williams, 1984: 5). Examples of attributes and variables from Geography and other Earth Sciences are seemingly unbounded in number and diversity, since these fields of investigation cover such a wide range of subject areas. Relevant attributes include such things as rock hardness, soil type, land use, ethnic origin and housing tenure, whereas stream discharge, air temperature, population size, journey time and number of employees are examples of variables. The terms attribute and variable are sometimes confused and applied interchangeably, although we will endeavour to keep to the correct terminology here.

Numerical data are commonly categorised according to their **scale of measurement**. There are four scales usually known as **nominal, ordinal, interval** and **ratio** and can be thought of as a sequence implying a greater degree of detail as you progress from the nominal to the ratio. However, strictly speaking, the nominal is not a scale of measurement since it only applies to qualitative attributes and therefore is an assessment of qualitative difference rather than magnitude between observations. The other three scales provide a way of measuring the difference between observations to determine whether one is smaller, larger or the same as any other in the set. Box 1.1a,b summarise the features of each of these measurement scales and show how moving from the interval/ratio to the nominal scale results in information being thrown away. Such collapsing or recoding of the values for data items may be carried out for various reasons but may be necessary when combining a qualitative data item, such as housing tenure, with a quantitative one, such as household income.

Box 1.1a Scales of measurement: location of a selection of McDonalds restaurants in Pittsburgh, USA.

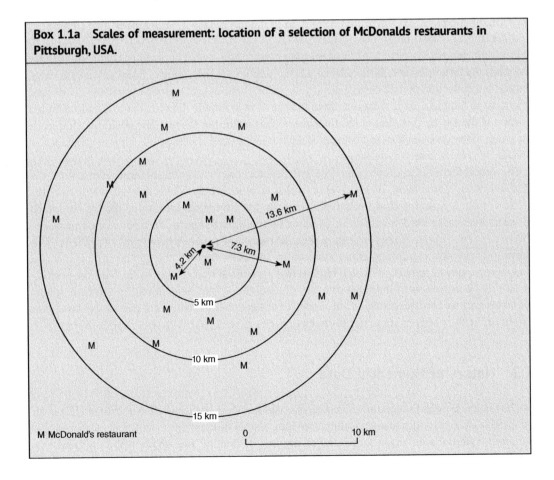

M McDonald's restaurant

Box 1.1b Definitions and characteristics.

- Nominal scale attributes record a qualitative difference between observations.
- Ordinal scale variables rank one observation against others in the set with respect to the ascending or descending data values of a particular characteristic. These are particularly useful when analysing information obtained from secondary sources where the raw data are unavailable.
- Interval scale variables record differences of magnitude between observations in a set where the measurement units have an arbitrary zero.
- Ratio scale variables record differences of magnitude between observations in a set where the ratio between the values of any two observations remains constant irrespective of measurement units, which have an absolute zero.

For most practical purposes, the interval and ratio scales are treated as equivalent with respect to various analytic procedures. The four scales of measurement are listed in order of increasing precision and it is possible to collapse the detail for a characteristic measured on the interval/ratio scales to the ordinal and to the nominal. This process results in a loss of detail as information is discarded at each stage, but this simplification may be necessary when analysing data from across different measurement scales. Box 1.1c illustrates that progressive loss of detail occurs with respect to the distance measurements between the centre of Pittsburgh and a selection of the MacDonald's restaurants in the city. Distance is measured on the ratio scale, with an absolute zero in this case indicating that a restaurant is located exactly at the centre. Sorting these raw data measurements in ascending order and assigning rank scores converts the information into an ordinal scale variable. There are two restaurants closest to the centre at 2.2 km from this point and the furthest is 18.4 km. Further loss of detail results from allocating a qualitative code to denote whether the restaurant is located in the inner, middle or outer concentric zone around the city centre. Although the detailed raw data of ratio/interval variables can be collapsed in this way, it is not possible to go in the other direction from qualitative attributes, through ordinal to interval/ratio scale variables.

Box 1.1c Collapsing between scales of measurement: distance of a selection of MacDonald's restaurants from centre of Pittsburgh, USA.

| Restaurant No. | Distance to centre | | Rank score | Concentric zone |
	Ratio variable	Sorted distance	Ordinal variable	Nominal attribute
1	17.5	2.2	1	1 (inner)
2	15.1	2.2	1	1 (inner)
3	14.2	2.8	3	1 (inner)
4	10.3	3.6	4	1 (inner)
5	6.0	4.2	5	1 (inner)
6	8.2	6.0	6	2 (middle)
7	9.2	6.4	7	2 (middle)
8	10.8	7.0	8	2 (middle)
9	4.2	7.2	9	2 (middle)

(Continued)

Box 1.1c (Continued)

| Restaurant No. | Distance to centre | | Rank score | Concentric zone |
	Ratio variable	Sorted distance	Ordinal variable	Nominal attribute
10	3.6	7.3	10	2 (middle)
11	2.2	8.1	11	2 (middle)
12	7.0	8.2	12	2 (middle)
13	2.2	9.2	13	2 (middle)
14	2.8	10.3	14	3 (outer)
15	11.2	10.8	15	3 (outer)
16	8.1	11.2	16	3 (outer)
17	6.4	11.2	16	3 (outer)
18	7.2	13.6	18	3 (outer)
19	7.3	13.9	19	3 (outer)
20	11.2	14.2	20	3 (outer)
21	13.9	15.1	21	3 (outer)
22	13.6	17.5	22	3 (outer)
23	18.4	18.4	23	3 (outer)

Most investigations will analyse a number of attributes and variables, and these are likely to be measured using one, two, three or all four of these scales of measurement. It is important to recognise the different scales, because some types of statistical quantity and technique are only used correctly with one measurement scale and not the others. In other words, statistical analysis is not just a question of 'pick and mix', more 'horses for courses' or fitting a 'round peg in a round hole'. An important distinction is between discrete and continuous measurements, which distinguishes nominal and ordinal data from interval and ratio scale variables. Discrete data values are when observations can only take certain values, usually integers (whole numbers), whereas theoretically any value, including negative ones for interval scale variables, is allowed with continuous measurements. Information technology has helped investigators to carry out research by using software to apply statistical analysis to numerical data, but we still usually need to 'tell' the software whether a given set of integers are nominal category labels for qualitative attributes, an ordinal sequence of integers or real data values on the interval or ratio scales. Investigators need to intervene to select appropriate types of analysis and not just tell the software to produce the same set of analyses for all their attributes and variables. Undoubtedly comments such as these may not be applicable in the future, but investigators will always be wise to confirm that software has made sensible decisions.

Investigations using attributes and variables involve holding the data in a numerical format. In the case of nominal attributes, this usually involves simple integer numbers such as 1, 2, 3 up to J, where J equals the number of categories, and there is a one-to-one correspondence between the number of categories and integers. However, in some cases, negative integers may be used. In surveys, people are often asked their opinion about a series of statements in terms of strongly disagree, disagree, neutral, agree or strongly agree. These qualitative labels may be linked to the numbers $-2, -1, 0, 1$ and 2, rather than 1, 2, 3, 4 and 5. The ordinal scale is based on ranking observations in order from smallest to largest, or *vice versa*. These are often converted into a rank score (first, second, third, etc.)

corresponding to either the largest or smallest value depending upon whether they are sorted in ascending or descending numerical order. These rank scores are usually stored as integers, with values duplicated if two or more observations are tied. Thus, there may be three observations tied at 15 and no 16 or 17, and the next observation is 18, although arguably the three observations could all be ranked 16. Interval and ratio scale variables are usually recorded to a certain number of decimal places, 2 or 3 for example, even if particular values happen to be integers (e.g. 5.0 km). Computers can store data values with far more precision, for example to 25 decimal places, but such detail is often unnecessary for geographical variables and you may only have captured the data values to one decimal place or only as whole numbers. An example will illustrate the problem. This book follows the convention of using the International System of Metric Units (kilometres, degrees Celsius, hectares, etc.) in scientific research. However, many people in the USA, the UK and some other countries will be unfamiliar with this system and will give distances in miles, temperatures in Fahrenheit and areas in acres. It is possible to convert between these two sets of scales, for instance 1 hectare equals 2.4691 acres (4 decimal places). If we apply this conversion to a specific acreage, we may end up with what appears to be a highly accurate number of hectares, but in reality, is an entirely spurious level of precision. For example, 45 acres divided by 2.4691 equals 18.225264266113861811996273 9 ha, which could be shortened to 18.2 ha without any significant loss of information.

Before leaving our consideration of the characteristics of numerical data, there are other types of numbers we may want to use in our analysis. Dates and times are another useful category of information that provides investigations with a temporal dimension. Two different types of date and/or time can be recorded depending upon how this temporal reference is obtained. One is recorded by the investigator and refers to the date and time when the specific data were collected. An example of this would be the day, month and year, and the start and end times when an individual survey questionnaire was completed or when a field measurement was recorded. The second is obtained from the data subjects themselves and is more common in investigations involving people. A survey of company Chief Executive Officers might be asked when it was founded, a survey of farmers might ask when they last acquired or disposed of land, or car drivers could be asked the time when they started their journey. In Earth Sciences, a date variable might be obtained by dating of lake sediment cores by means of microfossils for the purpose of environmental reconstruction or the age of trees might be obtained by counting the annual growth rings from cores bored in tree trunks.

Another type of numerical variable requiring special mention is the recording of financial values in a currency (e.g. US $, £ Sterling and € Euro), which are often needed in investigations concerned with people and institutions. In some respects, a currency is just another unit of measurement like metres or litres, which are expressed to two decimal places. Many currencies are made up of units with 100 subunits, for example £8.52 (i.e. 8 whole pounds and 52 pence), which is notionally like 8.52 m (i.e. 8 whole metres and 52 centimetres). However, collecting accurate monetary figures of this type is notoriously difficult despite their obvious importance. Individuals and households may not only be reluctant to give details of their income and expenditure but also may not know the precise figures. How many households in the UK could say **exactly** how much their total gross income was in the last six months? Or equally how many US corporations could report precisely how much they had spent on staff expenses in the last three months? We may decide to ask people to estimate expenditure or income to the nearest £1000, but when these are analysed they may produce results that seem more accurate than they really are, for instance an average household income of £19,475.67.

Finally, before moving on from looking at the characteristics of numerical data, the various spatial variables that are used in geographical investigations such as distance, length, surface area, volume and so on can be analysed on their own or combined with 'standard' non-spatial variables to create composite indicators, such as population per km^2, cubic metres of water discharged per second or point pollution sources per linear km of a river. Such indicators are similar to rates expressed as per

100 (percentages), 1000 (per mille) or 1,000,000 (parts per million) insofar as they standardise data values according to some fixed unit, which makes comparison easier. Comparing population density in different countries or the concentration of chemicals dissolved in water is easier to understand, if the variables representing these concepts are expressed as relative values rather than absolute figures, even if the latter are collected as raw data in the first place. Investigators should be ready to analyse not only the raw data that they collect but also additional variables that can be calculated from these, in conjunction with some information obtained from another published source.

1.3 Simplifying Mathematical Notation

We cannot avoid the fact that quantitative analytical techniques depend on the calculation of mathematical values and on underlying statistical theory. No doubt many students of Geography, Earth and Environmental Science would be pleased if this were not the case. It is likely that most students were drawn to studying these subjects because of an interest in topics and themes, such as those mentioned previously (asylum refugees, climate change and environmental hazards), rather than the prospect of encountering mathematical equations. Unfortunately, in order to understand what these statistical techniques do and when they should be used requires some knowledge of their underlying mathematical basis and how they help to extend our understanding of the substantive topics that really interest us. This inevitably raises the question of how to explain these mathematical principles without providing an excess of equations and formulae that would deter some readers from progressing beyond the first chapter and that others might choose to ignore. This text employs two strategies to try and overcome this problem: first, the remainder of this section provides a simple basic principles introduction to the mathematical notation and calculation processes that will be encountered in later chapters and second wherever possible putting equations and formulae to Boxes that incorporate shading and graphics to illustrate how the calculations work.

The notation used in mathematical equations can be thought of as a language, just as it is possible to translate from English to French, and French into German, mathematical notation can be 'translated' into English or any other language. The problem is that a typical equation expressed in prose will use many more words and characters than standard mathematical notation.

For example:

$Y = \sum X^2 - \sum X$ translates as   the value of Y equals the result of adding together each value of X multiplied by itself after subtracting the total produced by adding up all the values of X.

5 'words'   30 words

There is clearly a saving in terms of both words and characters (excluding spaces) between these two 'languages', with the equation requiring 5 'words' (8 characters) and the English translation 30 words (130 characters). Languages allow communication of concepts and information from one person to another, although speech recognition software extends this to between people and computers. When taking notes in a lecture or from published literature, students will often develop their own abbreviated language to represent familiar words or phrases. For example, hydrology may be shortened to hyd. or government to gov., which are easy to guess what they mean, but incr. and ↑ might be used interchangeably to denote increase. A language is most useful when both parties understand what the symbols and abbreviations mean. Some mathematical and statistical notation has migrated into everyday language. So, books, newspapers and other media will often use %, for instance with respect to the national unemployment rate, to indicate a percentage. Many people will understand what the % symbol means, if not necessarily how the figure was calculated. However, for many people, much mathematical and statistical notation is like a foreign language that sometimes uses letters from a familiar alphabet in unfamiliar ways.

The basic principle of a mathematical equation is that the value on the left side is produced once the calculations denoted on the right have been carried out. In one sense, it does not really matter what the numbers on the right-hand side are, since you can always perform the calculation and obtain a matching value on the left-hand side. In simple mathematics, the numbers do not have any intrinsic meaning, you could just as easily calculate 11 by adding 3 to 8, or subtracting 3 from 14, multiplying 5.5 by 2 or dividing 33 by 3. However, when using numerical data and applying statistical techniques in Geography and other subject areas, we usually work with numbers that mean something in relation to the attributes and variables for the different observations that we are analysing. If one of the variables in a survey of farms is the area of land managed by the farmer, then each observation (farm) could have a different value for this variable. Variations in these values between the different farms actually have meaning in the sense that they say something about farm size. They are not just a series of numbers, but signify something seemingly important, in this case that some farms are larger than others.

The four sections in Box 1.2 provide a straightforward guide to explain how to read the different mathematical symbols used in later chapters. First, some general symbols are presented to indicate labelling conventions for variables, attributes and statistical quantities. Second, the standard elementary mathematical operators indicate how to compute one number from another. Third, mathematical functions specify particular computations to be carried out on a set of numbers. Finally, comparison or relational operators dictate how the value on one side of an equation compares with that on the other.

Box 1.2 Guide to mathematical notation and conventions and their demonstration in boxes.

Symbol	Meaning	Examples
General		
Decimal places	Number of digits to the right of the right of the decimal point	3 decimal places = 0.123, 5 decimal places 0.00045
X, Y, Z	Attributes or variables relating to a statistical population	X represents the values of the variable distance to city centre measured in kilometres for each member of the statistical population
x, y, z	Attributes or variables relating to a statistical sample	x represents the values of the variable customers per day for each member of a sample
Greek letters	Each letter represents a different parameter calculated with respect to a statistical population	α (alpha), β (beta), μ (mu) and λ (lambda)
Arabic letters	Each letter represents a different statistic calculated with respect to a statistical sample	R^2, d, and s
()	Brackets used, possibly in a hierarchical sequence to subdivide the calculation into stages in a sequence – work outwards from innermost calculations	$10 = 4 + (3 * 2)$ $10 = 4 + ((1.5 * 2) * 2)$

(Continued)

Box 1.2 (Continued)

Symbol	Meaning	Examples
Mathematical operators		
$+$	Add	$10 = 4 + 6$
$-$	Subtract	$-2 = 4 - 6$
$*$	Multiply	$24 = 4 * 6$
/ or ——	Divide	$0.667 = 4/6$ $0.667 = \dfrac{4}{6}$
Mathematical functions		
X^2	Multiply the values specified variable by themselves once – the square of X	$5^2 = 25.00$, $9.3^2 = 18.49$
X^3	Multiply the values specified variable by themselves twice – the cube of X	$5^3 = 5 \times 5 \times 5 = 125.00$, $9.3^3 = 9.3 \times 9.3 \times 9.3 = 79.507$
X^n	Multiply the values specified variable by themselves n times – exponentiation of X to n	2^n (where n = 5) $= 2 * 2 * 2 * 2 * 2 = 16$, 9.3^n (where n = 5) $= 9.3 * 9.3 * 9.3 * 9.3 * 9.3$ $= 1470.08443$
$\sqrt{X}$	Determine the number which, when multiplied by itself, will produce X – the square root of X	$\sqrt{25.00} = 5$, $\sqrt{18.49} = 9.3$
$\sum X$	Add all the values of the specified variable	3 data values for X (4.6, 11.4, 6.1) $\sum X = 4.6 + 11.4 + 6.1 = 22.1$
$X!$	Factorial of the integer value X	$5! = 5 * 4 * 3 * 2 * 1 = 120$
$\lvert (X - Y) \rvert$	Take the absolute difference between the value of X and the value of Y (i.e. ignore any negative sign)	$\lvert(12 - 27)\rvert = 15$ (not -15)
Comparison or relational operators		
$=$	Equal to	$10 = 4 + 6$
$\neq$	Not equal to	$11 \neq 4 + 6$
$\approx$	Approximately equal to	$7 \approx 40/6$
$<$	Less than	$8 < 12$
$\leq$	Less than or equal to	$8 \leq 8$
$>$	Greater than	$5 > 3$
$\geq$	Greater than or equal to	$5 \geq 5$

The second strategy for introducing and explaining the mathematical background to the statistical techniques presented in this text is the use of boxes. These provide a succinct definition and explanation of the various techniques. They illustrate the calculations involved in these techniques by using a series of case study datasets relating to various countries and topics from Geography and the Earth Sciences. These datasets include attributes and variables referring to the spatial and aspatial (non-locational) characteristics of the observation units. Box 1.3 provides an example of the general design features of the Boxes used in later chapters, although the figures (data) are purely illustrative. Most Boxes are divided into different sections labelled (a), (b) and (c). The first of these is often a diagram or photographic image to provide a geographical context for the case study data. The second section is text describing the key features and application of the statistical concept. The third section defines the statistical quantity in question in terms of its equation, tabulates the data values for the observation units and details the calculations required to produce a particular statistic (see example in Box 1.3). Some Boxes vary from this standard format and datasets introduced in the early chapters may be re-used and possibly joined by other data in later chapters to enable comparisons to be made and to allow for techniques involving two or more attributes/variables (bivariate and multivariate techniques) to be covered.

Box 1.3   Guide to conventions in demonstration in boxes.

Population symbol: sample symbol:
The symbols used to represent the statistical measure covered in the box will be given in the title section, if appropriate.

(a) Graphic image – photograph, map or another diagram
Boxes introducing some of the case study data will often include a graphic image to provide a context for the data and to help with visualising the significance of the results.

(b) Explanatory text
This section provides a relatively short two- or three-paragraph explanation of how the particular statistical technique 'works', summary of how it should be calculated and interpreted.

(c) Calculation table
This section explains how to calculate statistical measure(s), in some instances comparing different methods. The first part provides the equation for the statistical quantity the elements of which are then used as column/row headings. Most of the table comprises a mixture of clear and shaded columns together with arrows to point out how one value turns into another because of addition, subtraction, multiplication or division.

The following example, although based on the calculation of a particular statistic (the variance) is provided simply to illustrate the general format of these computation tables.

Identification of individual observation units	$s = \sqrt{\dfrac{N(\sum x^2) - (\sum x)^2}{N(N-1)}}$		Definitional equation
	X	X^2	Column headings:
			X = the values of variable X
			X^2 = the values of X squared
1	3 →	9	• Arrows indicating that mathematical
2	5 →	25	operations applied to values in one column
3	6 →	36	produce new values in another column
			• Shading to lead the eye down a column of numbers to the calculations
$N = 3$	$\sum X = 14$	$\sum x^2 = 70$	Column totals
	$S^2 = \dfrac{3(70) - (14^2)}{3(3-1)}$		Numbers substituted into equation
	$S^2 = 3.33$		Result
Sample mean	$\dfrac{\sum X}{N} = 4.67$		Other useful statistics

---

1.4   Introduction to Case Studies and Structure of the Book

This text is intended as an introduction to statistical techniques and their application in contemporary investigations of the type typically undertaken by undergraduate students during the final year of study in Geography and other Earth Sciences. The book also provides an overview of how to undertake statistical analysis using computer software and although calculating statistics using calculators is outdated, some understanding of how they are computed helps to apply and interpret them correctly. This summary of software usage is not intended as a substitute for online help or other texts giving detailed instructions on how to carry out procedures in specific software (e.g. Bryman and Cramer, 1999). Instead, we outline the essential generic characteristics of analysing digital geographical and spatial datasets and communicating/visualising the results of such enquiries.

The book has six main sections. The present chapter is in the *First Principles* section and has introduced quantitative analysis and outlined the features of numerical data and elementary mathematical notation. Chapters 2 and 3, also in this section, focus on collecting and accessing data for your analysis and considers the differences between statistical populations and samples, and strategies for collecting data. Chapters 4 and 5, in a section titled *Exploring Geographical Data*, investigate ways of reducing substantial amounts of numerical data into more manageable chunks either as summary measures or as frequency distributions. Chapter 5 examines probability, defining research questions, stating hypotheses and introduces inferential statistics. Sections II and III are separated by a pair of diagrams and a short series of questions that help with deciding how to carry out your own research investigation.

These diagrams refer to the statistical techniques covered by the chapters in Sections III and IV. Section III, *Testing Times*, includes two chapters: Chapter 6 focuses on parametric statistical tests and Chapter 7 on non-parametric statistical tests. Each chapter has a similar structure and works through situations in which there are one, two, or three or more attributes and/or variables to be tested. Section IV, as its title, *Forming an Association or Relationship*, suggests goes beyond testing whether samples are significantly different from their parent populations or each other and examines ways of expressing and measuring relationships between variables in terms of the nature, strength and direction of their mathematical connections. Examination of the explanatory and predictive roles of statistical analysis of relationships covers situations with both bivariate (two variables) and multivariate (three or more variables). Section V, *Explicitly Spatial*, concentrates on exploring spatial aspects of geographical data (Chapter 10) and the analysis of spatial patterns and the distinctive features of modelling relationships using spatial data (Chapter 11). Finally, Section VI, *Practical Application*, has one chapter that uses project case studies to provide guidance on developing your own research project pointing out some of the pitfalls to avoid and some of the features to include for a more impressive, robust piece of work.

The attributes and variables in these datasets are, by and large entirely genuine 'real' data values, although details that would have allowed the identification of individual observations have been removed, and in some cases, the number of cases has been restricted in order to limit the size of the Boxes. The datasets can be regarded as illustrating diverse types of investigation that might be carried out by undergraduate students either during their independent studies or during group-based field investigations.

References

Bryman, A. and Cramer, D. (1999) *Quantitative Data Analysis for Social Scientists*, London, Routledge.

Getis, A. and Getis, J. (1966) Christaller's central place theory. *Journal of Geography*, **5**(5), 220–226. doi:10.1080/00221346608982415.

Hartshorne, R. (1958) The concept of geography as a science of space, from Kant and Humbolt to Hettner. *Annals of the Associations of American Geographers*, **48**(2), 97–108. doi:10.1111/j.1467-8306.1958.tb01562.x.

Johnston, R.J. (1979) *Geography and Geographers: Anglo-American Geography since 1945*, London, Edward Arnold.

Smith, C.H. (1986) A general approach to the study of spatial systems. Two examples of application. *International Journal of General Systems*, **12**(4), 385–400. doi:10.1080/03081078608934944.

Williams, R.B.G. (1984) *Introduction to Statistics for Geographers and Earth Scientists*, London, Macmillan.

Further Reading

Bryman, A. (1988) *Quantity and Quality in Social Research*, London, Unwin Hyman.

Clifford, N., Cope, M., Gilespie, T., French, S. and Valentine, G. (eds) (2010) *Key Methods in Geography*, 3rd edn, London, Sage.

Ebdon, D. (1984) *Statistics in Geography*, 2nd edn, Oxford, Blackwell.

Goodchild, M.F. and Longley, P.A. (1999) The future of GIS and spatial analysis, in *Geographic Information Systems: Principles, Techniques, Management and Applications* (eds P. A. Longley, M. F. Goodchild, D. J. Maguire and D. W. Rhind), 2nd edn, New York, John Wiley and Sons.

Hammond, R. and McCullagh, P.S. (1978) *Quantitative Techniques in Geography*, Oxford, Clarendon Press.

Haynes, R. (1982) *Environmental Science Methods*, London, Chapman and Hall.

Heiman, G.W. (1992) *Basic Statistics for the Behavioural Sciences*, Boston, Houghton Mifflin.

Kitchin, R. and Tate, N. (2000) *Conducting Research into Human Geography*, Harlow, Prentice Hall.

Limb, M. and Dwyer, C. (eds) (2001) *Qualitative Methodologies for Geographers*, London, Arnold.

McGrew, J.C.J. and Monroe, C.B. (2000) *Introduction to Statistical Problem Solving in Geography*, Boston, McGraw-Hill.

Moser, C. and Kalton, G. (1971) *Survey Methods in Social Investigation*, 2nd edn, London, Heinemann.

Rogerson, P.A. (2006) *Statistical Methods for Geographers: A Student's Guide*, Los Angeles, CA, Sage.

Rose, D. and Sullivan, O. (1996) *Introductory Data Analysis for Social Scientists*, 2nd edn, Buckingham, Open University Press.

Sachs, L. (1984) *Applied Statistics: a Handbook of Techniques*, New York, Springer-Verlag.

Simpson, L. and Dorling, D. (eds) (1998) *Statistics in Society*, London, Arnold.

Walford, N.S. (1995) *Geographical Data Analysis*, Chichester, John Wiley and Sons.

Williams, R.B.G. (1985) *Intermediate Statistics for Geographers and Earth Scientists*, London, Macmillan.

2

Geographical Data: Quantity and Content

> Chapter 2 starts to focus on geographical data, looks at the issue of how much to collect by examining the relationships between populations and samples and introduces the implications of sampling for statistical analysis. It also examines different types of geographical phenomena that might be of interest and how to specify attributes and variables. Examples of sampling strategies that may be successfully employed by students undertaking independent research investigations are explained in different contexts.

Learning Outcomes

This chapter will enable readers to

- Outline the relationship between populations and samples and discuss their implications for statistical analysis;
- Decide how much data to collect and how to specify the attributes and variables that are of interest in different types of geographical study;
- Make data collection plans for an independent research investigation in Geography and related disciplines using an appropriate sampling strategy.

2.1 Geographical Data

What are geographical data? Answering this question is not as easy as it might at first seem. The geographical part implies that we are interested in things – entities, observations, phenomena, etc. – immediately above, below and on, or indeed comprising, the Earth's surface. We are interested in where these things are, either as a collection of similar entities in their own right or as one type of entity occurring close to or at a distance from one or more different types. For example, we want to investigate people in relation to settlements; rivers to land elevation; farms to food industry corporations; or volcanoes to the boundaries of tectonic plates. The data part of the term 'geographical data' also raises some definitional issues. At one level, data can be thought of as a set of 'facts and figures' about 'geographical things'. However, some would argue that the reduction of geographical entities to a list of 'facts and figures' in this fashion fails to capture the essential complexity of the phenomena. Thus, while we can list a set of **data items** (attributes and variables) relating to a river, a farm, a settlement, a volcano, a corporation, etc., we can only understand the nature of such phenomena by looking at the complete entity. It is really accepting the cliché that the whole is greater than the sum of its parts.

Practical Statistics for Geographers and Earth Scientists, Second Edition. Nigel Walford.
© 2025 John Wiley & Sons Ltd. Published 2025 by John Wiley & Sons Ltd.
Companion website: www.wiley.com/go/PracticalStatistics2e

Carrying out research with geographical data is not just about using data that you have collected yourself but can also involve acquiring and then (re-)analysing data that someone or some other organisation has collected and made available to you. The collection or acquisition of geographical data raises two sets of issues: how much data should you collect or acquire to answer your research question(s), and how are you going to obtain the desired amount of data? This chapter examines the first of these questions, and Chapter 3 explores the ways of obtaining geographical data from **primary** and **secondary** sources and of locating where geographical 'things' are in relation to the Earth's surface. Chapter 1 suggested that from a theoretical perspective all geographical entities are unique. However, at each location there are many different geographical attributes and variables: these include physical (e.g. soil pH, underlying rock type and annual hours of sunshine (insolation)), human (e.g. land use, ownership and value) and spatial (elevation, aspect, etc.) characteristics. Many investigations are not interested in one occurrence of a particular type of entity at a specific location, but with comparing and contrasting at least two and often many more examples of the same or different types. Consequently, one of the first 'data quantity' issues needing to be addressed is how many occurrences of a particular type of entity should we include in our study. The second main issue relates to specifying a set of data items (attributes or variables) to collect so that we can answer our research questions. Taken together, selected entities and data items comprise what is commonly referred to as a **dataset**. In other words, this is a coherent collection of data (facts, opinions, measurements, etc.) for each geographical entity in a set, hence the word dataset. The remainder of this chapter examines these issues and shows how we can assemble datasets to answer research questions of the type asked by undergraduate students of Geography and other Earth Sciences.

2.2 Populations and Samples

One of the first issues to address when planning an investigation involving the collection of original data and/or the assembly of information from existing sources is to decide how many observation units to include in your dataset. At first sight, there seem to be two answers to this question that are the opposites of each other – obtain data relating to as many or as few observations as possible. Option 1 avoids the potential problems of leaving out some items that might be unexpectedly important, whereas option 2 saves time and money and cuts out unnecessary duplication and waste of effort. If we knew all the information, there was to know about the observations in advance: we could select the minimum number required to exactly reproduce these overall characteristics. However, if we were in this fortunate position of perfect knowledge, we might with some justification decide to abandon the investigation altogether. Imagine we knew in advance the winning numbers that were going to come up in the UK's National Lottery™. There would be two logical responses: everyone buys a ticket and marks it with the known set of winning numbers, with the result that over 50 million people aged 16 and over in the UK population would win an equal share of the prize fund; nobody buys a ticket because they realise that they are likely to win a few pence when the prize fund is divided between over 50 million people, given that only a minority percentage of each person's £2 stake goes into the prize fund. The winners would be the people who knew what everyone else was going to do and then did the opposite themselves.

Unfortunately, just as when playing The National Lottery™, when we carry out research, we do not normally have perfect knowledge in advance and therefore we need to come up with a strategy for selecting enough observations to provide sufficient insight into all the observations but not so many that we waste too much time and effort. When playing The National Lottery™, many people use numbers related to memorable dates or events, such as birth and wedding dates and number of grandchildren. Such numbers serve as no more of a reliable guide for choosing the winning numbers in a game of chance as they would for selecting entities for inclusion in a research study. The observations or individuals selected for inclusion in a research study where quantitative techniques are to be used should be a large enough group so that you can make reliable statements about all the observations of that type. They should be **representative** of the complete set.

There are some occasions when it is possible or desirable to collect data for all observations in a set. The term **census** is usually reserved for situations where a count of all members of a human or animal population is carried out, but the origin of the word census simply means 100 per cent of observations are enumerated. Often such censuses are carried out regularly by government or other national organisations; for example, many countries under United Nations guidance carry out a Population Census every 10 years or in some instances at shorter intervals. Despite the possible omission and the certainty of change in observation units, censuses perform the important function of providing a benchmark against which to compare the results when surveying a subset of observations on a later occasion. Thus, censuses may contain information that will help us to decide how many people or households to select for a more in-depth enquiry. Remote sensing carried out by satellites and, for smaller areas, by aircraft or drones can also be thought of as censuses, since they collect data in digital or analogue form for all of the surface in their overview. However, just as censuses of human and animal populations may accidentally omit some individuals, intervening clouds and shadows may mean that remotely sensed data have gaps in their coverage.

The collection or assembly of a dataset for research relates to a specific period of time, and if all observation units are counted, measured or surveyed, the data are likely to become out of date after or even before they have all been collected. The various phenomena that interest us are subject to alteration and change. Air and water temperature, people's age and location, and corporations' market valuation and workforce are theoretically and practically in an almost constant state of flux. The problem originates from the perpetual interaction between space and time. Only one geographical entity of a certain type can occupy a particular portion of space at a given moment in time. But geographical phenomena are a little like Russian dolls with one inside another. Suppose a person is in a room in their dwelling, there are three other rooms in that dwelling, with or without people in them, those rooms are behind the front door of an apartment on the fifth floor of a block with 12 floors, and it is one of five blocks in a housing development. The person, the room, the apartment, the floor, the block and the housing development are all geographical phenomena worthy of study. If we wanted to discover the socio-economic characteristics of people living in the development, should we attempt to interview everyone or just a subset of people?

Now imagine we wanted to know what it was like living in this development when it was built and first occupied 15 years ago. For most practical purposes, time is unidirectional, moving forwards and not backwards; nevertheless, we can imagine or visualise how that unit of space appeared in the past, relying on the accuracy of historical records to recreate or model the past using three-dimensional visualisation technology, but we cannot return to a previous time. There may be some people who have lived in the development since the start or others who worked in farming the land before it was built over, and we could ask them about what it was like then. The developer or local authority may hold some records about people who moved into the apartments when they were built that includes information about their employment or family circumstances. Using such diverse sources, it may be possible to reflect on what this portion of the Earth's surface was like before the development was built.

A complete set of observations is called a **population** in quantitative analysis, and a subset selected from a population is known as a **sample**. Some investigations in Geography and the Earth Sciences will use data for a population of observation units, such as in the case of analysing Agricultural or Population Census statistics covering, respectively, all farm holdings and all individuals/ households. The data for these farms, households and individuals will often be aggregated into spatial units such as parishes or electoral wards. Similarly, investigators of ice sheets or vegetation may analyse data captured remotely by satellite-mounted sensors for every square metre of their area of interest. However, it is unusual for academic researchers to have the resources to initiate an investigation that involves collecting original data on this scale for a statistical population, and, as these examples illustrate, analysis of all observation units usually involves the use of data from secondary sources. Often, and invariably in the case of undergraduate project work, a sample of observations is selected for analysis from which we reach conclusions.

There are five main reasons why collecting original data about an entire population is an unrealistic proposition for most researchers – cost, time, indeterminacy, unavailability and contamination of the entities (see Box 2.1). These difficulties invariably mean that, if the collection of new, original data is considered important for answering the research questions, a strategy for selecting one or

---

Box 2.1 Five reasons for sampling.

Time	Primary data collection is time-consuming, and the amount of time that can be assigned to this task has to be scheduled alongside all the other activities required for completing an investigation. An important part of project planning is to specify the amount of time available – days, weeks or months – to collect or acquire data from primary or secondary sources. Producing a Gantt chart is a useful way of visualising the different parts of a project and identifying possible overlaps between them. It also helps to specify milestones and outputs (deliverables).
Cost	Every means of data collection has direct and indirect costs. Administration of questionnaires by post, email, social media or face-to-face interview, the gathering of field specimens and samples, the surveying of land use, landforms, slope and altitude and the counting of vehicles, cyclists or public transport passengers, and species of fauna or flora all cost money in terms of travel and equipment expenses. Indirect or opportunity costs are those associated with not being able to carry out some other activity. For example, the loss of earnings by a student from not being in paid employment when collecting data versus the lower mark for a dissertation is associated with a smaller-than-ideal sample size and the consequent impact final degree outcome.
Indeterminacy of entities	Some populations contain an infinite or constantly changing number of observations, and thus, unless the required data can be collected instantaneously for those observations existing at a particular time, sampling is necessary. This problem is common when the units of observation are part of a flow, such as water in an ocean, lake or river, gases in the atmosphere, vehicles on a road or pedestrians in a shopping centre.
Unavailability of entities	Some observations in a population may not be available, even if they can in theory all be defined as existing now or in the past. If investigators can only find those observations that have been left behind, they cannot be certain whether those remaining were differentially privileged, allowing them to survive processes of decay and transformation. For example, historical documents, oral history interviewees and small-scale landscape features (e.g. mediaeval field systems or dew ponds) are those that have survived deterioration, destruction and obliteration and not necessarily all of entities that once existed. The researcher cannot feasibly decide whether those observations that remain are in some indeterminate way different from the population that existed formerly.
Contamination of entities	In some circumstances, the method and/or quantity of data collected can produce negative feedback with the observation units, causing their characteristics to change or be inaccurately measured. Some landforms or physical features may be disturbed by the approach of human researchers; for example, walking over a beach or dunes can alter the morphology and mineralogy of small- and micro-scale features. Similarly, interviewees can give mixed responses if they become confused or annoyed with excessive or pointless questioning.

more samples of observation units has to be devised. A sample, provided it is selected with careful planning, can give just as clear a picture of what the population as a whole is like, but at a fraction of the cost, time and effort. There are two main sampling strategies: one based on using probability to select observations at random and the other choosing them in a purposive or subjective non-random fashion. These are examined in the following sections, although the challenges involved with achieving a sampling strategy based on purely random selection should be recognised at the outset.

2.2.1 Probability Sampling Techniques

Most statisticians would regard sampling based entirely on probability as providing the ideal means of representing all the typical, diverse and idiosyncratic features of a statistical population. The basis of probability sampling is that chance, not some biased whim of the investigator or historical accident, determines whether a particular observation is included. Initially, it might seem perverse to base your sampling strategy on chance dictating which observations to include and exclude an investigation. However, the value of this approach is that it recognises the innate variability of 'real world' events. Many such events when considered in aggregate terms tend to display a particular pattern, but when focusing on an individual observation it will often be difficult to see where it fits into the pattern. For example, people living on the same housing estate or part of a city tend to have similar socio-economic characteristics in respect of attributes such as income, marital status, education level and occupation. However, this does not mean that these individuals are identical, simply that they are more similar to each other than they are to people living on a different housing estate or in another part of the city. Similarly, pebbles lying in the same area of a beach in respect of high and low mean water levels are likely to be similar to each other in terms of size, angularity, lithology and shape. Again, they will not be identical, but simply more like each other in these respects than pebbles from a different area of the beach. Samples of people and pebbles in these examples based on chance probability are likely to capture the innate variability of their respective populations more closely than one based solely on subjective judgement.

The 'gold standard' sampling strategy based on probability, known as **simple random sampling**, gives each and every observation in the population an equal chance of inclusion and exclusion at all stages in the sample selection process. The strategy depends on being able to identify all observations in the population and then selecting at random from this **sampling framework** the required number to make up the sample. If the population is finite in size, in other words has a fixed and unchanging number of observations within the time period covered by the investigation, their identification may be entirely possible. But, if the population is infinite or variable in size (see above), then producing a complete list is not possible. In these circumstances, the investigator can only realistically assume that those observations capable of being identified are not substantively different from those that are unknown. Observations are then selected at random from the list. Apart from using such imprecise and unsatisfactory manual methods to achieve this such as sticking a pin in the list of observations, the standard approach involves giving each observation a unique sequence number (e.g. 1–453) and using random numbers to select sufficient observations to make up the desired sample size (more about sample sizes later). Traditionally groups of 2, 3, 4 or however many digits as necessary were selected within the required range by moving diagonally, horizontally or vertically from a randomly determined starting point in a table of random digits from 0 to 9. Any sequences of digits forming a number larger than the total observations in the population would be ignored. A simpler approach is to use the random number function in software such as SPSS (originally Statistical Package for the Social Sciences), Minitab or MS Excel to generate the requisite random numbers. Table 2.1 illustrates the procedure for generating random numbers in a selection of software programs used for statistical analysis.

Table 2.1 Procedures for generating random numbers using a selection of software programs (population total observations = 150).

Software	Procedure
SPSS	In SPSS, select *Compute Variable* from the *Transform* menu item. Specify a Target Variable name (RANDNUMS) into which the list of random numbers will be inserted and apply the UNIFORM function with the total count of observations in the population in brackets. Note that it is necessary to open a Data Editor window with a column of non-blank numbers equal to the required sample size.

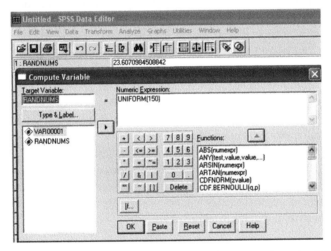

Minitab	In Minitab, select *Random data ...* followed by Integer from the *Calc* menu item. Specify a name (e.g. RANDNUMS) for the column into which the list of random numbers will be inserted, and enter how many random numbers are to be generated, and the minimum and maximum values of the observation identifiers.

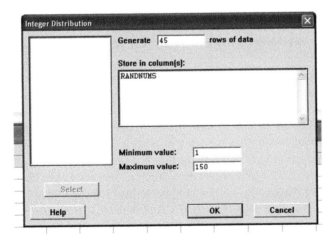

Table 2.1 (Continued)

Software	Procedure
MS Excel	Duplicate Excel *RANDBETWEEN* command in successive cells as necessary to equal the desired sample size. A new random number is returned every time the worksheet is calculated. Note that the function requires the Data Analysis Pack Add-in to be available.

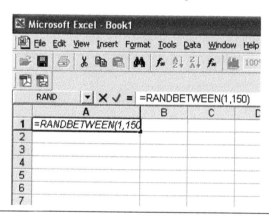

Whatever method is chosen to select a set of random numbers onto the corresponding unique numerical observation identifiers, there remains one further question to consider. There is nothing to prevent a particular random number from occurring more than once. Unfortunately, this produces the illogical situation of an observation being included in the sample several times. Simple random sampling in its purest form aims to give each observation in the population an equal chance of selection at all stages in the process; otherwise, some bias is present because the probability of an observation being selected changes. An observation left in the population towards the end of the sampling process has a greater chance of being selected than they all did at the start. It comes down to whether the selection process views sampled observations as being 'physically' transferred from a list of the total known observations onto a list of sampled items, or whether it copies the randomly selected identifiers from one list to another. In essence, there are two problems to deal with: what to do about duplicate random numbers that map onto the same observation more than once, and, if the answer is to allow items to be included only once, how do you assess the bias arising from the changing likelihood of inclusion at the second as each subsequent stage or iteration in the selection process? From a statistical analysis perspective, these problems relate to whether you should sample with or without replacement and are illustrated in Box 2.2.

Box 2.2 Simple random sampling with and without replacement.

Sampling with replacement means that once a randomly selected observation has been included in the sample, it is returned to the population and has a chance of being re-selected at any subsequent stage in the sampling process, if its identification number so happens to be chosen again. This means that the size of the population from which the sample is taken remains constant each time an observation is selected, but individual observations might feature more than once. Sampling without replacement effectively removes sampled observations

---

(Continued)

Box 2.2   (Continued)

from the population as if they are physically transferred into the sample. Thus, each time an observation is chosen the size of the population reduces by one and the chance of selection for each of the remaining items increases slightly. In practice, this problem is usually overcome by selecting additional random numbers to substitute for any duplicates.

Boxes 2.2a and 2.2b illustrate the implications of sampling with or without replacement with respect to selecting a sample of 20 observations from a population of 100 items. Only the stages selecting the first five observations and then the 19th and 20th are shown in order to limit the size of the tables. The 20 random numbers, restricted to integers between 1 and 100, were generated using statistical software (SPSS), which resulted in three duplicates (12, 64 and 70) when sampling with replacement. Sampling without replacement avoided this problem by requiring that, at any stage, if a random number selected had already been chosen then another was used as a substitute. This prevented any duplicates and resulted in the numbers 2, 71 and 84 being added to the initial list of 20 random numbers in place of the second occurrences of 12, 64 and 70. If a different set of random numbers were to be generated, then such duplication might not occur, but three duplicates occurred without any attempt to 'fix the result' and illustrated the problem.

Comparing the rows showing the percentage of the population sampled at each stage in Boxes 2.2a and 2.2b indicates that this remained constant at 1% when sampling with replacement but reduced by a steady 0.01% as every new observation was transferred into the sample when sampling without replacement. Thus, the first observation chosen had a 1 in 100 (1.00%) chance of selection, but by the time of the 20th this had reduced to 1 in 81 (1.23%). Provided that the size of the sample is small in relation to the number of observations in the population, which may of course be infinite, this change does not cause major problems in most academic research. However, suppose the sample size in this example was raised to 50, then the 50th observation would have a 1 in 51 chance of selection or would be 1.96% of the sampled population when sampling without replacement. This is nearly double what it was at the start of the process.

Box 2.2a   Sampling with replacement (includes duplicates in the final set).

Stage (iteration)	1	2	3	4	5	...	19	20
Population (remaining)	100	100	100	100	100	...	100	100
Accumulated sample size	1	2	3	4	5	...	19	20
% population sampled	1.00	1.00	1.00	1.00	1.00	...	1.00	1.00
Probability of selection	0.01	0.01	0.01	0.01	0.01	...	0.01	0.01
1st number selected	64	64	64	64	64	...	64	64
2nd number selected	–	39	39	39	39	...	39	39
3rd number selected	–	–	90	90	90	...	90	90
4th number selected	–	–	–	64	64	...	64	64
5th number selected	–	–	–	–	05	...	05	05
6th number selected	–	–	–	–	–	...	51	51

	1	2	3	4	5	...	19	20
7th number selected	–	–	–	–	–	...	92	92
8th number selected	–	–	–	–	–	...	18	18
9th number selected	–					...	73	73
10th number selected	–	–	–	–	–	...	70	70
11th number selected	–	–	–	–	–	...	88	88
12th number selected	–	–	–	–	–	...	52	52
13th number selected	–	–	–	–	–	...	73	73
14th number selected	–	–	–	–	–	...	48	48
15th number selected	–	–	–	–	–	...	70	70
16th number selected	–	–	–	–	–	...	92	92
17th number selected	–	–	–	–	–	...	34	34
18th number selected	–	–	–	–	–	...	12	12
19th number selected	–	–	–	–	–	...	23	23
20th number selected	–	–	–	–	–	...	76	76

Box 2.2b   Sampling without replacement (excludes from the final set).

Stage (iteration)	1	2	3	4	5	...	19	20
Population (remaining)	100	99	98	97	96	...	82	81
Accumulated sample size	1	2	3	4	5	...	19	20
% population sampled	1.00	1.01	1.02	1.03	1.04	...	1.22	1.23
Probability of selection	0.01	0.01	0.01	0.01	0.01	...	0.01	0.02
1st number selected	64	64	64	64	64	...	64	64
2nd number selected	–	39	39	39	39	...	39	39
3rd number selected	–	–	90	90	90	...	90	90
4th number selected	–	–	–	64		...		64
5th number selected	–	–	–	–	05	...	05	05
6th number selected	–	–	–	–	–	...	53	51
7th number selected	–	–	–	–	–	...	92	92
8th number selected	–	–	–	–	–	...	18	18
9th number selected	–	–	–	–	–	...	73	73
10th number selected	–	–	–	–	–	...	70	70
11th number selected	–	–	–	–	–	...	88	88
12th number selected	–	–	–	–	–	...	52	52
13th number selected	–	–	–	–	–	...	73	73

(Continued)

Box 2.2b	**(Continued)**							
14th number selected	–	–	–	–	–	...	48	48
15th number selected	–	–	–	–	–	...		70
16th number selected	–	–	–	–	–	...	92	92
17th number selected	–	–	–	–	–	...	34	34
18th number selected	–	–	–	–	–	...		12
19th number selected	–	–	–	–	–	...	23	23
20th number selected	–	–	–	–	–	...	76	76
1st replacement number	–	–	–	–	–	...		71
2nd replacement number	–	–	–	–	–	...		84
3rd replacement number	–	–	–	–	–	...		02

Most investigations in Geography and the Earth Sciences use sampling without replacement to avoid the ludicrous situation of potentially including an observation more than once. Respondents in a survey would surely be rather puzzled, reluctant or annoyed, if they were asked to answer the same set of questions twice, thrice or even more times. Provided that the **sampling fraction**, the proportion or percentage of the population included in the sample is small, but nevertheless large enough to provide sufficient observations for the intended analysis, then the issues raised by the reducing population size and consequently increasing the chance of selection with each stage (iteration) of the sampling process are unimportant. Figure 2.1 illustrates the relationship between the sampling fraction and population size. The lines depict sampling fractions of two sample sizes (40 and 100 observations) selected from a population total that increments by 1 observation for each data point graphed along the horizontal (X) axis. Thus, the first point plotted for each line represents the situation where the population and sample sizes are equal; thereafter, the sampling fraction decreases, as the sample total becomes an ever-smaller proportion of the population, which continues to diminish beyond the 2500 shown in Figure 2.1. The shape of the graph is similar for both sample sizes with an exponential decrease in sampling fraction as the population size increases. Once the population total reaches 2500, the sampling fractions are indistinguishable, varying between 1.54 and 3.85 per cent, although up to this point there is some variation. The two sample sizes, which would not be untypical of those used in undergraduate investigations, pass through the 10.0 per cent sampling fraction level at population totals of 400 and 1000. The dashed lines in Figure 2.1 plot the difference in the probability for the first and last observation selected for inclusion in the sample, which extends the more limited case study in Box 2.2b. The general form of the curves is like those for the decrease in sampling fraction with considerable difference between the 40 and 100 size samples up to a population total of approximately 750. In summary, sampling without replacement is entirely feasible when the population has at least 1000 observations and the sample fraction does not exceed 10.0 per cent. One formal approach to determining sample size involves applying the following formula:

$$n = \frac{CI^2 p(1-p)}{m^2}$$

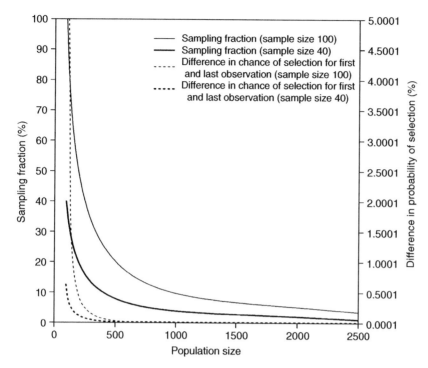

Figure 2.1 Relationship between sampling fraction and difference in probability for first and last sampled observations with population size.

where n is the sample size, CI is the confidence interval (see Chapter 6), p is the estimated occurrence of the variable of interest in the population (e.g. 0.3 of the population are pensioners) and m is the margin of error typically 0.05 (equivalent to level of significance (see Chapter 5)).

A sampling strategy based purely on simple random sampling is often difficult to achieve, irrespective of the sampling with or without replacement issue. There are frequently legitimate reasons for regarding the population of observations as being differentiated into a series of classes (i.e. the population is subdivided). The justification for defining these **strata** is usually a theoretical presumption that the observations of interest are fundamentally not all the same. Households living in a city may be subdivided according to the tenure of their accommodation, for example as owner occupiers, renting from a private landlord or renting from the local authority or housing association; soil and vegetation specimens may be taken from different types of location, for instance, according to aspect and altitude; medium-size enterprises may be selected on the basis of where they operate, such as rural, suburban or city centre areas. When such strata can be assumed to exist with respect to the entire population, **stratified random sampling** may be used to ensure that observations from each stratum are randomly selected in such a way that those in each are fully represented. This involves associating unique identification numbers to each observation in the different strata and then generating sets of random numbers corresponding to the ranges of the identifiers in the strata. In other words, a simple random sample is selected from each stratum.

However, there are several ways of achieving this: an equal number of observations can be selected at random from each stratum; observations can be chosen in the same proportion as each stratum's percentage of the total population; or after a simple random sample of observations has been selected from the whole population, it is subdivided according to the stratum to which they belong. Each approach has its strengths and weaknesses. Equalising the 'sub-samples' comes up against the problem that strata with few observations overall will be over-represented and those with large numbers will be under-represented. Sampling proportionately avoids this problem but requires knowledge about the number of observations in each of the strata. The riskiest of the three options is to take a single simple random sample and then subdivide, since this could easily result in no observations being selected from some strata, especially those with few members. Stratified random sampling is widely used, for example by organisations carrying out public opinion polls, and the decision on whether to use proportionate or disproportionate selection depends on the anticipated extent of variation between the strata. However, given that there is unlikely to be sufficient information available in advance of the survey to assess the extent of variation and that it will certainly vary between the different attributes and variables, then a proportionate approach is normally advised.

> Which strata might be suitable for a sample survey of farmers in your country? How easy would it be to obtain the information needed to identify the populations of farms in these strata?

The basic assumption of stratified random sampling is that there are sound theoretical reasons for subdividing the overall population into a series of strata. Equally, there are occasions when an investigator has no prior justification for stratifying the population in this way. Nevertheless, if the population comprises a very large number of observations, has a wide geographical distribution or is fluid in the sense that individuals flow in and out, subdivision into groups for the purpose of **cluster sampling** may be useful. The observations are allocated to mutually exclusive groups or clusters according to their geographical location or temporal sequence, for example households on streets or travellers passing through a train station during 30--minute time periods. A subset of clusters is selected at random (streets or time periods), and then either all observations or a simple random sample is chosen from each. This procedure assumes that the geographical location and/or temporal sequence of the observations do not produce any substantive differences between them in respect of the topic under investigation. This may be a risky assumption since many geographical phenomena exhibit Tobler's (1970: 236) first law of Geography: 'Everything is related to everything else, but near things are more related than distant things'. In other words, the same type of observations (i.e. from one population) that are located close together (e.g. living on the same street) are more likely to be similar to each other than they are to those that are further away. Since cluster sampling works by randomly excluding a substantial number of observations from consideration when selecting the sample, it should be seen as a way of helping to reduce the amount of time and money required to collect original data.

Cluster sampling has been used widely in Geography and the Earth Sciences, for instance, to select regular or irregular portions of the Earth's surface or 'slices' of time in which to obtain data relating to vegetation, soil, water, households, individuals and other phenomena. There are three problems with cluster sampling that are especially important in respect of Geography and the Earth Sciences. First, the focus of interest is often on discovering whether there is geographical (spatial) and/or temporal differentiation in the topic under investigation, which makes nonsense of assuming that where observations are located and when they occur is irrelevant to selecting a sample. Second, there is no way of being certain that the sampled observations are truly representative of the overall population, since the final sample is based on selection from a set of arbitrary clusters. These could be defined in many ways. Time may be divided into various combinations of temporal units (e.g. seconds, minutes, hours, months, years, decades, centuries or millennia), or into convenient epochs, for example when a political party held power, when an individual led an organisation or when the effects of a natural disaster persisted. Geographical space may be similarly dissected into portions on the basis of units of physical distance or area (e.g. metres, hectares and kilometres), of administrative and political convenience (e.g. census tracts, counties, regions, states and continents) or with cultural connotation (e.g. buildings, communities, gang territories and tribal areas). Third, the number of clusters and their 'sub-population' totals has a critical influence on the size of the sample. If clusters are defined to equalise the number of observations, then it is reasonably easy to determine sample size in advance, but random selection of clusters with uneven numbers of observations will produce an unpredictable sample size.

The principle of cluster sampling may be taken further by means of **nested sampling**, which involves successive subdivision and random selection at each stage down the tiers of a hierarchy. This system can be illustrated with the example of selecting areas of deciduous woodland to form a sample in which a count of flora species will be made. A regular grid of 100 km^2 placed over the country under investigation can be subdivided into 10 and 1 km units. First, a random group of the largest units is chosen, then another of the intermediate units and finally one of the 1 km^2. The areas of woodland lying in these squares constitute the sample. In addition to selecting a sample of observations with respect to a nested hierarchy of geographical or spatial units, it is also possible to select from a hierarchy of time slices. For example, two months in a year might be chosen at random, then five days in each month and then four hours on those days. The problems associated with cluster sampling also apply to nested sampling; in particular, the number of clusters chosen randomly at each stage strongly influences the eventual sample size and the extent to which it represents the population.

One further sampling strategy that introduces an element of chance is to use **systematic sampling**. The procedure involves selecting observations in a regular sequence from an initial randomly determined starting point, for example every 44th person encountered in a shopping mall, every house number ending in a 3 on a city's residential estates or every 15th individual met on a tourist beach, where the numbers 44, 3 and 15 were selected randomly. Systematic sampling is particularly useful when drawing a simple random sample is beyond the resources of the investigator and the total number of observations is difficult to identify and quantify. The method assumes that the observations themselves are distributed randomly in time and space, and since regular systematic selection starts from a random point in this distribution, the resulting sample should be reasonably representative. The problem is that it may be difficult to determine if the observations are randomly distributed and that no obscure, underlying pattern is present introducing undetected bias.

This difficulty is exacerbated in many geographical and Earth scientific investigations, since they are often underpinned by a belief that spatio-temporal variation is of crucial importance. They are seeking to explore the processes that have produced certain spatial and temporal patterns of observations. In these circumstances, it is hardly appropriate to sample on the basis that the observations are in fact randomly distributed, when the aim is to discover if this is the case. However, provided that the presence and potential impact of an underlying pattern in the population of observations is recognised and that the systematic sample does not coincide with this regularity, so that only one subset of observations are included and not the full range, then systematic sampling is a feasible option. The presence of such patterns only really becomes problematic when their potential impact on the sample is ignored.

The random component that gives strength to simple random sampling clearly diminishes moving through stratified, cluster, nested and systematic sampling. Boxes 2.3 and 2.4 illustrate the application of these strategies to human and physical geographic research topics. The illustration in Box 2.3 depicts the concourse of a major railway terminus in London with people passing through on their daily commute to and from work, *en route* to and from airports and seaports, when visiting shops in the station or for other reasons. There are many entrances and exits to the concourse, resulting in a constant flow of people arriving and departing, connecting with other modes of transport or using the station for various other reasons. The statistical population of observations (individuals) formed of the people on the station concourse in a 24-hr period, even if not strictly infinite, is fluid. Selecting a sample of people for inclusion in a survey using simple random sampling, where all people can be identified, is impractical. The examples of applying stratified, cluster, nested and systematic sampling to this situation are intended to be illustrative rather than prescriptive.

Box 2.3   Sampling strategies in transport geography: Liverpool Street Station, London.

Each of the alternatives to simple random sampling seeks to obtain a sample of observations (rail passengers) from the concourse at Liverpool Street Station in London. Passengers continually enter and leave the station concourse, which has a number of entrances and exits, and hence it is impossible to identify a finite population with unique identifier numbers, which rules out the option of simple random sampling.

Stratified random sampling:
The concourse has been divided into three zones based on what appear to be the main throughways for passenger movement and areas where people congregate (part of each stratum is shown in the image). An interviewer has been positioned in each of these zones. A set of 25 three-digit random numbers has previously been generated. Interviewers have been instructed to approach people and ask for their mobile or landline telephone number. If this includes any of the sequences of three-digit random number sequences, the person is interviewed for the survey.

Cluster sampling:
Seven zones in the concourse have been defined arbitrarily based on the entrances and exits. Two of these have been selected at random. Interviewers are positioned within each of these areas and select people by reference to a previously generated set of five random numbers between 1 and 20 that tells the surveyor the next person to approach once an interview has been completed (e.g. 4 denotes selecting the 4th person, 11 the 11th and so on).

Nested sampling:
The same two randomly selected zones used in the cluster sampling procedure are used for nested sampling. But now the surveyors attempt to interview all passengers and visitors within these areas during a randomly selected 15-min period in each hour.

Systematic sampling:
The team of four interviewers have been positioned across the concourse (circled locations on the image) and have been instructed to approach every 8th person for an interview. This number was chosen at random. The interviewers have also been told to turn to face the opposite direction after each completed interview so that interviewees come from both sides of the concourse.

Box 2.4 illustrates an area of lateral moraine of Les Bossons Glacier in France in the mid-1990s that was built up during the glacial advances during the so-called Little Ice Age in the 18th century. The glacier's retreat in recent times has exposed the lateral moraines, enabling an investigation to be carried out into the extent to which the constituent rock debris has been sorted. The material ranges in size from small boulders and stones to fine particulate matter, providing a partial clay matrix. There is a break of slope on the moraine, above which the angle of rest of the partially consolidated material is $55°$ and below the surface slopes gently at $12°$. The upper section is stable in the short term, although occasionally there are falls of small stones. Thus, the observations (stones and pebbles) forming the statistical population of surface material may be regarded as infinite and changeable as various depositional and erosional processes move them across the slopes of the moraine and into the outwash stream. The illustrated applications of stratified, cluster, nested and systematic sampling strategies to this problem represent four of many approaches to sampling material from the surface of the moraine.

Box 2.4 Sampling strategies in geomorphology: Les Bossons Glacier, French Alps.

The four alternatives to simple random sampling have been applied by overlaying the chosen study area with a 1-m grid (each small square represents 1 m^2). Each sampling strategy seeks to obtain a sample of observations (stones, pebbles and similar rock debris) from the surface of the moraine that can be measured with respect to such morphological variables as size, shape and angularity. The difficulties of identifying a finite population with unique identifier numbers make simple random sampling impractical.

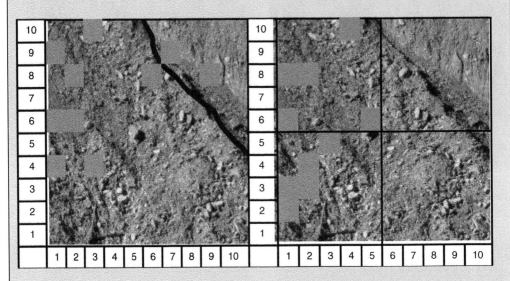

Stratified random sampling:
The study area has two strata according to slope angle: 20% of the surface area (20 m^2) is steeper (upper right). A sample of grid cells selected using two sets of paired random numbers covers 10% of the 100 m^2 with 20% (2 m^2) and 80% (8 m^2) in steeper and gentler areas.
7, 9; 9,8.
2,6; 2,8; 1,6; 3,10; 1, 9; 1,4; 6,8; 3,4.

Cluster sampling:
The study area has been divided into four equal-sized quadrants of 25 m^2. Two of these have been selected at random, and then 2 sets of 5 random number pairs within the range covered by each quadrant provide a 10% sample of grid cells across the study area.
2,8; 5,6; 1,6; 4, 10; 1,8.
1,3; 1,2; 2,3; 2,4; 3,5.

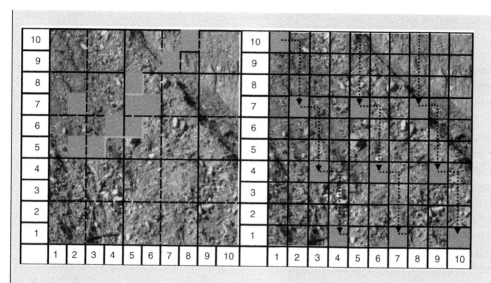

Nested sampling:
The study area has been covered by another grid with four whole squares of $16\,m^2$ and parts of 5 other similar units. These are further subdivided into $4\,m^2$ quadrants and then 1m squares. In two randomly selected $16\,m^2$ squares four pairs of random numbers and in one of the $16\,m^2$ squares two pairs have been selected. These provide the grid references for the sample of ten 1 m squares, representing a 10% sample of the study area.

2, 5; 3, 5; 2, 7; 4, 6.
5, 7; 5, 6; 5, 8; 6, 7.
8, 10; 7, 9.

Systematic sampling:
A random starting point (1, 10) within the grid of $1\,m^2$ squares has been selected. Then moving to the next column and down by a randomly determined number of squares (4) and so on across the columns provides a 10% sample of the overall study area.

What are the strengths and weaknesses of the sampling strategies shown in Boxes 2.3 and 2.4? List some other situations where they could be applied.

2.2.2 Subjective Sampling Techniques

An element of subjectivity exists in most sampling techniques, even in those based on probability examined in the previous section, although they are regarded as more representative of the parent population and free from investigator bias. For these reasons, they are considered preferable for collecting data that are to be analysed using quantitative statistical techniques. However, an alternative view is that, since we can rarely, if ever, realistically achieve the ideal of simple random sampling and the shortcomings of the alternatives render them incapable of providing a truly representative sample, we might as well make use of overtly subjective approaches. Subjective sampling techniques place less emphasis on obtaining a sample that is representative of a population, preferring to rely on the investigator's perception of what constitutes a selection of sufficiently diverse and typical observations. The difficulty with this approach is that what might seem like an entirely acceptable sample of observations to one person may not be regarded as such by another researcher.

One of the main differences between a sampling strategy based on some element of chance and one relying on subjectivity is the question of replication. If two researchers, A and B, each selected a simple random sample or at least a partially random sample from a population of observations and they were to carry out identical analyses, they should produce comparable results so that both reached the same conclusions. A subjective sampling strategy that relies on an individual investigator's or a research team's collective perception and interpretation of the population will inevitably result in an inability to replicate and verify results. Qualitative analysis does not demand the opportunity for replication but recognises the uniqueness of events and the investigator's role in influencing the research process. Proponents of qualitative research often maintain that the objectivity afforded when employing chance in a sampling strategy is more illusory than real, since the definition of a statistical population entails subjective judgement.

The sampling strategies illustrated in Boxes 2.3 and 2.4 for selecting samples of observations using an element of chance are indicative of the many situations in which it is difficult to define or identify a statistical population. An alternative option would be to employ **convenience sampling**, in other words, to select those observations that are accessible without excessive cost, danger or inconvenience to the investigator. Using convenience sampling to select stones and pebbles for measurement on the lateral moraine of Les Bossons Glacier might result in a daring researcher venturing onto the steeper, dangerous slopes to collect specimens, whereas a less intrepid investigator might only select material while traversing the gentler area near the outwash stream. Similar differences could arise when sampling passengers conveniently in the concourse of Liverpool Street station: one self-assured interviewer might approach anyone as a potential respondent to the survey, whereas a more reticent interviewer might only speak to people who appeared 'friendly' and willingly cooperated. Passengers who agree to take part in the survey may be regarded as self-selected volunteers, and the sample could end up comprising docile, amiable people with time on their hands, rather than a representative cross section of visitors to the concourse. These cases illustrate the two-way interaction between the investigator and the individual observations that comprise the population, which makes convenience sampling a risky option when the reactions of surveyors and observations cannot be standardised or at least deemed equivalent.

An alternative method, relying on the opinion of the investigator rather than the convenience of the observations, is **judgemental sampling**. This strategy requires that the investigator can obtain some initial overview of the population, for example by visual inspection of a terrain, by perusal of published statistics or by examination of a cartographic or remotely sensed image, to decide what the population is like. This assessment provides the basis for selection of a set of individual observations that the investigator judges will constitute a 'representative' sample. The main problem with this approach is that we are not particularly good at making unbiased judgements, especially from limited information, people are inclined to overestimate the number

Figure 2.2 Passengers on the concourse of Liverpool Street station, London.

of observations of some types and to underestimate others. An investigator casting his/her eyes over the scenes presented in Boxes 2.3 and 2.4 would have difficulty in deciding how many stones of varied sizes or how many passengers in different age groups to select for the respective samples. Figure 2.2 shows another image of people in the concourse of the same London station. Examine the image carefully, and see how difficult it is to make a judgement about basic characteristics of the population, such as the proportions of people who are male and female. Judgemental sampling is often used when only limited information can be found about the nature of the population, but even so it is important that its validity is checked and the investigator's judgement is as free from bias as possible.

Quota sampling can be viewed as the subjective equivalent of stratified random sampling as far as it presumes that the population subdivides into two or more separately identifiable strata. However, the similarity ends here, since in other respects quota sampling, by combining features of the convenience and judgemental approaches, introduces the negative aspects of both. The procedure for quota sampling starts from two decisions: how many observations to select and whether to split these equally or proportionately between the strata. The latter requires the researcher to estimate the proportions of observations in the different strata, which may be difficult given a lack of detailed knowledge about the population. Investigator judgement may inflate or deflate the number of observations chosen from some categories, or it may be part of the research design that certain strata are over-/under-represented. Whichever method is required, the investigator proceeds by filling the strata up to the predetermined limit with conveniently accessible observations. An investigator using quota sampling in the earlier transport geography and geomorphology projects might employ the same strata as outlined previously but select observations with reference to accessibility and judgement rather than chance. Deciding when sufficient observations in each stratum have been included is one of the main problems with quota sampling, which comes back to the problem of not having sufficient knowledge about the structure and composition of the population in advance.

One of the most common reasons for an investigator moving away from a strategy based on random selection towards one where subjectivity plays a significant part is the difficulties of adequately identifying and delimiting the population. A carefully designed strategy based on simple random sampling will still fail if the population is incompletely defined. Some members of a single population may be more difficult to track down than others. For example, farmers managing large-scale enterprises may be listed in trade directories, whereas those operating smaller or part-time concerns may be less inclined to pay for inclusion in such listings. Thus, the population of farmers is imperfectly defined. One way of partially overcoming this problem is to use an approach known as **snowball sampling**, which involves using information about the known members of the population to select an initial, random sample and adding to this with others that are discovered by asking participants during the data collection phase. The obvious disadvantage of this approach is that the composition of the sample is determined by a combination of the investigator's subjective judgement and the views and suggestions of those people who readily agreed to participate in the survey. For example, an investigator may be told about other potential respondents who are likely to cooperate or to have particularly 'interesting' characteristics, whereas others who may be reluctant to participate or considered 'ordinary' may not be mentioned. Increased use of social media as a way of delivering surveys has seen an expansion in snowball sampling as respondents are invited to share the questionnaire with their virtual, online friends.

Purposive sampling is another type of non-random approach often attractive in situations where it is difficult to identify the total population. It involves the researcher selecting participants in accordance with already established criteria. It represents another version of **judgemental sampling** because the approach relies on the researcher's judgement about the suitability of individuals or other entities for inclusion in the sample as a basis for achieving the desired research objectives (the purpose of the research). This inevitably introduces some subjectivity and bias into the selection process and is therefore rarely used with statistical analysis unless it forms part of a mixed methods strategy in which complementary quantitative and qualitative analyses are carried out.

2.2.3 Closing Comments on Sampling

The observations in geographical and Earth scientific investigations are often spatial units or features. In theoretical terms, such populations tend to be difficult to identify, because there are an infinite number of points, lines and areas that could be drawn across a given study area and hence a similar number of locations where measurements can be taken. The grid squares, used as basic spatial building blocks in Box 2.4, were entirely arbitrary and reflected a decision to use regular square units rather than any other fixed or irregular areas, to arrange these on the 'normal' perpendicular and horizontal axes rather than at any other combination of lines, at right angles to each other, and to define their area as 1 m^2. Selecting observations for inclusion in a sample as part of a geographical or Earth scientific investigation needs to distinguish whether it is the spatial units or the phenomena that they contain that are of interest (i.e. are the relevant observations). For example, an investigation into the extent of sorting in lateral moraine material needs to focus on the characteristics of the stones, pebbles and other measurable items rather than the gird squares, and a project on the reasons for passengers visiting a major rail station is concerned with individual people rather than their location in the concourse. However, location, in the sense of where something is happening, frequently forms a context for geographical and Earth scientific enquiries. Thus, the location of the lateral moraine as adjacent to the western flank of Les Bossons Glacier in the European Alps and the situation of Liverpool Street station close to the City of London and as a node connecting with North Sea ports and Stansted airport provide geographical contexts in these examples.

No matter how perfect the design of a particular sample is, unforeseen events can occur that lead to the data collected being less than perfect. Consider the example of a company running the bus service in a large city that decides to carry out a simple on-board survey of passengers to discover how many people use a certain route, their origins and destinations and their reason for making the journey. The survey is carefully planned to take place on a randomly selected date with all passengers boarding buses on the chosen route being given a card to complete during the course of their journey, which will be taken from them when they get off. Unfortunately, one of the underground lines connecting two rail termini is closed temporarily during the morning commuter period due to trains being unexpectedly out of service. An unknown proportion of commuters who usually take the underground between the rail stations decide to travel by buses on the route in the survey. Unless the survey team become alert to this unanticipated inflationary factor, the number of passengers normally travelling the bus route is likely to be exaggerated. This unfortunate change of circumstances is not as fanciful as it might at first seem since this situation arose when the author participated in such a survey on his normal commute through London when the first edition of this text was being written.

2.3 Specifying Attributes and Variables

This section could in theory be endless! The list of attributes and variables that are of interest to Geographers and Earth Scientists in their investigations is potentially as long as the proverbial piece of string. Any attempt to compile such a list would be futile and certainly beyond the competence of a single person or even outside the scope of an international 'committee' of Geographers and Earth Scientists. There have been attempts to produce lists of attributes and variables relating to a theme, such as the 'Dewdney' set of demographic and socio-economic indicators from the 1981 British Population Census (Dewdney, 1981) or Dorling and Rees' (2003) comparable counts from British Population Censuses spanning 1971–2001. However, in both these cases the researchers' efforts were constrained to those counts that were contained in the published statistics, and, while they might have had their personal wish list, this could not be realised because they were referring to data from secondary sources. Similarly, there are sets of indicators of environmental quality. Such lists may not be universally accepted, and the exclusion or inclusion of indicators is likely to be a contentious issue.

A comprehensive, all-embracing list of geographical and Earth scientific attributes and variables that we can call upon when undertaking an investigation may not exist; nevertheless, there are some general principles that can be established to promote more successful research endeavours. Section 2.2 suggests two alternative responses when faced with the question of how many observations to include in a sample – including the maximum or the minimum. Deciding what attributes and/or variables to collect could easily present a similar dichotomy. Collect as many as possible just on the off chance that they may be useful or seek the minimum necessary to answer the research question specified at the start of the investigation. The first approach is likely to lead to excessive levels of data redundancy as things are collected that are never used. Whereas the minimalist approach may answer the question but prevent any opportunity for serendipitous discovery and going back to collect more data is normally out of the question since you cannot recreate the situation exactly as it occurred on the first occasion.

At the start of an investigation, you not only need to have research question(s) that you want to answer, but also some idea of what kind of information would provide answers to these questions. In a different context, if you wanted to discover how many British cricketers had hit six sixes in an over of world-class cricket, then you would turn to Wisden (www.wisden.com), The Cricketers Almanack, and not to the records of the Lawn Tennis Association. This chapter started by defining geographical data as attributes and variables, called **data items**, about geographical or spatial 'things' – those phenomena that occupy a location under, on or near the Earth's surface. We need to distinguish between geographical entities or phenomena, such as cities, lakes, sand dunes, factories, glaciers, road networks, national parks and rivers and the information collected about them. In other words, a researcher needs to be clear about the units of observation and the attributes or variables used to express certain characteristics or quantities of these units. Some of these attributes or variables may apply to distinct categories of an entity. For example, we may be interested in a farm or urban green space, or similarly the volume of water in a lake or frozen in an ice sheet. The examples depicted in Boxes 2.3 and 2.4 have indicated that it would be feasible to collect data relating to some of the people on the station concourse or some of the material on the surface of the moraine and still reach valid conclusions about the phenomena as a whole (i.e. passengers in the train station and debris on the Bossons Glacier moraine). The next stage in planning our research involves deciding exactly what attributes and variables we will collect or acquire in respect of the geographical phenomena of interest.

Geographical phenomena (cities, lakes, sand dunes, etc.) are commonly separated into three types based on their spatial properties, namely points, lines and areas. Additionally, surfaces are sometimes regarded as a further category of geographical feature, although in other respects these may be treated as an extension of area features into three-dimensional entities. A lake can simply be considered as a two-dimensional surface of water, although we may also be interested in its subsurface characteristics. For example, microorganisms (flora and fauna) are often dispersed at various levels of concentration, and systematic sampling throughout the water body would enable this variability to be captured in a (sub-) surface model. In an equivalent way, people are dispersed at varying levels of density within urban settlements and these patterns may change over time. Quantifying the number of people per unit area (e.g. per hectare) allows surface models to be created. However, some features change from being points to areas and from lines to areas depending upon the **scale** at which they are viewed cartographically and the nature of the research questions being asked. For example, the city of Norwich may simply be viewed and located as a point on a map of all cities in the UK and in a study of urban population change be represented as a graduated symbol whose size depicts a certain level of population growth or decline between two census dates. Figure 2.3(a) illustrates a range of estimated population growth for selected English cities between 2001 and 2006 and indicates that Norwich's population increased between 5.0 and 9.9 per cent. However, such a study may choose to focus on population change within and between areas of the city in which case the detail of Norwich's urban morphology and the arrangement of its socio-economic and physical features are likely to be relevant. Figure 2.3(b) records estimated population growth and decline in Norwich's wards between 1991 and 2001 adapted to compensate for boundary changes together with selected features of the urban landscape (major roads and buildings). In the present context, the relevance of Figure 2.3 lies not in a discussion of the population changes themselves, but in reflecting on how geographical entities and research questions need to be related to the scale of the investigation.

Boxes 2.5, 2.6 and 2.7 illustrate some of the attributes and variables that might be associated with a selection of point, line and area geographical phenomena. Chapter 3 will examine how we can locate geographical phenomena by means of coordinates in a grid referencing system. However, for the present it is relevant to note that point features require one pair of coordinates (X and Y), lines a minimum of two pairs (X_{start}, Y_{start}; X_{end}, Y_{end}) and areas (polygons) at least three, and usually many more pairs to specify their location and boundary. Before discussing examples of point, line

(a)

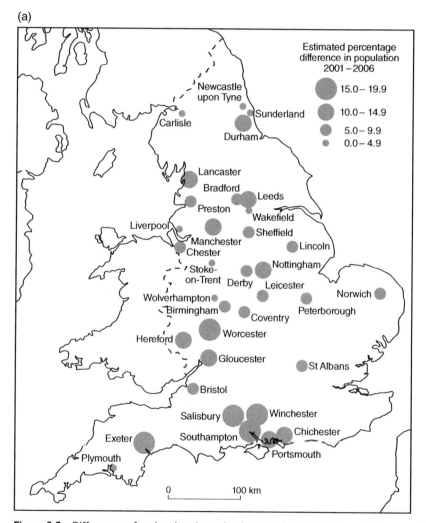

Figure 2.3   Differences of scale when investigating population increase and decrease. (a) Estimated population change in a selection of English cities, 2001–2006. (b) Estimated population change in Norwich, England, 1991–2001.

(b)

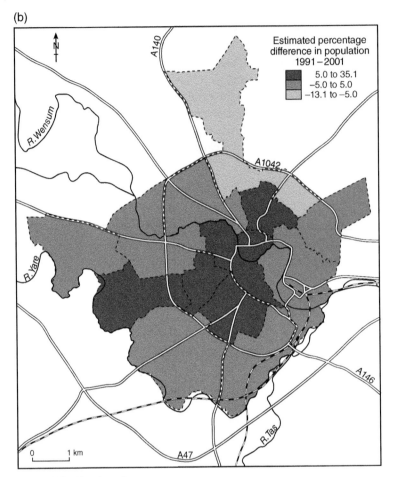

Estimated percentage difference in population 1991–2001

- 5.0 to 35.1
- −5.0 to 5.0
- −13.1 to −5.0

Figure 2.3   (Continued)

and area features, it should be noted that the attributes and variables associated with them could relate to their geometrical properties, their spatial dimensions (e.g. the elevation of points, the length of lines or the shape of areas) or to the features themselves. We have already referred to population density where the concentration of people is expressed as a ratio in respect of the area of land they occupy (e.g. persons per km^2). There are many occasions when we want to relate an attribute or variable associated with an entity to one or more of its spatial characteristics. If such measures as population density or discharge from a drainage basin are expressed in a standardised fashion, respectively, as persons per km^2 or cubic metres per second (CUMECs), they become comparable in different contexts and universally applicable.

2.3.1   Geographical Phenomena as Points

Points represent the simplest form of geographical feature and are, from a theoretical perspective, dimensionless, in the sense that they do not have a linear or areal extent: hence the importance of being aware of scale in an investigation. Nevertheless, we will often be interested not in a single isolated point, for example just the city of Norwich, but in the occurrence of some or all points of the same type. Rarely are such collections of points located on top of each other, but more commonly occur within a defined area or along a line and can therefore be related to these more complex types of features. For instance, we may be interested in the density or distribution of points in an area, or the distance between points. Most point features are land-based, and there are few located at sea, and some of those that might be treated as points, such as small rock outcrops may appear and disappear with the rise and fall of the tide. Although such features do not vanish when the sea covers them, they do illustrate the ephemeral or transitory nature of some types of geographical entity. While there may be some features at sea that could be regarded as points, for example oil platforms and wind turbines, although these have an areal extent, there are no geographical point entities at fixed locations in the Earth's atmosphere, although some point features anchored to Earth's surface do not extend some considerable distance into the atmosphere, such as telecommunication masts.

Another feature illustrating this characteristic is a skyscraper block, such as those shown in downtown Honolulu in Hawaii in Box 2.5a. Although they are viewed obliquely in the photograph from the 17th floor of one such building and clearly each has a footprint that occupies a physical area, the buildings can be treated as point features with a discrete location. Such areal extent is a less obvious characteristic of the example of point features as shown in Box 2.5b. These are some of the trees across the moraines of the Miage Glacier in the Italian Alps. Using the X, Y coordinate location of the skyscrapers and the trees, the spatial variables relate to the density of blocks or trees per unit area (e.g. per km^2 or per m^2) and to the distance between buildings or trees. In the case of skyscrapers, simple density measurements may be weighted according to number of floors on the basis that higher buildings with more stories will have more impact on the urban landscape. Similarly, the density of trees might be weighted according to their height or the width of the canopy.

Box 2.5 also includes some substantive attributes and variables relating to both types of point feature. These are illustrative because the details of which attributes and variables are required will depend upon the aims and objectives of the investigation. The skyscrapers provide an example of how something that might be regarded as a single-point feature may in fact be subdivided in terms of some of its characteristics. As a single geographical feature, each tower block has several floors, a number of lifts or elevators and a date of completion (age), although in respect of other characteristics, such as the use or ownership of the building, it may have multiple values. For example, a single block may be used entirely for offices and occupied by one corporation operating in the financial services sector or a hotel run by one company. In contrast, another building might be in multiple use, with a gym, shop and restaurant on the ground floor and residential apartments on the remaining floors. Some characteristics, such as the address or post (zip) code of the building, which might initially seem to have one value, could in fact have many different values. If such a tower block in the UK was occupied by several companies, they are likely to each have a different 'large user postcode'. Similarly, if it were an apartment block, then each dwelling unit would have its individual identifying number so that the mail for households was delivered correctly. These examples illustrate what is often referred to as hierarchical data, where one characteristic, such as occupier, may have one or many different genuine values. In some cases, we may have to deal with multiple hierarchies, such as where a skyscraper not only has multiple uses present (e.g. retailing on the ground floor, financial services on floors 1–10 and residential from 11 to 20), but each of these uses may also be subdivided, with different retail firms on the ground floor, several financial corporations and many households in the residential apartments. Some examples of attribute variables for the trees on the Miage Glacier moraines are included in Box 2.5b. These have been selected from the collection of forest mensuration variables that are of interest to those involved with timber management and production.

Box 2.5   Geographical features represented as points.

(a) Honolulu skyscrapers

Skyscrapers represent point features in an urban environment.

Spatial attributes/variables:
- X, Y location
- Distance to the nearest neighbour
- Density per km^2

Substantive attributes/variables:
- Use (e.g. residential, finance and leisure)
- Single/multiple occupancy
- Occupants' details
  - Company name
  - Number of employees
  - Turnover
- Address
- Postcode
- Number of floors

Trees represent point features in different types of environments.

Spatial attributes/variables:
- X, Y location
- Distance to the nearest neighbour
- Density per km^2
- Height

Substantive attributes/variables:
- Species
- Age
- Canopy cover

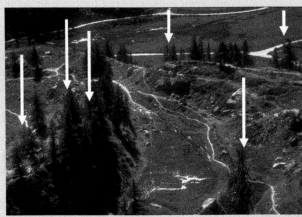

(b) Trees on the Miage moraines, Italian Alps

The list of attributes and variables shown in Box 2.5b does not include any that could be regarded as hierarchical (i.e. potentially having multiple values for the same individual specimen). Can you specify any additional attributes or variables associated with the trees that might be hierarchical?

2.3.2   Geographical Phenomena as Lines

Simple linear features follow a path through or across space and represent a connection between a minimum of one point to another. Often these are referred to, especially in the context of Geographical Information Systems (GIS), as 'start' and 'end' points, which implies a direction of flow along the line. However, this may have more to do with how the digital coordinate data representing the line were captured and held in the system (i.e. with which terminal point on a line was digitised first), than with the flow of something along the line. Just as the occurrence of a single point is rarely of interest, so it is also unusual to concentrate on one instance of a simple linear feature. Groups of lines can be compared to see if they follow the same bearing or direction, whether some are more sinuous, or some shorter or longer than others. It is also possible to determine the density of lines within an area and whether the lines display a clustered, regular or random pattern. Lines are also inherently land-based features, although again some linear features may describe paths over the surface of the Earth's seas and oceans, for example shipping lanes that are used to help navigation. But such features are not visible on the water's surface; instead, they are shown in an abstract form on navigation charts, and even the wake from passing vessels is transitory and soon dispersed by the waves, which are also ephemeral linear features moving across the water's surface.

Box 2.6a illustrates the multiple occurrences of a particular type of linear feature. The linear scratches formed by glacial ice dragging sub-surface moraine over a solid rock on a valley floor, referred to as striations, often occur in groups where the individual lines are not connected with each other. These striations are micro-features in the landscape and again illustrate the importance of scale in an investigation. In detail, such features as striations may be considered not only as having a linear extent but also depth and width as well as length. Box 2.5a includes some of the spatial variables associated with striations, such as orientation and length, which may be used to indicate the direction of flow of the glacier and the pressure exerted on the rock where they occur. Such features are therefore indicative of the erosive force of the glacier that created them as well as the quantity and hardness of the moraine being dragged along by the ice. Box 2.6a also includes some of the substantive attributes or variables that might be relevant in respect of the striations and the situation in which they occur, such as rock type and hardness.

A group of simple lines, whether straight or sinuous, may also form a network, which implies not only a direction of flow along the lines, but also that what flows along one line can pass to another. A flow along a line may be two-way (bidirectional) or constrained to one direction either physically (e.g. by a downward slope allowing the flow of water in a river) or artificially (e.g. by a 'No entry' sign in a transport network). The junctions where the lines meet are referred to as nodes, which denote locations in the network where the material in transit can proceed in two or more directions. The regulation of the flow at such nodes may be achieved in diverse ways depending on the type of network. For example, in a road transport network the flow of vehicles may be controlled by signals that prevent drivers from taking a right turn, whereas in drainage networks slope and the velocity of flow may prevent the flow of water from the main channel into a tributary. It is sometimes suggested that there are two main types of linear network known as closed and open. However, totally closed networks are extremely rare and most networks can be regarded as having entry and exit points. In a hydrological network, for example, water flows through from source to destination and there is rarely any significant backward flow, although this is not necessarily the case in other instances of open networks such as a metro system where there will be a two-way flow of passengers and trains along rail tracks going in opposite directions.

Box 2.6b shows some of the watercourses formed by glacial meltwater streams from the Lex Blanche Glacier in the Italian Alps. These flow across a flat area composed of a mixture of rocky debris brought down by the streams and boggy wet areas where silt has accumulated and vegetation becomes established. After flowing across this expanse, the streams combine and cascade through a narrow gap, towards the lower left in the photograph, and thus unified carry on down the valley. Some of the streams in the network flowing across this area have been highlighted in the photograph, although there is clearly a fairly high degree of bifurcation and braiding as the main channels divide and reunite as shown lower right in the photograph and is apparent in the distance. This example demonstrates how the notion of a hydrological network as being formed from a series of simple lines may be misleading. Even assuming that the main channels can be identified, as has been attempted with the annotation lines on the photograph, these will vary in width and depth, and in some places, the water is stagnant, spreading out into areas of surface water. Thus, the conceptualisation of the streams as lines is a matter of convention and one of scale. For the purposes of investigating the hydrology of a drainage basin, the linear model applies well, since it is possible to quantify the capacity of the water to transport diverse types of material (chemicals, particulate matter, etc.). However, from a geomorphological perspective the dynamics of the streams as geographical features need to be considered. Since these streams are fed by melting water from glacial ice, there will be seasonal and diurnal variation not only in flow velocity but also in the number of streams and the location of their main channel. Thus, the network of streams is not a static but a dynamic network.

Box 2.6   Geographical phenomena represented as lines.

(a) Striations on an erratic in the European Alps

Striations are micro-features on rocks in glaciated environments.

Spatial attributes/variables:
- Length
- Depth
- Angle
- Density per m^2

Substantive attributes/variables:
- Rock type
- Rock hardness

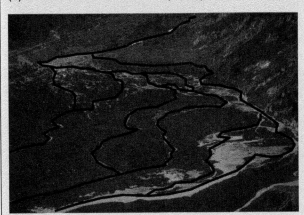

(b) Streams flowing from the Lex Blanche Glacier, Italian Alps

A **stream or river** represents an environmental unit in a network.

Spatial attributes/variables:
- Length
- Width
- Connectivity
- Cross-sectional profile

Substantive attributes/variables:
- Water velocity
- Concentration of suspended matter
- Water pH
- Water temperature
- Air temperature

The list of substantive attributes and variables shown in Box 2.6b (water velocity, concentration of suspended matter, etc.) relate to the water flowing through the streams. Assuming suitable field equipment was available to you, where would you set about measuring these variables and how often should such measurements be made?

2.3.3 Geographical Phenomena as Areas

The discussion and examples of point and line features have noted that whether these are treated as areas is a matter of the scale and the purpose of the investigation. However, there are many geographical features that are genuine areas with substantial physical extent, which cannot realistically be conceived as points or lines. The fundamental idea behind what makes up an area feature is that it occupies a portion of the Earth's that is undifferentiated in respect of at least one characteristic. Conceptualising areas in this way leads to a recognition that there are two ways in which such portions of the Earth's surface might be defined: by means of the boundary drawn around the area or by identifying where concentrations of values of the defining characteristic occur. In the first instance, the presence of a boundary line around a feature implies there is an abrupt change in the characteristic, whereas in the latter there is a gradation across the surface between high and low values. The farm boundaries annotated onto the photograph in Box 2.7a illustrate the former situation, since they denote where the land occupier alters from one person or company to another. Similarly, the boundaries of the London Boroughs in Box 2.7b signify where the responsibility for local government passes from one council to another. However, the upland landscape in Box 2.7c shows part of a slope where there is a change of vegetation type from improved grassland in the foreground to more mixed vegetation in the middle distance, where the grass grows in tussocks and is interspersed with other taller species. Drawing a boundary line between these two combinations of vegetation types to delimit zones is more problematic than in either of the previous types of area feature, since improved grass dominates the floral assemblage in some parts of the fuzzy border zone, whereas in others the tussocky grass and taller species are dominant. The dashed line annotations indicate two of the many possible boundaries that might be drawn between these vegetation zones.

Geographical features in the form of areas are sometimes examined individually, for example in studies of a particular mountain range or country. However, such large features are more often subdivided into smaller units, such as the individual mountains and valleys in a mountainous region or the local authorities in a country. An investigation may focus on a sample of such units and further divide these into their constituent components. In general, there has been a trend in recent years for researchers to concentrate on smaller units as the basis of their investigations, and in some fields of human geography, this has resulted in small numbers of individuals becoming the focus of enquiries. Where the investigation involves several areal units, their spatial characteristics, shape, physical area, distribution and juxtaposition may be of interest. Many area features of interest to Geographers and Earth Scientists are inherently land-based, particularly when conceived as delimited by a boundary line. Thus, although the territorial limits of countries may extend a certain distance from the mean tide line around the coastline, this does not result in a visible boundary out to sea. The various oceans and seas as well as lakes are themselves area features, although the water molecules, fauna and flora they contain are of course able to move between one of the 'liquid' areas and another. However, there are occasions when the marine environment contains area features. Spillage of oil from tankers can spread out across the surface of the water, thus forming an area of pollution. Although such features may be ephemeral, they will only remain until the salvage teams have secured the flow of oil leaking from the vessel and applied dispersal chemicals to aid the breakup of the area by microorganisms and wave action. The temporary existence of such an oil slick does not invalidate it as an area feature, since geographical features are rarely permanent, immutable fixtures of the Earth's landscape or seascape. An increasingly well-known example of such a marine 'area' is the Great Pacific Garbage Patch comprising debris from countries around the Northern Pacific (Dautel, 2009).

The three geographical area features in Box 2.7 include examples of spatial and substantive attributes or variables. The farm represented in Box 2.7a illustrates how one occurrence of an area feature might comprise several physically separate but substantively connected land parcels. This example also shows how a geographical feature of one type (i.e. a farm) may itself be subdivided into another type of entity, in this case fields, which are visible through the shading. Thus, our geographical data may not only be hierarchical in relation to the substantive attributes and variables, as illustrated with respect to the occupancy of the skyscraper blocks in Box 2.5a, but also in respect of the features themselves having constituent spatial components, potentially with distinctive characteristics. In this case, some of the fields on the farm may grow the same crop but have different types of soil. The example attributes and variables for the London Boroughs (Box 2.7b) indicate some of the diverse types of information that may be relevant in a particular investigation, spanning not only demographic and political attributes but also environmental indicators. Finally, the attributes and variables included in Box 2.7c represent some of those that might be used to help with defining the boundary between one vegetation zone and the other. In this, we could envisage covering the hill slope with a regular 1 m² grid and counting the numbers of different floral species or the soil pH in each square and using these characteristics to allocate the squares between each of the vegetation zones.

Box 2.7 Geographical phenomena represented as areas.

(a) A farm in the Auvergne, France

A **farm** represents an environmental and socio-economic unit.

Spatial attributes/variables:

- $X_1, Y_1 \ldots X_n, Y_n$
- Area
- Number of land parcels

Substantive attributes/variables:

- Soil types
  - Loam (%)
  - Sand (%)
  - Clay (%)
  - Etc.
- Crops grown (area)
  - Cereals (ha)
  - Potatoes (ha)
  - Vegetables (ha)
  - Etc.
- Diversification present/absent

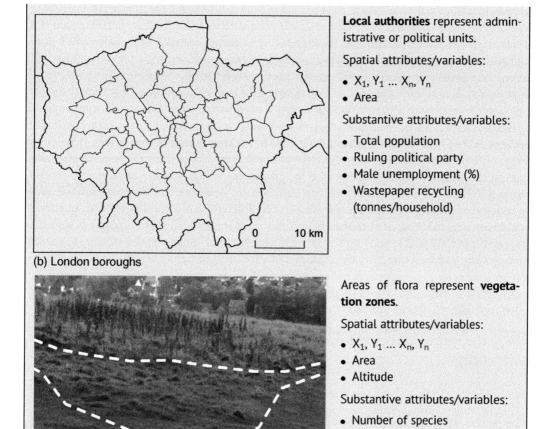

Local authorities represent administrative or political units.

Spatial attributes/variables:

- $X_1, Y_1 \ldots X_n, Y_n$
- Area

Substantive attributes/variables:

- Total population
- Ruling political party
- Male unemployment (%)
- Wastepaper recycling (tonnes/household)

(b) London boroughs

Areas of flora represent **vegetation zones**.

Spatial attributes/variables:

- $X_1, Y_1 \ldots X_n, Y_n$
- Area
- Altitude

Substantive attributes/variables:

- Number of species
- Rainfall (mm)
- Mean annual temperature (°C)
- Soil pH

(c) Upland landscape in Northern England.

2.3.4 Closing Comments on Attributes and Variables

This chapter has focused on two important issues with respect to the collection of geographical data – how much and what data to collect. These are not easy questions to answer, and part way through an investigation, it may well become clear that either not enough or too much data were collected or that the appropriate attributes and variables were not obtained or defined carefully enough. Unfortunately, it is rarely possible to go back and remedy these mistakes after the event and, as we shall see in Chapter 3, researchers are often constrained not by their ability to define exactly what data they need but by being unable to obtain this information in the time and with the resources available. Students will invariably need to undertake some form of sampling when carrying out an independent investigation, since the time and resource constraints are likely to be even more pressing than in funded academic research. The discussion of sampling strategies in this chapter has highlighted that there are many different approaches, and exercising careful consideration as to how the resultant data will be analysed is a crucial factor in determining where that chosen procedure will be satisfactory and achieve the desired outcomes. There has been a trend towards reducing the number of observation units or geographical features as we have referred to them; however, such a reduction in the amount of data obtained is not advisable simply based on saving time and money. While investigations following a qualitative as opposed to a quantitative approach may not seek to ensure that the sample of observations is statistically representative of the parent population, it remains the case that those included in the study should not overemphasise the idiosyncratic at the expense of omitting the typical. This is not to dismiss the importance of the unusual but to remind us of the need to include what is commonplace.

The specification of what information is going to be relevant is another important question facing students charged with the task of undertaking some form of independent investigation. It is all very well to say that the attributes and variables that you need should follow from the research questions that you are seeking to answer. However, if the questions are imperfectly or ambiguously defined then it is likely that unsuitable attributes and variables will be devised and the study will conclude by saying that if other data had been available then different answers to the research questions would have resulted. This is, of course, not helpful or informative. It is in the nature of research that one project will lead to and improve on another, but this is not an excuse for failing to specify attributes and variables carefully. The examples of attributes and variables given in Boxes 2.5, 2.6 and 2.7 are purely illustrative and not prescriptive. There are many alternatives that could have been included. Those included do indicate that in geographical studies we might need to take account of both the spatial and the substantive characteristics of what is being investigated.

References

Dautel, S.L. (2009) Transoceanic trash: international and United States strategies for the Great Pacific Garbage Patch. *Environmental Law Journal*, **3**(8), 181–208. https://digitalcommons.law.ggu.edu/gguelj/vol3/iss1/8.

Dewdney, J. (1981) *The British Census*, Headley, Brothers Ltd.

Dorling, D. and Rees, P.H. (2003) A nation still dividing: the British Census and social polarisation 1971–2001. *Environment and Planning A*, **35**, 1287–1313.

Tobler, W. (1970) A computer movie simulating urban growth in the Detroit region. *Economic Geography*, **46**(2), 234–240.

Further Reading

Anderson, J. 2021. Census 2021: why this year's poll could be the UK's last, according to leading ONS statistician. i Newspaper. https://inews.co.uk/news/uk/census-2021-uk-poll-this-year-last-statistician-ons-906876#:~:text=In%202023%2C%20the%20ONS%20will,considered%20scrapping%20the%202021%20survey (accessed 28/06/23).

Cochran, W.G. (1997) *Sampling Techniques*, 3rd edn, New York, John Wiley and Sons.

Diamond, I. and Jeffries, J. (1999) *Introduction to Quantitative Methods*, London, Sage.

Ebdon, D. (1984) *Statistics in Geography*, 2nd edn, Blackwell, Oxford.

Keylock, C.J. and Dorling, D. (2004) What kind of quantitative methods for what kind of geography? *Area*, **36**, 358–366.

Lynn, P. and Lievesely, D. (1991) *Drawing General Population Samples in Great Britain*, London, Social and Community Planning Research.

Marsh, C. (1982) *The Survey Method*, London, George Allen and Unwin.

Rogerson, P.A. (2006) *Statistical Methods for Geographers: A Student's Guide*, Los Angeles, Sage.

3

Geographical Data

Collection and Acquisition

Chapter 3 continues looking at the issues associated with geographical data but focuses on collection and acquisition. Having discovered in the previous chapter that it is usually not necessary and often impossible to collect data about all observations in a population, we examine some of the different tools available for collecting geographical and spatial data. It also explores the option of obtaining data from secondary sources including 'crowdsourcing' as a way of avoiding primary data collection altogether or at least in part. Examples of diverse ways of collecting data from global remote sensing to local scale questionnaire surveys are examined together with the ways of coding or georeferencing the location of geographical features. In other words, how to tie our data to locations on the Earth's surface.

Learning Outcomes

This chapter will enable readers to:

- Explain the differences between primary (original) and secondary (borrowed) sources of data;
- Select appropriate data collection tools and procedures for different types of geographical study;
- Recognise different ways of locating observation units in geographical space;
- Continue data collection planning for an independent research investigation in Geography and related disciplines.

3.1 Originating Data

Where do geographical data come from? There are many ways of obtaining geographical data depending on what research questions are being asked, and there are often several diverse types of data that can be used to address comparable topics (Walford, 2002). For example, information on land cover or land use patterns could be collected by means of a field survey, by the processing of digital data captured by orbiting satellites or by examining planning documents. Each method has its strengths and weaknesses; each is potentially prone to some form of error. However, as illustrated by this example, one common division of data sources is between whether they are primary or secondary, original or borrowed. This distinction is based on the relationship between the person or organisation originally responsible for collecting the data and the purpose for which they are

Practical Statistics for Geographers and Earth Scientists, Second Edition. Nigel Walford.
© 2025 John Wiley & Sons Ltd. Published 2025 by John Wiley & Sons Ltd.
Companion website: www.wiley.com/go/PracticalStatistics2e

about to be used. A researcher going into the field in their study area and identifying individual land parcels is able to take decisions on the spot about whether to record a particular parcel's use as agriculture, woodland, recreation, housing, industry, etc., as well as the various sub-categories into which these may be divided, but may reach different decisions to another investigator conducting a similar survey in the same location. The researcher using digital data sensed remotely by a satellite and obtained from an intermediate national agency, commercial organisation or free Internet resource has no control over whether the sensor is functioning correctly or exactly what wavelengths of electromagnetic radiation (EMR) are captured (Mather, 2004; Walford, 2002), but is dependent upon someone else having taken these decisions.

The primary versus secondary distinction is more transitory than real since a primary source of data for one person or organisation may subsequently be a secondary source for another. The word 'data' (singular datum) comes from the Latin meaning 'having been given', in other words something (facts and figures perhaps) that is just waiting to be collected or gathered up. This suggests that different people would interpret and understand the data used in an investigation in the same way. The problem is that often the very opposite is true. The difference between primary and secondary sources of data is not about the **methods** of data collection (e.g. questionnaire survey, field measurement using equipment, remote sensing, laboratory experimentation or observation and recording), but about the **relationship** between the person(s) or organisation undertaking the analysis and the extent of their control over the selection and specification of the data collected. Our investigator of land cover/use patterns could theoretically, if sufficient funds were available, organise the launch of a satellite with a sensor to capture reflected radiation in the desired sections of the electromagnetic spectrum. Data obtained from a secondary source, such as an archive or repository, a commercial corporation, a government department/agency or over the Internet, will have been collected according to a specification over which the present analyst has no direct control. Sometimes published journal articles, textbooks and official reports are referred to as secondary sources of data, but this fails to recognise that much of the material in these is argument, debate and discussion based on the author's interpretation of analytical results and other information. This clearly forms part of a literature review but not part of your own quantitative or qualitative analysis. In some cases, such publications will include numerical data and results that may be integrated into your own investigation, but this is not quite the same as extracting data from an Internet website where datasets can be downloaded and used in new analysis.

The choice between using data from primary or secondary sources in an investigation is often difficult and researchers often combine information from both sources because they can potentially complement each other and provide more insightful answers to research questions. Recognising that the difference between primary and secondary sources of data is a question of whether you, as a researcher, have control over key aspects of the data collection process. Data obtained from secondary sources may be less costly and time-consuming to acquire, but they may not be exactly what you need and their quality may be uncertain. The book titled *Use and Abuse of Statistics* (Reichman, 1961) showed that unscrupulous or thoughtless selection of data, analytical techniques and results had the potential to allow investigators to reach contradictory conclusions; in effect to argue that black is white and white is black. Obtaining data and statistics in the era of the Internet is even easier than when the *Use and Abuse of Statistics* was published, but this also makes it more important that those seeking to answer research questions check the reliability and authenticity of the data sources they use. The next section focuses on the different means of collecting data for research investigations in the geographical and Earth sciences.

3.2   Collection Methods

Many geographic and earth science investigations use a combination of methods to collect and acquire data about the observations of interest. Therefore, a second question to be addressed when planning an investigation relates to choosing the methods: how the data, the attributes and variables, are obtained from the observations making up the population or sample. This issue transcends the division between primary (original) and secondary (borrowed) sources of data, since most, even all, data at some time were **collected** from measurable or observable phenomena. The focus in this section is not on the primary/secondary division, but on the methods or means whereby the quantitative and qualitative attributes and variables are obtained from the observations. If the origin of the word data is 'having been given', how do you get animate or inanimate observations to give up the data they contain?

The different methods of collecting data can be classified in several ways and Box 3.1 illustrates some of the options. A useful starting point is to decide where the data are going to come from. There are three main options: to carry out field work and collect data by direct contact with people or the environment; to enter a laboratory either to process samples collected in the field, create data from mathematical or physical, analogue models; or to gather data indirectly through remote sensing devices, operated either by the researcher (e.g. drones) or by external organisations (e.g. satellite imagery), that measure some characteristics of the phenomena that act as a surrogate for the attributes and variables of interest. A second classification relates to the way in which data are stored, which divides broadly between analogue and digital formats. Analogue data are represented in an equivalent way to our perception of the information. For example, an analogue 12-hr clock or watch portrays the passage of time through a day with hour and minute hands and we refer to time using phrases such as '20 to 11' or '10 minutes past 12'. A digital clock or watch shows time as a numerical sequence and would represent these times as 10:40 and 12:10. The difference between analogue and digital data is partly related to how the recording or measuring device operates (e.g. mechanically or electronically) and partly to how it stores the data that are obtained. There is usually greater flexibility for analysing and transferring digital data, since such data can easily be moved between different storage media, for example from a digital camera or scanner to a computer. Transfer from text documents, photographic prints or handwritten questionnaires introduces another stage where errors can enter the process.

3.2.1   Field Observation, Measurement and Survey

A useful starting point for many geographical investigations is observation, measurement and survey of phenomena in the field. Most funded projects and undergraduate/postgraduate degree courses in these disciplines include a period in the 'field', despite increased competition for research funding and pressure on budgets in teaching departments. These visits provide the opportunity for students on taught courses to develop their skills of observation, measurement and survey and for researchers to collect data and for in-depth investigation of the area in a way that is simply not possible at a distance. There is a long tradition of using the field sketch as a device for getting students at various levels of geographical education to look at the world around them, to record the locations of and to think about the interrelationships between phenomena. This approach partly goes back to the philosophy that emphasised the connection between the human and physical domains in the study of Geography, because visualising the juxtaposition of phenomena and their proximity to one another helped to indicate how they might be connected.

Box 3.1 Selected methods and means of collecting data geographical and Earth science research.

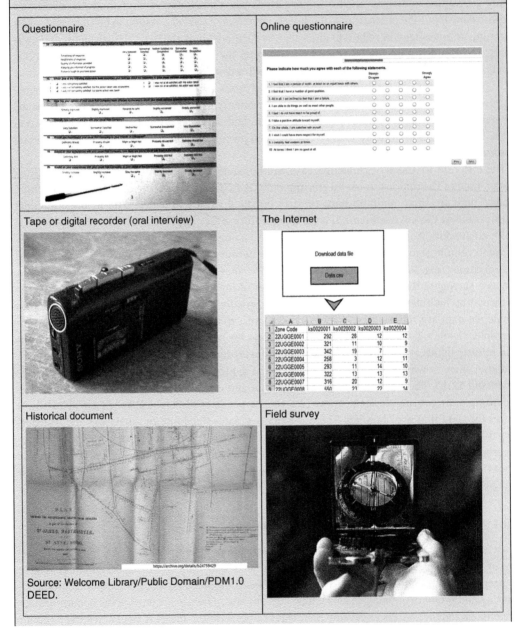

Source: Welcome Library/Public Domain/PDM1.0 DEED.

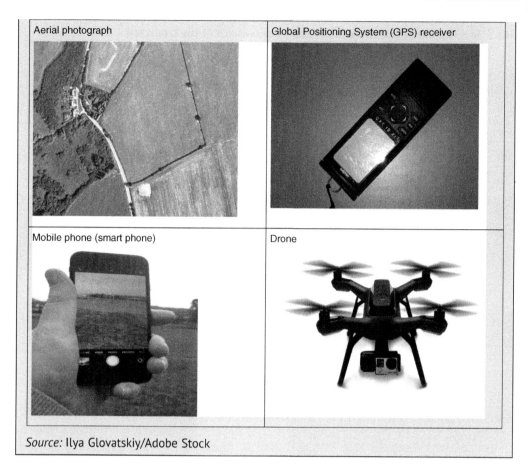

| Aerial photograph | Global Positioning System (GPS) receiver |
| Mobile phone (smart phone) | Drone |

Source: Ilya Glovatskiy/Adobe Stock

Box 3.2 attempts to reproduce such an experience using a photograph to simulate being in the field with a clear view of a glaciated landscape from an elevated vantage point looking down onto the outwash plain of the Bossons Glacier near Chamonix in the French Alps. The use of cross-hatch shading in a sketch can indicate areas with a particular type of surface feature or vegetation cover, such as the extensive coniferous woodland in the upper part of the photograph. Lines of distinct types can show the approximate boundaries of such areas and paths can be denoted with different line styles. Arrows pointing down slope together with labelling aligned with the contours can show variations in elevation. Returning to the notion that field sketches can indicate the juxtaposition of features, it is worth observing that the footpath created by visitors (tourists) to the area has been defined along the valley side away from the incised main channel an area where stream braiding occurs, and roughly along the line where there is a break of slope. Sketches of this type do not of course provide data capable of statistical analysis by the methods covered in later chapters. However, as an aid to understanding processes of human and physical geography, they provide a useful reminder of the principal visible relationships evident in the field.

The selection of sites or locations for field observation, survey and measurement is a key issue (Reid, 2003), which in some respects relates to the question of sampling strategy discussed in

Box 3.2a   The use of sketch maps to record key features of the landscape.

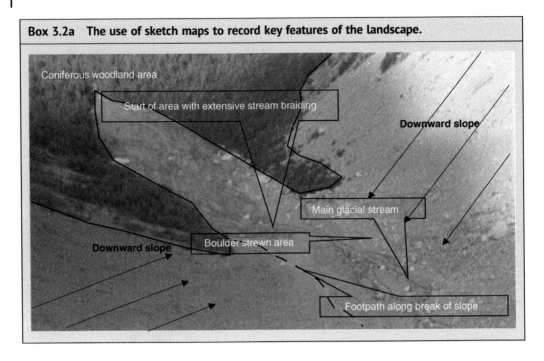

Box 3.2b   Sample sketch map of the area shown in Box 3.2a.

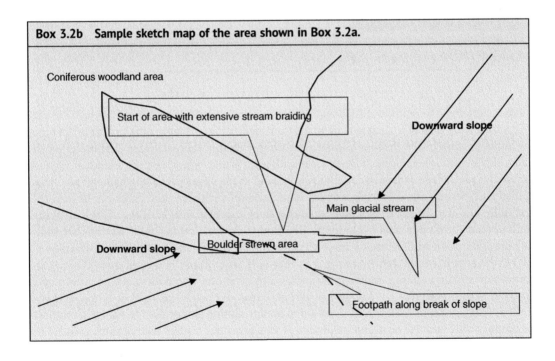

Chapter 2. One of the primary drivers that should guide investigators in their selection of suitable sites for field survey is to eliminate or 'hold constant' the effect of influences capable of distorting the outcome. However, selection of the ideal, perfect site where the cause-and-effect relationship between the variables will be manifest is rarely if ever achievable and the investigator should have recourse to the statistical techniques covered in later chapters in order to make statements about the probability that they have discovered how the variables are connected. Making measurements of variables in the field often involves the use of instruments, and just as with any other piece of technology, there are often more or less sophisticated versions of any particular type of equipment. Reid (2003) offers a basic equipment list that should enable students to observe and measure sufficient variables relevant to their own piece of research.

3.2.2 Questionnaire Surveys

3.2.2.1 Questionnaire Delivery

Questionnaire surveys and interviews represent one of the most used means of collecting data from people either in their role as individual members of the public, as living within families and households, or as representatives, stakeholders and actors in several types of organisation (e.g. commercial businesses, public agencies, non-government bodies, environmental interest and amenity groups). The essential purpose of a questionnaire is to ask people questions yielding answers or responses that can be analysed in a rigorous and consistent fashion. Questions can seek **factual** information, such as someone's age, period of time working with the same employer, level of educational qualification, years living at an address, previous residential address, participation in community social activities, or number of times visited a doctor in the last six months, etc. The list of topics is almost endless. Questions can also seek **attitudinal** information or people's **opinions** about various matters, such as whether they agree with government health policies, whether congestion charging should be introduced, whether the interval between domestic waste collection should be altered, whether 4×4 wheel drive vehicles should pay higher or lower licence fees, whether they anticipate taking a holiday overseas in the next 12 months, etc. Again, there is an almost infinite number of things about which people might express an opinion. When carrying out academic research using a questionnaire the key aim is to ensure that the questions are delivered in an unambiguous, consistent and reliable fashion without introducing any bias into the responses. A researcher needs to be confident that participants in such a survey have had the opportunity to answer quantitative and/or qualitative questions in a consistent way.

One of the issues to be decided when carrying out a questionnaire survey is how to administer or deliver the questions to the potential respondents. Over many years, the main methods of delivery have been telephone, face-to-face interview, drop-and-collect and the postal service, although these traditional means have been replaced in many cases by the Internet, including using one of the freely available survey websites (e.g. www.hostedsurvey.com, www.surveymonkey.com or https://www.google.co.uk/forms/about/). Some comments about the more traditional methods are useful as they shed light on issues that need to be thought about when carrying out any survey. Face-to-face and telephone questionnaires involve an interviewer reading the questions orally and recording the respondents' answers (interviewer-completion), whereas other methods including online surveys expect respondents to read and answer the questions unaided (self-completion). In general, telephone interviewing is less successful than face-to-face because the respondent can terminate the interview at any time simply by stopping the phone call, building a rapport between interviewer and interviewee is more difficult, and respondents may be more inclined to give misleading or untruthful answers. Social media platforms have offered an increasingly accessible approach for recruiting survey participants and for delivering questionnaires in recent years and have been especially attractive to younger researchers, who have experience with these platforms since an early age in many cases. When selecting a delivery method researchers should consider whether bias has occurred because some sections of the sampled population have been unintentionally disadvantaged from participating. For example, using a social media platform or online questionnaire may limit participation of people unfamiliar with or not possessing the necessary technology to enable completion of the survey. It is important to design the questionnaire or interview schedule and to select the method of delivery after you have decided what your research questions are and who you want as your survey respondents. Otherwise, you might end up including only those people you can contact because of your chosen delivery method and type of questionnaire. Traditional methods of questionnaire delivery, such as on-street interviewing, postal service or drop-and-collect have become less common but may still be appropriate in some circumstances.

Whatever method of delivery is employed, it is important to introduce potential respondents to the survey, for example by means of a letter and/or introductory section. This should explain:

- why you are seeking the information
- why they are being asked to take part
- what rights do they have to withdraw from the survey at any stage
- what confidentiality measures are in place
- how secure storage of the data obtained will be ensured.

Researchers should stress that it will not be possible to identify individual respondents by name in the data that are held on computer or on other media, and anonymity will continue in subsequent dissemination of findings, unless explicit permission is granted. Participants should also confirm that they have read and agree with taking part in the survey and have been informed of their right to withdraw. Universities, professional institutes, learned societies and other research organisations invariably have ethical codes of practice for you to follow, which will ensure that you carry out a survey in an ethical manner and not cause offence or bring the practice into disrepute. These ethical codes often require investigators to complete an ethics approval form, especially if the research involves direct contact with human or animal subjects. These procedures reflect the requirements of the General Data Protection Regulations (GDPR) introduced in the UK and many other countries. If research involves using data from secondary sources, such as population censuses or official surveys then ethical approval may not be necessary because the investigator is not having direct contact with human subjects and the statistical data from these sources are increasingly already in the public domain. Historical records of the British Population Census and other surveys are governed by the 100-year rule meaning that the original documents showing people's names may be viewed in The National Archives or as scanned images through its commercial partners.

Table 3.1 summarises the advantages and disadvantages of interviewer and self-completion forms of delivery. Some counterbalance each other, for example self-completion questionnaires are relatively inexpensive and less time-consuming, whereas face-to-face interviews are more costly of time and financial resources. However, the latter allows you to use more sophisticated and searching types of questions, whereas self-completion requires standard, easy to understand questions capable of completion in a short time period without any assistance. Self-completion questionnaires need to be unambiguous, whereas interview questionnaires allow questions to be explained, although this raises the problem of variation in the phrasing of questions by interviewers if several people are doing this. Surveys completed by an interviewer often produce a higher response rate,

Table 3.1 Advantages and disadvantages of interview and self-completion surveys.

Interviewer-completion		Self-completion	
Advantages	**Disadvantages**	**Advantages**	**Disadvantages**
Opportunity for clarification	Difficult to standardise phrasing	Standardised question format	Impossibility of clarification
Higher response rate	Possibly smaller samples	Potentially larger samples	Lower response rate
Possibility of complex questions	Excessive complexity	Simple question format	Produces only simplistic data
Opportunity for follow-up	Longer data collection time	Shorter data collection time	Difficulty of follow-up
Possibility of adding questions spontaneously	Potential of introducing inconsistency	Fixed question content	Impossibility of adding questions spontaneously
	Expensive	Inexpensive	
Allows for variable ability of respondents			Assumes the standard ability of respondents
Control question structure			Likelihood of more spoiled responses

although conversely might result in a smaller sample size, because potential respondents may be less likely to refuse a face-to-face or even telephone request to take part. Self-completion questionnaires achieve lower response rates, typically less than 40 per cent, but the overall sample size may be larger because of the ease of contacting more numbers of potential respondents.

3.2.2.2 Question Wording

There are some aspects of questionnaire design that vary according to the mode of delivery: for example, the layout of questions in an online survey and the instructions for completion need to be clear and completely unambiguous. However, irrespective of the mode of delivery there is a general need to ensure consistency in responses so that the answers obtained from one respondent may be validly compared with all the others. Alternative forms of questions seeking information on what is essentially the same topic can produce quite different responses. Consider the following questions and answers about where someone lives:

- Where do you live?   In a bungalow
- Where is your home?   England
- Where is your address?   On a main road

All of the answers given to these questions are entirely legitimate, but they do not actually provide the information that the investigator was seeking. A questionnaire is a form of communication between one person and another and provided that the researcher (interviewer) and respondent are communicating 'on the same wavelength' as far as understanding what information is wanted in answer to a question there should be no problem. Unfortunately, we cannot realistically make this assumption given the variability in people's oral and written communication skills and it is important to eliminate poorly defined questions.

We will now examine these three questions about where someone lives in a little more detail and explore the range of answers that might be given.

- Where do you live? Closer inspection reveals this is quite an open-ended question. It does not make clear whether respondents should reply with the type of dwelling or property they live in; with a geographical description of where they live, such as city centre, suburb, small town or village; with the name of the place such as Bute, Guildford, Fishguard, Happisburgh or Stratford-upon-Avon; or simply 'here' (i.e. in the location where the survey is taking place) or 'elsewhere' (somewhere else). A researcher receiving answers from across these four possibilities will be unsure how to analyse the data in a consistent way.
- Where is your home? Home is a social construct and not in itself a geographical location and is therefore likely to produce responses reflecting people's feelings of social identity rather than geography. People who have moved to another country for reasons of employment may feel that their home is where their parents or other relatives still live. Students who have gone away to university or college may consider the place and indeed dwelling where they were raised by one or both parents as home. Such feelings may subtly change if such students after graduation gain employment and remain in their university town, which may slowly assume the role of home. Older people who become unable to live independently may move into a residential care home, but it may take some time if ever for such a setting to truly take on the social meaning of their own home.
- Where is your address? This potentially sounds as though it might objectively focus on the geographical location of where somebody lives. However, as the answer suggested above 'on a main road', or alternatives such as in a cul-de-sac, on an unmade track, on a farm, in a dockyard, above a fish and chip shop and next to the A12 indicate the range of answers can also be potentially be difficult to analyse in a consistent way. The essential problem with this question is that it asks where someone's address is located not what the address is.

Chapter 2 mentioned that the UK and many other countries have postcode or zip code systems that are used to help with delivery and collection of postal and other items from residential and commercial addresses. Using such systems as a way of asking in a questionnaire survey what someone's address is helps to standardise responses, avoids GDPR issues and allows the researcher to subsequently identify where the group of addresses that share any given postcode are geographically located. The use of a postcode also potentially helps with linking the data obtained from a survey with other secondary sources of statistical information such as those from population censuses or administrative records (see Section 3.2.3).

3.2.2.3 Questionnaire Structure

A questionnaire, as a means of communication between researcher and respondent, sometimes with an interviewer acting as an intermediary, needs an overall structure that helps all parties in the venture to feel comfortable and reassured. Some questions will be easier for researchers to ask and for respondents to answer than others, and these often appear early on in a questionnaire and help with establishing some rapport. Most questionnaires seek some basic, demographic and socio-economic details about respondents, such as gender, age, residential location, education, employment status and ethnicity, because differences in these characteristics between one person and another may account for variations in attitudes and opinions concerning the topic that is the focus of the research. The importance of such details as explanatory factors accounting for these differences led to the questions seeking this information to be appear on the first page of a questionnaire. This occurred either because respondents could easily answer these questions or if the questionnaire ended prematurely there would potentially be some data available for analysis. An alternative argument is that these straightforward questions are simple factual ways of finishing a questionnaire that leaves both parties satisfied.

The overall structure of a questionnaire or interview schedule typically identifies and separates different questions by means of a sequence number and may also include sections to group together questions concerned with a topic, such as leisure activities and pursuits. This structure helps respondents and interviewers to progress through the whole document. It is also often necessary to include instructions that guide people through the process of completing a questionnaire. The most common reason for these instructions is to direct the self-completing respondent or interviewing researcher towards the next question or section that should be answered. This often occurs when there is a filter question that determines which question should be answered next. Filter questions are often dichotomous (only two mutually exclusive answers are possible) or with a limited range of options, such as those with answers 'Yes' or 'No', or 'Male' or 'Female', 'Owner occupier', 'Social sector renter' or 'Private sector renter' or 'In paid employment' or 'Not in paid employment'. These can conveniently be used to route respondents or interviewers through a questionnaire. This can be achieved by including an instruction such as:

- If answer Yes go to Section 3
- If answer No go to Section 4

The two groups of respondents separated by the filter question can be brought back together by an instruction at the end of Sections 3 and 4 stating that they should go to go to Section 5 next. This routing of respondents and interviewers through a questionnaire is not unlike giving someone directions about how to navigate between places on a map.

3.2.2.4 Questionnaire Design

Our examination of questionnaire survey design, delivery, completion, questions and structure concludes with reference to the example project outlined in Box 3.3a. The main aim of the project is to investigate if areas in cities and towns that have been redeveloped and regenerated after industrial decline are suitable for raising children or are only attractive to households without children. The project is notionally based in part of the redeveloped London Docklands area, which for the present purpose became housing and low-rise apartment blocks 15 years ago. The researcher wants to separate the sampled addresses (households) into two groups: first, those where at least one person has lived since the accommodation first became habitable; and second those where none of the residents were present at the start (i.e. no 'first generation' occupants). Slightly different versions of the questionnaire have been prepared for each type of household to make self-completion simpler. Published statistics obtained from a population census held 15 months ago recorded 1260 households in the area and the researcher decides a 10 per cent sample (126 addresses) split equally between the two types of address (household) is sufficient. The researcher is unable to determine which of these two types of household lives at any of the addresses and decides that the most suitable approach is to randomly sample double the required 126 addresses and to visit them with a fellow student for support and safety reasons. At each address visited when inviting people to take part in the survey and sign the participation document, the researcher will ask an initial filter question to determine which of the two types of household lives at the address. The corresponding version of the questionnaire is left for completion and return either by using a prepaid envelope for return by post or for collection by the researcher on an agreed date. The researcher and companion proceed in this way until 63 questionnaires (and envelopes if required) have been left at the two types of sampled addresses.

> Suppose the researcher had decided to create an online questionnaire accessible from a social media website but still wants to sample 126 addresses randomly split equally between both types of household. How might the procedure outlined above be adapted to obtain the information required from the two distinct types of address (household) (i.e. those where no present residents were there at the start and those where at least one original resident is still there)?

Box 3.3b does not show both versions of the questionnaire in full but illustrates a small selection of topics that are asked differently or identically in the two versions, or that are only included in one version. Box 3.3b focuses on factual information about the turnover of people in a household and specifically about when the last person joined or left a household. These questions appear on both versions of the questionnaire, although their context is slightly different. Households including at least one original resident could have gained or lost people (adults and/or children) at any time during the last five years (see upper part of Box 3.3b). The lower part of Box 3.3b, taken from the second version of the questionnaire, first establishes when the person who has lived at the address longest arrived and then a filter question asks if there is more than one adult now. The purpose is to establish if the adults moved as a household unit to their present address. If they did move as a household unit proceed to the questions in Box 3.3c, otherwise find out when individuals arrived or departed. Box 3.3c focuses on any anticipated future changes in household composition over the next five years. It appears on both versions of the questionnaire and uses a predefined set of categories to seek views on the likelihood that either the whole household or any one individual member will leave within the next five years. Such a series of categories should not be excessive but be as exhaustive as possible, although an 'Other' category has been included as a 'catch all' Asking questions about what might happen in the future is inevitably an inexact science, since people cannot know for certain how their circumstances will alter. However, this project is interested in both past and potential future turnover of residents in the London Docklands and therefore consideration of future changes was necessary. If an investigator is interested in probable

---

Box 3.3a   Example project illustrating issues of questionnaire design.

In recent years many cities and towns in Britain and other countries have experienced regeneration when areas that were no longer required for industry or transportation have been re-developed as offices or for residential, retail and leisure purposes. This process has been associated with industrial restructuring and economic globalisation, and has led in some cases to changes in the social and economic character of the areas and contrasts have emerged between residents of the surrounding areas of the redevelopment sites and those occupying the new residential accommodation.

The London Docklands Development started in the 1980s was the first major regeneration project of this type in the UK. Some 30 years on a researcher has decided to investigate 'population turnover' in this area and whether the early residents have remained or moved on. The researcher has decided to select a random sample of residential addresses and to distribute a drop-and-collect questionnaire to each of these. Some of the addresses are likely to be occupied by the same groups of residents who moved in at the start, while others will house a completely different group of people. In those addresses where at least one member of the original household is still present, the researcher wants to find out details of the changing composition of the household over the intervening period.

Box 3.3b Turnover of residents at address.

Asked of households with at least one original resident

1. How many people have **JOINED** your household since you first arrived, including those who do not live here now?

    Enter number of additional adults
    Enter number of additional children

2. In what year did the last new member of this household come to live at this address?

3. How many people have **LEFT** your household since you first arrived, including those who do not live here now?

    Enter number of departed adults
    Enter number of departed children

4. In what year did the last person leaving this household cease to live at this address?

Asked of households without any original residents

1. How many years has your household lived at this address?

    If there is more than one adult in the household, otherwise go to Question 3:

2. Did the adult(s) in your household live together at their previous address?

    Yes, please go to next set of questions in Box 3.3c   1

    No, please go to Section 3   2

3. How many people have **JOINED** your household since moving here?

    Enter number of adults
    Enter number of children

4. In what year did the last new member of this household come to live at this address?

5. How many people have **LEFT** your household since moving here?

    Enter number of adults
    Enter number of children

6. In what year did the last person leaving this household cease to live at this address?

---

future occurrences, then the questions should indicate that they are seeking information about the likelihood of some event and the answers obtained should be treated with due care. The 'future move' questions for whole households and for individuals have follow-on parts that attempt to discover the reasons for the indicated categorical response. The follow-on questions seek to understand the reason the whole household or an individual member of it would consider moving within the next five years. There are some obvious problems with this form of questioning. The respondent might only have stated one of the categories (e.g. move to another London Borough) because there is a realistic possibility that this might occur, but the follow-on question asks for the person to justify or rationalise this response. The reason given for moving might therefore be fairly imprecise, misleading and people might be tempted to give a standard response, even based on how they imagine other people might respond. Although this type of question is more suited to an interviewer-completed survey, it can be used in in self-completion questionnaires.

Box 3.3d illustrates another way of obtaining people's opinions and attitudes. It involves presenting respondents with a series of statements and asking them to express an opinion, in this case on a five-point scale from 'Strongly Agree' to 'Strongly Disagree' together with a 'Don't know' option.

Box 3.3c   Future turnover of residents at address.

Asked of ALL households

1. Is it likely that your complete household will move from this address within the next 5 years?

Unlikely that the household will move together. (Go to Question 3)	1
Yes, likely to move within London Docklands Development area. (Go to Question 2)	2
Yes, likely to move to another London Borough. (Go to Question 2)	3
Yes, likely to move elsewhere in the South East region. (Go to Question 2)	4
Yes, likely to move elsewhere in the UK or overseas. (Go to Question 2)	5
Yes, location unknown. (Go to Question 2)	6
Don't know. (Go to Question 3)	7

2. If YES, what would be the main reason for your household deciding to move within the next 5 years?

3. Is anyone currently living at this address likely to leave within the next 5 years?

Unlikely that anyone will leave.	1
Yes, child(ren) likely to go away to university/college	2
Yes, adult(s) or child(ren) likely to set up household within London Docklands Development area	3
Yes, adult(s) or child(ren) likely to set up household in another London Borough	4
Yes, adult(s) or child(ren) likely to set up household elsewhere in South East region	5
Yes, adult(s) or child(ren) likely to set up household elsewhere in the UK or overseas	6
Yes, location unknown	7
Don't know	8

Box 3.3d   Attitudes to living in the area.

1. Please rank each of the following statements on the scale: 1 = Strongly agree; 2 = Agree; 3 = Neutral; 4 = Disagree; 5 = Strongly disagree; 6 = Don't know.

a) There are not enough facilities for families in this area.	1	2	3	4	5	6
b) You don't need to be wealthy to live in this area.	1	2	3	4	5	6
c) This area should not be developed any further.	1	2	3	4	5	6
d) This area has a strong sense of community.	1	2	3	4	5	6
e) It is not necessary to have a car to live in this area.	1	2	3	4	5	6

This type of attitude battery question can elicit opinions quite successfully in both interviewer and self-completion surveys, they are designed to provoke a response, although some people are unwilling to express opinions at the extremes, which might exaggerate the number of responses in the middle of the range. However, there are issues over how many categories to use and questions force people to answer with one of the options. Some people's reaction to such questions is to probe the statements and conclude that 'on the one hand they agree and on the other hand they disagree', strongly or otherwise. This might precipitate people opting for the 'Neutral' choice or 'Don't know', when they have a split opinion on the subject. Another problem with attitudinal and factual questions, particularly in self-completion surveys, relates to whose opinion is obtained. In general, they assume that the person completing the questionnaire can answer on behalf of everyone to whom the survey is addressed (e.g. all residents in the household or all directors of a company). In the case of interviewer-completed questionnaires, it might be possible to determine the accuracy and comprehensiveness of the answer, for example, by interviewing all members of a household, although this might be quite difficult to achieve.

There is often a tension between what information is needed to answer the research questions and the feasible methods of data collection. Discovering the genuine motivations and reasons for people's actions and attitudes can be difficult to obtain, especially by means of a self-completed postal or Internet-delivered questionnaire, no matter how sophisticated the design. In some cases, these tensions can only satisfactorily be resolved by re-designing the overall approach to the research, for example by including both extensive but simplified questionnaires to produce a lot of superficial data accompanied by in-depth interviews to probe the topic with smaller number of interviewees. A carefully designed, interviewer-completed questionnaire can allow sufficient probing to obtain the required depth of information. It seems important not to let a questionnaire rule the scope and success of your research project. Compromises made about the questions asked will inevitably mean that the depth of understanding of the research topic becomes limited. One type of question frequently used to obtain survey data about people's behaviour, opinions or views involves using Likert scale questions. These can be designed in several ways but present respondents with a numerical scale, typically with 5, 7 or 9 options, for example ranging from complete disagreement to complete agreement with a set of statements (see above). A more nuanced approach could present people with a measurement scale (like on a ruler) and ask them to mark, or in the case of an online survey drag a pointer, to the position on the scale most closely reflecting their opinion. This section has focused on questionnaires at some length because they are such a commonly used method for collecting data about people, but there are other options for obtaining the data that will become the attributes and variables in your analysis which will now be examined.

3.2.3 Administrative Records and Documents

Investigations in Geography and the Earth Sciences are grounded in the world around us and therefore it is not surprising that commercial, public and voluntary sector organisations collect data as part of the administrative and operational procedures that would be relevant to undertaking research. Obtaining access to these data sources at the required level of detail can be difficult, but it is invariably worth exploring what might be available at the start of an investigation. These data will have been collected, processed and analysed by another organisation and therefore they clearly fall within the type of data known as secondary or borrowed sources, although how the raw data are collected in the first case is variable. It is worth describing in general terms how such data may be collected, created and processed before examining some examples. Many countries provide some form of financial support for members of their populations who are unable to work in paid employment, for example through physical or mental disability. The governments of such countries need to communicate the potential availability of such support and to collect details from people who regard themselves as entitled to claim the welfare benefit. This administrative process will often be carried out by public sector employees, referred to in the UK by the overarching term civil servants. Obviously, details about individuals recorded in this process are confidential, but statistics that have been suitably anonymised to prevent disclosure of personal information may be relevant to researchers concerned with this topic. There is general agreement that the demand for information increases as governmental and commercial organisations become more complex and the extent of their operations expands into new areas. This process has historically fuelled the demand for more data and particularly in the digital era has raised questions about the compatibility, consistency and confidentiality of data held by different organisations and the various parts of the state bureaucracies. It may also be argued that when administrative and operational data generated by these processes have been collected at public expense there is an obligation to make summary information available subject to confidentiality safeguards. This can include making the data records of individuals available provided that all variables are removed that might allow the person to be identified. Overall, it is not sufficient for administrative and operational data to be made accessible; they need to become usable to be useful.

Many of the obvious examples of such data relate to people rather than to features of the physical environment, although the latter are by no means absent. The former includes population registration systems concerned with births, marriages and deaths, which provide basic data about vital events in people's lives, and procedures for welfare assistance (e.g. free school meals, mobility allowance, housing benefit, etc.), which in its various forms potentially act as indicators of relative deprivation and disadvantage. In many countries, a periodic population census constitutes a major source of information about the demographic, economic and social condition of the population, and provides an example of how what is generally conducted as a statistically comprehensive questionnaire-based survey of a population at one point in time by government can become a secondary source of data for researchers across a wide spectrum of subject areas not only embracing the social sciences, but also the humanities and natural sciences. Population censuses have for many years been a fundamental source of information in geographical studies, since they provide one of the few long-term, reasonably consistent and authoritative statistical summaries of the condition of national, regional and in some cases smaller area populations, and a population represents both a country's key resource and its constituent character.

Population censuses and other surveys conducted by governments are in many respects different from other types of administrative data since they are carried out with the explicit purpose, at least in part, of answering research and policy questions. One of the definitive characteristics of administrative data is that their use for research investigations is a byproduct of the process for which the information was originally collected. Several implications follow from this. The data may not be ideally suited for the research questions that the investigator seeks to address. This can result in compromise either in respect of the research aims and objectives, or the analysis being conceptually flawed. The essential problem is the difficulty or impossibility of fitting data obtained and defined for administrative purposes to the requirements of an academic research project. Many of the datasets collected as a result or administrative or statutory processes involve holding names, addresses and other personal details of individuals, who expect their interests and privacy to be safeguarded. Thus, a fundamental requirement for treating such information as a secondary source of data for researchers is that records are stripped of details that could be used to identify individuals and organisations, and in most cases, this includes removing specific geographical location.

Another problem is administrative and statutory data sources may be held in software systems that do not easily allow information to be extracted in a format suitable for statistical analysis, although greater openness in database structures now makes this less of an issue than formerly. A third, more problematic issue, is that many administrative databases are designed to operate in 'real time', in other words to be subject to continual update and amendment. For example, as people's financial circumstances change, they might no longer be eligible for certain types of welfare payment and thus their status on the 'system' would need to be altered, and similarly other individuals might need to be added. The core of this problem is that things change and changes involving the connections between people and places can easily be disrupted, severed and rearranged. For example, most people during their lifespan will live at several addresses and some of the addresses that people move to will be on newly developed residential land and others may be created by building additional dwellings in the garden(s) of one or more formerly large single properties.

Suppose one residential address representing a detached house on a parcel of land with a large garden is redeveloped as three new, smaller houses on the same plot when the elderly resident of the house has moved into a care home because of ill health. What changes would need to be recorded in the address databases held by the local authority, utilities companies (gas, electricity, water, etc.) and by the Royal Mail, which is responsible for assigning postcodes? What operational issues might arise for these organisations in respect of the three new houses?

Administrative databases are therefore often dynamic in their nature, which raises issues from an academic research perspective. A typical research investigation involving the collection of original or new data, especially if carried out by a degree-level student with limited time and financial resources, can only collect data relating to a restricted period. For example, one intensive period of data collection during the vacation between the second and third years of study, although occasionally one or two repeat visits to an area may be possible. The resultant data will therefore be 'one-off', cross-sectional in nature and provide limited opportunities for examining change as it occurs. There are examples of longitudinal and panel surveys in many countries involving repeat questioning of the same set of participants often with additions made according to strict criteria to compensate for departures from the initial collection of survey respondents. One such dataset in the UK is the British Longitudinal Study linking a sample Population Census records since 1971 with other vital events data. In contrast, the records in an administrative database are potentially changing all the time, which raises the twin questions of when and how often to take a 'snapshot' of the data so that the analysis can be carried out. Academic research is rarely concerned with continual monitoring of phenomena, although clearly issues and locations are re-visited as time goes by. There comes a time in all academic research when the data collection stage ends and the analysis phase begins. Researchers are aware that during the analysis phase, additional data could be collected by someone else, which might produce different results. This is why it is important to be clear about the time data were collected when reporting results. The passage of time is a more worrisome issue when analysing data from administrative sources, because researchers know that someone running the ongoing system is capturing more up-to-date information while they are carrying out their analysis.

There are some similarities between administrative data sources and those examined in other sections of this chapter. The issues of accuracy and reliability of the data and the design of the data collection document are common and merit some attention. Most datasets collected by researchers will employ sampling (see Chapter 2) and thus the investigator accepts that some observations will have been omitted. Although this will inevitably introduce the possibility of sampling error into the analysis, the application of well-designed sampling strategy, ideally because of introducing some element of randomness into the selection process, should minimise this source of error. However, when carrying out analysis on data obtained from administrative sources, the investigator usually has little way of knowing how complete or partial the dataset is, and most importantly the reasons why some observations are omitted. Potentially the major difference between sample surveys and administrative datasets is that in the former case, omissions occur because of a random element in the sampling process, whereas in administrative data missing information is more likely to arise because certain categories of people have chosen to exclude themselves from the process. In the UK, the annual electoral register is created by a process administered by local authorities that involves householders completing a form online or returning a printed form by post. The resultant electoral register used to be a reliable way of selecting a sample of adults for inclusion in a questionnaire or other type of survey, but legislation was introduced that allowed people to request their name and address to be excluded from the public register. Omissions arising from such measures are likely to be far from unbiased and random. Although datasets generated from administrative processes can fill an important gap with respect to certain types of information or in obtaining data in sufficient quantity on some topics, their use in research investigations should always be treated with a certain amount of caution. Just because the data exist and are accessible does not mean they are necessarily reliable, even if the source seems reputable.

Most administrative data are obtained compulsorily or voluntarily using a printed or online form. The design of such a document in many respects raises the same issues involved in producing a clear, intelligible and easy-to-complete questionnaire. Clarity and simplicity are key features of administrative data collection forms and this often results in the information sought being even more streamlined and simplified than in a questionnaire survey. Although this facilitates the process of collecting the dataset, it does mean that its usefulness for answering research questions is likely to be diminished. Governments in most countries in the developed and developing world publish statistical abstracts and digests, some of which are brought together by larger organisations such as the United Nations, the Organisation for Economic Cooperation and Development, the World Bank and the European Union (Eurostat), which are often based on data obtained through administrative and operational processes. Increasingly these are available over the Internet from the different government websites or from those of international organisations. These information sources often provide a context for researchers, but their usefulness for localised investigations may be hampered by their lack of geographical detail.

3.2.4 Interviewing, Focus Groups and Audio Recording

The section concerned with questionnaires emphasised the importance of clarity in designing and administering a survey instrument, since it is a form of communication between the researcher and members of the public. Questionnaires that are completed with a researcher asking the questions can avoid many of the difficulties associated with self-completion if respondents do not understand the purpose of a question and what sort of information is being sought. There is an alternative approach that may be the main means of collecting data or used as a follow-up to a subset of questionnaire survey respondents. This involves a more conversational method of data collection, talking to people to obtain a richer, fuller set of information. One of the main problems with questionnaires is that they are designed under the assumption that it is understood in advance how people will answer the questions and that the answers given will be both truthful and meaningful. Even open-ended or verbatim response questions, which are often included in a questionnaire, may not be phrased in a way that allows the full range of interpretations and responses to be elicited. In other words, even the most carefully designed questionnaires will have a standardising, potentially limiting effect on data collection.

Consider the three seemingly clear and unambiguous questions contain in Box 3.4:

Box 3.4 Examples of seemingly clear and unambiguous questions.		

1) How often do you see a film?
  a) Once a week
  b) Once a month
  c) Once a year
  d) Less often
	2) What type of film to do prefer?
	  a) Horror
	  b) Comedy
	  c) Western
	  d) Science fiction
	  e) Other
		3) Area films less entertaining now?
		  a) Strongly agree
		  b) Agree
		  c) Neutral
		  d) Disagree
		  e) Strongly disagree

So how did you respond to these questions? Well, if your answer to question 1 was do you mean on TV, in a cinema or streamed, you are already aware of the problems associated with asking people simple questions with a limited set of fixed pre-coded answers. What if you cannot distinguish a preference for Horror and Science Fiction films? When asked your opinion about whether you think films are more or less entertaining than they used to be, how far back are you supposed to go? If a question prompts respondents to reply with a question of their own, this often indicates that it is not well designed or simply that the standardised rigidity of a questionnaire is not allowing people to answer in a way that is meaningful them, even if it is convenient for the investigator. Structured or semi-structured interviews are more conversational and seek to tease out how people make sense of their lives and the world around them. In many respects this approach is the antithesis of the statistical analyses examined in later chapters, since interviews are based on a two-way dialogue between interviewer and interviewee that are contextualised not only by their own past but also their present and possibly their future experiences, particularly if they are anxious to be somewhere else. In contrast, rigidly adhering to a prescribed questionnaire reflects only the investigator's past perception and does not usually admit the adaptation and flexibility that is possible as an interview progresses.

Social science researchers have increasingly adopted a mixed methods approach in their investigations combining a larger scale questionnaire survey with more in-depth interviewing and/or focus group with a smaller subset of respondents. Because the intention with qualitative forms of data collection is to be illustrative rather than representative and to seek corroboration rather than replication of information (Valentine, 2008), there is little need to use a random process for identifying potential interviewees, who might be chosen from respondents to a previously completed quantitative questionnaire survey or be chosen entirely separately. Fundamentally qualitative interviews are based on the principle that interviewees can articulate their 'subjective values, beliefs and thoughts' (Layder, 1993: 125) and communicate these to the researcher without loss or distortion of understanding. Using qualitative methods is as much about investigators getting to know themselves as about trawling data from other people. Researchers should therefore be aware of how they might be perceived by the people they are interviewing from a range of perspectives, such as gender, age, ethnic background, sexuality and wealth. This is often referred to as the positionality of the researcher in relation to research participants.

There are alternative approaches to selecting people to interview. Cold calling entails identifying potential interviewees from a list of 'candidates', such as the electoral register or organisation membership list, subject to ethical approval. Careful use of such lists can help to identify certain social groups, males and females, people living alone and even people of different ages by following fashions in first names. However, increasing concern over 'identity theft' and unwanted contact from salespeople has resulted in some people being reluctant to appear on a published list or register. Telephone and trade directories were previously a convenient way of selecting people to interview but have become less useful as mobile smartphones have superseded traditional landline telephony. Disadvantages associated with 'cold calling', its intrusiveness, indeterminate accuracy and potential to cause anxiety amongst vulnerable groups (e.g. elderly people and those living alone), limit its usefulness for research investigations. One way of overcoming these difficulties is to seek an introduction to members of the target social group through an intermediary, sometimes referred to as a 'gatekeeper', such as the leader of community or social organisations. Although this form of approach can result in contact with more interviewees, these intermediaries are likely to exercise their own judgement over who to introduce and the full range of members in the organisation may not be suggested. Thus, the investigator may achieve more seemingly successful interviews, but at the expense of only having approached a particularly cooperative or friendly group of people.

The 'snowballing' method of selecting people to interview (see Chapter 2) represents a permutation of this approach and is based on the idea of using an initially small number of participants to build up a whole series by asking each interviewee to suggest other people who might be willing to be interviewed, and for this second tier of interviewees to introduce other people, and so on until sufficient numbers have been obtained. The snowballing analogy refers to the idea of a snowball increasing in size as it rolls down a slope gathering more snow as it does so. Clearly the accretion of interviewees through 'snowballing' can suffer from the same problem of only receiving recommendations from friends of the initial contacts, who are of an amenable disposition or share similar opinions. This difficulty can be ameliorated by starting with a broad, perhaps random, selection of initial interviewees. One of the main strengths of this method is that the investigator can mention to new contacts that they have been given a person's name as a potential interviewee by an acquaintance, which can help to build trust between the different parties.

Whatever approach is used to recruit interviewees, the outcome is likely to be more successful if an initial contact can be made by post, telephone or email so that the interview can be scheduled at a mutually convenient date and time. This is especially important when seeking interviews with people in organisations, because you will usually be asking them to devote some of their time, possibly when they are at work, to help you carry out your research. Careful preparation is essential in this case, especially if the investigator seeks to interview people with certain or even various levels of responsibility or function in organisation, rather than someone in 'customer relations' or 'external affairs'. Initial contact with a named individual is usually more likely to be successful than simply a letter to the Chief Executive of a company or Director of a public organisation. It is often necessary to be flexible when making an initial approach to an organisation and the example of trying to interview farmers provides some useful insights into the pitfalls. Farm businesses can be run in various ways. In some cases, there might be a single farmer running the business as a sole proprietor, a husband and wife partnership, a group of family members (e.g. two brothers and their sons), a manager working in close conjunction with the owner(s) or a manager working as an employee of a limited company. Researchers should make themselves aware of the possibilities and decide who they want to interview in each case. For example, if a farm business is run by a husband and wife in partnership, should the researcher interview both together, separately, or just one of them? It will be important when making the initial approach to someone in the 'farm office' to establish how the business is run so that some measure of consistency can be achieved when conducting the interviews. This type of initial approach requires a considerable amount of planning, but invariably increases the response rate, although possibly at the expense of only having discussed the questions with co-operative individuals.

The conduct of interviews requires careful planning and each interview will be different even if the investigator uses the same question script with each interviewee. One of the most basic issues to decide is where the interviews are to take place. In many cases, the type of respondent will dictate location, thus an interview with someone in a company will often take place in their office, with a member of the public in their home and with a child in their school, most probably in the presence of a responsible adult. The issues associated with locating a series of interviews relate to establishing a comfortable and relaxing setting for the interviewee and a safe and secure environment for the interviewer. The interviewer should aim to establish a rapport with the interviewee at an early stage in the process and this can usually be achieved by courteous behaviour, by allowing the interviewee time to answer your questions and by being willing to explain the purpose of the investigation and to answer questions. Participants are more likely to be cooperative if they can see where the questions and the research are leading and they can relate this to what you stated as the purpose of the research at the start. An interview-based method is less rigid and formal than a questionnaire survey, but it is still important to prepare a set of topics, sometimes called 'lines of enquiry' in advance, which may in some circumstances be sent to interviewees before the interview takes place. While these are not intended to be as rigid and uniform as the questions on a questionnaire, they are intended to ensure that the issues you are researching get covered in the conversation with each respondent. A useful opening gambit in many interviews is to ask the interviewee to 'Tell me about ...', which provides a topic for discussion but does not constrain the interviewee's response. Recording interviews, with the respondents' permission is an especially useful way of enabling an interviewer to remain focused on the task of keeping the interview on track and listening to what the interviewee says. This makes it easier to ensure all the issues are covered and allows unexpected themes to be explored. Recording allows the researcher to revisit the interview on several occasions and to pick up the nuances of what occurred in a way that is impossible if simply relying on written notes. Modern digital recorders and smartphones help to avoid some of the difficulties associated with the older style tape recorders, such as the tape running out, and being smaller are generally less intrusive.

Even after completing a recorded interview, it is important to make some brief notes about issues that were discussed and the conduct of the interview itself. These will be useful should there be problems with the recording and will help an interviewer learn from the process so that subsequent interviews are even more effective. Analysis of the qualitative data obtained by means of interviews, whether written verbatim or recorded, involves the researcher teasing out the themes and making connections between the discussions conducted with different interviewees. This requires that the interviews be transcribed so that they can conveniently be compared with each other. Although modern digital recordings can be passed through voice recognition software to assist with the process of transcription, such software is still being perfected and some notes about the points discussed will always be useful. Ideally, interviews should be transcribed as soon as possible after they have taken place to clarify any points that are uncertain and to avoid a backlog of interviews potentially stretching back over several weeks or even months building up.

Interviewing on a one-to-one basis (i.e. interviewer and interviewee) is not the only way of engaging in qualitative methods of data gathering. Focus groups have become a productive way of eliciting information, since if they work well, they allow an interchange of opinions and views between participants. They have emerged as a popular means of gauging public opinion about political issues, but they are also well suited to generating data to help with addressing certain types of research questions. Although it is important for the researcher to be present, opinion is divided over whether it is advisable to have an independent facilitator to chair the event and to keep the conversation focused on the topic being investigated or whether the researcher can take on this role without biasing the outcome of the discussion. It might be tempting to treat a focus group as an encounter between people with differing or even opposing views on a topic. However, there are significant ethical issues associated with this strategy and from a practical perspective, it might be difficult to control if the discussion becomes heated and gets out of hand.

Many of the points raised previously in relation to one-to-one interviewing apply to focus groups. Additionally, it is necessary to organise a neutral location where the event will take place, which might involve hiring a venue that is conveniently accessible to all participants. At the start, the facilitator should state clearly the 'ground rules' for the discussion, for example that people should avoid interrupting others when they are speaking and should introduce themselves. It is even more important to record a focus group session, since there may be instances when different people are trying to speak and taking written notes would be almost impossible. It may be useful to organise some refreshments for participants at the start of the event, although consumption of these should not be allowed to interfere with the progress of the discussion. Transcription of the discussion should not only seek to provide an accurate record of what was said from start to finish, but also enable the contributions of individual participants to be tracked through the conversation. This has the potential to allow researchers to explore whether some people maintain or change their opinions in the light of contributions from others.

The previous discussion of approaches to data collection has, perhaps tacitly assumed, that the activity takes place in a mutually convenient space, such as an interviewee's workplace or home, in front of a computer or smartphone with a respondent self-completing a questionnaire, on the pavement of a road in a city or town centre with a researcher asking a respondent from a printed questionnaire or as a focus group meeting in a shared space such as a community hall or library. However, a range of alternative approaches involving movement of participants with or without a researcher and often involving some form of technical device such as smartphone with Global Positioning System (GPS) tracker or simple digital recorder have been used. These methods have opened new research directions and are felt to yield data from participants that are in some way 'more genuine' and immediate. For example, a researcher seeking information about people's attitudes toward a certain type of environment, such green spaces in urban centres, can be elicited in that context rather than removed from it. Similarly, participants can be allowed to move through an area on foot, cycle or other form of transport and record audio and photographic data about things they encounter according to a guiding principle provided by the researcher. Participants may be accompanied on these journeys by a researcher, in which case they are often called 'go alongs' (Carpiano, 2009; Jones and Evans, 2012; Kusenbach, 2003), which is a mobile interview that allows the researcher's initial 'lines of inquiry' script to be augmented spontaneously as new situations are encountered during the journey.

3.2.5 Crowdsourced Data

Researchers have for centuries sought new ways and types of technology for capturing data about the objects of their research to test and challenge existing and new theories. The continuing development of information technology in its diverse forms has contributed to the emergence of a new source of data during the early years of the 21st century. Important drivers facilitating the growth of such 'crowdsourced data' or 'user generated content' are the ease of creating content on the Internet using Web 2.0, the ubiquity of mobile devices capable of storing feature and operator location, online mapping applications and open access to remotely sensed imagery (See et al. 2016). The term crowdsourced data includes both volunteered geographic information (VGI) and contributed geographic information (CGI). The distinction between these two sources is that VGI refers to data that are deliberately created and shared, whereas CGI suggests that those producing the data have not explicitly given permission or are not aware of the content being shared. Harvey (2013) argued that the VGI and CGI distinction reflected the difference between participants opting-in to a study on a voluntary basis and people opting-out because they do not wish to contribute to the research. This is not unlike traditional data collection involving observation of people, vehicles and so on that are on open display as they move around public spaces but has the advantage of speedy and repeated data creation. People's motivation for participating in the creation of crowdsourced data is many and varied ranging from an altruistic desire to contribute content to a freely available map of the Earth (OpenStreetMap) (Jokar Arsanjani et al. 2015), assisting with large-scale scientific experiments, recording observations of plant or animal species (e.g. RSPB Big Garden Birdwatch, https://www.rspb.org.uk/get-involved/activities/birdwatch/) to posting geotagged photographic images of places people are visiting on social media. Here there is a similarity with administrative type of data considered in a previous section, where data collected or created for one purpose may in the hands of scientific investigators help to answer other research questions.

These developments have occurred in response to a research interest in ever finer-grained and up-to-date, perhaps near real-time information to investigate the spatio-temporal patterns of our daily lives than can be achieved 'with traditional methods, such as field observation and questionnaires' (Bubaloa, van Zanten and Verburg, 2019: 102). The research questions potentially addressable and the opportunities for analysing crowdsourced data are still emerging. Crowdsourcing data presents a new set of challenges for researchers, especially if the project involves retaining voluntary participants over a significant period. West and Pateman (2016) noted three points in particular: participants' motivation; individual circumstances and demographics; and raising awareness of the project. Box 3.5 provides an example of VGI data obtained from participants in a research project and the tracklogs for journeys made over 7 days. There are already

---

Box 3.5a   Example of volunteered geographic information (tracklogs).

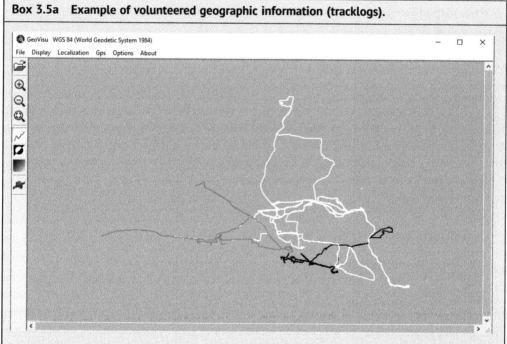

The image shows the sets of three tracklog navigation routes in a city on the south coast of England provided by older age voluntary participants in project. These are a subset of all the tracklogs captured using GPS receivers of the type illustrated in Box 3.1. The tracklogs record routes of various lengths and complexities, and for each individual volunteer, they covered all journeys by any mode of transport over 7 days. The GPS receivers recorded the coordinates of the routes in latitude and longitude.

Box 3.5b   GPS captured data and navigation route for individual tracklog.

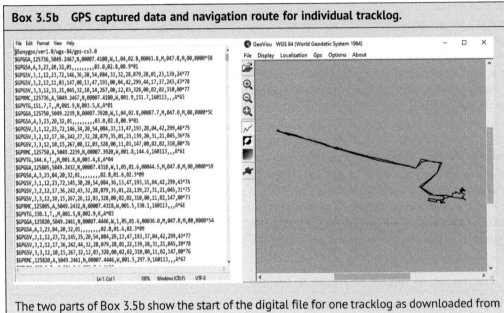

The two parts of Box 3.5b show the start of the digital file for one tracklog as downloaded from the GPS receiver and its routes is plotted on the right side.

several examples of where crowdsourced data obtained from social media websites (e.g. geo-tagged photos from Flickr.com) and other forms of VGI have been used in geographical research (Barros et al., 2020; Derdouri and Osaragi, 2021; Han et al., 2021; Hu et al., 2020; Negrini and Walford, 2022). There are already also instances, albeit unpublished ones, of undergraduate students assembling data through social media platforms, subject to ethical approval procedures, that go beyond the launching of online questionnaires into the realms of crowdsourced data via discussion groups.

3.2.6   Remotely Sensed Collection Methods

Remote sensing is defined as 'the observation or gathering information about, a target by a device separated from it by some distance' (Crackell and Hayes, 1993: 1). Such methods are generally accepted to include aerial photography and satellite imagery, thus defined remote sensing can include other approaches such as observation and recording without the target's knowledge, although such approaches in respect of human subjects clearly raise ethical issues. There are several texts available, many of which have multiple editions, that comprehensively cover the principles of remote sensing (e.g. Curran, 1985; Lillesand, Kiefer and Chipman, 2015; Mather, 1987; Richards, 2022). This section is not intended as a substitute for such thorough coverage but simply aims to provide a brief overview of satellite remote sensing including LIDAR (light detection and ranging) and aerial photography. These systems capture quantitative data in the form of EMR as a way of measuring characteristics of the Earth's surface together with its atmosphere and oceans. These are processed often using multivariate statistical techniques covered in later chapters of this text. The rapidity and regularity of collecting copious quantities of data are two of the main advantages of remote sensing, although the set-up costs for developing, launching and maintaining the sensor devices are high. This is offset by the length of time sensors may be in service continuing to supply large quantities of data. Recently cheaper platforms such as drones, which have increasingly joined the equipment inventories of university Geography and Earth Sciences departments, have allowed faculty academics and in some cases students the opportunity to operate remote sensing devices for their research projects. In summary, the various permutations of data collection devices available under the heading 'remote sensing' allow large quantities of data gathering across wide areas and over many years, which results in the unit cost of the data being low in comparison with alternative methods of obtaining the same quantity of information.

The history of remote sensing of the Earth's environment from airborne sensors dates back more than 150 years to early experiments with cameras used to take photographs from balloons. During the First World War cameras were used in aircraft to take photographs for reconnaissance purposes and between the World Wars such aerial photographic surveys were being used to supplement traditional land-based surveys for map making. Although further development occurred with diverse types of film during WWII, it was the launch of the first Earth Observation satellite in 1960 and the subsequent images of the Earth from space that set the trend in remote sensing for the closing decades of the twentieth and opening decades of the 21st centuries. The early images of the Earth from space confirmed for the public and scientists alike that 'The Earth was … an entity and its larger surface features were rendered visible in a way that captivated people's imagination' (Crackell and Hayes, 1993: 4). The arrival of data from satellite-mounted remote sensors helped to initiate the era of modelling global environmental systems and focused a spotlight on the impact of humankind's activities on the Earth.

The data (variable) captured by many remote sensors is typically EMR, although some systems record ultrasonic waves, and before examining the characteristics of the data obtained by this means it is helpful to review the nature of EMR. The principal source of EMR reflected from the Earth's surface and its features is the Sun, which is detected by sensing devices mounted on platforms directed towards the surface. EMR travels at the speed of light in a wavelike fashion creating the electromagnetic spectrum, which provides a scale or metric for measuring this phenomenon. The length of the EMR waves varies along this scale, which is often categorised into ranges, such as the near-infrared or visible sections respectively 0.7–3.0 μm and 0.4–0.7 μm. A proportion of the EMR reaching the Earth's atmosphere is attenuated (reduced) by absorption, reflection and scattering in certain parts of the spectrum. The EMR reaching the Earth's surface is reflected differentially by contrasting terrestrial features. For example, the amount of EMR reflected in certain sections of the spectrum by healthy and drought-stressed (desiccated) vegetation is different. Sensors directed towards the Earth passively record the level and intensity of EMR in one or more sections of the spectrum producing a collection of digital data with characteristic highs and lows. Processing these data allows the pattern to be determined and the terrestrial features and their characteristics to be interpreted. The following reflectance pattern indicates healthy green leaves:

- high level of near-infrared 0.75–1.35 μm
- medium level green light 0.5 μm
- low levels of red and blue 0.4–0.6 μm

The high level of near-infrared indicates the presence of photosynthetic activity and the medium level of green light the presence of chlorophyll. Variability in the ability of the Earth's atmosphere to absorb EMR along the spectrum means that certain sections are favoured for remote sensing.

- visible or optical (wavelength 0.4–0.7 μm)
- near (reflected) infrared (wavelength 0.7–3.0 μm)

- thermal infrared (wavelength 3–14 μm)
- microwave (wavelength 5–500 mm)

Therefore, rather than attempting to detect EMR along the full range of the spectrum, most sensors are set or 'tuned' to capture emitted EMR in specific ranges. One of the main differences between traditional aerial photographs and satellite images is the format in which the data are stored. In the case of aerial photographs, the data were held in an analogue format, corresponding to what the human eye can see, so that a photograph of a tree looks like a tree. The EMR is captured by focusing light onto a photosensitive target, such as a film emulsion, and is then processed with chemicals to produce the visible image. In contrast, sensors on satellites collect data in a digital format and an image of a tree is a series of digital numbers representing the EMR reflected in different sections of the spectrum by the tree. The digital image comprises a series of cells, called pixels, arranged in a regular grid that each possesses an intensity value in proportion to the EMR detected by the sensor. It is possible to convert between analogue and digital data by scanning a photograph, where the scanner is akin to the remote sensing device on an aircraft or satellite except that it is obtaining digital data from a previously processed photographic image rather than from EMR emitted by the terrestrial features themselves. Modern aerial photography also uses digital devices to capture images in much the same way as smartphone cameras are now routinely used to take our holiday and other pictures, whereas formerly we used cameras with film that was printed as images on photographic paper. An important characteristic of remote sensing is that the process does not automatically record either the location or the attributes of the geographical features. In the case of analogue data, it is a case of interpreting a photograph to determine what is present, whereas digital data requires processing in order to represent recognisable features.

3.2.6.1 Satellite Imagery

Most satellite-mounted sensors collect digital rather than analogue (photographic) data and these, alongside other spacecraft, have become increasingly important for Earth observation since the 1960s. There have been numerous remote sensing satellites launched since the start of the Earth observation era in 1960 and a growing number of countries have become involved individually or collaboratively in operating 'space programmes'. Some notable examples are the LANDSAT (USA), SPOT (FRANCE) and NOAA (USA) (formerly TIROS) series. Conventionally the digital imagery collected by these systems was expensive and likely to lie beyond the resources of student projects and this remains the case for some of the very recent, high spatial and spectral resolution data. However, several of the organisations responsible for these satellite series have now started to allow free access to some of the imagery. Such is the importance and detail contained in these data that the Ordnance Survey, Britain's national mapping agency, now uses this source rather than digitising from aerial photographs and field survey as the primary means of obtaining data for cartographic purposes. The organisation believes that 'as the accuracy increases, this method [remote sensing] could replace ground survey and aerial photos' (Ordnance Survey, 2007). Here we concentrate on reviewing the essential characteristics of digital satellite imagery as a means of obtaining information about geographical features for statistical analysis.

The first environmental remote sensing satellite launched in April 1960 has been followed by many others, some of which form part of a series (e.g. the National Aeronautics and Space Administration's (NASA) LANDSAT and France's SPOT series), others are 'one-offs'. The three main distinguishing characteristics of a satellite are its orbit, the spectral resolution and the spatial resolution and viewing area. The orbits of satellites vary in their relationship with the Earth and its position in the solar system in respect of their nature and speed. Geostationary satellites are located in a fixed position in relation to the Earth usually over the equator. Polar-orbiting satellites trace a near circular orbit passing over the North and South Poles or these regions at different altitudes and speeds according to satellite series. These characteristics define the temporal or repeat frequency at which data for a given portion of the Earth's surface are captured. The combination of a satellite's orbit and the Earth's rotation shifts the swath or scanning track of the sensing device's perpendicular view westwards so that the entirety of the surface is covered over a relatively brief period of time. The number of satellites operating in this fashion now means that there is likely to be at least one sensor in position to capture unexpected environmental events, such as the Asian tsunami on 26 December 2004 or the fire at the Buntsfield Oil Storage depot Hemel Hempstead in 12 December 2005.

Our discussion of EMR referred to the fact that some sections of the spectrum are preferentially absorbed by the atmosphere, including by clouds, gases and particulate matter, leaving certain wavelengths to reach the Earth's surface and be reflected for detection by satellite-mounted sensors. Most systems are multi-spectral and the sensors are 'tuned' to capture data in more than one section of the EMR spectrum or bandwidth and to avoid those ranges that tend to be absorbed. The clear advantage of multi-spectral as opposed to single-band systems is that the interplay between the radiance data values for the different bandwidths for each pixel is captured simultaneously and can be processed in a combinatorial fashion, which aids with interpreting the features located on the ground surface. However, a high spectral resolution can result in an increased signal-to-noise ratio, which reduces accuracy. Satellite-mounted sensors, as with most types of measuring instruments may operate imperfectly, and are subject to a certain amount of variation in the signal they receive on account of technical faults. This variability can be quantified by means of the signal-to-noise ratio, which can be reduced by increasing the bandwidth or by using a series of individual sensors to detect radiance in each scan line element, the so-called 'push broom' system.

Spatial resolution, although important, is rather more difficult to define and in its simplest form refers to the circular area on the ground, assuming no rectangular framing system is in place, viewed by the sensor at a given altitude and point in time. This is commonly referred to as the Instantaneous Field of View (IFOV), although alternative definitions of spatial resolution exist that take account of the sensor's ability to detect and record radiance at discrete points and for features to be identified irrespective of their relative size (for a full discussion see Campbell, 2007; Lillesand, Kiefer and Chipman, 2015; Mather, 1987). Spatial resolution should not be confused with pixel size, which refers to a rectangular or more typically square area on the ground surface that is covered by one cell in the digital image. Each of the pixels in the regular grid stores the radiance values emitted in certain sections of the EMR spectrum by the terrestrial features wholly or more commonly partially covered by each cell on the ground. If a given cell (pixel) straddles two contrasting features, for example bare rock and healthy vegetation the radiance values will reflect both these characteristics, whereas if a cell lies over a single, spectrally undifferentiated surface, such as concrete or bare soil, then the data will be more closely aligned with the spectral imprint of this feature. Reducing the size of the pixels will potentially enable greater surface detail to be determined, as the smaller pixels approach the scale of local variations in the ground surface. However, even seemingly uniform features, such as bare soil or concrete, have small-scale variations in the radiance values that might confound the spectral message. The spatial resolution of satellite imagery has reduced over the years, and careful selection from the range of available sources is required to ensure that the data employed are fit for purpose. Small is not necessarily beautiful when choosing satellite imagery for certain types of application, notably where the feature under investigation possesses minor variation over the extent of the surface. Many small pixels in this situation will result in a high degree of data redundancy, which is wasteful of data storage and processing time, and gives rise to a spurious sense of detailed variation in something that is inherently relatively constant over a large area.

The types of sensors just outlined are usually described as passive sensors, which simply means that they capture the EMR that originates from the Sun and reflected from the Earth's surface. In contrast, LIDAR is an example of an active sensor that inputs energy into the system, in this case light from a rapidly firing laser. The laser is directed towards the ground surface and terrestrial features, such as vegetation, buildings and valleys, which reflect light that is detected by a sensor. The time taken for the reflected light to be detected by the sensor is used to calculate the distance to the surface features and thereby their elevation above the ground. A GPS determines the X, Y and Z locations of the reflected light energy in conjunction with an Inertial Measurement Unit (IMU) that records the orientation of the aircraft considering its pitch, roll and yaw. The light energy returned to the sensor produces a waveform showing peaks and troughs in the amount or intensity of energy, which represent features on the ground. LIDAR data are typically stored as a set of discrete points in a 'point cloud'. Like the EMR data captured by passive remote sensors, LIDAR point data will usually need to be classified to identify the type of feature that reflected the light. Unlike satellite-mounted remote sensors, LIDAR systems can be mobile (e.g. on the roof of a road vehicle) or stationary on the ground (e.g. on a tripod) from which the laser is 'fired' horizontally at a vertical surface such as a building or cliff. Some of the main advantages of LIDAR sensors are that they can detect small differences in complex surfaces, provide 3-D depth in visualisation without further processing and can operate at night or during bad weather.

3.2.6.2 Aerial Photography

The arrival of Earth observation by means of satellite-mounted sensors started a slow process of decline in the importance of aerial photography, although some resurgence of interest arose in the 1990s when these were seen at the time as providing greater detail than the data available from satellite systems. However, aerial photography remains significantly challenged as a means of collecting data about geographical features, although digital camera technology has again led to some resurgence. Two of the clear advantages of an analogue aerial photograph are that the photographic image can be held in the hand and viewed optically, and pairs of overlapping photographs can be examined stereoscopically to determine the relative elevation of terrestrial features.

The three main distinguishing characteristics of aerial photographic images relate to the type of camera and film, and the position of the camera in relation to the horizontal. Typically, a single lens reflex (SLR) camera was used with either 35 or 70 mm format, although 23 cm format was also available, where format refers to the size and shape of the negative images produced. The focal length of the lens, the distance between its centre when focused on infinity and the film plane, influences the field of view of the camera, which in turn affects the scale of the photograph. Increasing the focal length of a camera at a fixed altitude results in an increase in the distance on the photograph between any given pair of points at fixed locations on the ground, this changes the scale of the image. For example, two electricity pylons located 100 m apart on the ground might be measured at 1 cm apart on a photograph with one focal length (i.e. a scale of 1:10,000), whereas with a different focal length the distance on the photograph increases to 1.25 cm (i.e. a scale of 1:8000). However, the question of scale in relation to aerial photography is not as simple as it might seem from this example. Suppose we extend the previous example to consider the situation of there being a series of pylons stretching diagonally across the photograph. Each pylon is exactly 100 m from its neighbour on the ground, but on the aerial photograph if the pair in the middle is 1 cm apart those near the corners will be separated by a greater distance. Thus, the scale of the photograph is not uniform across the entire area. This characteristic arises because the photograph has been produced from a camera with a perspective or central rather than orthographic projection. Techniques are available to apply geometrical correction to aerial photographs, but those using scanned images of the aerial photographs to capture digital data representing the geographical features (see below) should be alert to this issue. The implication of this variation in scale is that aerial photographs have two scales, one relating to the average for each photograph or the whole mission, and the other to each individual point at a particular elevation.

The third differentiating characteristic of aerial photographs relates to the types are film available. This is like distinguishing between satellite-mounted sensors that are 'tuned' to detect radiance in one or more discrete bandwidths. The main types of film used in small and medium format aerial photography are panchromatic (black and white), near-infrared black and white, colour and false colour near-infrared (Curran, 1981). The position of the film plane in the camera relative to the horizontal is divided into vertical and oblique. The former is preferred since it does not introduce distortion or displacement provided pitching, rolling and height changes of the aircraft are avoided. Photographs above 5° from the vertical are referred to as oblique, which are subdivided according to whether horizon is included or excluded from the field of view, respectively, known as high and low.

The process of converting between the multi-scale environment of an aerial photograph and the single-scale, orthographic projection of a map is known as photogrammetry. This involves the integration of the geometrical properties of the photograph (focal length, displacement, etc.) in relation to its centre. Photogrammetric processing of successive, regularly spaced, partially overlapping aerial photographs can produce three-dimensional data and enable the creation of a digital elevation model (DEM) of the surface. Such models include X, Y and Z values, which represent genuine measurements of altitude from which interpolated figures can be obtained over the entire surface. Aerial photographs can also be scanned and information about geographical features digitised from the images, although it is important to be aware of displacement, distortion and scale issues. The success of this process depends on transforming locations on the photograph to match the ground coordinate system using a minimum of three control points that can be precisely located on the photograph and on an existing map or in the field by using a GPS device. The digital datasets created by scanning aerial photographs are in some respects like the digital images obtained from satellite-mounted sensors. The scanning process creates an image of pixels of a certain density, usually measured in dots per inch (dpi), and with each pixel holding a data value representing the radiance (EMR) detected by the scanner. The number of greyscale or colour shades that can be stored is related to the storage capacity of the scanner (e.g. an 8-bit scanner stores 256 colours).

3.3 Locating Phenomena in Geographical Space

The main thing that distinguishes geographical data from other types of data is that the information can be located in relation to where it occurs on the Earth's surface. Therefore, irrespective of whether we are referring to analogue or digital data, there needs to be some way of linking the features (entities or observations) to which the data relate and by associating their attributes and variables to some location in respect of the Earth's surface. How this connection is achieved depends on the way the data are collected and how the geographical component of the data is to be analysed. Chapter 2 introduces the idea that as Geographers and Earth Scientists we are interested in the things, variously referred to as entities, observations and phenomena that exist at some location in relation to the Earth's surface and the relationships between these features. Recording where one observation in a population is located in respect of the others, and where all members of that population are in comparison with other populations of similar or distinct groups of observations is important. Fundamentally the individuals in a population possess variation in their spatial location (i.e. they do not occur on top of each other) and populations of similar observations may be compared with each other. This focus on locations requires that some means of recording where the constituent members of populations are located needs to be considered when planning an investigation. Sometimes the need to preserve people's anonymity in surveys means that their exact location, for example in terms of an address, should not be recorded and measures are required for generalising an address to an area where there are more addresses so that those relating to survey respondents are 'lost' and cannot be discovered.

The essence of the problem is to have some way of fixing the location of populations of observations and their individual constituent members in the short, medium or long term. Geographical entities exist at a fixed location at a given point in time, although on another occasion in the past or the future they may be located elsewhere or have a different form. The car park shown in Figure 3.1 serves London's Stansted Airport: the arrival and departure of cars associated with their drivers' and passengers' air travel aptly illustrate the geodynamic nature of most spatial phenomena. Sometimes there will be an abundance of empty spaces and at others, the majority will be occupied. This example illustrates not only the transient nature of most geographical entities but also in respect of its individual components (i.e. the occupancy of parking spaces) can change over time. Over a medium time period, the car park can be regarded as a fixed geographical feature assuming no enlargement or internal restructuring has occurred (e.g. changing the layout of roads or parking spaces). However, the internal spatial pattern of vehicles changes in connection with their arrival and departure, which produces a transient occupation of parking spaces. However, over a longer time period the car park's overall size and internal layout may be altered, thus potentially resulting in a change to its outer boundary and the network of roads connecting separate parking areas could be reconfigured.

Figure 3.1 Long Stay Car Park at Stansted Airport, London. *Source:* GeoInformation Group (adapted).

The importance of these issues from the perspective of locating geographical features relates to how the data associated with geographical features can be connected with locations on the Earth's surface. The location of a spatial feature is often specified by means of a **georeferencing system**. At a regional or national scale, this usually takes the form of a regular grid with a system of numerical coordinates, which defines the extent of national, regional or local space (geography) as though it was a flat (Euclidean) surface. The geometry of space is quantified by a regular, usually square, grid of X and Y coordinates, a Cartesian grid, that enables the position of phenomena to be determined. At a global scale, coordinates of longitude (meridian) and latitude (parallel) normally locate the position and thereby determine the distance between phenomena and reflect the shape of the Earth as a spheroid, not a perfect sphere. In the case of longitude and latitude, the units of measurement are degrees, minutes and seconds which are defined in relation to the geometrical properties of the Earth (see Figure 3.2a), whereas the British National Grid, introduced by the Ordnance

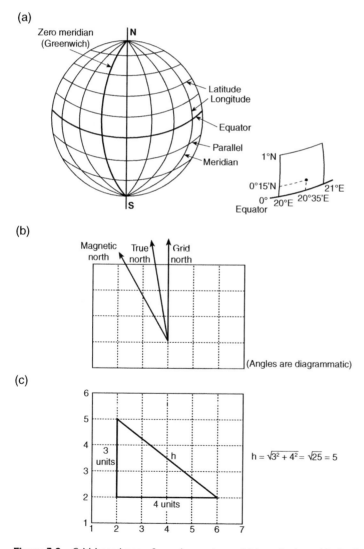

Figure 3.2 Grid-based georeferencing systems. (a) Longitude and latitude geographic coordinate system; (b) Connection between a national grid and the global longitude/latitude geographic coordinate system; (c) Calculation of distance between two georeferenced points using Pythagoras' theorem.

Survey in 1936, is based on the Airy 1830 ellipsoid. It uses the metric system and therefore enables distances to be determined in kilometres or metres directly rather than involving transformation from degrees and minutes. National grids can be linked with the global longitude/latitude system as illustrated in Figure 3.2b. An important feature of regular square grids is that distances can be calculated between phenomena not only in the directions of the horizontal and vertical planes of the X and Y axes but also diagonally. Any pair of points within the grid can form the endpoints of the hypotenuse of a right-angled triangle and its length can be obtained by using Pythagoras' theorem (Figure 3.2c).

Chapter 2 introduced the notion that geographical features could be regarded as points, lines or areas, and that these could be conceptualised in two main ways: in the case of areas as a portion of the Earth's surface surrounded by definite boundary; or by identifying variation in the concentration or density of some characteristic or variable. Following on from this distinction there are two main ways of georeferencing where features are located. The former case lends itself to capturing a series of points and lines, the latter formed of smaller segments or arcs, containing vectors of pairs of X and Y coordinates, as examined in Section 2.3.1. Except in the case of there being just one area feature of interest, the nodes at the end of each line represent the points where lines bounding two or more areas meet. If area features are conceived as being formed by variations in the concentration of one or more characteristics, the surface can be divided into a series of regular or irregular units (e.g. squares or triangles) and values of the characteristic(s) recorded for each of these units indicate where a particular feature is located. Population density provides an example of how this system works: each square in the regular grid has a value representing the concentration of people per unit area. Across the surface covered by the grid groups of neighbouring squares with relatively high values indicate a city or town occurs at that location, whereas at the other extreme a large area of squares with low values and an occasionally higher figure suggests a sparsely populated rural area of villages and small towns.

The referencing of geographical features to the Earth's surface by means of numbered vertical and horizontal lines in a regular grid involves using pairs of numerical X and Y coordinates to locate where they are to be found and to define their spatial or geometrical form and shape. Treating geographical features as composed of points, lines and areas, their location and geometry can be constructed from sets of X and Y coordinates. However, if they are conceptualised as being constructed from variations in the concentration or density of a particular characteristic in a regular square grid then it is the spatial units (squares) holding the data values that are located by means of row and column numbers of georeferenced coordinates. The metric of the grid provides a means of relating distances on the ground to those on a digital or printed map and thus defines the scale of representation.

Both approaches represent formal ways of recording the location of data about geographical features and phenomena. However, there are alternatives that apply when the investigation is less concerned with spatial characteristics of the features and more focused on using a qualitative methodology with respect to the geographical component of the data. A research question seeking to explore the lifestyles of adolescents in urban areas is likely to be less concerned with determining the precise location of these individuals and more interested in the types and characteristics of places where they 'hang out'. These provide a spatial or geographical context for an adolescent lifestyle that might vary in accordance with different ethnic groupings and gang membership. In these situations, it is less relevant to know the precise boundaries of where such groups congregate than the types of environments where their behaviours can flourish. Despite these contrasts between quantitative and qualitative approaches to locating geographical phenomena, both are concerned with similar spatial questions, such as the juxtaposition of individuals and the grouping of phenomena 'in space'. Such a view reflects a philosophical position that considers space as a 'container' in which phenomena or entities occur. The quantitative approach conceives spatial relationships as capable of direct analysis whereas the qualitative one argues that the fuzziness of location and uniqueness of places makes this unrealistic. A focus of where phenomena occur is a distinguishing characteristic of investigations in Geography and the Earth Sciences, even including such abstract, virtual locations as cyberspace. Later chapters exploring the application of statistical techniques in these fields will recognise the variety of ways in which location can be incorporated into your analysis.

Georeferencing by means of coordinates provides a potentially accurate and precise way of locating and describing the form and shape of geographical features. Ready access to the GPS via mobile geospatial technologies (e.g. smartphone and satellite navigation) allows people to easily, but unknowingly, determine where they are or to find another location on the Earth's surface by means of grid references. However, the notion of people using grid coordinates in everyday conversations seems some way off and both formal and informal languages exist enabling people to say where they are and to describe geographical phenomena. The simplest way of referring to a location is to use its name, which has the advantage of familiarity and the disadvantage of imprecision. Unless an investigator's research questions are concerned with place names, their main function in most studies is to indicate the general area to which the study relates. Possibly more useful than a name is some form of index comprising letters and/or numbers (alphanumeric), particularly where this forms a commonly used system, such as in the case of postal or zip codes, or the standard indexation applied to administrative areas (counties, districts and wards) by British governmental statistical agencies. A useful feature of these coding systems is that they are often hierarchical, thus enabling a group of smaller or lower units to be associated with a common larger or higher level area.

These spatial coding systems provide a means of precisely and uniquely identifying locations in space and the things they represent can be regarded as geographical entities, nevertheless, other than in a very simple way, they are unimportant from the perspective of statistical analysis. However, capturing the location of phenomena as georeferenced coordinates enables further types of geospatial analyses to be carried out. In July 2013 what3words (https://what3words.com/about) was launched that has assigned three words to each 3-by-3 m square across the Earth as a non-numeric way of geographical location. The extent to which spatial form and location are important aspects of an investigation will vary according to the research questions asked and the techniques for analysing these characteristics are examined in the closing chapters of this text.

References

Barros, C., Moya-Gómez, B. and Gutiérrez, J. (2020) Using geotagged photographs and GPS tracks from social networks to analyse visitor behaviour in national parks. *Current Issues in Tourism*, 23(10), 1291–1310. DOI:10.1080/13683500.2019.1619674.

Bubaloa, M., van Zantena, B.T. and Verburg, P.H. (2019) Crowdsourcing geo-information on landscape perceptions and preferences: a review. *Landscape and Urban Planning*, 184, 101–111.

Campbell, J. (2007) *Map Use and Analysis*, Oxford, Wm C Brown.

Carpiano, R.M. (2009) Come take a walk with me: the "go-along" interview as a novel method for studying the implications of place for health and well-being. *Health and Place*, 15(1), 263–272. DOI:10.1016/j.healthplace.2008.05.003.

Crackell, A.P. and Hayes, L.W.B. (1993) *Introduction to Remote Sensing*, London, Taylor and Francis.

Curran, P.J. 1981. Remote Sensing: The Role of Small Format Light Aircraft Photography. Geog. Papers, No. 75, Dept. of Geography, Univ. of Reading.

Curran, P.J. (1985) *Principles of Remote Sensing*, New York, Longman.

Derdouri, A. and Osaragi, T. (2021) A machine learning-based approach for classifying tourists and locals using geotagged photos: the case of Tokyo. *Information Technology and Tourism*, 23(4), 575–609. DOI:10.1007/s40558-021-00208-3.

Han, S., Liu, C., Chen, K., Gui, D. and Du, Q. (2021) A tourist attraction recommendation model fusing spatial, temporal, and visual embeddings for Flickr geotagged photos. *ISPRS International Journal of Geo-Information*, 10, 20. DOI: 10.3390/ijgi10010020.

Harvey, F. (2013) To volunteer or to contribute locational information? Towards truth in labeling for crowdsourced geographic information, in *Crowdsourcing Geographic Knowledge* (eds D. Sui, S. Elwood and M. Goodchild), Dordrecht, Springer. DOI:10.1007/978-94-007-4587-2_3.

Hu, L., Li, Z. and Ye, X. (2020) Delineating and modeling activity space using geotagged social media data. *Cartography and Geographic Information Science*, 47(3), 277–288. DOI:10.1080/15230406.2019.1705187.

Jokar Arsanjani, J., Zipf, A., Mooney, P. and Helbich, M. (2015) *OpenStreetMap in GIScience; Lecture Notes in Geoinformation and Cartography*, Cham, Springer International Publishing.

Jones, P. and Evans, J. (2012) The spatial transcript: analysing mobilities through qualitative GIS. *Area*, 44, 92–99. DOI: 10.1111/j.1475-4762.2011.01058.x.

Kusenbach, M. (2003) Street phenomenology: the go-along as ethnographic research tool. *Ethnography*, 4(3), 455–485.

Layder, D. (1993) *New Strategies in Social Research*, Cambridge, Polity Press.

Lillesand, T.M., Kiefer, R.W. and Chipman, J.W. (2015) *Remote Sensing and Image Interpretation*, 2nd edn, New York, John Wiley and Sons.

Mather, P.M. (1987) *Computer Processing of Remotely Sensed Images*, Chichester, John Wiley and Sons.

Mather, P.M. (2004) *Computer Processing of Remotely-Sensed Images – An Introduction (with CD)*, 3rd edn, Chichester, John Wiley and Sons.

Negrini, C. and Walford, N.S. (2022) A stroll in the park, a view of water: quantifying older people's interaction with 'green' and 'blue' spaces in urban areas. *Journal of Applied Geography*, 149, 102808. DOI: 10.1016/j.apgeog.2022.102808.

Reichman, W.J. (1961) *Use and Abuse of Statistics*, Harmondsworth, Penguin.

Reid, I. (2003) Making observations and measurements in the field: an overview, in *Key Methods in Geography* (eds N. J. Clifford and G. Valentine), London, Sage.

Richards, J.A. (2022) *Remote Sensing Digital Image Analysis*, Springer. DOI: 10.1007/978-3-030-82327-6.

See, L., Mooney, P., Foody, G., Bastin, L., Comber, A., Estima, J., Fritz, S., Kerle, N., Jian, B., Laakso, M., Liu, H.-Y., Milčinski, H., Nikšič, M., Painho, M., Pődör, A., Olteanu-Raimond, A.-M. and Rutzinger, M. (2016) Crowdsourcing, citizen science or volunteered geographic information? The current state of crowdsourced geographic information. *ISPRS International Journal of Geo- Information*, 5(5), 55. DOI:10.3390/ijgi5050055.

Valentine, G. (2008) Living with difference: reflections on geographies of encounter. *Progress in Human Geography*, 32(2), 323–337. DOI: 10.1177/0309133308089372.

Walford, N.S. (2002) *Geographical Data: Characteristics and Sources*, Chichester, John Wiley and Sons.

West and Pateman (2016) Recruiting and retaining participants in citizen science: what can be learned from the volunteering literature? *Citizen Science: Theory and Practice*, 1(2), 1–10. DOI: 10.5334/cstp.8.

Further Reading

Asher, H. (1998) *Polling and the Public. What Every Citizen Should Know*, 4th edn, Washington, CQ Press.

Batty, M. (2013) Big Data, smart cities and city planning. *Dialogues in Human Geography*, 3(3), 274–279. DOI:10.1177/2043820613513390.

Billari, F.C. and Zagheni, E. (2017) Big data and population processes: a revolution? in *SIS 2017. Statistics and Data Science: New Challenges, New Generations. 28–30 June 2017 Florence (Italy).* Proceedings of the Conference of the Italian Statistical Society (eds A. Petrucci and R. Verde), Firenze University Press, pp. 167–178.

Elwood, S. (2008) Volunteered geographic information: key questions, concepts and methods to guide emerging research and practice. *GeoJournal*, 72(3–4), 133–135. DOI:10.1007/s10708-008-9187-z.

England, K. (2002) Interviewing elites: cautionary tales about researching women managers in Canada's banking industry, in *Feminist Geography in Practice: Research and Methods* (ed. P. Moss), Oxford, Blackwell.

Fowler, F. (1995) *Improving Survey Questions: Design and Evaluation*, Thousand Oaks, Sage.

Gibson, P. and Power, C. (2013) *Introductory Remote Sensing: Principles and Concepts*, 2nd edn, London, Routledge, p. 216.

Goodchild, M.F. and Glennon, J.A. (2010) Crowdsourcing geographic information for disaster response: a research frontier. *International Journal of Digital Earth*, 3(3), 231–241. DOI:10.1080/17538941003759255.

Hetz, R. and Luber, J. (1995) *Studying Elites: Using Qualitative Methods*, London, Sage.

Kirkby, M.J., Nadem, P.S., Burt, T.P. and Butcher, D.P. (1992) *Computer Simulation in Physical Geography*, Chichester, John Wiley and Sons.

Meijering, L., Osborne, T., Hoorn, E. and Montagner, C. (2020) How the GDPR can contribute to improving geographical research. *Geoforum*, 117, 291–295. DOI:10.1016/j.geoforum.2020.05.013.

Misra, A., Gooze, A. and Watkins, K. (2014) *Transportation Research Record: Journal of the Transportation Research Board, No. 2414*, Washington, Transportation Research Board of the National Academies, pp. 1–8. DOI:10.3141/2414-01.

Oppenheim, A.N. (1992) *Questionnaire Design, Interviewing and Attitude Measurement*, London, Continuum.

Oskamp, S. and Schultz, P.W. (2004) *Attitudes and Opinions*, 3rd edn, Mahwah, Erlbaum.

Sheehan, K.B. (2018) Crowdsourcing research: data collection with Amazon's Mechanical Turk. *Communication Monographs*, 85(1), 140–156. DOI:10.1080/03637751.2017.1342043.

Skelton, T. (2001) Cross-cultural research: issues of power, positionality and 'race', in *Qualitative Methodologies for Geographers* (eds M. Limb and C. Dwyer), London, Arnold.

Slocum, T. (1990) The use of quantitative methods in major geographical journals, 1956-1986. *Professional Geographer*, 42, 84–94.

Smith, F. (2003) Working in different cultures, in *Key Methods in Geography* (eds N. Clifford and G. Valentine), London, Sage.

Spielman, S.E. (2017) The potential for big data to improve neighborhood-level census data, in *Seeing Cities Through Big Data* (eds P. Thakuriah, N. Tilahun and M. Zellner), Springer Geography, Cham, Springer. DOI:10.1007/978-3-319-40902-3_6.

Sudman, S. and Bradburn, N. (1982) *Asking Questions: A Practical Guide to Questionnaire Design*, San Francisco, Jossey-Bass.

Section II

Exploring Geographical Data

4

Statistical Measures (or Quantities)

> Chapter 4 concentrates on the basic descriptive measures or quantities used to compare and contrast sets of numbers in their role as measuring quantifiable differences between attributes and variables. Attention is directed towards summary measures referring to numbers, such as the median, mean, range and standard deviation, and to the location of geographical phenomena, such as mean centre and standard distance. This chapter establishes the characteristics of the basic measures and quantities that are subject to statistical testing.

Learning Outcomes

This chapter will enable readers to:

- Explain the purpose of spatial and non-spatial descriptive statistics for comparing and contrasting numerical data;
- Identify the appropriate summary measures to use with variables recorded on different measurement scales;
- Start to plan how to use descriptive statistics in an independent research investigation in Geography, Earth Science and related disciplines.

4.1 Descriptive Statistics

Descriptive statistics are one of two main groups of statistical techniques, and the other is known as inferential statistics or hypothesis testing. Descriptive statistics are about obtaining a numerical measure (or quantity), or a frequency count to describe the characteristics of attributes and variables of observations in a dataset. We will look at frequency distributions in Chapter 5. The principles of hypothesis testing and inferential statistics are examined in Chapter 6, but for the present, they can be defined as a group of analytical techniques that are used to find out if the descriptive statistics for the variables and attributes in a dataset can be considered as important or significant. Descriptive statistics are useful in their own right, and in most geographical investigations, they provide an initial entry point for exploring the complex mass of numbers and text that make up a dataset. For this reason, the topics and techniques covered in this and the next chapter are now commonly labelled exploratory statistics (analysis), or in the geographical

Practical Statistics for Geographers and Earth Scientists, Second Edition. Nigel Walford.
© 2025 John Wiley & Sons Ltd. Published 2025 by John Wiley & Sons Ltd.
Companion website: www.wiley.com/go/PracticalStatistics2e

context exploratory spatial statistics (analysis). Some research questions can be adequately answered using descriptive statistics on their own. However, in many cases, this proves insufficient because of an unfortunate little problem known as **sampling error**, to which we will return a number of times in this text.

Descriptive statistics can be produced for both population and sample data, when they relate to a population they are known as **parameters**, whereas in the case of a sample they are simply, but perhaps slightly confusingly, called **statistics**. The purpose of descriptive statistics is to describe individual attributes and variables in numerical terms. We are all familiar with descriptions of people, places and events in books, and Box 4.1a reproduces the first sentence from Jane Austen's novel *Pride and Prejudice but* try to follow the narrative by reading the words when they have been jumbled (Box 4.1b). It is worse than looking at a foreign language, since the individual words are familiar, but their order makes little sense. Even if the words are sorted into alphabetical order (Box 4.1c), they would not convey the meaning intended by the original author. However, we can see that Jane Austen used the word 'a' four times and both 'in' and 'of' twice. We can make a similar comparison with a set of numbers. In Box 4.1d, the numbers generated by the UK's National Lottery™ over 26 Saturdays in 2003 have been listed in the order in which they appeared. The range of possible numbers for each draw went from 1 to 49. Were some numbers selected more often than others? Were some numbers omitted? Were numbers above 25 picked more or less often than those below this middle number between 1 and 49? It is difficult to answer these questions until the numbers are presented in numerical order and Box 4.1e enables us easily to see that 10 and 12 appeared more often than any other number (7 times each), 40 is the only number not to appear in the draw during this 26-week period and that 86 of the numbers were below 25 and 90 were above. This example illustrates that the simple process of reorganising the numbers into the sequence from lowest to highest allows us to derive some useful information from the raw data. Nevertheless, the sorted sequence still includes 182 numbers (26 weeks × 7 numbers) and there is scope for reducing this detail further to obtain more summary information from the raw data. Indeed, an alternative name for the process of producing descriptive statistics is **data reduction**.

> Which combination of six main numbers would seem to offer the best chance of winning the jackpot prize? Why would using this set of numbers fail to guarantee winning the jackpot?

There are two main, commonly used groups of summary measures or quantities known as **measures of central tendency** and **measures of dispersion**. These are examined in detail later in this chapter. The first group includes the mode, median and mean (commonly called the average). The main measures of dispersion are the range, interquartile range, variance and standard deviation. The two sets of measures provide complementary information about numerical measurements, in the first case a central or typical value, and in the second how 'spread out' (dispersed) the data values are. The choice of which measure to use from each group depends largely upon the measurement scale of the attribute or variable in question. Most analyses will involve using two or three of the measures from each group according to the nature of the data. In addition to summarising the central tendency and dispersion of a numerical distribution, it may be important to examine two further characteristics, namely its **skewness** and its **kurtosis** ('peakedness'). Measures of skewness describe whether the individual values in a distribution are symmetrical or asymmetrical with respect to its mean. Measures of kurtosis indicate the extent to which the overall distribution of data values is relatively flat or peaked.

Box 4.1   Comparison of data reduction in respect of textual and numerical data.

(a) Original words in English sentence
'It is a truth universally acknowledged that a single man in possession of a good fortune must be in want of a wife'.
Pride and Prejudice by Jane Austen

(b) Jumbled words
Of man a in single a fortune be good universally wife that acknowledged want it truth is of possession a in that a.

(c) Words sorted alphabetically
A a a a acknowledged be fortune good in in is it man must of of possession single that truth universally want wife

(d) 182 Lottery numbers

08/02/03: 43, 32, 23, 37, 27, 41, **35,**
22/02/03: 30, 19, 42, 33, 38, 44, **31,**
01/03/03: 29, 31, 45, 44, 22, 24, **35,**
15/03/03: 47, 28, 04, 25, 38, 24, **11,**
22/03/03: 45, 10, 47, 49, 08, 02, **09,**
29/03/03: 33, 42, 05, 15, 35, 21, **26,**
12/04/03: 45, 25, 03, 16, 05, 43, **23,**
19/04/03: 48, 08, 38, 31, 13, 10, **14,**
26/04/03: 35, 27, 21, 09, 48, 33, **18,**
03/05/03: 07, 03, 49, 46, 06, 29, **37,**
10/05/03: 26, 32, 18, 08, 38, 24, **31,**
17/05/03: 48, 15, 36, 12, 08, 22, **37,**
31/05/03: 22, 25, 02, 09, 26, 12, **21,**
14/06/03: 46, 43, 37, 03, 12, 29, **11,**
21/06/03: 36, 20, 45, 12, 39, 44, **49,**
28/06/03: 23, 17, 35, 10, 01, 29, **36,**
05/07/03: 10, 18, 06, 19, 22, 43, **08,**
12/07/03: 15, 13, 21, 12, 09, 33, **43,**
19/07/03: 09, 34, 03, 17, 10, 48, **36,**
26/07/03: 27, 49, 16, 01, 12, 26, **23,**
02/08/03: 45, 28, 49, 47, 05, 26, **08,**
09/08/03: 41, 25, 14, 30, 19, 11, **38,**
16/08/03: 01, 09, 42, 45, 12, 10, **27,**
23/08/03: 42, 25, 15, 10, 18, 48, **11,**
30/08/03: 18, 38, 43, 44, 26, 33, **02,**
06/09/03: 21, 25, 39, 01, 27, 42, **17**

(e) Sorted lottery numbers

01	01	01	01	**02**	02	02	03
03	03	03	04	05	05	05	06
06	07	**08**	**08**	08	08	08	08
09	**09**	09	09	09	09	10	10
10	10	10	10	10	**11**	**11**	**11**
11	12	12	12	12	12	12	12
13	13	**14**	14	15	15	15	15
16	16	**17**	17	17	**18**	18	18
18	18	19	19	19	20	**21**	21
21	21	21	22	22	22	22	**23**
23	23	23	24	24	24	25	25
25	25	25	25	**26**	26	26	26
26	26	**27**	27	27	27	27	28
28	29	29	29	29	30	30	**31**
31	31	31	32	32	33	33	33
33	33	34	**35**	**35**	35	35	**35**
36	**36**	36	36	**37**	**37**	37	37
38	38	38	38	38	38	39	39
41	41	42	42	42	42	42	**43**
43	43	43	43	43	44	44	44
44	45	45	45	45	45	45	46
46	47	47	47	48	48	48	48
48	49	49	**49**	49	49		

Note: 'Bonus Ball' numbers are shown in bold.

4.2   Spatial Descriptive Statistics

One thing setting Geography and other Geo (Earth) Sciences apart from many other subjects is an interest in investigating variations in the **spatial location** of geographically distributed phenomena as well as their attributes and variables. In some respects, the spatial location of geographical phenomena can be considered as 'just another attribute or variable', but in practice where phenomena are located in respect of each other and the patterns thus produced are often of interest in their own right. When measuring differences between non-spatial phenomena, we concentrate on the values of individual or groups of attributes and variables, but usually ignore the position of each observation in the set. Thus, a non-spatial analysis of sampled households is likely to have little interest in distinguishing between, for example, whether any of the households live next to each other or how far apart any pair of households live. Spatially distributed phenomena have the extra dynamic that their position can also be quantified. Haggett (1979) highlighted the importance of spatial location when he described where people might decide to position or locate themselves on a beach and the thoughts underpinning these decisions. Most measurable entities, even if infinitely small or large, are located in space, what distinguishes 'spatially aware' disciplines is that this spatial distribution itself is not dismissed as a chaotic, random occurrence, but as capable of interpretation and in the case of Geography capable to revealing the outcome and impact of human–environment interaction.

Suppose a researcher seeks to understand the reasons for differences in the quality of households' accommodation and well-being. Respondents in each sampled household might be interviewed and information obtained about the number of rooms and their size, the availability of various amenities (e.g. central heating, air conditioning, internet access, mains gas and electricity), the entrance floor level, the number of people in the household, the presence/absence of a swimming pool, etc. Analysis of these data might reveal certain non-spatial associations or connections; for instance, a large living area per person and the presence of a swimming pool may both indicate relative financial and social advantage. However, if analysis of the survey data reveals households possessing these and other similar combinations of characteristics live close together, then perhaps we can also conclude that this is a relatively wealthy suburban area. In other words, the physical distance between households and the patterns they display may add to the analysis of their socio-economic characteristics. In this situation, we can start to say something about the characteristics of the places people occupy as well as the people themselves.

Spatial patterns may be interesting in themselves, but perhaps more importantly for geographers is that they indicate the outcome of a process of competition for space amongst the phenomena of interest. In some plant communities, for example, there may be a mix of species that flower at different times of the year and so it does not matter too much if there is some variation in height, provided that tall, medium and short plants do not flower at the same time. Similarly, if a gardener tries to grow too many plants in a limited space, some will thrive, whereas others will decline and possibly die. In a commercial context, a farmer growing fruit bushes and trees outdoors (e.g. apples, cherries, blackcurrants, and gooseberries) will try to ensure that they are planted an equal distance apart in order to maximise their exposure to sunlight, rainfall, soil nutrients and to make harvesting the fruit easier. However, suppose the farmer decides to grub up the fruit bushes and trees and to sow the field to grass in order to graze cattle. Grazing animals are usually allowed relatively unconstrained movement over a field of grass or are occasionally moved between fenced-off sections: their competing demand for forage is controlled not by tethering the animals at different locations in the field, but by limiting the overall number within this bounded portion of space and allowing the animals to roam freely in search of fresh pasture. Competition for space is therefore rarely, if ever, an unrestricted process, since there is normally some form of controlling or managing function in place, which may be either 'natural' (e.g. competing demand for exposure the sun's radiation) or 'human' (e.g. restricting demand for access to fresh grass).

There are three main ways in which phenomena can be distributed across any bounded portion of space: regular, clustered or random. Figure 4.1 is an aerial view of part of the long stay car park at London Stansted airport and different areas of the image illustrate these three patterns. In some parts of the car park, the cars are parked in a regular fashion either in pairs next to each other or singly with one or two spaces in between. In other parts, there are cars parked in small clusters with between four and eight cars in each. However, taking an overall view the cars would seem to be parked in a fairly random fashion with some close together and others further apart, some

Figure 4.1 Spatial distribution patterns. *Source:* GeoInformation Group (adapted).

forming small groups and others on their own. Figure 4.1 presents a snapshot of the spatial pattern of cars in this car park, but thinking about the process whereby cars arrive and depart from an airport car park and how those managing the car park open and close different sections for new arrivals helps in understanding how the pattern may have arisen. It provides an interesting example of the dynamic interrelationships between space and time.

> Where might drivers arriving at this section of the car park their cars? How will length of stay in the car park influence the spatial pattern? How might the way the operator of the car park influence the spatial pattern by controlling car arrivals and departures?

This interest in geographical patterns has led to the development of a whole series of spatial statistics, sometimes called **geostatistics** that parallel the standard measures used with non-spatial data. Descriptive spatial statistics aim to describe the distribution and geometrical properties of geographical phenomena. For example, lines (e.g. roads) can be straight or sinuous; areas may be large or small, regular (e.g. rectangular city blocks) or irregular (e.g. UK address postcodes). Spatial statistics may not be relevant in all geographical investigations, but explanations of the standard measures are usefully complemented by descriptions of their spatial equivalents.

The division of statistical procedures into descriptive and inferential categories also applies to spatial statistics. Numerical measures and frequency counts produced to describe the patterns created by spatial phenomena either relate to one event, as in the aerial snapshot of car parking shown in Figure 4.1, or to a series of events on different occasions tor the purpose of revealing change over time, which could be illustrated by continuous filming of vehicles arriving and departing the car park. Geographical phenomena are usually divided into three basic types when examining spatial patterns, namely points, lines and areas. However, the question of scale can sometimes confuse this simple classification, especially between points and areas. Figure 4.2 shows the complete aerial photograph from which the extract in Figure 4.1 was taken. Zoomed out this distance, the individual cars in certain parts of the car park can still be identified, but in the large central area, they have

Figure 4.2 The blurring of individual points into areas. *Source:* GeoInformation Group (adapted).

virtually merged into one mass and just about transformed into an area, although the circulation routes between the parking spaces enable lines of cars to be identified. Clearly few geographers or Earth scientists spend their time analysing the patterns produced by cars in a parking lot, but nevertheless this example illustrates that the scale at which spatial phenomena are considered not only affects how they are visualised but also how their properties can be analysed.

Measures of central tendency and dispersion also form the two main groups of summary descriptive geostatistics and include equivalent descriptors to their non-spatial counterparts. The details of these are examined later in this chapter. These measures focus in the first case on identifying a central or typical location within a given collection of spatial phenomena and in the second on whether the features tightly packed together or are dispersed around this location. They are sometimes referred to as **centrographic techniques,** and they provide a numerical expression of the types of spatial distribution illustrated in Figure 4.1.

4.3 Central Tendency

4.3.1 Measures for Non-spatial Data

We have already seen that a certain amount of descriptive information about the numerical measurements for a population or sample can be obtained simply by sorting them into ascending (or descending) order. However, calculating a single quantity representing the central tendency of the numerical distribution reveals more. There are three main measures of central tendency known as the mode, median and mean, and each is a number that typifies the values for a non-spatial attribute or variable. The procedures for calculating these measures are outlined in Box 4.2a with respect to a variable measured on the ratio/interval scale, water temperature in a fluvioglacial stream measured at half-hourly intervals during daytime. The lowest of the central tendency measures is the mean (9.08); the mode is highest (11.40) with the median (10.15) falling between them. Thinking about the variable recorded in this example and the time period over which measurements were taken, this particular sequence is not surprising. A daytime rise in air temperature after sunrise warms the water melting from the glacier, which is sustained by the warmth from the mid-day and early afternoon sunshine, when the peak temperature occurs. Recording of water temperature stopped shortly after this peak in comparison with the time interval between sunrise and the start of recording in the morning. It took five hours to reach the peak (09.00–14.00 hr), but recording stopped three and a half hours afterwards (17.30 hr). In other words, the length of the period over which the data measurements were captured has possibly had an impact on the statistical analysis.

> What lessons might the researcher take from this apparent relationship between the change in air and water temperature during the day when next going into the field to record the temperature of glacial meltwater streams?

> Why are there three measures of central tendency?

Each measure has its relative strengths and weaknesses and should be used with attributes or variables recorded according to the different scales of measurement. If a large number of values are possible, then the mode is unlikely to provide much useful information. At one extreme, it might simply record that each value occurs only once or, slightly more promisingly, that 1 out of 50 values is repeated twice or perhaps even three times. In other words, the mode is only really helpful when dealing with nominal attributes that have a limited range of values representing the

Box 4.2a   Central tendency measures for non-spatial data: sample site for the measurement of water temperature in a fluvioglacial stream from Les Bossons Glacier, France.

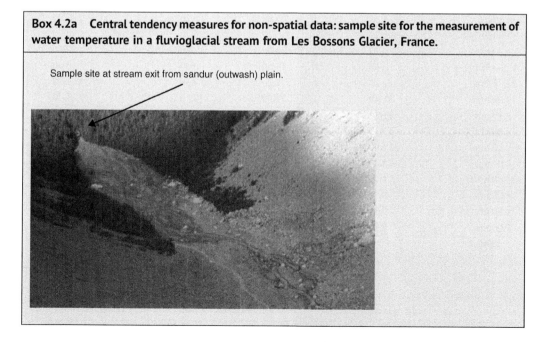

Sample site at stream exit from sandur (outwash) plain.

Box 4.2b   Calculation of the mode, median and mean.

Mean – population symbol: μ; sample symbol: $\bar{x}$

The measures of central tendency are perhaps some of the most intuitive statistics (or parameters) available. Their purpose is to say something about the typical value in an attribute or variable and allow you to quantify whether the central values in two or more sets of numbers are similar to or different from each other. Each of the main measures (mode, median and mean) can be determined for variables measured on the interval or ratio scale, ordinal variables can yield their mode and median, but only the mode is appropriate for nominal attributes. So, if you have some data for two or more samples of the same category of observations, for instance samples of downtown, suburban and rural households, then you could compare the amounts of time spent travelling to work.

The mode is the most frequently occurring value in a set of numbers and is obtained by counting. If two or more values appear the same number of times, the question arises as to whether there are two modes (i.e. both values) or one, midway between them. If no value occurs more than once, then a mode cannot be determined. The median is the value that lies at the mid-point of an ordered set of numbers. One way of determining its value is to sort all the data values into either ascending or descending order and then identify the middle value. This works perfectly well if there is an odd, as opposed to even, number of values, since the median will necessarily be one of the recorded values. However, if there is an even number, the median lies halfway between the two observations in the middle and will not be one of the recorded values if the two middle observations are different. Calculation of the arithmetic mean involves adding up or summing all the values for a variable and then dividing them by the total number of values (i.e. the number of observations).

The methods used to calculate the mode, median and mean are illustrated below using half-hourly measurements of water temperature in the fluvioglacial stream from Les Bossons Glacier near Chamonix in the European Alps (see Box 4.2a).

	Mode	Median		Mean	$\bar{x} = \dfrac{\sum x}{n}$
	x	x	Sorted x		x
09.00 hrs	4.30	4.30	4.30		4.30
09.30 hrs	4.70	4.70	4.70		4.70
10.00 hrs	4.80	4.80	4.80		4.80
10.30 hrs	5.20	5.20	5.20		5.20
11.00 hrs	6.70	6.70	6.70		6.70
11.30 hrs	10.10	10.10	7.80		10.10
12.00 hrs	10.50	10.50	8.80		10.50
12.30 hrs	11.20	11.20	9.30		11.20
13.00 hrs	**11.40**	11.40	10.10		11.40
13.30 hrs	11.80	11.80	10.20		11.80
14.00 hrs	12.30	12.30	10.50		12.30
14.30 hrs	11.90	11.90	11.10		11.90
15.00 hrs	**11.40**	11.40	11.20		11.40
15.30 hrs	11.10	11.10	11.40		11.10
16.00 hrs	10.20	10.20	11.40		10.20
16.30 hrs	9.30	9.30	11.80		9.30
17.00 hrs	8.80	8.80	11.90		8.80
17.30 hrs	7.80	7.80	12.30		7.80
					$\sum x$ 163.50
			$\dfrac{10.10 + 10.20}{2}$		$\dfrac{163.50}{18}$
	Mode = 11.40		Median = 10.15		Mean $\bar{x}$ = 9.08

labels attached to the raw data. For example, respondents in a survey may be asked to answer an attitude battery question with 'Strongly Agree', 'Agree', 'Neutral', 'Disagree' or 'Strongly Disagree' as the possible pre-defined responses. Suppose further that these responses have been assigned numerical codes from 1 to 5, respectively. It is entirely feasible that the mode will be one of the extreme values (1 or 5, 'Strongly Agree' or 'Strongly Disagree'). Hence, the notion of central tendency with respect to the mode relates it being the most common nominal category. The values 1–5 in this example do not imply any order of magnitude in measurement, and there is no requirement that 'Strongly Agree' should have been labelled 1, 'Agree' as 2, etc. It is simply a matter of arbitrary convenience to allocate the code numbers in this way – they could just as easily have been completely mixed up ('Strongly Agree' as 3, 'Agree' as 5, 'Neutral' as 4, 'Disagree' as 1 and 'Strongly Disagree' as 2). The median is, by definition, more likely to provide a central measure within a given set of numbers. The main drawback with the median is its limited focus on either a single central value or on the two values either side of the mid-point. It says nothing about the values at either extreme. Suppose there are two groups of five data values (19, 22, **24**, 26 and 27; 3, 13, **24**, 37 and 45), median value in both sets is 24, but the spread is very different.

The arithmetic mean is probably the most useful measure of central tendency, although its main feature is both a strength and weakness. All values in a set are taken into account and given equal importance when calculating the mean, not just those at the extremes or those in the middle. However, if the values are asymmetrical about the mean, for example there are a few outlying values at either the lower or upper extreme, these can 'pull' the mean in that direction away from the main group of values. Trimming values at both extremes, for example by ignoring the lowest and highest 5 or 10% of values, may reduce this effect, but the decision to do so represents a subjective judgement on the part of the investigator, which may or may not be supported from either a theoretical or statistical perspective. Another advantage of the mean over the median is that it can easily be calculated from totals (i.e. total count of observations and total sum of data values). For example, the 2001 UK Population Census reports that there was a total of 3,862,891 persons living in 1,798,864 privately rented furnished housing units in England, easily allows you to calculate that there were 0.47 persons per dwelling unit under this form of tenure (1,798,864/3,862,891) (on average). The equivalent median value cannot be determined in this fashion. The arithmetic mean possesses two further important properties. First, if the mean is subtracted from each number in the set, then these differences will add up to zero; and second, if these positive and negative differences are squared and then summed, the result will be a minimum (i.e. the sum of the squared differences of any number apart from the mean would be larger). Therefore, unless the median and mean are equal, the sum of the squared differences from the median would be greater than those from the mean. You might feel these properties of the mean are just a quaint feature of the statistic, although intuitively they suggest that the mean really does encapsulate the central tendency of a set of data values.

4.3.2 Measures for Spatial Data

The equivalent spatial statistics to the median and the mean for phenomena represented graphically as points are known as the **median centre** and the **mean centre**, both of which can be weighted according to the numerical values associated with attributes or variables measured with respect to the phenomena occurring at the points. The calculation of the median and mean centres, and the weighted equivalent of the latter, is given in Box 4.3a using the example of the location of Burger King food outlets in Pittsburgh. Box 4.3a indicates that the distribution of the companies' outlets was concentrated in the downtown area, but that there were also some outlying, suburban enterprises. The question is whether the overall centre of the corporation's enterprise in the city is in the urban core or otherwise. The mean and median centres are relatively close together in this example and can be regarded as providing a reasonable summary of the centre of the spatial distribution and as indicating that the Burger King has good coverage across the city as a whole.

The median centre shares the same problem as the (non-spatial) median, since it focuses on the centre of the spatial distribution and hence ignores possibly important outlying occurrences. Figure 4.3 illustrates how the median centre can occur with a range of X and Y co-ordinates at the intersection of the shaded stripes. Nevertheless, it is a useful intuitively simple location for the centre of a spatial distribution. The mean centre also has some issues. A few observations located at the extremes of a distribution can lead to the mean centre being drawn in their direction. An alternative to the Median centre, especially in American texts is the Centre of Minimum Travel, which is located at the point of minimum total distance between all of the points within a given set. The mean centre lies within a rectangular area with its dimensions defined by boundary lines drawn north–south and west–east through most northerly, southerly, westerly and easterly points in the distribution. This is known as the Minimum Bounding Rectangle and defines the extreme extent of the distribution. Consequently, if a few or even one point lies at some distance from the remainder, its extreme X and/or Y co-ordinates will drag the mean centre away from where the majority of phenomena are located. Another seemingly anomalous result can occur if the distribution of points includes two or possibly more distinct clusters each with a similar number of points. The mean centre could fall between the groups. Again, if the points form a crescent or

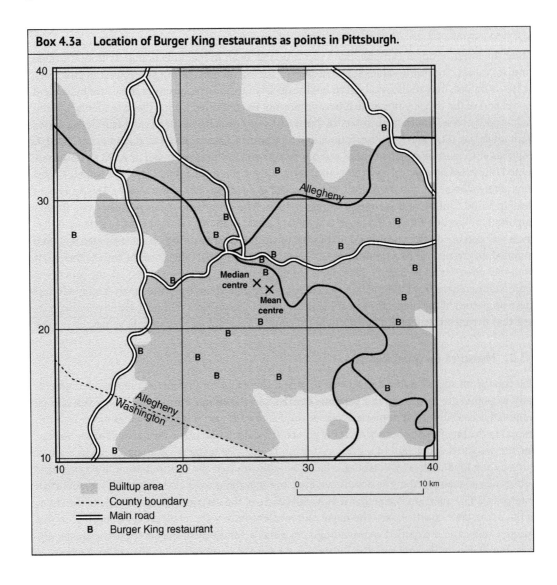

Box 4.3a Location of Burger King restaurants as points in Pittsburgh.

Box 4.3b Calculation of the mean and median centres of Burger King restaurants in Pittsburgh.

Mean centre – population symbol: $\overline{X}, \overline{Y}$; sample symbol: $\overline{x}, \overline{y}$

Measures of central tendency for spatially distributed phenomena serve much the same purpose as their non-spatial counterparts, but here it refers to a physically central location in the distribution. Calculation of the mean and/or median centre for the distributions of two or more spatial phenomena within a given area enables you to determine if their centres are relatively close or distant. Commercial enterprises competing with each other for customers may seek to open outlets in locations where they can attract each other's clientele. In this case, you might expect the mean centres of each corporation's outlets to be reasonably close to each other. In contrast, if enterprises obtained a competitive advantage by operating further way from rival companies, there would be more separation between their mean centres.

The median and mean centres are normally calculated from the X and Y coordinate grid references that locate the phenomena in Euclidean space. The median centre is determined by interpolating coordinate values midway between the middle pairs of X and Y data co-ordinates. The median centre lies at the intersection of the two lines drawn at right angles to each other (orthogonal lines) through these points and parallel with their corresponding grid lines. This procedure partitions the point phenomena into four quadrants (NW, NE, SW and SE). However, this procedure will not necessarily assign equal numbers of points to each quadrant. An alternative method involves dividing the points into four equal sized groups in each of the quadrants and then determining the median centre as where two orthogonal lines between these groups intersect. The problem is that many such pairs of lines could be drawn, and hence, the median centre is ambiguous.

The mean centre is far more straightforward to determine and simply involves calculating the arithmetic means of the X and Y co-ordinates of the data points. The mean centre is located where the two orthogonal lines drawn through these 'average' co-ordinates intersect. The mean centre is thus unambiguously defined as the grid reference given by the mean of the X and Y co-ordinates.

Calculation of the mean and median centres is illustrated using points locating Burger King outlets in Pittsburgh, denoted by subscript B for X and Y co-ordinates (see Box 4.3a).

	$\overline{x} = \dfrac{\sum x_B}{n}$	$\overline{y} = \dfrac{\sum y_B}{n}$	Median	
	x_B	y_B	Sorted x_B	Sorted y_B
1	11	27	11	10
2	14	10	14	15
3	16	18	16	16
4	19	23	19	16
5	21	17	21	17
6	22	16	22	18
7	22	27	22	19
8	23	19	23	20
9	23	28	23	20
10	26	20	26	22
11	26	24	26	23
12	26	25	26	24
13	27	16	27	24
14	27	25	27	25

(Continued)

Box 4.3b   (Continued)

$\bar{x} = \dfrac{\sum x_B}{n}$	$\bar{y} = \dfrac{\sum y_B}{n}$	Median	
x_B	y_B	Sorted x_B	Sorted y_B
15	27	32	27	25
16	32	26	32	26
17	36	15	36	27
18	36	35	36	27
19	37	20	37	28
20	37	22	37	28
21	37	28	37	32
22	38	24	38	35
	$\sum x_B = 583$	$\sum y_B = 497$		
	$\dfrac{583}{22}$	$\dfrac{497}{22}$		
	$\bar{x}_B = 27$	$\bar{y}_B = 23$	Median $x_B = 26.0$	Median $y_B = 23.5$
Mean centre = 27.0, 23.0			Median centre = 26.0, 23.5	

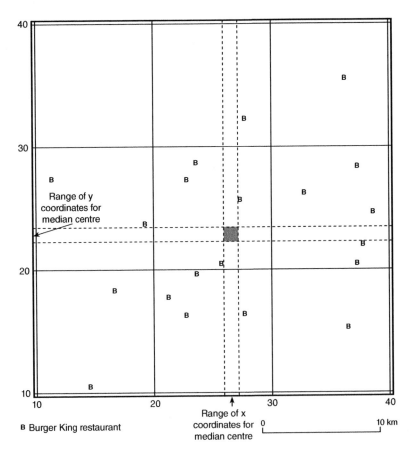

Figure 4.3   Square area containing alternative locations for the median centre.

circular-shaped group, the mean centre is likely to fall in space between the tips of the crescent or in the circle where the phenomena themselves are not located. These problems do not negate the value of the mean centre as a descriptive measure of the central tendency of spatial distributions, they simply re-emphasise the issue relating to the application of all statistical techniques – 'caveat investigator', let the researcher beware!

Weighting the location of the mean centre according to the values of a measured attribute or variable can enhance its usefulness. The distribution of spatial phenomena may be of interest in its own right, but the underlying processes responsible for producing the spatial distribution are likely to be associated with one or more variables having differentially high and low values at the location of the different points. Consequently, we are usually not only interested in the distribution of the phenomena, but also the values of variables and attributes associated with each point, line or area. For example, if we were interested in finding out the mean centre of Burger King's customer base in Pittsburgh, a team of researchers could be positioned at each restaurant and count the number of people entering the premises on one day, assuming the corporation were not willing to release these data for reasons of their commercial sensitivity. Each point in the distribution would then be weighted by multiplying its X and Y co-ordinates by the corresponding number of customers, then totalling the results and dividing them by the total number of customers (see Box 4.4). This process takes into account the size of some activity or characteristic occurring at the point locations, which can be thought of as a height measurement (Z value). Weighting the mean centre can also be achieved by applying a mathematical function that connects several variables together, for example the customers entering the Burger King premises during expressed as a percentage of the total number of people living or working within a 50 m radius.

One of the strengths of the arithmetic mean over the median was that it could be calculated when only the total number of observations and the sum total of a variable are known. Similarly, the mean centre and its weighted equivalent can be estimated even if the co-ordinates for the individual points are missing. This may be useful if the phenomena can be aggregated into counts per regular grid square. Suppose that the exact X and Y co-ordinates of the Burger King restaurants in Pittsburgh were unknown, but there was a count of the number per 10×10 km grid squares (see Box 4.5a). Conventionally, all the points within a square are assigned grid references representing one of the corners, typically the lower left, or the centre of the square, as in this case. In comparison

---

Box 4.4 Calculation of the weighted mean centre of Burger King restaurants in Pittsburgh.

Weighted mean centre – population symbol: $\bar{X}_w, \bar{Y}_w$; sample symbol: $\bar{x}_w, \bar{y}_w$

The weighted mean centre takes into account the 'centre of gravity' in a spatial distribution, since it is a measure that combines the location of spatial phenomena with the values of one or more attributes or variables associated with them. If all of the spatial entities had the same data values, then the weighted mean centre would occur at exactly the same location as the 'basic' mean centre. However, if, as an extreme example, one entity in a set dominates the data values, then the weighted mean centre will be drawn towards that single case.

Calculation of the weighted mean centre involves multiplying the X and Y co-ordinates for each point by the corresponding data value(s) or possibly in more complex cases by the results of a mathematical function. The outcome of these calculations is a weighted X coordinate and a weighted Y coordinate, which together constitute the grid reference of the weighted mean centre. This point is located where two lines at right angles to each other (orthogonal lines) drawn through these 'average' weighted co-ordinates intersect.

(Continued)

Box 4.4   (Continued)

The methods used to calculate the weighted mean centre are illustrated below using points locating Burger King outlets in Pittsburgh, denoted by subscript B for X and Y co-ordinates (see Box 4.3a).

Note: the figures for daily customers are hypothetical.

	$\bar{x} = \sum \frac{x_B}{n}$	$\bar{y} = \sum \frac{y_B}{n}$		$\frac{\sum x_B w}{\sum w}$	$\frac{\sum y_B w}{\sum w}$
	x_B	y_B	w	$x_B w$	$y_B w$
1	11	27	1,200	13,200	32,400
2	14	10	7,040	98,560	70,400
3	16	18	1,550	24,800	27,900
4	19	23	4,670	88,730	107,410
5	21	17	2,340	49,140	39,780
6	22	16	8,755	192,610	140,080
7	22	27	9,430	207,460	254,610
8	23	19	3,465	79,695	65,835
9	23	28	8,660	199,180	242,480
10	26	20	7,255	188,630	145,100
11	26	24	7,430	193,180	178,320
12	26	25	5,555	144,430	138,875
13	27	16	4,330	116,910	69,280
14	27	25	6,755	182,385	168,875
15	27	32	1,005	27,135	32,160
16	32	26	4,760	152,320	123,760
17	36	15	1,675	60,300	25,125
18	36	35	1,090	39,240	38,150
19	37	20	1,450	53,650	29,000
20	37	22	2,500	92,500	55,000
21	37	28	1,070	39,590	29,960
22	38	24	1,040	39,520	24,960
	$\sum x_B = 583$	$\sum y_B = 497$	93,025	2,283,165	2,039,460
	$\frac{583}{22}$	$\frac{497}{22}$		$\frac{2,283,165}{93,025}$	$\frac{2,039,460}{93,025}$
	$\bar{x}_B = 27$	$\bar{y}_B = 23$		$\bar{x}_w = 24.5$	$\bar{y}_w = 21.9$
Mean centre	27.0, 23.0			Weighted mean centre	24.5, 21.9

with Box 4.3a, the details of exactly where each outlet is located have been lost with their positions generalised to within a series of 100 km² squares. Both the mean centre and the weighted version relating to such aggregate distributions of data points are clearly less accurate than those for phenomena whose exact grid co-ordinates are available. However, they represent useful alternative measures for analysing data obtained from secondary sources.

Unfortunately, mean centres based on aggregate counts of spatially distributed phenomena within areal units such as grid squares suffer from the problem that such units are arbitrarily

Box 4.5a   Allocation of Burger King restaurants to an arbitrary grid of squares in Pittsburgh.

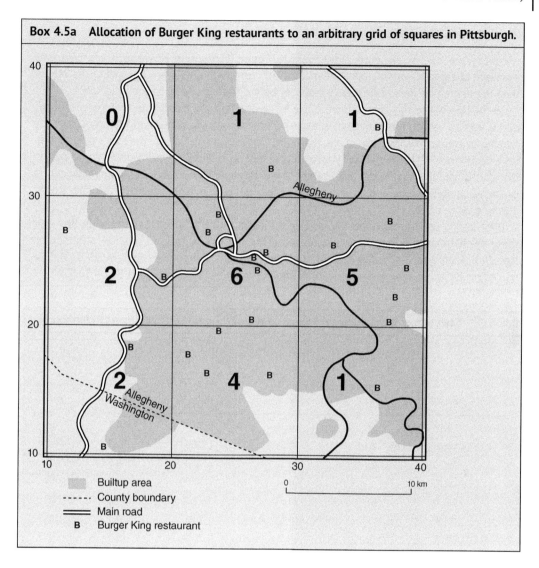

Box 4.5b   Calculation of the aggregate mean centre and the aggregate weighted mean centre of Burger King restaurants in Pittsburgh.

Mean centre – population symbol: $\overline{X}, \overline{Y}$; sample symbol: $\overline{x}, \overline{y}$

Weighted mean centre – population symbol: $\overline{X}_W, \overline{Y}_W$; sample symbol: $\overline{x}_W, \overline{y}_W$

The aggregate mean centre treats the spatial units, usually grid squares, in which entities are located as the primary unit of analysis. This results in some loss of detail but may be useful where either the exact georeferenced locations of entities cannot be determined or there are so many occurrences that it is not necessary to take each one into account individually. Since the aggregate mean centre generalises the spatial distribution of entities, local concentrations of points can be obscured and its location will depend on the origin (e.g. bottom left corner or centre point) of the superimposed grid. If the number of entities per grid square is fairly similar across the whole study area, then the grid unit representing the aggregate mean centre is likely

(Continued)

Box 4.5b (Continued)

to contain the 'true' mean centre of the individual point distribution. However, if, as an extreme example, one grid square includes the majority of cases with only a few in some of the others, then the aggregate mean centre will probably be a different square to the 'true' one.

Calculation of the aggregate mean centre involves using the X and Y grid references for the grid squares in which the individual entities occur. Multiply the chosen grid references by the number of entities within each square, then sum these values and divide the totals by the number of squares covered by the study area. This results in X and Y co-ordinates to a grid square that constitutes the aggregate mean centre. A weighted version of the aggregate mean centre can be calculated, which involves multiplying the count of points per square by the values of an attribute or variable.

The method used to calculate the aggregate mean centre is illustrated below in Box 4.5c using the number of Burger King restaurants in the six 10 × 10 km grid squares in Pittsburgh (see Box 4.5a) and grid references for the mid-point of the squares (e.g. 10.5, 20.5). The procedure for calculating the aggregate weighted mean centre is shown in Box 4.5d.

Box 4.5c   Calculation of the aggregate mean centre of Burger King restaurants in Pittsburgh.

$$\bar{x} = \frac{\sum x_B n}{\sum n} \qquad\qquad \bar{y} = \frac{\sum y_B n}{\sum n}$$

	Square	Pts/sq n	x_B	$x_B n$	y_B	$y_B n$
1	10,10	2	10.5	21.0	10.5	21
2	10,20	2	20.5	21.0	20.5	41
3	10,30	0	30.5	0.0	30.5	0
4	20,10	4	10.5	82.0	10.5	42
5	20,20	6	20.5	123.0	20.5	123
6	20,30	1	30.5	20.5	30.5	30.5
7	30,10	1	10.5	30.5	10.5	10.5
8	30,20	5	20.5	152.5	20.5	102.5
9	30,30	1	30.5	30.5	30.5	30.5
	$\sum n = 22$			$\sum x_B n = 481$		$\sum y_B n = 401$
				$\frac{481}{22}$		$\frac{401}{22}$
				$\bar{x} = 21.9$		$\bar{y} = 18.2$
Aggregate mean centre				21.9, 18.2		

---

defined. This arbitrariness arises from three decisions that have been taken about the nature of the grid squares: the size of the units; the position of the grid's origin; and its orientation. Figure 4.4 illustrates the impact of the first two of these choices with respect to placing a regular square grid over the location of Burger King fast-food outlets in Pittsburgh. Three alternative grid square sizes – 500, 1000 and 1500 m – and three different origins – x_1, y_1; x_2, y_2; and x_3, y_3 – are shown. A change in orientation would involve rotating the grid(s), for example through

Box 4.5d Calculation of the aggregate weighted mean centre of Burger King restaurants in Pittsburgh.

$$\bar{x} = \frac{\sum x_B w}{\sum w} \qquad\qquad \bar{y} = \frac{\sum y_B w}{\sum w}$$

	Sq	Pts/sq n	Weight/sq w	x_B	$x_B w$	y_B	$y_B w$
1	10,10	2	8590	10.5	90,195.0	10.5	90,195.0
2	10,20	2	5870	20.5	61,635.0	20.5	120,335.0
3	10,30	0	0	30.5	0	30.5	0
4	20,10	4	18,890	10.5	387,245.0	10.5	198,345.0
5	20,20	6	45,085	20.5	924,242.5	20.5	924,242.5
6	20,30	1	1005	30.5	20,602.5	30.5	30,652.5
7	30,10	1	1675	10.5	51,087.5	10.5	17,587.5
8	30,20	5	10,820	20.5	330,010.0	20.5	221,810.0
9	30,30	1	1090	30.5	33,245.0	30.5	33,245.0
			93,025		$\sum x_B w = 1{,}898{,}263$		$\sum y_B w = 1{,}636{,}413$

$$\frac{1,898,263}{93,025} \qquad\qquad \frac{1,636,413}{93,025}$$

$$\bar{x} = 20.4 \qquad\qquad \bar{y} = 17.6$$

Aggregate weighted mean centre   20.4, 17.6

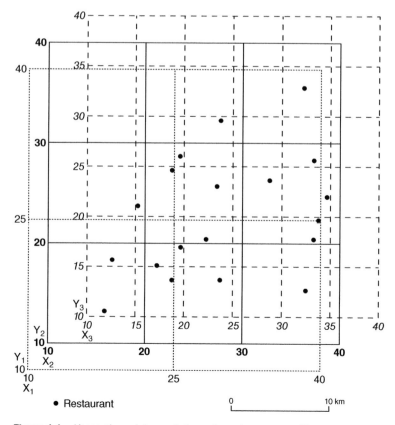

• Restaurant   0   10 km

Figure 4.4 Alternative origins and sizes of regular square grid.

45°C. Although calculations of the mean centres in each of these cases are not included, it will be apparent that decisions about the size and origin of regular grids can have a profound impact on aggregate spatial counts and hence on the location of the mean centre. There is, of course, nothing to dictate that a regular square, hexagonal or triangular grid has to be used, and once irregular, variable sized units are admitted, the arbitrariness becomes even more obvious. This example illustrates the **Modifiable Areal Unit Problem** (MAUP), which is a common difficulty encountered when attempting to analyse spatial patterns. We will return to the MAUP later in this text.

4.3.3 Distance Measures for Spatial Data

An alternative approach to analysing spatial patterns is to summarise distributions by reference to the **mean distance** between the set of spatial entities. Once this matrix of distances between each pair of points in the set has been determined, these measurements can be treated like any other variable that is quantified on the ratio scale and the overall mean or the **nearest neighbour mean distance** can be calculated by summation and division. In general terms, a small mean distance implies that the collection of phenomena is more tightly packed together in comparison with a

---

Box 4.6a   Mean distance for Burger King restaurants in Pittsburgh.

Mean distance – population symbol: $\bar{D}$; sample symbol: $\bar{d}$

The physical distance that separates each pair of entities can be treated as a variable much like 'standard' thematic attributes and variables to calculate measures of central tendency. The mean, and for that matter the modal and median, distance between phenomena provide measures of how far apart, or close together, they are. Each of these measures potentially suffers from the same problems as their non-spatial counterparts. Insofar as particular entities lying at the extremes of spatial distributions can distort the results, by suggesting a larger distance between entities is the norm rather than the exception.

The overall mean distance is obtained by determining the distance measurements between every pair of entities in the set, adding them up or summing the distances and then dividing by the total number of distance measurements (not the total number of points). The nearest neighbour mean distance is calculated by summing the distance between each point and its nearest neighbour and dividing it by the total number of points. The distance measurements can be calculated using Pythagoras' theorem, although standard statistical and spreadsheet software is not well suited to this task.

The upper part of the table below illustrates how to calculate the distance between a selected subset of the pairs of points locating Burger King restaurants in Pittsburgh (see Box 4.3a) and indicates the complexity of the computational problem. In this example, there are 22 points giving rise to 231 pairs $(1 + 2 + 3 + 4 + \ldots + 19 + 20 + 21)$, and the calculations are shown for the distance between points 1 and 2, 2 and 3, 3 and 4, to 22 and 21. The calculations between 1 and 3, 1 and 4, 1 and 5, etc., or 2 and 4, and 2 and 5, 2 and 6, etc., and so on are not considered. The calculations shown produce the diagonal in the matrix in Box 4.6c, which includes the 231 distance measurements resulting from pairing each Burger King restaurant with all others in the set. The overall modal, median and mean distance between the Burger King restaurants in Pittsburgh can be determined only when all the distance measurements have been calculated and sorted into ascending (or descending) order (see Box 4.6d).

Box 4.6b   Calculation of distances between selected pairs (1 and 2, 2 and 3, 3 and 4, etc.) of Burger King restaurants in Pittsburgh, USA.

Calculation of point-to-point distances using Pythagoras' theorem

Pt	Pt	Start	End	Start	End			Distance =
m	n	x_m	x_n	y_m	y_n	$(x_m - x_n)^2$	$(y_m - y_n)^2$	$\sqrt{(x_m - x_n)^2 + (y_m - y_n)^2}$
1	2	11	14	27	10	9.0	289.0	17.3
2	3	14	16	10	18	4.0	64.0	8.2
3	4	16	19	18	23	9.0	25.0	5.8
4	5	19	21	23	17	4.0	36.0	6.3
5	6	21	22	17	16	1.0	1.0	1.4
6	7	22	22	16	27	0.0	121.0	11.0
7	8	22	23	27	19	1.0	64.0	8.1
8	9	23	23	19	28	0.0	81.0	9.0
9	10	23	26	28	20	9.0	64.0	8.5
10	11	26	26	20	24	0.0	16.0	4.0
11	12	26	26	24	25	0.0	1.0	1.0
12	13	26	27	25	16	1.0	81.0	9.1
13	14	27	27	16	25	0.0	81.0	9.0
14	15	27	27	25	32	0.0	49.0	7.0
15	16	27	32	32	26	25.0	36.0	7.8
16	17	32	36	26	15	16.0	121.0	11.7
17	18	36	36	15	35	0.0	400.0	20.0
18	19	36	37	35	20	1.0	225.0	15.0
19	20	37	37	20	22	0.0	4.0	2.0
20	21	37	37	22	28	0.0	36.0	6.0
21	22	37	38	28	24	1.0	16.0	4.1
22		38		24				

larger mean distance. But how small is small and how large is large? The answer partly depends on the size of the study area and how it is defined, which again raises issues associated with the MAUP. What might seem a short mean distance within a relatively large area may appear as a long distance in a small area. Box 4.6a illustrates the procedure for calculating the mean distance between the Burger King outlets in Pittsburgh. The overall mean distance is 13.0 km and the nearest neighbour mean is 4.4 km, with modal and median values of 11.2 and 12.0 km, respectively. This variation partly reflects the presence of some outlets outside the central downtown area. For comparison, just focusing on the central downtown 100 km^2 grid square the equivalent measures are a mean of 4.3 km with the median and mode both 5.0 km, which suggests the outlets are much closer together downtown than in the study area overall.

Box 4.6c Distance matrix for pairs of Burger King restaurants in Pittsburgh, USA.

Pts		1	2	3	4	5	6	7	8	9	10	11	12	13	14	15	16	17	18	19	20	21	22
	Co-ords	27	10	18	23	17	16	27	19	28	20	24	25	16	25	32	26	15	35	20	22	28	24
1	11																						
2	14	17.3																					
3	16	10.3	8.2																				
4	19	8.9	13.9	5.8																			
5	21	14.1	9.9	5.1	6.3																		
6	22	15.6	10.0	6.3	7.6	1.4																	
7	22	11.0	18.8	10.8	5.0	10.0	11.0																
8	23	14.4	12.7	7.1	5.7	2.8	3.2	8.1															
9	23	12.0	20.1	12.2	6.4	11.2	12.0	1.4	9.0														
10	26	16.6	15.6	10.2	7.6	5.8	5.7	8.1	3.2	8.5													
11	26	15.3	18.4	11.7	7.1	8.6	8.9	5.0	5.8	5.0	4.0												
12	26	15.1	19.2	12.2	7.3	9.4	9.8	4.5	6.7	4.2	5.0	1.0											
13	27	19.4	14.3	11.2	10.6	6.1	5.0	12.1	5.0	12.6	4.1	8.1	9.1										
14	27	16.1	19.8	13.0	8.2	10.0	10.3	5.4	7.2	5.0	5.1	1.4	1.0	9.0									
15	27	16.8	25.6	17.8	12.0	16.2	16.8	7.1	13.6	5.7	12.0	8.1	7.1	16.0	7.0								
16	32	21.0	24.1	17.9	13.3	14.2	14.1	10.0	11.4	9.2	8.5	6.3	6.1	11.2	5.1	7.8							
17	36	27.7	22.6	20.2	18.8	15.1	14.0	18.4	13.6	18.4	11.2	13.5	14.1	9.1	13.5	19.2	11.7						
18	36	26.2	33.3	26.2	20.8	23.4	23.6	16.1	20.6	14.8	18.0	14.9	14.1	21.0	13.5	9.5	9.8	20.0					
29	37	26.9	25.1	21.1	18.2	16.3	15.5	16.6	14.0	16.1	11.0	11.7	12.1	10.8	11.2	15.6	7.8	5.1	15.0				
20	37	26.5	25.9	21.4	18.0	16.8	16.2	15.8	14.3	15.2	11.2	11.2	11.4	11.7	10.4	14.1	6.4	7.1	13.0	2.0			
21	37	26.0	29.2	23.3	18.7	19.4	19.2	15.0	16.6	14.0	13.6	11.7	11.4	15.6	10.4	10.8	5.4	13.0	7.1	8.0	6.0		
22	38	70.7	62.1	22.8	19.0	18.4	17.9	16.3	15.8	15.5	12.6	12.0	12.0	13.6	11.0	13.6	6.3	9.2	11.2	4.1	2.2	4.1	

Box 4.6d **Sorting and summation of all distances to determine mean, modal and median distance measures between Burger King restaurants in Pittsburgh, USA.**

$$\text{Mean distance } \bar{d} = \sum \frac{d_B}{n}$$

Sorted distances (d)	d	d	d	d	d	d	d	d	d
1.0	5.1	7.1	9.2	11.2	12.6	14.3	16.3	19.4	26.9
1.0	5.1	7.1	9.4	11.2	12.7	14.4	16.6	19.4	27.7
1.4	5.4	7.2	9.5	11.2	13.0	14.8	16.6	19.8	29.2
1.4	5.4	7.3	9.8	11.2	13.0	14.9	16.6	20.0	33.3
1.4	5.7	7.6	9.8	11.2	13.0	15.0	16.8	20.1	62.1
2.0	5.7	7.6	9.9	11.2	13.3	15.0	16.8	20.2	70.7
2.2	5.7	7.8	10.0	11.4	13.5	15.1	16.8	20.6	
2.8	5.8	7.8	10.0	11.4	13.5	15.1	17.3	20.8	
3.2	5.8	8.0	10.0	11.4	13.5	15.2	17.8	21.0	
3.2	5.8	8.1	10.0	11.7	13.6	15.3	17.9	21.0	
4.0	6.0	8.1	10.2	11.7	13.6	15.5	17.9	21.1	
4.1	6.1	8.1	10.3	11.7	13.6	15.5	18.0	21.4	
4.1	6.1	8.1	10.3	11.7	13.6	15.6	18.0	22.6	
4.1	6.3	8.2	10.4	11.7	13.6	15.6	18.2	22.8	
4.2	6.3	8.2	10.4	**12.0**	13.9	15.6	18.4	23.3	
4.5	6.3	8.5	10.6	12.0	14.0	15.6	18.4	23.4	
5.0	6.3	8.5	10.8	12.0	14.0	15.8	18.4	23.6	
5.0	6.4	8.6	10.8	12.0	14.0	15.8	18.4	24.1	
5.0	6.4	8.9	10.8	12.0	14.1	16.0	18.7	25.1	
5.0	6.7	8.9	11.0	12.0	14.1	16.1	18.8	25.6	
5.0	7.0	9.0	11.0	12.1	14.1	16.1	18.8	25.9	
5.0	7.1	9.0	11.0	12.1	14.1	16.1	19.0	26.0	
5.0	7.1	9.1	11.0	12.2	14.1	16.2	19.2	26.2	
5.1	7.1	9.1	11.2	12.2	14.2	16.2	19.2	26.2	
5.1	7.1	9.2	11.2	12.6	14.3	16.3	19.2	26.5	

$$\sum d = 2993.8$$

$$\frac{2993.8}{231}$$

Mode = 11.2   Median = 12.0 $\bar{d} = 12.96$

> What can you conclude about the distribution of Burger King restaurants in Pittsburgh from this information?

4.4 Dispersion

4.4.1 Measures for Non-spatial Data

We have seen that there are various ways of quantifying the central tendency or typical values in a set of data, but it is also useful to know the spread of data values. Are they reasonably similar to this typical value or are they rather different? This information can be obtained by calculating a statistical quantity representing the dispersion of a numerical distribution. Imagine sitting in a stadium and looking down at a football match, the 22 players (and the referee) are all located somewhere within the 90 by 45 m rectangle that defines the pitch (minimum dimensions). At kick off, when the match starts, the players will be fairly evenly distributed around the pitch in their allotted positions. However, if one side is awarded a free kick, the large majority of the players will rush to cluster together in the goal area attempting to defend or score depending on which side they are on. In other words, sometimes the players will be distributed in a relatively compact way, and at other times, they will be more dispersed.

There are three main measures of dispersion known as the range, variance and standard deviation, all of which are numerical quantities indicating the spread of values of a non-spatial attribute or variable. These measures are interpreted in a relative fashion, in the sense that a small value indicates less dispersion than a large one, and *vice versa*. The range is simply calculated as the difference between the smallest and largest values. Alternatively, various percentile ranges can be used that omit an equal percentage of values at the upper and lower ends of the distribution. For example, excluding both the largest and smallest 25% of data values and calculating the difference between the quantities at either end of the remainder produces the interquartile range. The procedures for obtaining the variance and standard deviation measures involve rather more complicated calculations than simply sorting the data values into ascending order and are presented in Box 4.7a with a variable measured on the ratio/interval scale. Nevertheless, the basic purpose in each case is the same, namely, to determine whether the data values are close together or spread out.

The reason for having several measures of dispersion is essentially the same as why there are different measures of central tendency. Each has advantages and disadvantages and should be used for attributes and variables recorded according to the different scales of measurement. Intuitively, the range seems the most sensible measure, since it reports the size of the numerical gap between the smallest and largest values. Unfortunately, this may be misleading, since extreme outlier values may exaggerate the spread. Discarding the top and bottom 10 or 25% helps to overcome this problem, but the choice of exclusion percentage is entirely arbitrary. A further problem with using the absolute or percentile range is that it fails to take into account the spread the values between the upper and lower values. We are none the wiser about whether the data values are clustered, regularly or randomly dispersed along the range. It could be that there are one or two very small values and all the rest are at the upper end of the range, or *vice versa*.

The variance and the standard deviation help to overcome this problem by focusing on the differences between the individual data values and their mean. However, this implies that these dispersion measures are not suitable for use with attributes recorded on the nominal or ordinal scales. If most of the data values are tightly clustered around the mean, then the variance and standard deviation will be relatively small even if there are a small number of extremely low and high values as well. A similar result will occur if the values are bimodally distributed with most lying at **each** end of the range and comparatively few in the middle around the mean. Conversely, the variance and standard deviation will be moderately large, if the data values are regularly spaced across the

Box 4.7a Measures of dispersion for water temperature measurements in a fluvioglacial stream from Les Bossons Glacier, France.

Variance – population symbol: σ^2; sample symbol: s^2
Standard deviation – population symbol: σ; sample symbol: s

The variance and the standard deviation (its close relation) are two of the most useful measures of dispersion when your data are recorded on either the interval or ratio scales of measurement. They measure the spread of a set of numbers around their mean and so allow you to quantify whether two or more sets of numbers are compact or spread out irrespective of whether they possess similar or different means. The concept of difference (deviation) is important when trying to understand what information the variance and standard deviation convey. A small variance or standard deviation denotes that the numbers are tightly packed around their mean, whereas a large value indicates that they are spread out, or at least some of them are. So, if you have data for two or more samples of the same type of observations, for instance samples of soil from the A, B and C horizons, then you can see whether or not the spread (dispersion) of particle sizes is similar or different.

The standard deviation is closely related to the variance, but one of its main advantages is that its value is measured in the same units as the original set of measurements. So, for example, if these measurements are of water temperature in °C or of journey time in minutes, their standard deviation is expressed in these units. The standard deviation is obtained by taking the square root of the variance. There are two main ways of calculating the variance and the standard deviation. The method on the left in the calculations table below involves simpler computation, since it is based on the sum of the X values and the sum of the squared X values. The second method uses the sum of the squared differences (deviations) of the X values about their mean and, although this entails slightly more complicated calculations, it provides a clearer illustration of what the standard deviation measures. These methods are illustrated using sample data that represent measurements of water temperature (X) in the fluvioglacial stream flowing from Les Bossons Glacier, France, where the stream leaves the sandur plain taken at half-hourly intervals between 09.00 and 17.30 hrs (see Box 4.2b).

The calculations show that the sample with a mean of 9.08°C has a variance of 7.78 and a standard deviation of 2.79°C – thankfully the two methods for deriving the standard deviation produce the same result.

Box 4.7b Calculation of the variance and standard deviation measures of dispersion for water temperature measurements.

	$s = \sqrt{\dfrac{n(\sum x^2) - (\sum x)^2}{n(n-1)}}$		$s = \sqrt{\dfrac{\sum (x - \bar{x})^2}{(n-1)}}$		
	x	x^2	x	$(x - \bar{x})$	$(x - \bar{x})^2$
09.00hrs	4.30	18.49	4.30	-4.78	22.88
09.30hrs	4.70	22.09	4.70	-4.38	19.21
10.00hrs	4.80	23.04	4.80	-4.28	18.35
10.30hrs	5.20	27.04	5.20	-3.88	15.08
11.00hrs	6.70	44.89	6.70	-2.38	5.68

(Continued)

Box 4.7b (Continued)

		$s = \sqrt{\dfrac{n(\sum x^2)-(\sum x)^2}{n(n-1)}}$		$s = \sqrt{\dfrac{\sum (x-\bar{x})^2}{(n-1)}}$	
		x	x^2	x	$(x-\bar{x})$	$(x-\bar{x})^2$

		x	x^2	x	$(x-\bar{x})$	$(x-\bar{x})^2$
11.30hrs		10.10	102.01	10.10	1.02	1.03
12.00hrs		10.50	110.25	10.50	1.42	2.01
12.30hrs		11.20	125.44	11.20	2.12	4.48
13.00hrs		11.40	129.96	11.40	2.32	5.37
13.30hrs		11.80	139.24	11.80	2.72	7.38
14.00hrs		12.30	151.29	12.30	3.22	10.35
14.30hrs		11.90	141.61	11.90	2.82	7.93
15.00hrs		11.40	129.96	11.40	2.32	5.37
15.30hrs		11.10	123.21	11.10	2.02	4.07
16.00hrs		10.20	104.04	10.20	1.12	1.25
16.30hrs		9.30	86.49	9.30	0.22	0.05
17.00hrs		8.80	77.44	8.80	-0.28	0.08
17.30hrs		7.80	60.84	7.80	-1.28	1.65
$n = 18$	$\sum x$	163.50 $\sum x^2$	1617.33	$\sum x$ 163.50	$\sum (x-\bar{x})^2$	132.21

$n = 18$

Variance $\quad s^2 = \dfrac{(18(1617.33)-(163.50)^2}{18(18-1)} \qquad s^2 = \dfrac{132.21}{18-1}$

$\quad\quad\quad\quad s^2 = 7.78 \qquad\qquad\qquad\qquad s^2 = 7.78$

Standard deviation $\quad s = \sqrt{\dfrac{18(1617.33)-(163.50)^2}{18(18-1)}}$

$\qquad\qquad\qquad s = \sqrt{\dfrac{18(1617.33)-(163.50)^2}{18(18-1)}} \qquad s = \sqrt{\dfrac{132.21}{18-1}}$

$\qquad\qquad\qquad s = 2.79 \qquad\qquad\qquad\qquad s = 2.79$

Sample Means $\quad \dfrac{\sum x}{n} \quad \dfrac{163.50}{18} \quad 9.08 \qquad \dfrac{\sum x}{n} \quad \dfrac{163.50}{18} \quad 9.08$

range. In the highly unlikely event that all data values are identical, then the range, variance and standard deviation measures will all equal zero and hence denote an absence of variation of the variable. In practical terms, the main difference between the variance and the standard deviation is that the latter is a measure in the same units as the original data values, which makes its interpretation easier. Thus, Box 4.7a the standard deviation of 2.79 for melt water stream temperature is in °C, whereas the variance of 7.78 is in squared °C, which does not make much sense.

4.4.2 Measures for Spatial Data

The dispersion or spread of spatial phenomena, especially those located as points, can be quantified by a measure known as **standard distance**. This is essentially the spatial equivalent of the standard deviation. It quantifies the extent to which the individual data points are clustered around the mean centre in two-dimensional space. We have seen that the standard deviation concerns the spread of a single set of data values; in contrast, the standard distance conjointly focuses on the

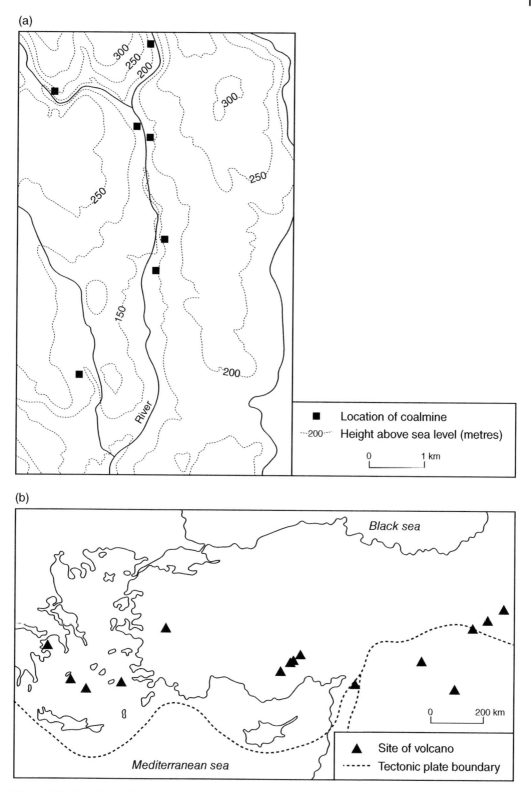

Figure 4.5 X or Y co-ordinates constrained within relatively narrow ranges. (a) co-ordinates spread out (dispersed) along north-south (Y) axis, compact on west-east axis; (b) co-ordinates spread out (dispersed) along west-east (X) axis, compact on north-south axis.

two-dimensional spread of the X and Y co-ordinates of a set of data points. A spatial distribution of points is located in two dimensions, X and Y, and it is possible for one set of co-ordinates to be clustered tightly around their mean and for the other to be more spread out. It follows from this characteristic that a set of points has two axes at right angles to each other: If the spread of co-ordinates on the X and Y dimensions are very similar, the points will approximate to a circular pattern. Figure 4.5(a) illustrates how a distribution of points depicting the positions of former coal mines within a relatively confined approximately north–south-oriented valley in the South Wales can lead to a narrower spread of X co-ordinates in comparison with Y co-ordinates. Figure 4.5(b)

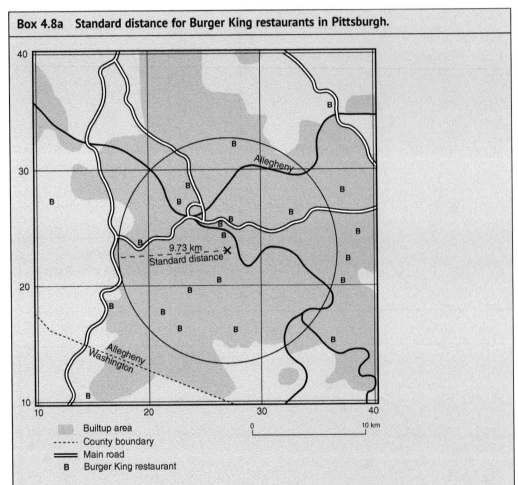

Box 4.8a   Standard distance for Burger King restaurants in Pittsburgh.

Builtup area
County boundary
Main road
B   Burger King restaurant

0   10 km

The notion that all points representing the location of a set of geographical phenomena will occur exactly on top of each other is just as unrealistic as to imagine that the values of any non-spatial variable for a given collection observations will be the same. The standard distance measure enables us to quantify how spread out the points is just as the standard deviation measures the dispersion of a set of numerical values. The larger the standard distance then the greater the dispersion amongst the points.

The standard distance is calculated from the X and Y co-ordinates of the points and does not require you to calculate the distances between all the pairs of points as described in Box 4.7. The procedure given below shows calculation of the variances for the X and Y co-ordinates, and these are summed and the square root of the result produces the standard distance. Examination of the X and Y variances indicates that the X co-ordinates are more dispersed than the Y co-ordinates.

Box 4.8b Calculations for standard distance.

Standard distance – population symbol: S_d; sample symbol: s_d

$$s_d = \sqrt{\frac{\sum (x - \bar{x})^2}{(n-1)} + \frac{\sum (y - \bar{y})^2}{(n-1)}}$$

Point	x	$(x-\bar{x})$	$(x-\bar{x})^2$	y	$(y-\bar{y})$	$(y-\bar{y})^2$
1	11	−15.5	240.25	27	4.4	19.36
2	14	−12.5	156.25	10	−12.6	158.76
3	16	−10.5	110.25	18	−4.6	21.16
4	19	−7.5	56.25	23	0.4	0.16
5	21	−5.5	30.25	17	−5.6	31.36
6	22	−4.5	20.25	16	−6.6	43.56
7	22	−4.5	20.25	27	4.4	19.36
8	23	−3.5	12.25	19	−3.6	12.96
9	23	−3.5	12.25	28	5.4	29.16
10	26	−0.5	0.25	20	−2.6	6.76
11	26	−0.5	0.25	24	1.4	1.96
12	26	−0.5	0.25	25	2.4	5.76
13	27	0.5	0.25	16	−6.6	43.56
14	27	0.5	0.25	25	2.4	5.76
15	27	0.5	0.25	32	9.4	88.36
16	32	5.5	30.25	26	3.4	11.56
17	36	9.5	90.25	15	−7.6	57.76
18	36	9.5	90.25	35	12.4	153.76
	$\sum x = 583$		$\sum (x-\bar{x})^2 = 1335.50$	$\sum y = 497$		$\sum (y-\bar{y})^2 = 749.32$

$$\frac{1335.50}{22} \qquad \frac{749.32}{22}$$

$$60.61 \qquad 34.06$$

$$s_d = \sqrt{60.61 + 34.06}$$

Standard distance $\qquad s_d = 9.73$

---

shows how points representing the location of volcanoes associated with the predominantly west to east aligned tectonic plates in the eastern Mediterranean region results in wide spread of X co-ordinate values and the Y co-ordinates in a narrower range.

Box 4.8a shows the calculation of the standard distance for the points representing Burger King restaurants in Pittsburgh and produces the value 9.73 km. This value on its own does not allow us to conclude whether these restaurants are dispersed or compact, you would want to compare this result with another fast-food chain in the same city, with Burger King's outlets in other cities, or possibly with some theoretical figure perhaps denoting a perfectly circular distribution. Unfortunately, if comparing the standard distance values for study areas that are substantially different in physical area, then an adjustment should be introduced to compensate. Neft (1968), for example, used the radius of his study area countries to produce a relative measure of population dispersion. Standard distance can be used as the radius of a circle drawn around the mean centre of a distribution of points and thus provide a graphic representation of the dispersion of the points.

4.5   Measures of Skewness and Kurtosis for Non-spatial Data

Skewness and **kurtosis** are statistical quantities concerned with the shape of the distribution of data values and so in some respects belong with discussion of frequency distributions in Chapter 5. However, since each is a quantity capable of being calculated for 'raw' data (i.e. not presented as a frequency distribution) and, following on from central tendency and dispersion (the first and second moments of descriptive statistics), they are briefly discussed here. The principle underlying skewness in respect of statistics is closely allied to the everyday word skewed, which indicates that something is slanted in one direction rather than being balanced. For example, a political party seeking re-election may be selective in the information it chooses relating to its current period in office in order to put a 'spin' or more favourable interpretation on its record. If information or statistics are skewed, this implies that, for whatever reason, there is a lack of symmetry or balance in the set of data values.

Kurtosis may most simply be translated into everyday language as 'peakedness' and therefore provides a method of quantifying whether one set of numbers is more or less peaked than another. Just as a mountainous landscape may be described as undulating if the height of the mountains is low relative to their surface area, in contrast with a deeply dissected upland region with steeper slopes. From the numerical point of view, it is a question of there being a difference in the ratio of the maximum height to radius of the mountain, assuming each peak to be perfectly circular when viewed from above (plan view).

Following on from this analogy, Box 4.9a shows typical shapes of two types of volcanic cone that are approximately circular on the horizontal plane but have rather different cross sections. In one case, Mauna Loa on Hawaii, the cone has been formed from relatively free-flowing lava that was able to spread extensively over the existing surface during successive eruptions and build up what is sometimes called a 'shield cone'. In contrast, Mount Teide on Tenerife in the Canary Islands was created during a more violent series of eruptions of more viscous lava that constructed a conical volcano. Lines have been superimposed along the profiles of the volcanoes to illustrate that one is relatively flat and the other more peaked (see Box 4.9a). The calculations involved with obtaining

---

Box 4.9a   Calculation of skewness and kurtosis using examples of contrasting volcanic cones in Hawaii and Tenerife.

Mauna Loa (shield volcano)

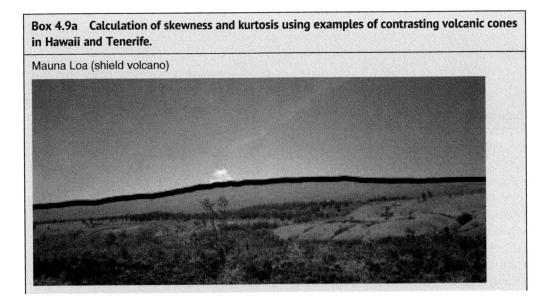

Teide (Tenerife)

Skewness is most useful when you are dealing with data recorded on the interval or ratio scales of measurement. It can be viewed as measuring the degree of symmetry and the magnitude of the differences of the values of a variable around its mean. A skewness statistic of zero indicates perfect symmetry and differences of equal magnitude on either side of the mean, whereas a negative value denotes that more numbers are greater than the mean (negative skewness) and a positive quantity the reverse. A large skewness statistic signifies that the values at one extreme exert a disproportionate influence. Both symmetrical and severely skewed distributions can display varying degrees of kurtosis depending on whether the values are tightly packed together or are more spread out. A kurtosis statistic of 3 indicates that the distribution is moderately peaked, whereas a set of data values possessing a relatively flat shape will produce a statistic less than 3 and those with a pronounced peak will be greater than 3. As with the other descriptive quantities, skewness and kurtosis are commonly used to help with comparing different sets of numbers to see if they are characteristically similar or different. Calculating the skewness and kurtosis statistics for two or more samples of data enables you to see whether or not the distributions of their values have a similar shape.

The sum of the differences between the individual values and their mean always equals zero and the square of the differences is used in calculating the variance (see Box 4.7), so the cube of the deviations provides the basis of the skewness measure. These cubed differences are summed and then divided by the cubed variance of the values multiplied by the number of data points. Following a similar line of argument, calculation of kurtosis involves raising the differences to the fourth power and then dividing the sum of these by the variance raised to the power of four multiplied by the number of observations. These calculations are shown below in Boxes 4.8b and 4.9c with respect to samples of elevation measurements for two contrasting volcanic cones, Mauna Loa in Hawaii and Mount Teide on Tenerife. The classical broad, reasonably symmetrical profile of the Mauna Loa shield volcano results in a skewness and kurtosis statistics of −0.578 and 2.009, respectively. The profile of Mount Teide is more conical (peaked in statistical terms) and slightly less symmetrical than Mauna Loa, which produces skewness and kurtosis values of −0.029 and 1.673.

Box 4.9b Calculation of skewness for samples of spot heights on Mauna Loa, Hawaii (x_L), and Mount Teide, Tenerife (x_T).

(Note: standard deviation for Mauna Loa elevations (s) = 73.984; and standard deviation for Mount Teide elevations (s) = 363.354).
Skewness – population symbol: β_1; sample symbol: b_1.
Kurtosis – population symbol: γ_2; sample symbol: g_2.

	Mauna Loa $b_1 = \dfrac{\sum(x-\bar{x})^3}{(n-1)s^3}$				Mount Teide $b_1 = \dfrac{\sum(x-\bar{x})^3}{(n-1)s^3}$		
N	x_L	$(x_L-\bar{x}_L)$	$(x_L-\bar{x}_L)^3$	N	x_T	$(x_T-\bar{x}_T)$	$(x_T-\bar{x}_T)^3$
1	3840	−146.44	−3,140,358.0	1	2600	−572.7	−187,817,497.0
2	3852	−134.44	−2,429,883.8	2	2700	−472.7	−105,609,182.0
3	3864	−122.44	−1,835,565.8	3	2800	−372.7	−51,761,668.0
4	3889	−97.44	−925,149.3	4	2900	−272.7	−20,274,953.0
5	3901	−85.44	−623,711.5	5	3000	−172.7	−5,149,038.3
6	3913	−73.44	−396,093.8	6	3100	−72.7	-383,923.6
7	3938	−48.44	−113,661.2	7	3200	27.3	20,391.2
8	3962	−24.44	−14,598.3	8	3300	127.3	2,063,905.9
9	3974	−12.44	−1925.1	9	3400	227.3	11,746,621.0
10	3986	−0.44	−0.1	10	3500	327.3	35,068,535.0
11	4011	24.56	14,814.4	11	3600	427.3	78,029,650.0
12	4023	36.56	48,867.3	12	3700	527.3	146,629,965.0
13	4035	48.56	114,508.1	13	3717	544.3	161,273,450.0
14	4047	60.56	222,104.6	14	3700	527.3	146,629,965.0
15	4059	72.56	382,025.0	15	3600	427.3	78,029,650.0
16	4072	85.56	626,343.1	16	3500	327.3	35,068,535.0
17	4088	101.56	1,047,533.9	17	3400	227.3	11,746,621.0
18	4072	85.56	626,343.1	18	3300	127.3	2,063,905.9
19	4059	72.56	382,025.0	19	3200	27.3	20,391.2
20	4047	60.56	222,104.6	20	3100	−72.7	−383,923.6
21	4035	48.56	114,508.1	21	3000	−172.7	−5,149,038.3
22	4023	36.56	48,867.3	22	2900	−272.7	−20,274,953.0
23	4011	24.56	14814.4	23	2800	−372.7	−51,761,668.0
24	3986	−0.44	−0.1	24	2700	−472.7	−105,609,182.0
25	3974	−12.44	−1925.1	25	2600	−572.7	−187,817,497.0

$\sum x_L = 113{,}356 \quad \sum(x_L-\bar{x}_L)^3 = -5{,}618{,}013.1 \qquad \sum x_T = 79{,}317 \quad \sum(x_T-\bar{x}_T)^3 = -33{,}600{,}939.0$

$$b_1 = \frac{5{,}618{,}013.1}{24(404{,}965.8)} \qquad\qquad b_1 = \frac{33{,}600{,}939.0}{24(47{,}972{,}275)}$$

$$b_1 = -0.578 \qquad\qquad\qquad b_1 = -0.029$$

Box 4.9c Calculation of kurtosis for samples of spot heights on Mauna Loa, Hawaii (x_L), and Mount Teide, Tenerife (x_T).

(Note: standard deviation for Mauna Loa elevations (s) = 73.984; and standard deviation for Mount Teide elevations (s) = 363.354).
Kurtosis – population symbol: γ_2; sample symbol: g_2.

	Mauna Loa				Mount Teide		
	$g_2 = \dfrac{\sum (x - \bar{x})^4}{ns^4}$				$g_2 = \dfrac{\sum (x - \bar{x})^4}{ns^4}$		
N	x_L	$(x_L - \bar{x}_L)$	$(x_L - \bar{x}_L)^4$	N	x_T	$(x_T - \bar{x}_T)$	$(x_T - \bar{x}_T)^4$
1	3,840	−146.44	459,874,025.8	1	2,600	−572.7	107,559,324,269.5
2	3,852	−134.44	326,673,582.4	2	2,700	−472.7	49,919,348,352.4
3	3,864	−122.44	224,746,679.3	3	2,800	−372.7	19,290,538,323.2
4	3,889	−97.44	90,146,548.1	4	2,900	−272.7	5,528,574,182.1
5	3,901	−85.44	53,289,906.6	5	3,000	−172.7	889,135,929.0
6	3,913	−73.44	29,089,126.0	6	3,100	−72.7	27,903,563.8
7	3,938	−48.44	5,505,750.6	7	3,200	27.3	557,086.7
8	3,962	−24.44	356,783.5	8	3,300	127.3	262,776,497.6
9	3,974	−12.44	23,948.7	9	3,400	227.3	2,670,241,796.4
10	3,986	−0.44	0.0	10	3,500	327.3	11,478,632,983.3
11	4,011	24.56	363,842.5	11	3,600	427.3	33,343,630,058.2
12	4,023	36.56	1,786,589.4	12	3,700	527.3	77,320,913,021.0
13	4,035	48.56	5,560,511.1	13	3,717	544.3	87,784,364,145.9
14	4,047	60.56	13,450,656.0	14	3,700	527.3	77,320,913,021.0
15	4,059	72.56	27,719,736.4	15	3,600	427.3	33,343,630,058.2
16	4,072	85.56	53,589,919.4	16	3,500	327.3	11,478,632,983.3
17	4,088	101.56	106,387,540.5	17	3,400	227.3	2,670,241,796.4
18	4,072	85.56	53,589,919.4	18	3,300	127.3	262,776,497.6
19	4,059	72.56	27,719,736.4	19	3,200	27.3	557,086.7
20	4,047	60.56	13,450,656.0	20	3,100	−72.7	27,903,563.8
21	4,035	48.56	5,560,511.1	21	3,000	−172.7	889,135,929.0
22	4,023	36.56	1,786,589.4	22	2,900	−272.7	5,528,574,182.1
23	4,011	24.56	363,842.5	23	2,800	−372.7	19,290,538,323.2
24	3,986	−0.44	0.0	24	2,700	−472.7	49,919,348,352.4
25	3,974	−12.44	23,948.7	25	2,600	−572.7	107,559,324,269.5
$\sum x_L = 99,661$		$\sum (x_L - \bar{x}_L)^4 = 1,501,060,350.0$		$\sum x_T = 79,317$		$\sum (x_T - \bar{x}_T)^4 = 704,367,516,272.0$	

$$g_2 = \frac{1,501,060,350.1}{24(29,961,099.8)}$$

$$g_2 = 2.009$$

$$g_2 = \frac{704,367,516,272.0}{24(17,430,924,527.8)}$$

$$g_2 = 1.683$$

the skewness and kurtosis measures are illustrated with elevation measurements for these differing shapes in respect of 5-km-long cross sections whose centres correspond with the peaks of the cones (Box 4.9b and 4..9c). Both cones are reasonably symmetrical across this distance, although the measure of kurtosis reflects the difference in profile.

From a computational perspective, there is no reason measures of skewness and kurtosis should not be calculated for spatial as well as for non-spatial data. For example, the skewness statistic could be calculated for all the pairs of measurements between Burger King restaurants in Pittsburgh and some commentary supplied about whether they are more or less than the mean distance and whether the magnitude of the differences is larger in the case of the latter or former. However, the skewness or kurtosis of spatial distributions is rarely examined and their calculations have not been carried out here.

4.6 Closing Comments

We started this chapter by thinking about how authors of fiction describe people and events and we saw how these descriptions rely on readers having an understanding of the language in which they have been written. Change the order of the words and they either become nonsense or have a meaning the author had not intended. We moved on to explore some of the descriptive tools available for understanding the meaning of numerical and other types of data collected for research purposes. In some respects, this chapter has been all about 'playing with numbers', seeing how we can describe the essential characteristics of a set of numbers. The four main elements of this description are central tendency, dispersion, skewness and kurtosis. Most of these elements of description can be quantified in several different ways, for example by taking the most typical or frequently occurring (mode), the middle (median) or mean (average) value in the case of central tendency. In the case of measuring dispersion, the different types of range (absolute and percentile) provide an intuitive impression of how spread out the values of a variable are, but in practice the variance and standard deviation are more useful quantities. In most cases, there are spatial statistics equivalent to these more 'standard' non-spatial versions. Although this chapter has tended to focus on spatial phenomena as points, similar measures exist for lines and areas, and in some instances, linear and areal feature may be located spatially as a point by means of a pair of X and Y co-ordinates. Such points may relate to the mid-point of a line or the spatial or weighted centroids of an area (polygon). In these cases, the measures outlined in this chapter can reasonably be used to describe, for example, a collection of areas.

When carrying out quantitative data analysis, the intention is not just to observe how different statistical quantities vary between one set of numbers and another. There needs to be a transition from regarding the numerical values as not just a set of digits but as being important in respect of the variable or attribute they are measuring. So, for example, when looking at the values for the temperature of Les Bossons Glacier melt water stream, the purpose is not simply to regard them as 'any old set of numbers' that happen to have a mean of 9.08 and a standard deviation of 2.79, there are many other sets of numbers that can produce these results. The important point is that the numbers are measurements of a variable relating to a particular geographical and temporal context. The statistical descriptive measures are not simply the outcome of performing certain mathematical computations but are tools enabling us to say something potentially of interest about the diurnal pattern of temperature change associated with the melting of ice from this specific glacier. When planning research investigations and specifying the attributes and variables that will be relevant, we should think ahead to the types of analysis that will be carried out, and without knowing what the values will be in advance, thinking about how to interpret the means, variances and standard distances that will emerge from the data.

References

Haggett, P. (1979) *Geography: A Modern Synthesis*, 3rd edn, London, Harper and Row.

Neft, D. (1968) Statistical analysis for areal distribution. Regional Science Research Institute, monograph series, No. 2. Regional Science Research Institute, Philadelphia, 1966. S4.75. vii and 172 pp. *Journal of the American Statistical Association*, **63**(322), 726–728.

Further Reading

Bachi, R. (1963) Standard distance measures and related methods for spatial analysis. *Papers of the Regional Science Association*, **10**, 83–132.

Harris, R. (2016) *Quantitative Geography: The Basics*, Sage Publications Ltd.

Mitchell, A. (2008) *The ESRI Guide to GIS Analysis, Volume 2: Spatial Measurements and Statistics*, Redlands, ESRI Press.

Rogerson, P.A. (2006) *Statistical Methods for Geographers: A Student's Guide*, Los Angeles, Sage.

5
Frequency Distributions, Probability and Hypotheses

Frequency distributions are an alternative way of unlocking the information contained in a set of data in the form of absolute and relative frequency counts. These are linked to the summary measures or quantities in the underlying distribution of data values examined previously. Frequency distributions also provide a starting point for understanding how mathematically defined probability distributions form the basis inferential statistical techniques. Three probability distributions are covered relating to situations where the outcome of events arising from chance is either dichotomous (e.g. Yes/No and Male/Female), discrete or integer (e.g. the 10 rock hardness categories, travel mode (train, car, bus, bicycle, etc.)), or continuous (e.g. commuter travel time, water temperature, and nitrate concentration). The chapter introduces the broad principles of statistical inference and hypothesis testing.

Learning Outcomes

This chapter will enable readers to:

- Recognise the relationships between frequency counts and summary measures;
- Identify the three main probability distributions and their association with attributes and variables recorded on different measurement scales;
- Connect the concepts of summary measures, frequency distributions and probability in the context of statistical analysis;
- Formulate simple Null and Alternative hypotheses;
- Continue planning how to use frequency distributions to summarise data in an independent research investigation in Geography, Earth Science and related disciplines.

5.1 Frequency Distributions

Each of the descriptive statistics examined in Chapter 4 was a single number that yielded some information about the whole set of data values from which they were calculated. The mean, median and mode represent a central or typical value, the range, variance and standard deviation indicate the spread of values, while skewness and kurtosis signify their symmetry and 'peakedness', respectively. A frequency distribution counts all the values in a set: a univariate frequency distribution focuses on those for one variable; a bivariate distribution for two variables together and a multivariate frequency distribution for three or more variables, although in the latter case, it can be difficult to visualise and interpret the information. Sometimes, particularly when using secondary data from official, governmental sources, these can only be obtained in the form of a frequency distribution.

In simple terms, a frequency distribution is a count of the data values allocated between classes or categories that span their full range. These counts can be expressed in two ways: the absolute counts showing the number of observations in the classes, or the relative counts usually shown as a percentage or proportion of the total either 100 per cent or 1.0, respectively. The cumulative absolute and relative frequencies are also useful, especially if the classes or categories are ordered in some way (e.g. commuting times in 30-minute intervals), in which case the count, percentage or proportion of observations in successive ordered classes are added together or accumulated. This makes it easier to discover the count, percentage or proportion of people who, for example, commuted up to 90 minutes to work.

Absolute and relative frequency distributions can be presented in the form of tables or as various types of graphs, although multivariate distributions can be difficult to interpret in both tabular and graphical format. Tables showing univariate distributions can include the absolute, relative and cumulative frequencies. The main graphical alternative is the histogram in which the heights of the columns projecting from the horizontal X axis are scaled according to the absolute or relative frequency on the Y axis. An alternative method used in respect of a continuous variable is a frequency polygon, which is produced by a line connecting the mid-values of each class of data values and sometimes extended beyond the lowest and highest classes containing observations to meet the horizontal axis at the mid-points of next lower and upper class.

Frequency distributions for nominal attributes can be produced without any further processing of the data values since the data are held in the form of categories. However, you might want to change nominal attribute categories, for example by combining detailed types of housing tenure into a smaller number of summary groups. For instance, owning a property outright and purchasing with a mortgage might be collapsed into an owner occupier category. The example of two categories of fast-food restaurant does not lend itself to this type of adjustment, although if the location of all restaurants in Pittsburgh had been recorded, then if would have been possible to collapse the detailed types into categories representing the cultural background of the food available (e.g. European, Latin American, and Oriental). However, when dealing with ordinal or continuous (interval or ratio) variables, the data values need to be separated into classes before examining their distribution. Box 5.1a illustrates these aspects of frequency distribution tables and histograms with reference data used previously, namely the numbers of fast-food restaurants operated by Burger King in Pittsburgh to which has been added the corresponding data for McDonald's (Box 5.1b) and the half-hourly temperature of the fluvioglacial stream at the exit from the outwash plain of the Bossons Glacier near Chamonix in the French Alps (Box 5.1c). The continuous variable has been divided into three mutually exclusive classes using two decimal places since the original data were recorded to one decimal place so that there is no question about which category should receive any observation. This is important since it overcomes the problem of deciding whether to allocate a value of, for example, 4.99 to the 0.00–4.99 or 5.00–9.99 class, whereas if the data value was 4.997, there would be some doubt over which of the two classes was appropriate. This distribution is clearly skewed towards to uppermost range (10.00–14.99), which contains 9 or 50.0 per cent of observations. The frequency distribution for the fast-food restaurants in Pittsburgh is simpler to interpret since there are only two categories, and when the data were collected, McDonalds dominated the local market for this type of outlet with 65.1 per cent of the total.

Box 5.1a Frequency distributions and histograms for nominal attributes and data values of continuous variables (interval scale measurements).

The use of categories when collecting nominal scale attributes makes it simpler to produce frequency distributions from this type of data. The categorisation of the fast-food restaurants operated by two companies (Burger King and McDonald's) in Pittsburgh provides an illustration below. Their absolute and relative frequencies can easily be produced and visualised as a histogram because the two types of restaurants have been defined as part of the data collection process.

The data values of ordinal, interval and ratio scale variables need to be grouped into classes to produce frequency distributions. The process is illustrated using the data values for the temperature of the water in the fluvioglacial stream at the exit from the outwash plain of the Bossons Glacier near Chamonix in the French Alps that was introduced previously. The 18 data values have been sorted in ascending order and then classified into ranges or classes, sometimes referred to as 'bins', 0.00–4.99, 5.00–9.99 and 10.00–14.99. These are equal range classes (i.e. each is 5°C wide).

The methods used to produce frequency distributions and histograms for types of fast-food restaurant and water temperature are shown in Boxes 5.1b and 5.1c, respectively.

Box 5.1b Frequency distributions and histograms for nominal attributes and data values of continuous variables (interval scale measurements).

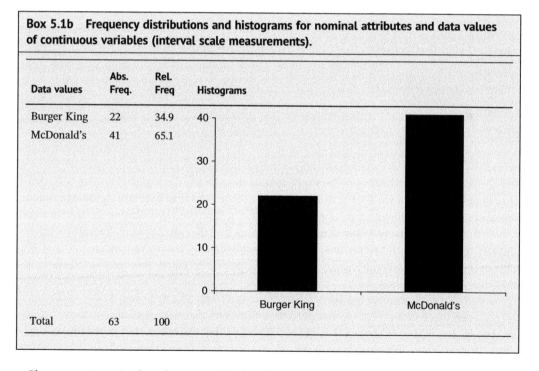

Data values	Abs. Freq.	Rel. Freq	Histograms
Burger King	22	34.9	
McDonald's	41	65.1	
Total	63	100	

Classes or categories for a frequency distribution table or histogram can be produced in several diverse ways. Box 5.1c illustrates the simplest approach, using equal ranges. Alternatives include dividing the data values into percentile ranges, identifying natural breaks or delimiting with respect to the standard deviation. Box 5.2a illustrates how these distinct types of classification

Box 5.1c Classification of data values of continuous variables (interval scale measurements).

Data values *x*	Abs. Freq.	Rel. Freq
4.30	3	16.7
4.70		
4.80		
5.20	6	33.3
6.70		
7.80		
8.80		
9.30		
10.10	9	50.0
10.20		
10.50		
11.10		
11.20		
11.40		
11.40		
11.80		
11.90		
12.30		
Total	18	100

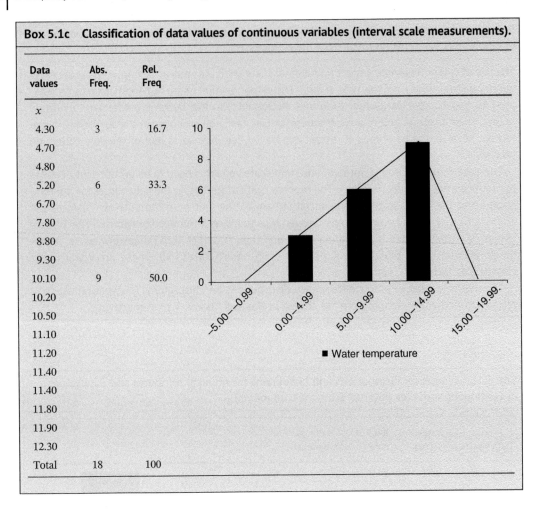

alter the frequency distribution of the water temperature in the fluvioglacial stream. Although this example only deals with a limited number of data values, the different methods produce contrasting frequency histograms. Not surprisingly, the quartile classification produced a flat with distribution with some 25 per cent of values in each of the four classes. The slightly higher numbers in the lowest and uppermost classes is simply an artefact of the procedure where the total observations do not divide exactly into four groups with an equal number. The clearest natural break in the set of data values occurred between observations four and five (5.20 and 6.70 – difference 0.75) with another between measurements 12 and 13 (10.50 and 11.40 – difference 0.90). Classes based on units of the standard deviation from the mean produced a more distinctive distribution with a clear modal range of 9.08–11.86 (i.e. the mean plus one standard deviation). In each case, the class boundaries do not occur at simple integers when compared with the regular, equal size ranges used in Box 5.1c. Examination of the four histograms reveals how the same set of data values can be allocated to classes in various ways to produce quite different frequency distributions and thus provide alternative information about the temperature of the in this particular fluvioglacial stream.

Box 5.2a   Alternative methods for classifying data values of continuous variables (interval scale measurements).

There are several alternative ways of classifying ordinal, interval and ratio scale variables so that they can be examined as frequency distributions. Classification of the data values into quartiles relates to the interquartile range measure of dispersion and separates the values into four groups each containing 25% of the observations. If, as in the example here, the total does not divide into classes with equal numbers of cases, then some will contain the extras. Classification according to where 'natural breaks' occur in the set of data values represents a more subjective method, although statistical software can also identify them. It locates the class boundaries where there are abrupt steps in the ranked sequence of values. Definition of classes in terms of units of the standard deviation about the average of the data values provides a more rigorous approach to the task, since it takes account of dispersion about the mean. The process is illustrated using the dataset values for water temperature in the fluvioglacial stream from Les Bossons Glacier near Chamonix in the French Alps. The 18 data values have been sorted in ascending order and then classified into data ranges or classes, sometimes referred to as 'bins', 0.00–4.99, 5.00–9.99 and 10.00–14.99. These are equal range classes (i.e. each is 5°C wide).

These methods are illustrated with respect to the variable recording half-hourly water temperature in the fluvioglacial stream at the exit from the outwash plain of the Bossons Glacier in Boxes 5.2b, 5.2c and 5.2d.

Box 5.2b   Percentile classes for data values of continuous variables (interval scale measurements).

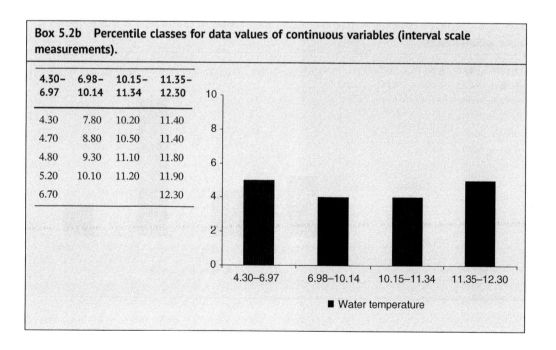

4.30–6.97	6.98–10.14	10.15–11.34	11.35–12.30
4.30	7.80	10.20	11.40
4.70	8.80	10.50	11.40
4.80	9.30	11.10	11.80
5.20	10.10	11.20	11.90
6.70			12.30

Box 5.2c   Natural break classes for data values of continuous variables (interval scale measurements).

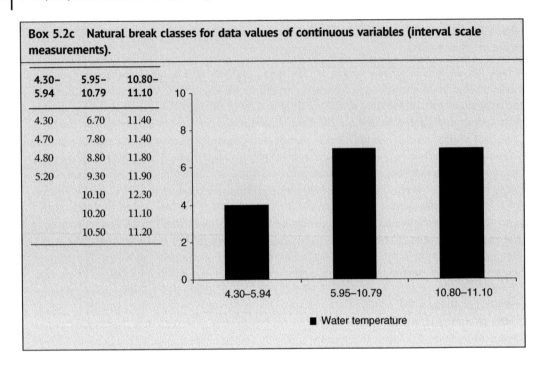

4.30–5.94	5.95–10.79	10.80–11.10
4.30	6.70	11.40
4.70	7.80	11.40
4.80	8.80	11.80
5.20	9.30	11.90
	10.10	12.30
	10.20	11.10
	10.50	11.20

Box 5.2d   Standard deviation-based classes for data values of continuous variables (interval scale measurements).

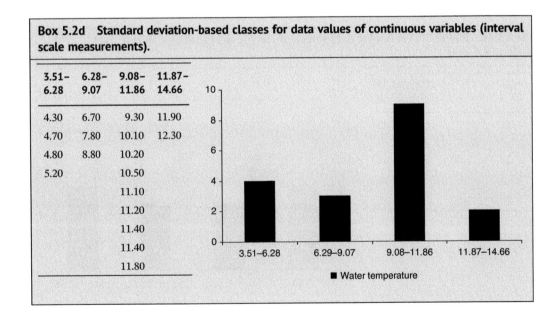

3.51–6.28	6.28–9.07	9.08–11.86	11.87–14.66
4.30	6.70	9.30	11.90
4.70	7.80	10.10	12.30
4.80	8.80	10.20	
5.20		10.50	
		11.10	
		11.20	
		11.40	
		11.40	
		11.80	

5.2 Bivariate and Multivariate Frequency Distributions

The univariate frequency distributions examined in the previous section focus on one attribute or variable. Therefore, we have no idea whether the Burger King and McDonald's restaurants share the same pattern in relation to distance from the centre of Pittsburgh. Similarly, we have no way of telling whether water temperature changes in relation to the time of day. Examination of bivariate frequency distributions will help to answer these questions. In some cases, it might be useful to explore how three or even more variables are associated with each other by viewing their multivariate frequency distribution. For example, is the occurrence of the two types of fast-food restaurant in different zones moving outwards from the central business district of Pittsburgh also connected with variations in land values? The water temperature in the stream might be linked to not only the time of day but also the amount of cloud cover. Before considering the characteristics of bivariate frequency distributions, it is worth noting that often the aggregate data used in geographical research from secondary sources exist in the form of bivariate or multivariate frequency distributions, with location acting as one of the variables or dimensions. For example, many of the statistics from the UK national Population Census are held as tables of aggregate absolute counts of people or households with different demographic and socio-economic characteristics. The presentation of data in the form of aggregate statistics helps to preserve the confidentiality of the raw observations, since the identity of individuals becomes subsumed with others in each of the geographical units.

The creation of a bivariate frequency distribution is like dealing a standard, shuffled pack of 52 playing cards according to the suit (spades, clubs, diamonds or hearts) and whether a picture (Ace, King, Queen and Jack) or a number (2–10). Figure 5.1 illustrates the outcome of this process where the two variables are the suit and value of the cards. Dealing the cards in this way distributes them into piles according to two criteria (suit and high or low value), and assuming no cards are missing from the pack, the outcome can be confidently predicted in advance: there will be eight piles four each containing nine numbered cards (2–10) and another four with the picture cards. However, when the 'observations' are not playing cards but geographical phenomena in a population or sample the outcome is less certain, although the total number to be distributed may be

Figure 5.1   Separation of playing cards into suits and high/low values.

known. There are three main ways of representing bivariate or multivariate frequency distributions: as a **contingency table** (cross-tabulation), as a **graph** (chart) or as a **thematic map**. We will return to visualisation through mapping later in this text and for we now focus on tables and graphs.

Contingency tables can be produced from two or more nominal attributes without any further processing, since the data are already categorised. However, when producing a table from ordinal or continuous (interval or ratio) variables or from one or more of these in combination with a nominal attribute(s), the raw data values will need to be classified using one of the methods already examined (equal classes, percentiles, units of the standard deviation, etc.). Box 5.3a illustrates key features of bivariate frequency distribution tables with reference to the data with which we

Box 5.3a   Bivariate contingency tables from nominal attributes and classified data values of continuous variables (ordinal and interval scale measurements).

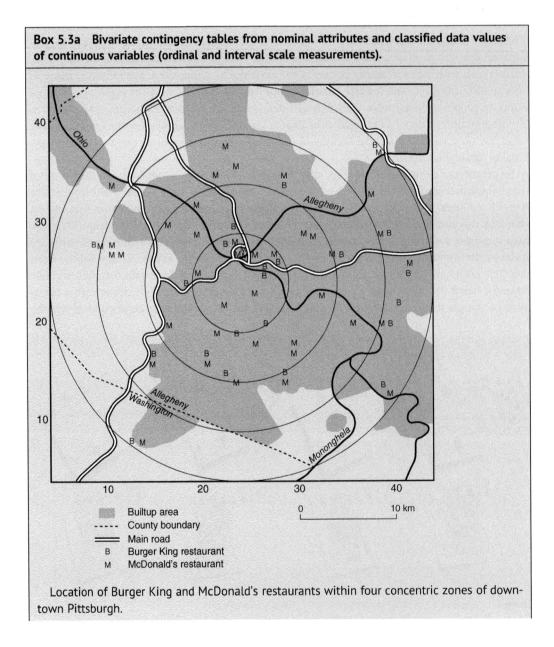

Location of Burger King and McDonald's restaurants within four concentric zones of downtown Pittsburgh.

The diagram indicates that Burger King and McDonald's restaurants are distributed spatially throughout Pittsburgh and that both companies have outlets in the four concentric zones moving outwards from the downtown area. However, it is not immediately obvious whether one type of restaurant dominates any zone, and if so to what extent. Trying to answer these questions involves interpreting the bivariate absolute and relative frequency distributions.

Absolute and relative frequency cross-tabulations provide a way of finding how many and what percentage (or proportion) of observations possess different combinations of characteristics. The boxes below reveal that there were very similar numbers of Burger King and McDonald's restaurants (eight and nine, respectively) in the outermost zone (15.0–19.9 km). However, these absolute figures accounted for 36.4% of the total in the case of Burger King outlets, but only 21.4% of McDonald's. This result indicates Burger King had a greater presence in the suburban areas of the city. McDonald's seems to dominate the fast-food restaurant sector in Pittsburgh, and overall, the outlets for both types are reasonably evenly distributed across the four zones with almost a quarter of the total, between 21.9 and 28.1%, in each.

Absolute frequency and expected counts, and relative percentage frequencies together with the calculations for these are given in Boxes 5.3b, 5.3c, 5.3d, 5.3e and 5.3f.

Box 5.3b   Absolute frequency distribution.

	0.0–4.9 km	5.0–9.9 km	10.0–14.9 km	15.0–19.9 km	Marginal column total
Burger King	5	5	4	8	22
McDonald's	9	13	11	9	42
Marginal row total	14	18	15	17	64

Box 5.3c   Expected absolute frequency distribution.

	0.0–4.9 km	5.0–9.9 km	10.0–14.9 km	15.0–19.9 km	Marginal column total
Burger King	$(14 \times 22)/64$ $= 4.8$	$(18 \times 22)/64$ $= 6.2$	$(15 \times 22)/64$ $= 5.2$	$(17 \times 22)/64$ $= 5.8$	22
McDonald's	$(14 \times 42)/64$ $= 9.2$	$(18 \times 42)/64$ $= 11.8$	$(15 \times 42)/64$ $= 9.8$	$(17 \times 42)/64$ $= 11.2$	42
Marginal row total	14	18	15	17	64

Box 5.3d   Row percentage relative frequency distribution.

	0.0–4.9 km	5.0–9.9 km	10.0–14.9 km	15.0–19.9 km	Marginal column total
Burger King	$(5 \times 100)/22$ = 22.5	$(5 \times 100)/22$ = 22.7	$(4 \times 100)/22$ = 18.2	$(8 \times 100)/22$ = 36.4	100.0
McDonald's	$(9 \times 100)/42$ = 21.4	$(13 \times 100)/42$ = 31.0	$(11 \times 100)/42$ = 26.2	$(9 \times 100)/42$ = 21.4	100.0
Marginal row total	$(14 \times 100)/64$ = 21.9	$(18 \times 100)/64$ = 28.1	$(15 \times 100)/64$ = 23.4	$(17 \times 100)/64$ = 26.6	100.0

Box 5.3e   Column percentage relative frequency distribution.

	0.0–4.9 km	5.0–9.9 km	10.0–14.9 km	15.0–19.9 km	Marginal column total
Burger King	$(5 \times 100)/14$ = 35.7	$(5 \times 100)/18$ = 27.8	$(4 \times 100)/15$ = 26.7	$(8 \times 100)/17$ = 47.1	$(22 \times 100)/64$ = 34.4
McDonald's	$(9 \times 100)/14$ = 64.3	$(13 \times 100)/18$ = 72.2	$(11 \times 100)/15$ = 73.3	$(9 \times 100)/17$ = 52.9	$(41 \times 100)/64$ = 65.6
Marginal row total	100.0	100.0	100.0	100.0	100.0

Box 5.3f   Overall percentage relative frequency distribution.

	0.0–4.9 km	5.0–9.9 km	10.0–14.9 km	15.0–19.9 km	Marginal column total
Burger King	$(5 \times 100)/64$ = 7.8	$(5 \times 100)/64$ = 7.8	$(4 \times 100)/64$ = 6.3	$(8 \times 100)/64$ = 12.5	$(22 \times 100)/64$ = 34.4
McDonald's	$(9 \times 100)/64$ = 14.1	$(13 \times 100)/64$ = 20.3	$(11 \times 100)/64$ = 17.2	$(9 \times 100)/64$ = 14.1	$(41 \times 100)/64$ = 65.6
Marginal row total	$(14 \times 100)/64$ = 21.9	$(18 \times 100)/64$ = 28.1	$(15 \times 100)/64$ = 23.4	$(17 \times 100)/64$ = 26.6	100.0

are already becoming familiar. It shows that there are several distinct types of tables that can be produced, each allowing a slightly different interpretation of the information obtained from the same raw data. Relative frequency cross-tabulations can contain row, column or overall percentages and the absolute counts (Box 5.3b) and may be accompanied by the expected frequency counts. These are the number of observations that would occur in each cell in the table if they were distributed in proportion to the row and column totals. Calculating expected frequencies will often result in decimal values (e.g. 4.8 in the Burger King 0.00–4.9 km cell in Box 5.3c). The observed absolute frequency count for this cell was five, very close to the expected value. However, the McDonald's and 15.00–19.9 km shows an observed count of 9 and an expected one of 11.2, indicating there were '2.2' fewer restaurants than expected. The importance of the differences between these two types of frequency will become apparent when we examine how to test to see whether the observed counts are significantly different from the expected ones in Chapter 7. Box 5.3a illustrates two-way or two-dimensional contingency tables; however, multivariate (i.e. with three or more dimensions) can be produced. These can become quite complex to summarise, and visualisation of multivariate distributions is best carried out using graphs.

> Describe some other interpretations of the information contained in the absolute and relative cross-tabulations apart from those referred to in Box 5.3a. There are at least three more things you could mention.

There are many ways of representing bivariate and multivariate frequency distributions graphically and the choice of method depends upon the nature of the data to be presented more than the visual appearance of the result. Again, the focus here will be on using graphs to illustrate the characteristics of bivariate and multivariate frequency distributions from the perspective of helping to understand statistical methods rather than as tools for communicating results visually. The choice of graph depends upon the nature of the data being examined, in other words whether the data values refer to nominal attributes or continuous measurements on the ordinal, interval or ratio scales. Box 5.4a illustrates bivariate graphing with a clustered column chart or histogram in respect of the fast-food restaurants in Pittsburgh that is equivalent to the contingency table in Box 5.3b, which has been included to emphasise the point. A clustered column graph is an extension of the univariate histogram and may be used with nominal attributes or categorised continuous variables. The two frequency distributions can be viewed side by side and their shapes compared. In this example, the number of McDonald's is nearly double the number of Burger King restaurants in the first three zones, whereas the two types are nearly equal in the outermost zone.

One of the most common methods for displaying a bivariate distribution of unclassified continuous variables is as a scatter graph (also called a scatterplot and scattergram). Box 5.4c includes a scatter graph for water temperature at two sampling points on the fluvioglacial stream flowing from Les Bossons Glacier in France. The variables, usually assigned to the shorthand labels X and Y, are plotted on the horizontal and vertical axes of the graph, respectively. These are scaled according to the measurement units of the variables, °C in this case, and provide a visual impression of the spread of values and how the data values for one variable relate to those of the other. The water temperature measurements on the fluvioglacial stream have been recorded at the same time of day at each location. Thus, there is a 'hidden' variable in the data connected with the time when the temperature measurements were taken. If they had been made randomly throughout the day and the time recorded independently, then insufficient data might have been available for certain periods, which would have limited the opportunity to examine whether temperature varies between the locations at the same time of day. This together with differences in the ranges of data values has helped to produce the slightly curious shape of the scatter graph. Scatter graphs do not provide a count of the number of observations possessing certain values. However, it is feasible to overlay the graph with a grid to represent the cells in a contingency table, which is equivalent to grouping the raw data values into classes. Such a grid has been superimposed on the scatter graph in the lower part of Box 5.4c, and the corresponding absolute frequency cross-tabulation appears on Box 5.4d. The data values for both sets of temperature measurements have been allocated to groups using the classification scheme originally applied to one of the sample locations (see Box 5.1b), and because those recorded near the glacier snout have a narrower range, there are some empty cells in the table.

> How could the classification schemes for the water temperature data values at the two locations be altered to avoid so many empty cells in the table?

Box 5.4a Bivariate graphical display with nominal attributes and classified continuous variables (interval and ratio scale measurements).

Sample site at stream exit from outwash plain.

Sample site on stream near glacier snout.

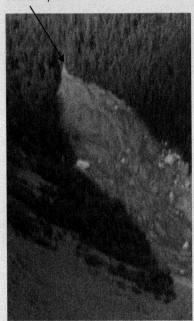 

The clustered histogram or column graph for fast-food restaurants in Pittsburgh (Box 5.4b) has been created with the frequency counts of the restaurants on the vertical axis and the four distance zones on the horizontal. The graph indicates whether the two companies' outlets possess a similar or different frequency distributions. The McDonald's restaurants show a rising then falling trend over the set of distance zones, whereas Burger King dips in the middle two zones and reaches a peak furthest from the city centre.

The scatter graph plots water temperature for the two locations on the fluvioglacial stream at half-hourly intervals throughout the day when the field work was carried out. The X and Y axes have been scaled to cover the same spread of units (0.0–$14.0\,°C$); although data values for the measurements made near the stream's exit from the outwash plain were more dispersed (spread out around the mean). The values on the X axis range between 4.0 and 8.0, which is much narrower than those on Y (4.0–12.0). Lower-temperature values tend to occur together at both locations, although the high values near the stream's exit were accompanied by only moderate figures at the glacier snout. The overlain grid used to indicate classification of the temperature recordings further emphasises this point.

A clustered column chart for type and zone of fast-food restaurant and a scatter graph of water temperature together with a bivariate classification of the values are given in Boxes 5.4b and 5.4c and 5.4d.

Box 5.4b   Clustered column bar graph of fast-food restaurants by distance from central Pittsburgh.

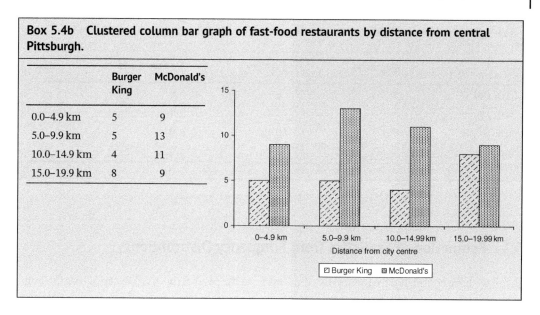

	Burger King	McDonald's
0.0–4.9 km	5	9
5.0–9.9 km	5	13
10.0–14.9 km	4	11
15.0–19.9 km	8	9

Box 5.4c   Scatter graph of raw data values of water temperature at glacier snout and outwash exit of Les Bossons Glacier, France.

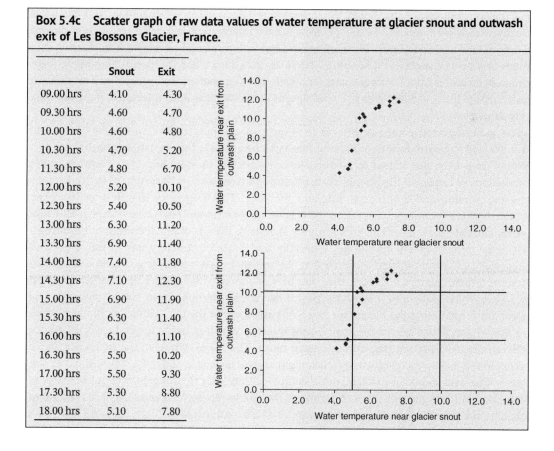

	Snout	Exit
09.00 hrs	4.10	4.30
09.30 hrs	4.60	4.70
10.00 hrs	4.60	4.80
10.30 hrs	4.70	5.20
11.30 hrs	4.80	6.70
12.00 hrs	5.20	10.10
12.30 hrs	5.40	10.50
13.00 hrs	6.30	11.20
13.30 hrs	6.90	11.40
14.00 hrs	7.40	11.80
14.30 hrs	7.10	12.30
15.00 hrs	6.90	11.90
15.30 hrs	6.30	11.40
16.00 hrs	6.10	11.10
16.30 hrs	5.50	10.20
17.00 hrs	5.50	9.30
17.30 hrs	5.30	8.80
18.00 hrs	5.10	7.80

Box 5.4d	Contingency table of water temperature at glacier snout and outwash exit.			
	0.00–4.99	5.00–9.99	10.00–14.99	Exit
0.00–4.99	3	0	0	5
5.00–9.99	2	3	0	11
10.00–14.99	0	10	0	2
Snout	5	13	0	18

5.3 Estimation of Statistics from Frequency Distributions

So far, we have considered frequency distributions from the perspective that we can access the individual raw data values for the full set of (sampled) observations (i.e. details of fast-food outlets and the water temperature measurements). These values are either nominal categorical attributes or have been classified into such groups for the purpose of producing an absolute or relative frequency distribution. This process of classification discards details in the raw data to produce an overview of how the observations are distributed. When analysing data from secondary sources, this detail has usually already been removed by the 'publisher' of the information and so the summary measures of central tendency and dispersion examined in Chapter 3 cannot be calculated from the raw data values to provide a statistical description. There are occasions when it is useful to estimate these summary measures from the frequencies, although the estimated measures (e.g. mean or standard deviation) are likely to differ from the 'true' figures that could have been calculated, if the raw data were available.

The main descriptive measures used in respect of continuous data values are the mean, variance and standard deviation. Each of these can be estimated for variables where the observations have been distributed as frequencies into discrete classes. Box 5.5a illustrates these calculations in respect of the frequency distribution of farm sizes in Government Office Regions covering London and South East England for 2002. This frequency distribution has been obtained from a secondary source, and the classes are those employed by the National Statistics Office. The data show that farm area is highly skewed to the left of the distribution indicating a very large number of small holdings, especially under 2.0 ha, and very few in the larger size categories (3.3% over 300.0 ha). Notice that the unequal class widths (e.g. 10, 50 and 100 ha) produce a distribution with decreasing numbers of farms in the classes up to 50 ha followed by a rise, which leads into a second series of decreasing columns from 100 ha onwards. The calculations estimating the mean (see Box 5.5b) assume that the data values within each class are evenly distributed across the entire class interval, which is likely to produce a difference between the values of the true and estimated mean. The left-skewness of the frequency distribution in this example suggests that farms might also be skewed in the same way within the size classes. The data source in this instance has provided not only the frequency count of farms in each class, but also the total area of farmland that these represent. Therefore, it is possible to calculate the true mean, which turns out to be 43.2 ha compared with an estimate of 44.7 ha (excluding 700 and over ha class).

Box 5.5a   Estimation of summary descriptive statistical measures from frequency distribution of farm sizes in London and South East England.

Mean – population symbol: μ; sample symbol: $\bar{x}$

Variance – population symbol: σ^2; sample symbol: s^2

Standard deviation – population symbol: σ; sample symbol: s

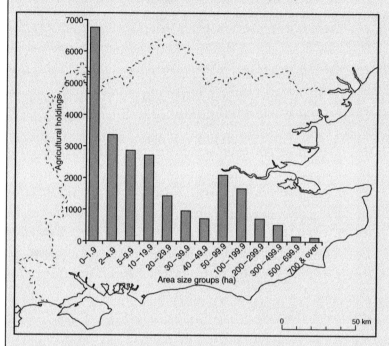

Frequency distribution of farm area in London and South East England 2002.

The mean as a measure of central tendency conveys something about the typical value of a continuous variable, whereas the variance and standard deviation indicate how dispersed all the data values are around the mean. The usual ways of calculating these statistics (adding up the values and dividing by the number of observations for the mean and summing the squared differences from the mean and dividing by the number of observations for the variance) are not possible when the data are held in the form of a frequency distribution. An estimate of the mean can be calculated by multiplying the mid-value of each class by the corresponding number of observations, then adding up the results and dividing the sum by the total number of observations. Class mid-values are obtained by dividing the width of each class by 2 and adding the result to lower class limit (e.g. 4.9–2.0/2 + 2.0 = 3.45). The variance may be estimated by subtracting the estimate of the overall mean from the mid-value of each class, squaring the results and summing before dividing by n (or n–1 in the of sample data). The square root of the variance provides the standard deviation.

The methods used to calculate estimates of the mean, variance and standard deviation for the frequency distribution of farm (holding) areas in South East England are illustrated below in Box 5.5b, where the letters f and m denote the frequency count and class mid-value, respectively, for the farm area variable (x). Note that the uppermost open-ended class (700 ha and over) has been omitted from the calculations to enable comparison with the published statistics.

Box 5.5b   Calculation of estimates for the mean, variance and standard deviation from data in a frequency distribution.

Size class	Class mid-value m	Count f	$\bar{x} = \dfrac{\sum fm}{n}$ mf	$s^2 = \sqrt{\dfrac{\sum\left(f(m-\bar{x})^2\right)}{n}}$ $(m-\bar{x})$	$f(m-\bar{x})^2$
0.0–1.9	0.95	6,759	6,421.1	−43.75	12,937,148.44
2.0–4.9	3.45	3,356	11,578.2	−41.25	5,710,443.75
5.0–9.9	7.45	2,872	21,396.4	−37.25	3,985,079.50
10.0–19.9	14.95	2,715	40,589.3	−29.75	2,402,944.69
20.0–29.9	24.95	1,433	35,753.4	−19.75	558,959.56
30.0–39.9	34.95	961	33,587.0	−9.75	91,355.06
40.0–49.9	44.95	724	32,543.8	0.25	45.25
50.0–99.9	74.95	2,103	157,619.9	30.25	1,924,376.44
100.0–199.9	149.95	1,668	250,116.6	105.25	18,477,374.25
200.0–299.9	249.95	721	180,214.0	205.25	30,373,972.56
300.0–499.9	399.95	523	209,173.9	355.25	66,003,940.19
500.0–699.9	599.95	155	92,992.3	555.25	47,786,897.19
		23,990	$\sum x = 1,071,985.5$	$\sum\left(f(x-\bar{x})^2\right) = 190,252,536.98$	

$$\frac{1,071,985.5}{23,990}$$

Mean $\qquad\qquad \bar{x} = 44.7$

Variance $\qquad\qquad s^2 = \dfrac{190,252,536.98}{23,990}$

$$s^2 = 7930.5$$

Standard deviation $\qquad\qquad s = \sqrt{\dfrac{190,252,536.98}{23,990}}$

$$s = 89.1$$

In this example, the skewness of the distribution and the use of different class intervals across the entire spread of data values are likely to have limited the difference between the estimated and true means. The calculations used to estimate the variance and standard deviation in Box 5.5b assume that there is no dispersion amongst the observations (farms) in each class and that the area of each one equals the class mid-value (i.e. the within class variance is zero). The effect of the highly skewed distribution of farm areas can be discovered by collapsing the classes so that they are equal (as far as the original classes allow) (see Box 5.5c), except for the lowest category, which is 100 ha. The effect is to inflate the estimated mean to 76.1 ha with the variance and standard deviation reducing slightly. Since the original estimated mean is close to the known true mean in this example, it may be concluded that the unequal class interval employed in assembling the frequency distribution helps to overcome the underlying skewness in the raw data values.

Box 5.5c   Calculation of estimates for the mean, variance and standard deviation from data in a frequency distribution with approximately equal size classes.

Size class	Class mid-value m	Count f	$\bar{x} = \dfrac{\sum fm}{n}$ mf	$s^2 = \sqrt{\dfrac{\sum\left(f(m-\bar{x})^2\right)}{n}}$ $(m-\bar{x})$	$f(m-\bar{x})^2$
0.0–99.9	49.95	20,923	1,045,103.9	−26.15	576,690.19
100.0–299.9	199.95	2,389	477,680.6	123.85	57,581,021.81
300.0–499.9	399.95	523	209,173.9	323.85	66,003,940.19
500.0–699.9	599.95	155	92,992.3	523.85	47,786,897.19
		23,990	$\sum x = 1{,}824{,}950.5$	$\sum\left(f(x-\bar{x})^2\right) = 148{,}338{,}606.78$	

$$\dfrac{1,824,950.5}{23,990}$$

Mean $\qquad\qquad \bar{x} = 76.1$

Variance

$$s^2 = \dfrac{148,338,606.78}{23,990}$$

$$s^2 = 6{,}183.4$$

Standard deviation

$$s = \sqrt{\dfrac{148,338,606.78}{23,990}}$$

$$s = 78.6$$

Source: Department of the Environment, Food and Rural Affairs (2003).

5.4   Probability

People encounter probability throughout their everyday lives without necessarily thinking about it and how it affects them. Our experience of probability is so commonplace that we often have some difficulty when conceptualising what it means in a more formal fashion. Texts dealing with statistical techniques often resort to explaining the concept with reference to tossing coins and other similar devices where there is clearly an element of chance involved in the action. At an early age, many children encounter chance events when playing board games that involve throwing dice. They soon learn that when a number between 1 and 6 lands uppermost, then it is possible to move forward by that number of spaces and at each throw of the die the same number does not appear. With experience, they learn that there is an equal chance they could move forwards by one, two, three, four, five or six spaces. In other words, they learn by practice the probabilities associated with the outcome of the activity. There is a 1 in 6 (0.167 (16.7%)) chance that a specified number will show uppermost at each throw of the die. Some games involve throwing two dice at the same time, such as Backgammon and Shut-the-Box, which means there are more possible outcomes, 1 and 1, 1 and 2, 1 and 3, ... 6 and 6: there are 36 outcomes.

> Using this information try to work out mathematically the probability of getting a 3 and 5 (or any other specific combination) when throwing two dice at the same time.
> Hint: should you add or multiply the probability of getting a specific number with one die?

One such board game with a long pedigree is Snakes and Ladders (or Chutes and Ladders) as represented in a slightly simplified format in Figure 5.2. If the number on the die results in the player's token landing on the head of a snake, then it has to 'fall down the snake' to its tail and resume its progress towards the Finish from that space at the next turn. However, a player landing on a square at the foot of a ladder can 'climb' to the space at its top and continue from there. The number of squares on a board of Snakes and Ladders is normally 8×8, 10×10 or 12×12 and the ladders and snakes are of various lengths, some of which will cascade or promote the player a small number of rows, whereas others will cause a more precipitous fall or dramatic ascent. The probabilities associated with this game are therefore more complex than one player simply having the luck to throw higher numbers on the die than his/her opponents and thereby overtake them in the race from Start to Finish. Nevertheless, it is the probabilities associated with each throw of the die that is of interest here.

This description of playing Snakes and Ladders begs one important question. How does a player of this or any other game that involves throwing a die **know** that on each throw it can fall with one and only one numbered side uppermost, and that each of the six numbers has an equal probability of facing upwards. There are two ways of answering this question: either by 'thinking about' the physical characteristics of the die, the action of throwing it and the possible range of outcomes when it lands; or by spending a lot of time experimentally throwing a die and carefully recording the number of times each digit faces upwards. The first approach entails theorising about the event and, assuming the die is unweighted and unbiased, reaching the conclusion that it is extremely unlikely to land on one of its edges or corners. Therefore, it must land with one face hidden from view lying on the surface where it has landed, four facing outwards at right angles to each other to the sides and the sixth visible appearing uppermost. There is an equal chance (probability) of any one of the numbered faces in this position. This is known as *a priori* probability, since the

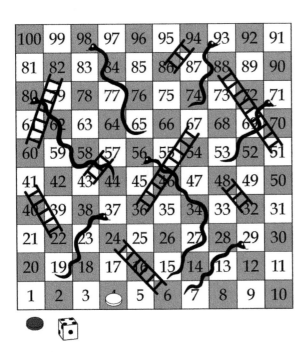

Figure 5.2 Typical layout of snakes and ladders board.

probability of a particular outcome is known or assessed in advance without ever actually throwing a die. The alternative way of determining probability involves experimentally throwing the die repeatedly, and recording the number facing upwards on each occasion will produce a frequency distribution of the range of outcomes. Overall, if this experiment is carried out enough times, the frequency distribution will show each of the six possible outcomes having occurred an equal or very nearly equal number of times. This type of probability is referred to as *a posteriori* because it is determined empirically after the (throwing) events have taken place.

In this example, both types of probability should produce the same answer. So, if you were to take time off from reading this chapter and to throw a die, what is the probability that a two will appear on the uppermost surface? How many different ways can you write the answer to this question?

The action of throwing a die produces discrete or integer outcomes, since it is possible to obtain a 1, 2, 3, 4, 5 or 6 only a whole number of times. For example, a six cannot be thrown on 0.45, 2.36 or any other decimal or fractional number of occasions. In other words, the type of event or observation represented by throwing a die is a nominal attribute with more than six categories. It would be quite feasible to produce a die where the faces were different colours (nominal categories) (e.g. red, blue, green, yellow, orange and purple). Likewise, the six faces could be numbered with the values 0.75, 1.25, 2.50, 3.75, 4.25 and 5.50 instead of 1, 2, 3, 4, 5 or 6. It would be difficult to use such a die in a game of Snakes and Ladders, how can a player's token move 3.75 squares on the board. However, if the coloured or decimally numbered die was thrown on many occasions, the number of outcomes associated with each colour or decimal value would still be an integer or whole number even though the labels of the nominal categories were different. Few, if any research projects in Geography and the Earth Sciences, involve throwing dice, except perhaps to obtain a random sample. Nevertheless, investigators often analyse events that have a discrete number of possible outcomes and/or situations where the values of the variables being measured are decimals.

A project will often involve a mixture of attributes and variables quantified according to these distinct types of measurement. The probabilities associated with these 'research' attributes and variables can also be thought about theoretically or empirically. Other things being equal, the proportion of males and females in paid employment should be the same as their proportions in the population of normal working age (currently 16–67 years in the UK). Assuming these proportions are 0.55 for males and 0.45 for females of working age, a random sample of 600 people in work aged 16–67 should contain 270 women and 330 men (see Table 5.1). This probability is based on the known proportions of the two genders in the parent population. If a die was to be thrown 600 times, you would expect a 1 to face upwards 100 times, a 2 facing upwards 100 times and the same frequency count (100) for the 3, 4, 5 and 6. However, in both cases, the observed or empirical results might be different from the expected and Table 5.1 also shows the hypothetical cases of slightly more males and less females appearing in the sample, and more or less than 100 observed occurrences of some of the numbered faces of the die. The expected outcomes are based on *a priori* probabilities since they can be asserted with the empirical results (the sample and the die throwing) having been obtained. In principle, similar arguments can be advanced in respect of the data values of continuous variables, although these relate to decimal measurements rather than integer outcomes. Thus, other things being equal, the temperature just below the surface of a water body on a day when the sun is shining will be higher than on days when it is raining. In other words, the temperature measurements in these two situations

Table 5.1 Observed and expected frequency distributions for dichotomous and discrete outcomes.

Gender split in working age population	Expected outcome Count (%)	Observed outcome Count (%)	Throwing a die	Expected outcome Count (%)	Observed outcome Count (%)
Male	330 (55.0)	338 (56.3)	1	100 (16.6)	96 (16.0)
Female	270 (45.0)	262 (43.7)	2	100 (16.7)	97 (16.2)
			3	100 (16.7)	107 (17.8)
			4	100 (16.6)	103 (17.2)
			5	100 (16.7)	100 (16.6)
			6	100 (16.7)	97 (16.2)
Total	600 (100.0)	600 (100.0)		600 (100.0)	600 (100.0)

can be put in order by reference to theoretical prior probability about the effect of thermal incoming electromagnetic radiation from the sun (i.e. asserting in advance they should be higher on a sunny day and lower on a cloudy day). Unfortunately, it is more difficult to say by how much the set of measurements on the sunny day are likely to be higher than those on the cloudy day, although this may be determined from past empirical results (i.e. *a posteriori* probability).

In summary, from a statistical perspective, three types of outcomes for different events are possible:

- dichotomous, where there are two possibilities (e.g. male or female gender; yes or no answer)
- integer or discrete, where there are more than two possibilities, but each is indivisible (e.g. rock hardness category, age group, housing tenure)
- continuous, where there are decimal or fractional possibilities resulting from measurement (e.g. annual gross income of £23,046.35, temperature of 15.6°C, farm area 46.3 ha).

The outcomes of these diverse types of events can be expressed in the form of frequency distributions, and the likelihood of obtaining a particular distribution can be determined by reference to mathematically defined probabilities. We have already seen that the probability of a specific outcome when throwing a die once is 0.1667 and that if the die was thrown repeatedly (even an infinite number of times), the frequency distribution would show each outcome occurred 16.67 per cent of the total. But what is needed is a distribution of probabilities that tells us whether a particular mean, median, standard deviation, variance or frequency distribution of observations provided by our research data is what would be expected. Statistical testing relies on probability distributions to act as a 'sounding post' or 'benchmark' against which to test results obtained from data collected empirically. The three main probability distributions are the Binomial, the Poisson and the Normal. The frequency distribution associated with each of these corresponds to what would be produced if the values of the variables and attributes occurred randomly. We will examine the characteristics of these probability distributions before exploring the tests that are available to help us to decide if the statistics or frequency distributions produced from empirical, research data are significantly different from those obtained from a probability distribution.

5.4.1 Binomial Distribution

Words starting with the letters 'bi' usually indicate that they have something to do with two or twice. For example, bisect means to cut into two pieces or sections and bilateral, as in bilateral talks, refers to discussions between two organisations or countries. The second part of binomial ('nomial') looks a little like nominal, which we have seen relates to categorical attributes where the values correspond to names or labels. It is not surprising that the binomial distribution is associated with those situations in which there are two nominal categories or outcomes to an experiment or event. The binomial distribution was devised by James Benouilli in the 17th century and applies to a series of similar events where two (dichotomous) outcomes are possible each time. Box 5.6a outlines how the binomial distribution is calculated using the now familiar example of Burger King restaurants in Pittsburgh, whose presence or absence in the six 10 km grid squares covering the city is shown in Box 5.6a.

The form and shape of the frequency distribution varies according to the number of events and the probability of each outcome. It is easy to determine the number of events, but deciding in advance on the expected probabilities of each outcome is more problematic. We cannot assume there is an equal (50/50) chance of each outcome and must turn to either theoretical reflection or empirical precedent. The frequency distribution produced from binomial probabilities has a

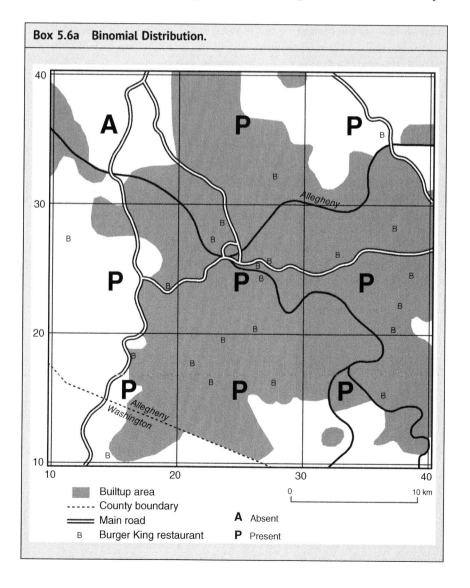

Box 5.6a   Binomial Distribution.

- Builtup area
- ----- County boundary
- ═══ Main road
- B   Burger King restaurant

- **A** Absent
- **P** Present

0 — 10 km

symmetrical shape if the probabilities of each outcome are identical (50/50 or 0.5/0.5). A skewed or asymmetrical distribution will result if there is distinct difference in the two probabilities, for example 75/25 or 0.75/0.25. The mean, variance, standard deviation and skewness statistics can be calculated for a frequency distribution produced from binomial probabilities. The mean refers to the average number of outcomes of one or other of the two types from a certain number of events. The variance and standard deviation relate to the dispersion of the distribution around the mean and skewness to the degree of symmetry.

Box 5.6b   Calculation of Binomial probabilities.

Binomial distribution probability: $p(X) = \dfrac{N!p^X q^{N-X}}{X!(N-X)!}$

The binomial distribution is used to determine the probability of each of two possible outcomes in a series of events. In this example, the outcomes are the presence or the absence of a Burger King restaurant in a set of 10 km grid squares superimposed over the built-up area of Pittsburgh. The probability of the occurrence of the event under investigation is denoted by p in the formula and probability of non-occurrence by q. In this example, the probabilities of occurrence and non-occurrence of a Burger King in any one grid square are equal (0.5). Across the complete set of N (9) grid squares, at one extreme, they might each include at least one of these restaurants, whereas at the other none of them does. The result of counting the number of squares with and without at least one Burger King outlet is shown as a frequency table below. It reveals that 83.3% or 5 squares include a Burger King and 16.7% do not.

The binomial distribution in this example determines the probability that a certain number of squares, referred to as X in the Binomial formula, will contain a Burger King: the number of squares in which the event occurs. The calculations in Box 5.6c provide the Binomial probability for the occurrence of at least one Burger King restaurant in 2, 3 and 5 squares. Now you might ask why it is useful to find out these probabilities when the empirical data on their own provide the answer – there are five squares with a restaurant and four without. The answer is that the binomial distribution has indicated the probable number of squares with a restaurant if they were distributed randomly. The probability of there being five squares with a Burger King is quite small (0.08 or 8 times in 100), certainly compared the probabilities for two (0.23) or three (0.31).

Box 5.6c   Calculation of Binomial probabilities.

$$p(X) = \dfrac{N!p^X q^{N-X}}{X!(N-X)!}$$

Burger King	Count	$p(2) =$	$p(3) =$	$p(5) =$
0 (absent)	1	$\dfrac{6!0.5^2 0.5^{6-2}}{2!(6-2)!}$	$\dfrac{6!0.5^3 0.5^{6-3}}{3!(6-3)!}$	$\dfrac{6!0.5^5 0.5^{6-5}}{5!(6-5)!}$
1 (present)	5	$\dfrac{720(0.25)(0.0625)}{2(24)}$	$\dfrac{720(0.125)(0.125)}{6(6)}$	$\dfrac{720(0.03125)(0.5)}{144(1)}$
	6	$p(2) = 0.23$	$p(4) = 0.31$	$p(5) = 0.08$

5.4.2 Poisson Distribution

Poisson is French for fish, but this does not help us to discover that it was a French mathematician, Siméon-Denis Poisson, who developed the Poisson distribution in 1830. Unlike the binomial distribution, there is nothing in the word 'Poisson' to suggest why the Poisson Distribution should refer to the probabilities associated with events having a discrete number of integer (whole) outcomes or that the number of possible outcomes increases as their probability reduces. This describes the situation where the outcomes are listed in ascending order (0, 1, 2, 3, ..., n) and the Poisson Distribution determines the probabilities associated with each of an ordered series of 'counted' outcomes. Two important applications relate to counting the number of times a particular outcome occurs within a given collection of spatial units or intervals of time. These units are treated as 'events' in such situations. Whereas the binomial distribution requires a fixed, finite number of outcomes to an event (e.g. Yes or No answers to a question), the Poisson does not impose an upper limit and theoretically an outcome could occur a large number of times within any individual spatial or temporal unit. However, one common feature of both distributions is that they deal with variables measured according to the ordinal scale, in other words a ranked or sorted sequence of outcomes.

Box 5.7a illustrates that the Poisson distribution is defined mathematically as a series of discrete terms, each representing an integer outcome in the sequence. The first provides the probability of a zero outcome (i.e. none of the phenomena being counted as present), the second one, the third two and so on to the maximum considered necessary. The calculations show that probabilities become smaller as the sequence progresses. The probability that zero or one outcome will occur in each unit of space or time is high, but there is a comparatively low chance that four, five or more occurring. Standard descriptive statistics (mean, standard deviation and skewness) can be calculated for a frequency distribution produced from Poisson probabilities. These are derived from one of the key elements in the series of Poisson terms λ (*lambda*), which equals the mean and variance. The standard deviation is the square root of λ, and skewness is the reciprocal of this statistic. The value of λ is usually obtained from empirical data and varies from one application to another.

There are two issues that should be noted. The value of λ, the average number of outcomes that have occurred across the group of spatial or temporal units, or other types of events, is usually used to calculate the Poisson probabilities. However, this average density of outcomes could easily vary across the overall study area (or over time). Suppose a researcher was interested in the spread of independent retail outlets across a county and decides to superimpose a regular grid of 1 km squares over the area. The number of shops is likely to vary between sparsely and densely populated areas: urban neighbourhoods might have a few shops competing for business, whereas villages, if they are lucky, might have one facility. The mean or average number of shops per 1 km square is likely to be higher or lower in distinct parts of the county. One solution would be to calculate separate means for the high- and low-density areas and then combine these by weighting according to the number of squares of the two types. However, this introduces the problem of deciding how to define the physical extents of these areas as groups of individual grid squares.

Another way of dealing with this issue would be to change the size of the spatial units, for example from 1 km squares to 2.5 km, 5.0 km or even larger squares. Each increase in area would mean that the number of squares would decrease, or conversely their number would increase if their size was reduced. Given that the number and location of outlets is fixed, these changes in size of unit will impact on the frequency distribution. A reduction in grid square size (e.g. to 0.25 km squares) is likely to increase the number of squares without any or only one shop and the count with three, four or more would decrease. Changing the size of the spatial or temporal units is still likely to produce a skewed frequency distribution, but the average number of outcomes per unit will alter, which results in a different value for λ.

Box 5.7a   Poisson Distribution.

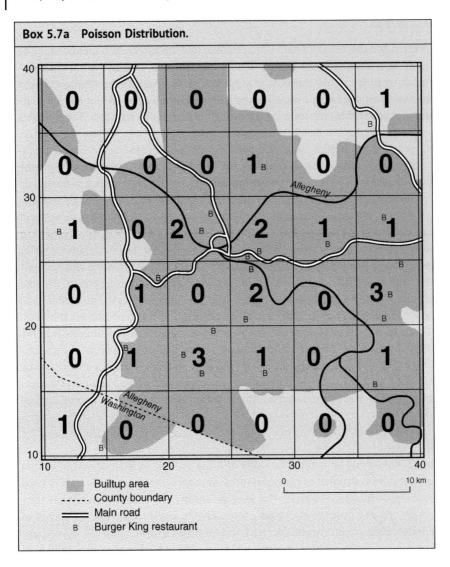

- ▨ Builtup area
- ----- County boundary
- ═══ Main road
- B   Burger King restaurant

0   10 km

Box 5.7b   Calculation of Poisson probabilities.

Sequence of terms in Poisson Distribution: $e^{-\lambda}$; $\lambda e^{-\lambda}$; $\dfrac{\lambda^2 e^{-\lambda}}{2!}$; $\dfrac{\lambda^3 e^{-\lambda}}{3!}$ ... $\dfrac{\lambda^x e^{-\lambda}}{N!}$

    The Poisson Distribution is used to discover the probability of the number of times that a discrete outcome will occur in a sequence. The present example complements the application of the binomial distribution in Box 5.6a by providing Poisson probabilities for the number of Burger King restaurant in a set of 5 km grid squares superimposed over the built-up area of Pittsburgh. Each square might contain 0, 1, 2, 3, 4, ... N restaurants and the Poisson Distribution uses a series of terms to represent each of these integer outcomes. The terms include a common element $(e^{-\lambda})$ where the constant e represents the base of natural logarithms (2.7182818) and λ is the average number of outcomes per unit. This equals 0.611 in this example (22/36). The lower part or denominators of the terms in the Poisson Distribution are the factorials of the integer number of outcomes that are of interest ($3! = 3 \times 2 \times 1 = 6$). These integers have been limited to 0 through 5 in the calculations in Box 5.7c.

The probability calculated for each term is multiplied by the total number of outcomes restaurants to produce the expected frequency distribution if Burger King outlets were distributed at random across the 36 grid squares. These are shown to 1 decimal place below but can easily be rounded to facilitate comparison with the empirical results. The Poisson Distribution indicates there should be 20 squares with zero Burger King restaurants and 12 with one, whereas the empirical survey counted 21 and 10, respectively. Looking further at the frequency distribution suggests that two squares with three outlets is a rather unexpected outcome given that the Poisson probability indicates there should be one.

Box 5.7c Calculation of Poisson probabilities.

Burger King	Count of squares	Count of B Kings	Poisson term		*P*	Est. count
0	21	0	$e^{-\lambda}$	$2.718^{-0.611}$	0.54	19.5
1	10	10	$\lambda e^{-\lambda}$	$(0.611)2.718^{-0.611}$	0.332	11.9
2	3	6	$\dfrac{\lambda^2 e^{-\lambda}}{2!}$	$\dfrac{(0.611^2)2.718^{-0.611}}{2}$	0.101	3.7
3	2	6	$\dfrac{\lambda^3 e^{-\lambda}}{3!}$	$\dfrac{(0.611^3)2.718^{-0.611}}{6}$	0.021	0.7
4	0	0	$\dfrac{\lambda^3 e^{-\lambda}}{4!}$	$\dfrac{(0.611^4)2.718^{-0.611}}{24}$	0.003	0.1
5		0	$\dfrac{\lambda^3 e^{-\lambda}}{5!}$	$\dfrac{(0.611^5)2.718^{-0.611}}{144}$	0.000	0.0
	36	22			1.000	36.0

5.4.3 Normal Distribution

The normal distribution, also referred to as the Gaussian distribution, was first specified by the German mathematician Carl Gauss. Unlike the previous Binomial and Poisson Distributions, it is used where variables are measured on the interval or ratio scales. We have already seen how a frequency distribution can be produced by counting the number of discrete dichotomous or integer outcomes from a series of events. Each additional occurrence increases the number of outcomes in a particular category by one unit each time. However, it is not immediately obvious how each measurement of a variable on the interval or ratio scale, for example the concentration of chemical pollutants in water or the minutes spent by a person commuting to work, creates a frequency distribution. At one extreme, the pollutant's concentration for each water sample or the travel time for each commuter could be a different value. Conversely, they could all be the same in both situations. Neither of these extreme situations is likely to occur. Each additional measurement might be the same as one already recorded, very similar or dissimilar. The normal distribution is based on the recognition that individual measurements in a set when taken together will tend to cluster around the mean or average rather than occur at the extremes of the range. In other words, while measured variables will normally include some comparatively high and low values, there will be relatively few of them and the majority will tend towards the mean. In its ideal form, the normal distribution traces a symmetrical, 'bell-shaped' curve with a convex peak and tails extending outwards towards plus and minus infinity (Figure 5.3). The points of inflection on the curve where it changes from convex to concave occur at one standard deviation from the mean in either direction (i.e. mean minus 1 standard deviation and mean plus one standard deviation). Since the standard deviation quantifies dispersion in terms of the original measurement scale units of the variable, unlike the variance, these points can be marked on the horizontal axis of X.

Application of the Binomial and Poisson Distributions involved calculating the probabilities associated with discrete outcomes (e.g. the number of grid squares with zero, one, two, three, etc., Burger King restaurants in Box 5.7a). Application of the normal distribution seeks to achieve a similar purpose, namely, to calculate the probability of obtaining a value for the variable X within a certain range. Since values of X are likely to cluster around the mean as a measure of central tendency, the probability of obtaining a value close to the mean will be higher than for one further away. The heights of the normal curve for different values of the variable, known as *ordinates*, represent the frequency of occurrence of different values of the variable, X. While from a theoretical perspective, the symmetrical, bell-shaped form of the normal curve can be created by inserting numerous different values of X into the equation given in Box 5.8, provided that the mean and standard deviation are known, there is usually a more restricted number of X values available when working with empirical data. Researchers rarely need to carry out the computations associated with the normal curve in order to make use of its important properties, although checking if their data values indicate that the variable follows the normal distribution is important. Several of the statistical tests examined in later chapters should only be applied under the assumption that the sampled data values are drawn from a population that follows the normal distribution, although in practice it is often acceptable if the empirical data approximate the normal distribution.

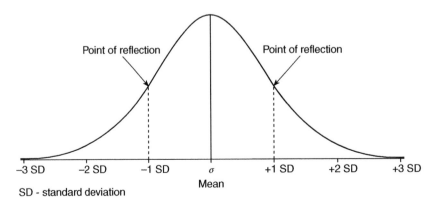

Figure 5.3 The normal distribution curve.

Box 5.8   Normal Distribution.

Normal distribution: $Y = \dfrac{1}{\sqrt{2\pi}\sigma} e^{-\left((X-\mu)^2/2\sigma^2\right)}$

   The normal distribution provides a way of determining the probability of obtaining values of a continuous variable measured on the interval or ratio scale. The equation for the normal distribution calculates the Y frequencies or ordinates of a collection of data values for a variable (X) with a mean (μ) and standard deviation (σ^2), which are known. If the values of the variable follow the normal distribution, plotting the ordinates will produce a symmetrical, normal curve. The equation can be broken down into three groups of terms: the constants π (3.1416...) and e (2.7128...); the mean (μ) and standard deviation (σ^2); and the known data values of the variable (X), for which ordinates are required.

   The mean of the data values in the normal distribution corresponds to the point on the horizontal axis of X values that lies under the central peak of the normal curve. A separate set of values for the same variable might result in a new mean, higher or lower than the first, but which still forms the centre of a symmetrical, bell-shaped normal curve. In other words, the position of the mean on the horizontal axis for a particular measured variable can be thought of as itself variable depending upon the individual data values that have been recorded. Imagine a series of symmetrical curves whose peak lies vertically above the same mean value and that these curves are themselves relatively more or less spread out (dispersed). The mean and standard deviation are thus of crucial importance in determining the normal curve's position and spread on the horizontal axis. This variability in the form of the normal curve produced from different sets of data for the same variable would limit the usefulness of the normal distribution were it not for the possibility of standardising the raw data.

What do we mean by standardisation?
   Percentages are standardised values between 0 and 100: converting raw data into percentages involves multiplying each value by 100 and dividing by the sum of all the data values. Suppose the sum of the raw data values is 350 and one of them is 125, the percentage is 35.7% (125 × 100/350).
   If another value is 45, what is it as a percentage of 350?
   A percentage can be converted into a proportion (values between 0.0 and 1.0) by moving the decimal point two places to the left (35.7% become 0.357).

   Standardisation in relation to the normal distribution expresses the differences between each X value and the overall mean as a proportion of the standard deviation. The results are known as Z values. In some ways, this procedure is like expressing raw data values as a percentage of the total. Percentages are useful for making comparisons between different pieces of information; for example it is more informative to know that there are 15 per cent unemployed working age adults in Area A and 4 per cent in Area B, rather than that Area A has 7500 and Area B has 12,000. It might be implied from the absolute counts that unemployment is more problematic in Area B, whereas the lower percentage indicates it is of less importance.

What do you conclude about the numbers of working age adults in areas A and B from this information?

Standardising the raw data values of the variable X results in positive and negative Z scores, respectively, indicating that the original value was less than or greater than the mean. These Z values are measured in units of the standard deviation, which can be plotted on the horizontal axis. Thus, a Z value of +1.5 indicates that it is one-and-a-half times the standard deviation. Provided the frequency distribution of the original values of X (see Section 5.1 for classification of raw data values to produce frequency distributions) appears to show that they follow a normal curve, the equation for the normal distribution can be simplified by substituting Z for a term on the right-hand side, which allows all other terms involving the standard deviation to equal 1.0 (see Box 5.9a). Box 5.9a shows the calculations involved in producing the standardised normal curve for the sampled water temperature measurements taken in the fluvioglacial stream from Les Bossons Glacier near Chamonix.

Probability statements in respect of the normal distribution focus on the chance (probability) of obtaining certain X values, or their Z score equivalents, within or outside certain ranges or sections of the horizontal axis. The two types of probability statement that can be made are best illustrated by means of a pair of questions:

- What proportion or percentage of X values can be expected to occur with a specified range (e.g. between −1.75 standard deviations less than the mean and +2.0 standard deviations greater than the mean)?
- What is the probability that any one X value or Z score will fall within a certain range (e.g. greater than +1.75 standard deviations from the mean)?.

Such questions are phrased a little differently to those relating to integer attributes, which are concerned with the probability of the frequency of discrete binary (dichotomous) or integer outcomes: for example, is the probability of 83 people answering Yes to a question in a questionnaire survey with 125 respondents more or less than would be expected if they answered randomly? Although worded differently, questions about probabilities associated with the normal distribution share the same basic purpose of trying to discover the probability of getting the result or outcome

---

Box 5.9a *Z* **scores and the standardised normal distribution.**

Z scores: $Z = \dfrac{X - \mu}{\sigma^2}$; *Z* standardised normal distribution: $Y = \dfrac{1}{\sqrt{2\pi}} e^{-(Z^2/2)}$

There are two steps involved in producing a standardised normal curve: first, standardisation of X values to Z scores and then insertion of these scores into the standardised normal distribution equation. Z scores are calculated relatively easily by subtracting each value in turn from the overall mean and dividing by the standard deviation. Calculations for the ordinates (Y values) are a little more complicated but involve the constants π (3.1416...) and e (2.7128...) and the Z scores.

The methods used to calculate Z scores and Y ordinates for the water temperature measurements at the snout of Les Bossons Glacier near Chamonix are shown below in Box 5.9b and the plot of the standardised normal curve for these data is included in Box 5.9c. The data values have been sorted (ordered) in respect of X, which leads to the standardised Z scores going from the lowest to highest value. The Z and Y (ordinate) values are used to plot the curve. Although this does not exactly conform to the classical symmetrical, bell-shaped curve, its shape provides a reasonable approximation.

Box 5.9b Calculations for ordinates of standardised normal curve.

	x (sorted)	$Z = \dfrac{X-\mu}{\sigma^2}$	$Z =$	$Y = \dfrac{1}{\sqrt{2\pi}}e^{-(z^2/2)}$	$Y =$
09.00 hrs	4.10	$\dfrac{4.10-5.66}{0.98}$	-1.67	$\dfrac{1}{\sqrt{2(3.1418)}}2.7128^{-(-1.67^2/2)}$	0.140
09.30 hrs	4.60	$\dfrac{4.60-5.66}{0.98}$	-1.17	$\dfrac{1}{\sqrt{2(3.1418)}}2.7128^{-(-1.17^2/2)}$	0.285
10.00 hrs	4.60	$\dfrac{4.60-5.66}{0.98}$	-1.17	$\dfrac{1}{\sqrt{2(3.1418)}}2.7128^{-(-1.17^2/2)}$	0.285
10.30 hrs	4.70	$\dfrac{4.70-5.66}{0.98}$	-1.07	$\dfrac{1}{\sqrt{2(3.1418)}}2.7128^{-(-1.07^2/2)}$	0.318
11.00 hrs	4.80	$\dfrac{4.80-5.66}{0.98}$	-0.97	$\dfrac{1}{\sqrt{2(3.1418)}}2.7128^{-(-0.97^2/2)}$	0.352
17.30 hrs	5.10	$\dfrac{5.10-5.66}{0.98}$	-0.67	$\dfrac{1}{\sqrt{2(3.1418)}}2.7128^{-(-0.67^2/2)}$	0.451
11.30 hrs	5.20	$\dfrac{5.20-5.66}{0.98}$	-0.57	$\dfrac{1}{\sqrt{2(3.1418)}}2.7128^{-(-0.57^2/2)}$	0.479
17.00 hrs	5.30	$\dfrac{5.30-5.66}{0.98}$	-0.47	$\dfrac{1}{\sqrt{2(3.1418)}}2.7128^{-(-0.47^2/2)}$	0.505
12.00 hrs	5.40	$\dfrac{5.40-5.66}{0.98}$	-0.37	$\dfrac{1}{\sqrt{2(3.1418)}}2.7128^{-(-0.37^2/2)}$	0.527
16.00 hrs	5.50	$\dfrac{5.50-5.66}{0.98}$	-0.27	$\dfrac{1}{\sqrt{2(3.1418)}}2.7128^{-(-0.27^2/2)}$	0.544
16.30 hrs	5.50	$\dfrac{5.50-5.66}{0.98}$	-0.27	$\dfrac{1}{\sqrt{2(3.1418)}}2.7128^{-(-0.27^2/2)}$	0.544
15.30 hrs	6.10	$\dfrac{6.10-5.66}{0.98}$	0.33	$\dfrac{1}{\sqrt{2(3.1418)}}2.7128^{-(0.33^2/2)}$	0.535
12.30 hrs	6.30	$\dfrac{6.30-5.66}{0.98}$	0.53	$\dfrac{1}{\sqrt{2(3.1418)}}2.7128^{-(0.53^2/2)}$	0.491
15.00 hrs	6.30	$\dfrac{6.30-5.66}{0.98}$	0.53	$\dfrac{1}{\sqrt{2(3.1418)}}2.7128^{-(0.53^2/2)}$	0.491
13.00 hrs	6.90	$\dfrac{6.90-5.66}{0.98}$	1.13	$\dfrac{1}{\sqrt{2(3.1418)}}2.7128^{-(1.13^2/2)}$	0.299
14.30 hrs	6.90	$\dfrac{6.90-5.66}{0.98}$	1.13	$\dfrac{1}{\sqrt{2(3.1418)}}2.7128^{(1.13^7/2)}$	0.299
14.00 hrs	7.10	$\dfrac{7.10-5.66}{0.98}$	1.33	$\dfrac{1}{\sqrt{2(3.1418)}}2.7128^{-(1.33^2/2)}$	0.234
13.30 hrs	7.40	$\dfrac{7.40-5.66}{0.98}$	1.63	$\dfrac{1}{\sqrt{2(3.1418)}}2.7128^{-(1.63^2/2)}$	0.150

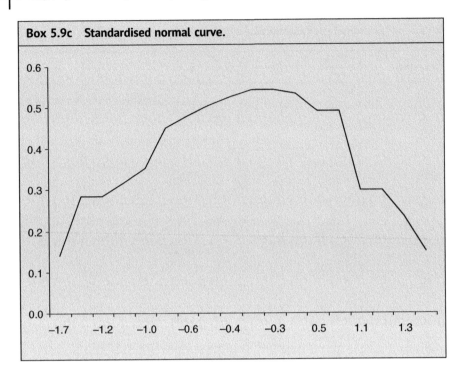

Box 5.9c   Standardised normal curve.

that has been obtained from your empirical data. For example, your field measurements show the cliffs had retreated from your sampling points by a mean of 2.47 m in the previous 12 months, whereas another student the previous year had recorded a mean of 2.15 m over an equivalent period. Is the difference between the two figures more or less than would be expected to occur at random or indicative increased coastal erosion.

Traditionally, the normal distribution probabilities associated with both types of question were obtained by consulting a table of probabilities for different values of Z. These tables, which have appeared as appendices in numerous texts on statistical techniques, can be presented in slightly different formats, but their essential purpose is to link the proportions of the area under the normal curve with different Z score values. These proportions provide the probabilities because the area under the normal is treated as finite, despite the tails of the distribution approaching but never reaching the horizontal axis at each extreme. The area under the curve is assigned an arbitrary value of 1.0 representing the total probability of all values. It follows from this that a vertical line drawn from the peak of the curve to the point on the horizontal axis representing the mean, where the Z score equals 0.0, bisects the area into two equal portions each containing 0.5 or 50 per cent of all values. The example in Figure 5.4 will help to show how this works. The shaded area in Figure 5.4 is composed of two parts, one on each side of the central vertical line: the part on the left goes from −1.75 to 0.0 and the one on the right from 0.0 to +1.75 on the Z score scale. The proportions of the total area under the curve associated with these two sections from a Z table are both 0.4199 and 0.4199. Summing these proportions indicates that the whole of the shaded area accounts for 0.8398, which indicates generically how to answer the first type of question above: in this case, 83.98 per cent of all values occur within the range between the specified Z scores. This also implies that there is 0.8398 probability of any one value lying within this range. Thus, we have answered the second type of question. Suppose interest was focused beyond this Z value range

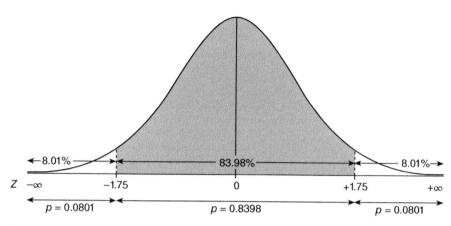

Figure 5.4 Probabilities and the standardised normal curve.

towards the negative and positive ends of the distribution. Since the area under the curve represents 100 per cent of values, the unshaded zone contains (100.00–83.98) per cent of values and there is 0.1602 (16.02 per cent) probability of any one value occurring in this region.

The standardised normal curve for the worked example in Box 5.9c approximates to the ideal symmetrical bell-shaped form, although the match is far from perfect. There are some geographical and geological variables, which tend to follow the normal distribution (e.g. administrative area population totals and river basin discharge), but there are many others, which are manifestly non-normal when viewing their frequency distributions. This problem can be addressed by a process of **normalisation,** which involves transforming or re-calibrating the raw data values. Variables having a strong positively skewed distribution (i.e. a high proportion of values at the lower end of the range and few large values) may be transformed so that they display a more normal shape by transforming into logarithms provided there are no negative or zero values. This is known as the log-normal transformation. Conversely, variables with strong negatively skewed distributions may be normalised by squaring the raw data values. One of the problems with these transformations is that interpretation of analytical results may be difficult because the variables are no longer recorded in a way that relates to the phenomena under investigation.

5.5   Inference and Hypotheses

Previous chapters have already referred to answering **research questions** as a way of describing what researchers are trying to do when carrying out an investigation. Such questions may be theoretical or empirical in their origin based on:

- thinking about why things occur and how they are connected in the world around us;
- seeking to discover some new 'factual' information about people, places and physical environments.

The difference between the two approaches is very similar to the distinction made between *a priori* and *a posteriori* probability examined earlier in this chapter, where we saw that the probability of the head side of a coin lying uppermost when it was tossed could be discovered in two ways: by thinking about the properties of the coin, namely that it has two flat surfaces one of which will land uppermost; or by carrying out a number of coin tossing experiments or trials tossing and recording the number of times each outcome occurred. The aim of our research, as students of Geography, Earth Science and related subjects, is to investigate phenomena from a global to a local scale and to establish whether our findings are reliable and how different phenomena relate to each other.

Irrespective of whether research questions originate from theoretical reflection or empirical observation, the process of undertaking an investigation in a rigorous and scientific manner requires that we specify the concepts relating to the phenomena of interest to us. Sustainability is currently a 'hot topic', but what does this phenomenon mean? For many years, student essays on the subject have invariably defined the term by quoting from the Brundtland Commission Report of the United Nations (United Nations 1993), which asserted that 'Sustainable development is development that meets the needs of the present without compromising the ability of future generations to meet their own needs'. But how can we tell whether actions undertaken at the present time are going to have a damaging impact on future generations? And even if society avoided taking such actions, what is to stop some other unforeseen event occurring that will have a harmful effect in the future? Who would have thought in the summer of 2019 that economies and societies would be disrupted by the global COVID-19 pandemic and that as of the 08.20 hr Central European Summer Time on 19 July 2023, 6,951,677 deaths and 768,237,788 confirmed cases of COVID-19 would have been reported to the World Health Organisation (WHO, 2023)?

The sustainable development example illustrates the importance of translating definitions of concepts into practical and precise terms that can be incorporated into hypotheses capable of being tested statistically. Careful specification of these terms allows us to identify the attributes and variables that need to be measured or quantified. The purpose of this process is to make some inference about what might happen or why something has happened. In other words, we are trying to reach some conclusion about the phenomena of interest. Recall that Chapter 2 suggested what is recorded or measured and how these recordings and measurements are achieved will inevitably affect the outcome of a piece of research. Vague definition and inconsistent application of concepts will often produce results that are difficult or impossible to interpret. It sometimes appears that conclusions from research are common sense, if this is the case from 'good' research; surely, they are nonsense from 'bad' research.

Most inferential statistical tests follow a standard procedure that involves the investigator specifying two diametrically opposed hypotheses. These are called the Null Hypothesis and the Alternative Hypothesis, which are represented by the symbols H_o and H_1. Most quantitative research investigations involve specifying several pairs of Null and Alternative Hypotheses to accumulate a series of answers to the research question(s). Some multi-stage investigations will include linked sets of hypotheses, so that the decisions taken with respect to one set will influence the direction of the research and the hypotheses tested at a later stage. An investigation into whether the present actions of a company or other organisation are sustainable may look at several distinct types of action. For example, it could examine recycling of waste materials (paper, plastic, computer consumables, etc.), type of fuel used in company vehicles (petrol, diesel, electricity, biofuel, etc.) and use of electricity in buildings (e.g. energy efficient lighting and equipment left on standby). The decisions taken in respect of Null and Alternative Hypotheses relating to this first stage of the research are likely to dictate which pairs of hypotheses will be appropriate at the second stage, and so on to a third stage and beyond.

The exact wording of any pair of Null and Alternative Hypotheses will obviously depend on the research question(s) being investigated; nevertheless, it is possible to offer some guidance on their generic format. In most cases, the focus of these hypotheses can be simplified to consideration of a quantifiable difference in the values of a statistical measure between a population and a sample, or between two or more samples to indicate whether they might have come from the same population. The statistical measure may be one of those already encountered (mode, median, mean, standard deviation, etc.) or those yet to be examined (e.g. correlation and regression coefficients). The wording of the Null Hypothesis is deliberately cautious in order to give rigour to the process and to reduce the opportunity for reaching inappropriate conclusions. Caution is exercised by always assuming at the outset that the Null Hypothesis is correct, which maximises the opportunity for accepting it and minimises the risk of incorrectly rejecting it. The penalty for reaching such an erroneous decision is rarely life-threatening in geographical, geological and environmental research, but this may not be the case in other fields of research. When the COVID-19 vaccine became available, the public were informed of very rare side effects, what this means in terms of statistics is that the probability of someone having such a side effect is extremely small. It may help to think of the Null Hypothesis as being put on trial and the test providing evidence enabling the jury (you the researcher) to reach a decision on its acceptance or rejection.

Continuing with the example of investigating the sustainability of an organisation's actions, the following pair of hypotheses might apply:

H_o The amount of material recycled by organisations that have introduced a policy promoting recycling during the last year is the same as in those that have not taken this action. Any observed difference is the result of chance and is not significant.

H_1 The observed difference in the quantity of material recycled is significantly greater in organisations that have incorporated recycling as part of company policy compared with those that have not done so.

Several different variables could be used to measure the quantity of recycling, for example tonnes of recycled materials per employee, per 100 m^2 floor space or per £1000 of turnover. A sample survey of organisations stratified according to whether they had introduced a policy to increase recycling might be carried out from which the mean (average) tonnage of recycled materials per employee, per 100 m^2 of floor space or per £1000 of turnover was calculated. A suitable test could be applied to reach a decision on whether the difference between the two mean values was significant or not. If the test suggested that Null Hypothesis should be accepted, the recycling policy seems to have been a waste of time, since it produced no significant gain in the number of recycled materials. It may have had other benefits, but these have not been assessed. This hypothesis is deliberately cautious because it avoids reaching the conclusion that introducing such a policy will inevitably lead to an increase in recycling, although acceptance of the Alternative Hypothesis, which suggests that the two things are linked, does not confirm that this connection necessarily exists.

The decision about whether to accept the Null or the Alternative Hypothesis enables an investigator to *infer* or tease out a conclusion in relation to the research question(s). Hence, the term inferential statistics is used to describe those techniques enabling researchers to infer conclusions from their data. All statistical tests involve reaching a decision in relation to the Null and Alternative Hypotheses: if H_o is accepted, then H_1 is necessarily rejected, and *vice versa*. They are mutually exclusive, and there is no 'third way'. The decision on which of these hypotheses to accept is made by reference to the probability associated with a test statistic calculated from the empirical data. The value of the test statistic depends on the number of observations (outcomes to use previous terminology) and the data values. There are many different statistical tests available that are often named using a letter, such as the Z, t and χ^2 tests, although some are also identified by reference to the statistician who first devised it, for example the Mann–Whitney U test.

The main reason for using inferential statistical tests is that some degree of uncertainty exists over the data values obtained for a sample of observations and we cannot be sure of the extent to which this **sampling error** is present. There are several reasons for sampling error to occur and it can be thought of as the sum of the error contained in a set of data that is present because the values have been measured for a sample rather than for the entire population of observations. Chapter 2 explores the reasons why most investigations are based on one or more samples rather than a population. If there is a large amount of sampling error present in a dataset, for example such that the mean or standard deviation is much larger or smaller than in the parent population, then we might make erroneous decisions about the Null and Alternative Hypotheses. In contrast, if there is little sampling error, the key features of the data are likely to remain and valid inferences may still be reached. We need to know how much sampling error is present, but as we cannot answer this question for certain all that can be done is to assume there is a random amount present. Our examination of the binomial, Poisson and normal distributions has shown that they provide the probability of a random outcome occurring in relation to dichotomous and ordinal integer and continuous outcomes. Comparing what they and other similar distributions predict and what our data contain therefore provides a means of discovering if the amount of sampling error present is random or otherwise.

Given that the probability of having obtained a particular value for the test statistic indicates whether or not to accept the Null Hypothesis, you need a criterion that enables you to reach a decision on acceptance or rejection. Having a standard criterion also means that any investigator analysing the attributes and variables in each dataset to answer a research question would reach the same conclusion. When students on a statistical techniques module are all given the same dataset to analyse they should, in theory at least, independently make the same decision about the Null and Alternative Hypotheses when applying a particular statistical test to a variable. This agreement is achieved by using a standard **level of significance** for statistical tests so that different people following the same standard would reach the same decision about the Null and Alternative Hypotheses. The most common level of significance is a probability of 0.05 or 5 per cent. This means that there is a 5 in a 100 chance of having obtained the test statistic because of chance or randomness. Some investigations might require a more stringent criterion to be applied, for example 0.01, 0.001 or 0.0001. Again, there is a parallel with the judge and jury system for trying criminal cases in British courts. If the case relates to a particularly serious crime, for example murder or rape, the judge might direct that the jury should return a verdict on which they are all agreed (i.e. unanimous), whereas in lesser cases, the judge might direct that a majority verdict (e.g. 10 for guilty and 2 for not guilty) would be acceptable.

> What are the probabilities 0.01, 0.001 and 0.001 expressed as percentages? How are they written if reported as a ratio?

Thus far, statistical testing has been discussed from the perspective of investigating whether different samples might have come from the same parent population or whether one sample is significantly different from its parent population. However, these questions can be reversed or 'turned on their head' to determine whether the values of statistical quantities (e.g. measures of central tendency and dispersion) calculated from sample data provide a reliable estimate of the corresponding figure in the parent population. The characteristics of a parent population are often difficult to determine for the reasons outlined in Chapter 3, and in these circumstances, the researcher's aim is often to use a sample as a surrogate for the population. Since it is not possible to be certain about whether statistics calculated from a sample are the same as would have been obtained if the population of observations had been analysed, the results are expressed within **confidence limits.** These limits constitute a zone or range within which an investigator can be 95, 99, 99.9 per cent or even more confident that the true, but unknown **population parameter** corresponding to the **sample statistic** lies. There will remain a small chance, in the latter case 0.01 per cent, that the population parameter falls outside this range, but this may be regarded as too small to cause any concern. Again, reflecting on the analogy with a court of law, it is not feasible to ask several juries to try the same case simultaneously and for the judge to take the average outcome so it may be sufficient to accept a majority verdict in certain less serious types of case but not in those where a narrower margin of error is more appropriate.

5.6 Connecting Summary Measures, Frequency Distributions and Probability

This chapter has explored three related aspects of statistical analysis. The presentation of attributes and variables in the form of frequency distributions provides a convenient way of visualising the pattern formed by a given set of data values in tabular and graphical formats. However, the form and shape of these distributions can be influenced by how the raw data values are categorised. This can relate to the classification schemes used for continuous and nominal data values, which may be re-defined in diverse ways. Frequency distributions not only complement single value descriptive statistical measures as a way of summarising attributes and variables (see Chapter 4), but these measures can also be obtained or estimated for such distributions.

The juxtaposition of discussion about frequency and probability distributions in this chapter reflects the important connection between describing the characteristics of a set of data values and the likelihood (probability) of having obtained those values during empirical data collection that includes an unknown amount of sampling error. Some other probability distributions will be introduced later; however, the binomial, Poisson and normal examined here illustrate the role of probability in statistical tests and how to discover the probabilities associated with random outcomes from diverse types of events. Finally, the principles of hypothesis testing and the general procedure for carrying out statistical tests have been examined. The Null and Alternative Hypotheses provide a common framework for testing whether data collected in respect of samples help to answer research questions that are phrased in diverse ways. Researchers will rarely have access to data for all members of a population and inferential statistics offer a way of confidently reaching conclusions about a parent population from a sample of observations. All statistical tests make assumptions about the characteristics of sample data, for example that observations have been selected randomly from the population or that variables follow the normal distribution. The reason for making these assumptions is to ensure that the probability of a calculated test statistic can be determined by reference to the known probabilities of the corresponding distribution so that by reference to the chosen level of significance a standard decision on acceptance or rejection of the Null Hypothesis can be reached.

There is a wide range of statistical tests available. The choice of test to use in different circumstances depends on the question being asked, the characteristics of the variable or attribute (i.e. the measurement scale employed), the number of observations, the degree of randomness in selecting the sample, the size of the sample and whether the data can be assumed to follow one of the standard probability distributions. If the more stringent assumptions demanded by certain tests are believed not to apply, for example in relation to complete random selection of sampled observations and normality, then other less demanding tests may be available to answer the same type of question or if the sample is sufficiently large certain assumptions may be relaxed. Switching between one test and another in this way might require an investigator to convert variables into attributes, for example by categorising a continuous variable into discrete, unambiguous classes. A simple, check list style guide to selecting the appropriate statistical test is provided between this chapter and the next, which examines the details of different tests. This takes the form of a checklist that students can use to help with examining the characteristics of their data so that you can choose the correct type of test. A project dataset will often include a combination of nominal, ordinal and interval/ratio attributes and variables and the analysis focuses on considering one in relation to another. For example, whether men and women fear distinct types of crime when in a shopping centre, whether arable or pastoral farming produces a greater build-up of nitrates in rivers and streams and whether organisations with or without a recycling policy are more sustainable. It is important that the check list is treated as a starting point for planning the data collection and statistical analysis in your research. All too often investigators fired with enthusiasm for their topic will rush into the data collection phase without thinking ahead to how the data collected will later be analysed and how they will be used to answer the research questions.

Each statistical test should be applied to address a specific type of question, for example to determine whether two samples of the same type of observation from different locations (e.g. soil samples from the side slopes and floor of a valley) are significantly different from each other in respect of the variable(s) under investigation (e.g. pH, particle size, and organic content). Nevertheless, there is a common sequence of stages or tasks involved in applying most tests as shown in Box 5.10. Most of the statistical tests examined in the next two chapters follow this sequence and any exceptions will be highlighted.

Box 5.10   Common sequence of stages in application of statistical tests.

- State Null and Alternative Hypotheses.
- Select the appropriate type of statistical test.
- State level of significance to be applied.
- Calculate appropriate test statistic (usually obtained from statistical or spreadsheet software).
- Decide whether there is any certain, undisputed reason for believing that the difference being tested should be in one direction or the other (e.g. whether one sample should necessarily have a greater mean than another); known as a one-tailed test.
- Determine the probability associated with obtaining the calculated value of the test statistic or a larger value through chance (usually provided by statistical software, but not necessarily by spreadsheet programmes).
- Decide whether to accept or reject the Null Hypothesis. (Note: if Null Hypothesis is rejected, Alternative Hypothesis is necessarily accepted).

References

United Nations (1993) Our Common Future: Brundtland Report, http://www.worldinbalance.net/intagreements/1987-brundtland.php.

World Health Organisation (2023) WHO Coronavirus (COVID-19) Dashboard, https://covid19.who.int/ (accessed 21/07/23).

Further Reading

Diamond, I. and Jeffries, J. (1999) *Introduction to Quantitative Methods*, London, Sage.

Ebdon, D. (1984) *Statistics in Geography*, 2nd edn, Oxford, Blackwell.

Harris, R. (2016) *Quantitative Geography: the Basics*, Sage Publications Ltd.

Rogerson, P.A. (2006) *Statistical Methods for Geographers: A Student's Guide*, Los Angeles, Sage.

Statistical Analysis Planner and Checklist

The selection of appropriate techniques to use when carrying out research is not an easy task, and there are many pitfalls for novice and even experienced researchers. Deciding what analysis to carry out involves delving into a large and possibly confusing collection of statistical techniques. Nowadays, there is a range of proprietary software, such as Minitab, SAS, SPSS and MS Excel, as well as freely available shareware (e.g. R) and websites on the internet that enable researchers to apply various techniques. Most of these resources include useful help information and have made the process of producing statistical results and getting an answer relatively easy. However, these resources rarely provide an overview of how to plan your statistical analysis so that you can choose the techniques that best fit the aim of your research.

This short interlude between the previous five introductory chapters and the following six explanatory chapters is intended to help you find your way through this complex collection of statistical techniques and to identify those that could be used in your research. Over the next two pairs of pages, the entry-level techniques covered in the following chapters have been divided into two broad groups: those dealing with hypothesis testing and those exploring associations and relationships. There are diagrams on these pages that summarise the connections between the techniques and how decisions about the number of samples and whether you want to analyse spatial patterns lead you towards applying different statistical techniques. The idea is that by considering your answers to a short checklist of questions, which is included afterwards, you will be able to follow a route through these diagrams to discover which techniques you should consider using. Referring to the index, you can then find explanations and worked examples of these techniques being used in Geography, Earth and Environmental Sciences.

Number of samples **Spatial / nonspatial**

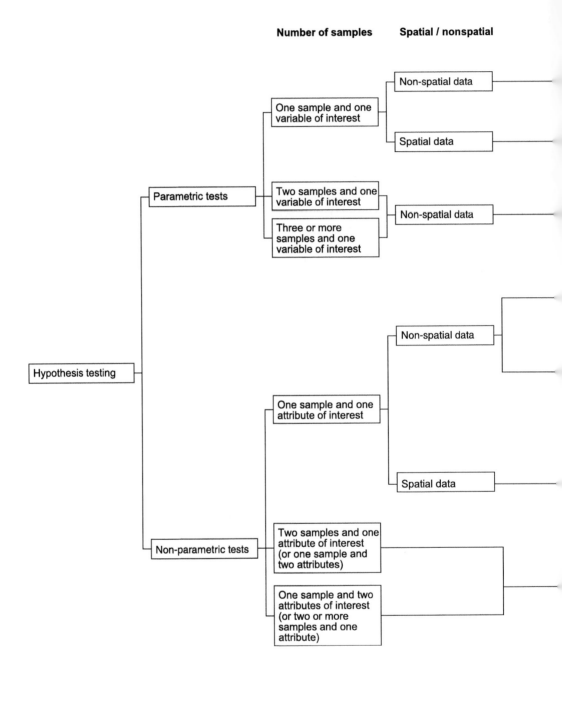

Scale of measurement   **Statistical procedure**

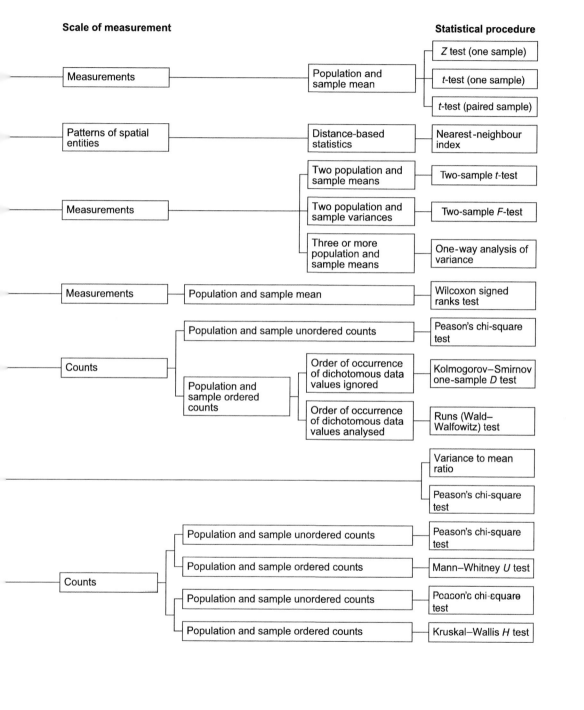

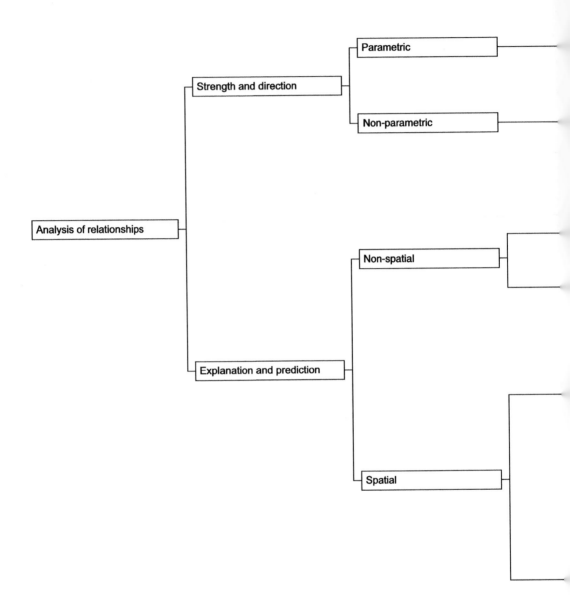

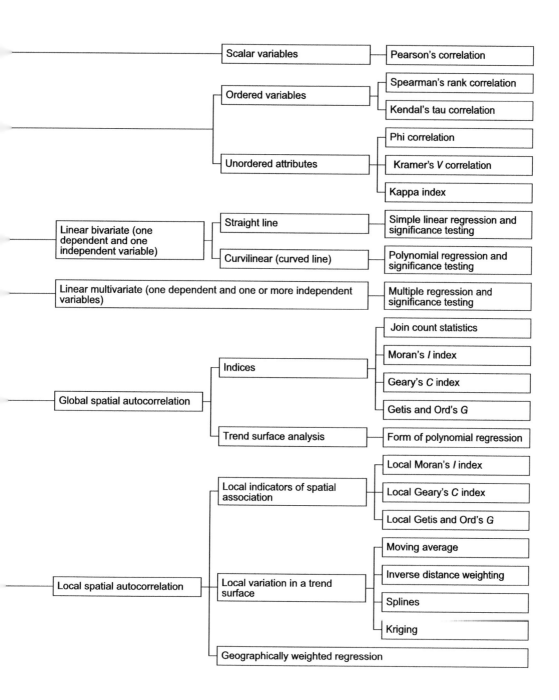

Scalar variables		Pearson's correlation
Ordered variables		Spearman's rank correlation
		Kendal's tau correlation
Unordered attributes		Phi correlation
		Kramer's *V* correlation
		Kappa index
Linear bivariate (one dependent and one independent variable)	Straight line	Simple linear regression and significance testing
	Curvilinear (curved line)	Polynomial regression and significance testing
Linear multivariate (one dependent and one or more independent variables)		Multiple regression and significance testing
Global spatial autocorrelation	Indices	Join count statistics
		Moran's *I* index
		Geary's *C* index
		Getis and Ord's *G*
	Trend surface analysis	Form of polynomial regression
Local spatial autocorrelation	Local indicators of spatial association	Local Moran's *I* index
		Local Geary's *C* index
		Local Getis and Ord's *G*
	Local variation in a trend surface	Moving average
		Inverse distance weighting
		Splines
		Kriging
	Geographically weighted regression	

Checklist of questions for planning statistical analysis

What is the objective of the statistical analysis?	a) To test hypotheses and/or estimate population parameters for attributes/variables b) To explore relationships between attributes and/or variables of interest
Is the intention to estimate population parameters with a certain degree of confidence and/or compare sample statistics with population parameters for normally distributed variables?	a) Yes – use parametric techniques b) No – use non-parametric techniques
How many samples provide the attributes and/or variables of interest? Are the samples related to each other (e.g. samples of males and females in the same household) or completely separate (e.g. hospitality businesses in five separate seaside towns)? Remember you may select subsamples from your original complete dataset.	a) 1 b) 2 c) 3 or more
Is the aim of the statistical analysis to examine the spatial patterns and location of phenomena? Are you interested in the spatial pattern of the features, or do you also want to investigate the spatial variation in the measured variable?	a) Yes b) No
What measurement scale has been used for the attributes and/or variables of interest? Do you want mix analysis of variables on separate scales?	a) Nominal (counts) b) Ordinal (counts or numerical measurements) c) Scalar (ratio and interval) (numerical measurements) d) Mixture
What is the purpose of exploring the relationship between attributes and/or variables of interest?	a) To determine its strength and direction b) To explain the relationship and make predictions

Planning the statistical analysis of data for a research project often involves carrying out a number of different types of analyses, and the results obtained from one type may mean that you decide to do some more analysis of your data. This means that you might need to cycle through these questions a number of times in preparation for different stages of your research. However, what is almost impossible is to return to the same set of observations under the same circumstances and collect some more data. This is why planning data collection carefully in the first place is so important.

Section III

Testing Times

6

Parametric Tests

Parametric tests comprise one of the main groups of statistical techniques available to students of Geography, Earth and Environmental Science and related disciplines. The choice of test to use for any specific piece of statistical analysis depends on the interplay between three issues: how many samples are being examined, the number of variables and attributes being tested and whether measures of central tendency or dispersion are the focus of attention. The chapter concludes by computing confidence limits to estimate population parameters from sample data as a way of indicating how 'good' data from a sample data are at providing information about its parent population.

Learning Outcomes

This chapter will enable readers to:

- Apply parametric statistical tests to variables measured on the ratio or interval scales;
- Reach decisions on Null and Alternative Hypotheses according to appropriate levels of significance;
- Undertake parametric procedures with non-spatial data;
- Plan statistical analyses using parametric tests relating to spatial and non-spatial data as part of an independent research investigation in Geography, Earth Science and related disciplines.

6.1 Introduction to Parametric Tests

Students can be introduced to the array of statistical techniques that are available for use in their projects and investigations in several different ways. The choice involves navigating a path through the many and varied tests and other techniques in a way that leads students to an endpoint where they feel confident about which statistical procedures to use in their own research projects and to know enough about the core techniques that they can understand the reporting of quantitative analysis in journal articles and books. Having specified that this chapter will concentrate on parametric tests and the next on non-parametric ones, the question remains of how to subdivide each of these two major groups into 'bite size chunks'. The approach adopted here is to group the tests according to whether one, two or three or more variables are under scrutiny and within these sections to examine procedures separately that compare means and variances or standard deviations.

Practical Statistics for Geographers and Earth Scientists, Second Edition. Nigel Walford.
© 2025 John Wiley & Sons Ltd. Published 2025 by John Wiley & Sons Ltd.
Companion website: www.wiley.com/go/PracticalStatistics2e

One key assumption of parametric statistical tests is that the variables under analysis follow the normal distribution in the parent population. This assumption of normality in the parent population means that the values of a variable display the same, or at least very nearly the same, frequency distribution and therefore probabilities are associated with the normal curve (see Chapter 5). The general purpose of testing a set of values for a variable in a sample in these circumstances is to determine whether their distribution and the descriptive statistics (notably the mean, variance and standard deviation) calculated from them are significantly different from the normally distributed parent population. Parametric tests are only applicable to continuous variables measured on the interval or ratio scales. However, many datasets of variables collected for a particular project are likely to include some that follow the normal distribution and others that do not. The analysis will include variables from both groups and therefore involve a combination of parametric and nonparametric techniques. Variables that are normally distributed may need to be combined with others that fail to meet this criterion or with attributes describing characteristics measured on the nominal scale. Ways of classifying the data values of continuous variables into groups or categories so that they can be analysed alongside other types of data were examined in Chapter 5.

The focus of this chapter is on different parametric statistical tests that progress through different situations according to the number of samples under investigation (one, two or three plus samples). Table 6.1 provides a summary of the types of comparison examined by the statistical tests in the subsections of this chapter. For example, the next section examines the Z and t parametric tests as applied to a single variable measured with respect to one sample of observations selected from a population. You might wonder why cells in the columns headed 'two variables' and 'three variables' are empty. Why might it not be appropriate, for example, to test the means of two, three or more variables for a single sample with respect to their corresponding parameters in the same parent population? There is no reason at all, but this would not imply that they were tested together but rather that they were examined concurrently or sequentially. What for example would be the point of comparing the mean length of commuter journey with a distance between home and workplace. The data values for one variable would be in hours/minutes and the other kilometres. Each of the procedures covered in this chapter relates to examining a single variable, although in any project the same techniques will usually be applied to more than one variable. Joint analysis of two, three or more variables only makes any sense if the research questions and hypotheses are about whether the variables are related to or associated with each other. In other words, whether the pairs of values display a pattern in the way they vary along their respective measurement scales. A separate set of statistical techniques are required to address such questions which are explored in Chapters 8 and 9.

Table 6.1 Subsections in Chapter 6 with parametric statistical tests for one, two and three or more samples.

	One variable	Two variables	Three variables
One sample	Section 6.2 6.2.1 comparison of sample and population mean 6.2.2 comparison of differences between pairs of values for a sample measured twice or divided into two parts		
Two samples	Section 6.3 6.3.1 comparison of sample and population means 6.3.2 comparison of sample and population variances		
Three or more samples	Section 6.4 6.4.1 comparison of sample and population means		

This introductory section started with a question: how to help students navigate through the spread of statistical techniques used in quantitative analysis. A supplementary question in relation to students of Geography, Earth and Environmental Sciences concerns how to introduce the group of techniques known as spatial statistics. Should these be relegated (or elevated) to a separate chapter or included alongside those similar quantitative but nonetheless fundamentally different procedures for statistically analysing non-spatial data? The latter option suggests that although spatial statistics have some distinctive features their essential purpose of enabling researchers to determine whether spatial patterns are likely to have resulted from random or non-random processes is like the purpose of non-spatial techniques. Spatial statistics treat the location of the phenomena under investigation, defined in terms of their position within a regular grid of numerical coordinates, as a variable alongside other thematic attributes and variables (see Chapter 2). Additional variables can be computed from the quantified location of entities such as their distance apart, mean density, sinuosity, length, shape and connectivity. Examination of spatial statistics with reference to these additional variables either on their own or in conjunction with thematic variables and attributes occurs in Section V (Explicitly Spatial). The remainder of this chapter, as indicated in Table 6.1, covers non-spatial, parametric statistical tests relating to one, two or three or more samples.

6.2 One Variable and One Sample

It would be very unusual for a project dataset to contain only one variable, although a single sample of observations is quite common, especially in an investigation where time and other resources are limited. Our exploration of parametric tests therefore starts by looking at the relatively straightforward situation of asking whether the data values of one continuous interval or ratio scale variable measured for a sample of observations are representative of their parent population, whose values are assumed to follow the normal probability distribution. Even when data for several variables have been collected and the aim is to examine relationships between them, exploring one variable at a time is often a useful starting point. The differences under scrutiny in this situation are between the descriptive statistics (mean, variance and standard deviation) for the sample and the equivalent known or hypothesised quantities for the parent population. We have already seen that the statistical characteristics of a population, its mean, variance, standard deviation, etc. are often difficult or impossible to determine and so it is necessary to hypothesise what these might be by reference to theory. This is similar to determining the *a priori* probabilities associated with tossing a coin, throwing a die or some other similar event considered in Chapter 5, which indicate the theoretical mean, variance and standard deviation of a series of trials (coin tosses, die throws, etc.). Thus, if a die was thrown 120 times, we would expect that each face would land facing upwards 20 times (i.e. 20 ones, 20 twos, 20 threes, 20 fours, 20 fives and 20 sixes) giving a mean of 3.5 dots (420/120) and their variance and standard deviation would be 2.94 and 1.71 respectively. These situations deal with discrete or integer outcomes (i.e. 1, 2, 3, 4, 5 or 6 dots). When dealing with continuous variables it is sometimes more difficult to theorise about what the population parameter might be. However, it may be possible to hypothesise about the direction of change: for example, the effect of irradiation from the sun during daytime can be expected to increase ambient temperature and result in some melting of glacial ice leading to an increase in the flow rate of a melt water stream. It would therefore be possible to hypothesise that water temperature would be higher in the afternoon than in the early morning.

Some investigations divide a single sample into two discrete parts, for example research on new treatments for medical conditions often divides a sample of people into two groups: those receiving the new drug and those receiving a placebo (control group). This division of one sample is not the same as selecting two different samples and comparing them, since the basis of the Null and Alternative Hypotheses is that the sampled phenomena are divided into two parts or measured twice to test if the new drug has or has not had a quantifiable beneficial effect. A similar division of a sample may also be useful in geographical research. A random sample survey of **all** individuals in a town might divide the respondents into males and females when examining attitudes towards a proposed new shopping centre. A random sample of **all** occurrences of a particular plant species might be divided into those above or below 250 m altitude. The point is that the whole sample was selected at random and then divided into two parts, which is different from random sampling two sets of observations. A single sample can also be measured before and after an event, for example soil samples might be weighed before and after heating to a certain temperature to determine if there is a significant change in the quantity of organic matter present.

6.2.1 Comparing a Sample Mean with a Population Mean

Data values for a sample of observations always include an unknown quantity of sampling error, which can cause the sample mean to under- or over-estimate the true population mean. When carrying out an investigation using simple random sampling with replacement (see Chapter 2), the sample of observations selected is but one of a potentially infinite number of samples that could have been obtained. If there was sufficient time and resources available, and the characteristics of interest of the observations were stable over time, many random samples could be selected and the means of any one variable (let us call it *X*) could be calculated. The individual sample means of *X* are likely to be a little different from each other, but they could be added together and divided by the number of samples to estimate the mean of *X* in the population. This value is known as the **grand mean** or the 'mean of the means'. In theory, the more sample means are added to this calculation, the closer the grand mean comes to the true population mean. Nevertheless, the individual means are still likely to be different from each other and there will be some dispersion or spread in their values: it is extremely unlikely they will all be the same. This dispersion can be quantified by a statistic called the **standard error of the mean**, which you can think of as the standard deviation of the sample means. Suppose many samples were selected and the mean of X calculated for each, and they were viewed as a frequency distribution, this would be the **sampling distribution of the mean** and it would follow the normal curve if the distribution of data values in the parent population is normal. However, if the population values are left- or right-skewed the sampling distribution of the mean would be approximately normal. In addition, the more observations there are in the samples the closer the sampling distribution of the mean approximates to the normal distribution irrespective of the form of the frequency distribution of population values. In practice an infinite number of samples cannot feasibly be selected from a population to find the true standard error of the mean, therefore its value has to be estimated from the data provided by just one sample.

These features define the Central Limit Theorem, which provides the theoretical basis of the first of our parametric tests, known as the *Z* test. When simple random sampling of an independent set of observations has been carried out, the Central Limit Theorem applies and the probabilities associated with the normal distribution can be used to answer the question: what is the probability that the difference between the sample mean and population mean has occurred through chance? The *Z* test statistic is calculated using a similar equation to the one given in Section 5.4.3 for converting raw data values into *Z* scores (see Box 6.1a). Because the sample mean has been obtained from a set

Box 6.1a   The Z test.

Z test statistic: $\dfrac{|\bar{X}-\mu|}{\sigma/\sqrt{n}}$

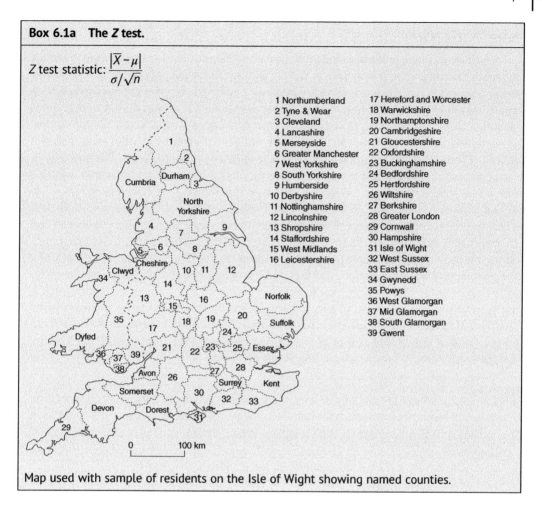

1 Northumberland	17 Hereford and Worcester
2 Tyne & Wear	18 Warwickshire
3 Cleveland	19 Northamptonshire
4 Lancashire	20 Cambridgeshire
5 Merseyside	21 Gloucestershire
6 Greater Manchester	22 Oxfordshire
7 West Yorkshire	23 Buckinghamshire
8 South Yorkshire	24 Bedfordshire
9 Humberside	25 Hertfordshire
10 Derbyshire	26 Wiltshire
11 Nottinghamshire	27 Berkshire
12 Lincolnshire	28 Greater London
13 Shropshire	29 Cornwall
14 Staffordshire	30 Hampshire
15 West Midlands	31 Isle of Wight
16 Leicestershire	32 West Sussex
	33 East Sussex
	34 Gwynedd
	35 Powys
	36 West Glamorgan
	37 Mid Glamorgan
	38 South Glamorgan
	39 Gwent

Map used with sample of residents on the Isle of Wight showing named counties.

Box 6.1b   Application of the Z test.

The Z test requires that values for the population mean (μ) and standard deviation (σ) can be specified by the investigator. The equation for the Z test statistic is a simple adaptation of the one used to convert individual data values into Z scores outlined in Chapter 5. The upper part of the Z statistic equation (the numerator) calculates the difference between the sample mean and the population mean. The lower part (the denominator) is the population standard deviation divided by the square root of the number of observations in the sample, which produces a value that is an unbiased estimate of the standard error of the mean ($s_{\bar{x}}$) when the population can be regarded as infinitely large. The distribution of the Z statistic follows the standardised normal curve and so the probabilities associated with different values of Z can be found to reach a decision on the Null and Alternative Hypotheses. The Z test is used to discover the probability of having obtained the sample mean rather than the population mean value. This probability helps an investigator decide whether the difference between the sample and population means has arisen through chance or whether there is something unusual or untypical about the set of sampled observations.

(Continued)

Box 6.1b   (Continued)

The worked example refers to a random sample of residents on the Isle of Wight (an island off the south coast of England), who were presented with a map showing the names of 53 counties in England and Wales. Each respondent was asked to rank the top five counties where they would like to live if they had a free choice. These ranks for the 30 counties (23 were not chosen) identified by the 26 respondents were aggregated to produce a mean score per county. For example, the sum of the ranks for Devon was 46, which produces a mean rank of 1.769 (46/30). These average scores are listed in the columns headed x_n below. The subscript n has been used to indicate that the names of the counties were shown on the map in contrast with a similar survey that was carried out with maps just showing the county boundaries that will be introduced later. How do we know what the population mean and standard deviation, which are necessary to apply the Z test, might be in this case? Each survey respondent had 15 'marks' to distribute across 53 counties (5, 4, 3, 2 and 1 ranks). If these were distributed randomly, the population mean score per county for Isle of Wight residents would be 0.283 (15/53). The population standard deviation is, as usual, a little more difficult to determine and has been estimated by generating random total scores per county within the range 0–26, representing the extremes of non-selection (0) and any one county as fifth favourite by all respondents (26). This produced a population standard deviation of 0.306. The Z test results indicate that, on the basis of the sample, Isle of Wight residents have distinct (non-random) preferences for certain counties in England and Wales as places to live.

The key stages in carrying out a Z-test are:

- *State Null Hypothesis and significance level*: the difference between the sample and population mean with respect to the preferences of Isle of Wight residents for living in the counties in England and Wales has arisen through sampling error and is not significant at the 0.05 level of significance.
- *Calculate test statistic (Z)*: the calculations for the Z statistic are given below.
- *Select whether to apply a one- or two-tailed test*: in this case, there is no prior reason to believe that the sample mean would be larger or smaller than the population mean.
- *Determine the probability of the calculated Z*: the probability of obtaining $Z = 2.163$ and $p = 0.015$.
- *Accept or reject the Null Hypothesis*: the probability of Z is <0.05, therefore reject the Null Hypothesis. By implication, the Alternative Hypothesis is accepted, recognising that this might be an erroneous decision 5 times in 100.

Box 6.1c   Assumptions of the *Z* test.

There are five main assumptions:

- Random sampling should be applied.
- Data values should be independent of each other in the population (i.e. the data value of the variable for one observation should not influence the value for another).
- The data values of the variable in the population should follow the normal curve (or at least approximately). Near normality may be achieved if the sample is sufficiently large, but this raises the question of how large is sufficiently large.
- The test ignores any difference between the population and sample standard deviations, which may or may not contribute to any difference between the means that are under scrutiny.
- The need to know the population standard deviation might be difficult to achieve.

Box 6.1d   Calculation of the *Z* test statistic.

Counties	Rank sum	x_n	Counties ...	Rank sum	x_n ...
Avon	11	0.423	Kent	19	0.423
Bedfordshire			Lancashire		
Berkshire			Leicestershire		
Buckinghamshire	2	0.077	Lincolnshire		
Cambridgeshire			Merseyside		
Cheshire	4	0.154	Mid Glamorgan		
Cleveland			Norfolk		
Clwyd	5	0.192	North Yorkshire	4	0.154
Cornwall	38	1.462	Northamptonshire	1	0.038
Cumbria	8	0.308	Northumberland	1	0.038
Derbyshire			Nottinghamshire		
Devon	46	1.769	Oxfordshire	11	0.423
Dorset	33	1.269	Powys		
Durham			Shropshire	3	0.115
Dyfed	2	0.077	Somerset	10	0.385
East Sussex	3	0.115	South Glamorgan		
Essex	7	0.269	South Yorkshire		
Gloucestershire	14	0.538	Staffordshire	1	0.038
Greater London	21	0.808	Suffolk	2	0.077
Greater Manchester			Surrey	27	1.038
Gwent			Tyne & Wear		
Gwynedd	4	0.154	Warwickshire	5	0.192
Hampshire	19	0.731	West Glamorgan		
Hereford & Worcestershire	3	0.115	West Midlands		
Hertfordshire	3	0.115	West Sussex	3	0.115
Hertfordshire			West Yorkshire		
Humberside	11	0.731	Wiltshire	5	0.192

$$\sum x = 12.115$$

Mean
$$\bar{x} = \frac{12.115}{30} = 0.404$$

Standard error of the mean
$$s_{\bar{x}} = \frac{\sigma}{\sqrt{n}} = \frac{0.306}{\sqrt{30}} = 0.056$$

Z statistic
$$\frac{|\bar{x} - \mu|}{\sigma/\sqrt{n}} = \frac{|0.404 - 0.283|}{0.056} = 2.163$$

probability
$$p = 0.015$$

of observations chosen at random, if the probability of difference is less than or equal to the level of significance (e.g. less than or equal to 0.05 or 5 per cent; or 0.01 or 1 per cent; etc.) then the Alternative Hypothesis would be accepted. Conversely, if the probability is greater than the level of significance, the Null Hypothesis should be regarded as valid.

Figure 6.1 helps to clarify the situation. It shows three samples of observations (A, B and C) selected from a normally distributed parent population and the data values in these three sets on the horizontal axis (X_1, X_2 and X_3) which yield different means for variable X. The number

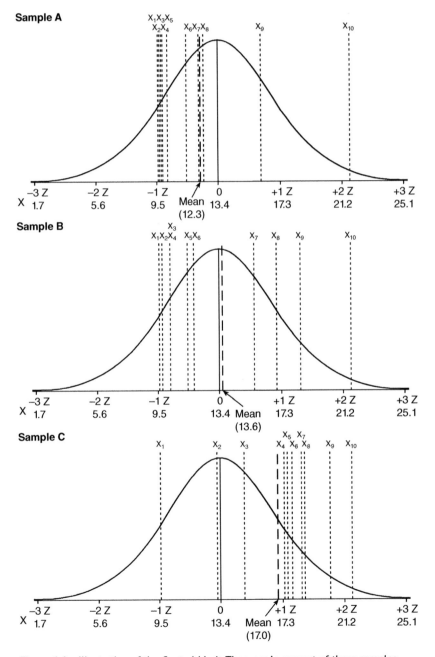

Figure 6.1 Illustration of the Central Limit Theorem in respect of three samples.

of data values in the samples has been limited in the interests of clarity. According to the Central Limit Theorem, these are just three of an infinite number of samples that could have been selected. The mean for sample A is less than the population or grand mean, the mean for sample B is a little higher and for sample C is a lot higher, the corresponding differences are −1.1, 0.2 and 3.6. The means have been converted into Z scores and the probabilities associated with these scores may be obtained from Z distribution tables (see Chapter 5) or from suitable statistical analysis software. In this case, the Z test values are 0.884, 0.162 and 2.903 and the corresponding probabilities for samples A, B and C are respectively 0.812, 0.436 and 0.002. These probabilities indicate that with reference to the 0.05 level of significance, the difference between the mean for sample C and the population or grand mean is larger than might be expected to have occurred by chance. This implies that a Z test value as large as 2.903 might be expected to occur with only two samples in 1000. The usual levels of significance correspond to certain Z scores or standard deviation units, thus 0.05 is +/−1.96, 0.01 is +/−2.57 and 0.001 is +/−3.303. So, the difference between the mean of sample C and the population mean is one of those unusual cases in one of the tails of the normal distribution beyond 1.96 standard deviations from the mean. Because there was no theoretical basis for believing that a sample should be more or less than the population means, the level of significance probability (0.05) is split equally between the tails with 0.025 in each. The probability of the Z test statistic for sample C is a lot less than 0.025 and the difference is therefore considered statistically significant, whereas those between samples A and B, and the population mean are not, because they are so small that they could easily have occurred by chance in more than 5 samples in 100. Their differences (−1.1 and 0.2) may be expected to occur with respect to 81 (0.2) and 43 (0.6) samples in 100.

Why is it useful to be 95 per cent sure that the difference between the mean of sample C and the population mean is significant, in contrast with the non-significant differences of samples A and B? This information is useful in two ways: first, it implies that in sample C the observations are untypical of the population with respect to their data values for the variable; and second that samples A and B are more representative of the population, at least as far as variable X is concerned. Suppose that by chance sample C included a disproportionately high number of recent additions to the parent population, for example new households that had come to live on a housing estate during the last five years. An alternative interpretation can now be made of the statistically significant difference between the sample and the parent population with respect to variable X. Suppose that variable X measured gross annual household income, we might now conclude that over the five-year period new households had contributed to raising the economic status of the estate.

Reflect further on this example to suggest other ways of interpreting these results.

All statistical tests make certain assumptions about the characteristics of the data to which they can be correctly applied. Some of the assumptions should be more strictly adhered to than others. The assumptions relating to the Z test are given in Box 6.1c and are so stringent as to make application of the test a rarity. The demands of the Z test are such that its application in the social and 'softer' natural sciences, such as Geography and Environmental Science, is less than in other areas. It might be regarded as the 'gold standard' of statistical testing and the possibility of relaxing some of the assumptions allows it to be used in certain circumstances. It also forms a useful starting point for examining other less demanding tests.

One important drawback of the Z test is that it requires an investigator to know the population standard deviation so that the standard error of the mean can be calculated. Although the population mean may be hypothesised or determined by reference to other information sources, it is often impossible to know the value of the population standard deviation since this has to be calculated from the raw data values. Fortunately, an alternative test exists for use in these situations, known as the t test or more formally Student's t test after William Gosset the statistician who published it under the pseudonym Student in 1908. It substitutes the sample standard deviation into the standard error equation in order to estimate this statistic, which is then used to produce the *t* test statistic in a similar fashion to calculating Z. This change potentially introduces additional sampling error that varies according to sample size, since inequality between the sample and population standard deviations will produce different values for the standard error. This implies that the shape of the t distribution curve, which although symmetrical about the mean like the (standardised) normal curve varies slightly with sample size. The difference between the standard error calculated from a small sample and that from the parent population is likely to be larger than for a large sample with more observations.

The probabilities associated with different Z values are independent of sample size. Thus, the Z test statistic and the associated probability obtained for sample C in Figure 6.1 (2.903 and 0.002, respectively) would be obtained irrespective of sample size. However, from a theoretical perspective there are as many *t* distribution curves as there are sample sizes. In practice the larger the sample size, the closer the t distribution curve matches with the standardised normal curve. Small sample sizes lower the peak of the t distribution curve and raise its tails in comparison with the Normal Curve. Samples with larger numbers of observations are likely to produce an estimate of the standard error that is closer to the population value, because it is calculated from a higher proportion of observations in the population, and in these circumstances, the *t* distribution provides a closer approximation to the Normal Curve. These characteristics cause the probabilities associated with different values of t to depend on the sample size, whereas with the Z statistic they are fixed. For example, the probability associated with a Z statistic of 1.96 is always 0.05 (0.025 in each tail of the distribution), whereas the probability of a t statistic of 1.96 is 0.056 when there are 50 sampled observations, 0.062 with 25 and 0.082 if there are 10. Thus, the t test statistic needs to be larger than the Z test equivalent to achieve the same level of significance. Figure 6.2 helps to clarify the

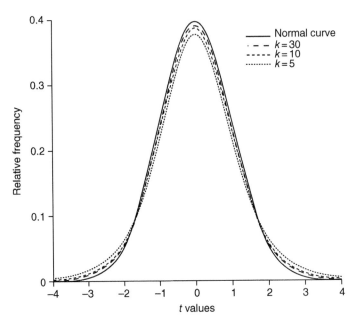

Figure 6.2 Comparison of the Normal and *t* distribution curves.

differences between the Z and t distributions. There are three t distribution curves corresponding to different sample sizes (k). These can be compared with the solid line of the standardised normal curve. A higher value of the t statistic is required for smaller sample sizes to achieve the same level of significance (0.05), 2.045 when there are 30 observations, 2.145 with 10 and 2.262 with 5. The t value associated with a probability of 0.05 approaches the equivalent Z values as sample size increases and is the same (to 3 decimal places) when sample size equals 4425 observations.

This link between sample size and t distribution probabilities introduces the concept of **degrees of freedom** (df). This refers to the number of data values in a sample that are free to vary subject to the overall constraint that the sample mean, standard deviation or any other statistic calculated for any variable are determined. This concept often causes some confusion at first. If you have been out in the field or laboratory and collected some data for a sample of observations, you have carefully entered the data into statistical software and produced the sample mean of your variables, surely there is no freedom for any of them to vary. Consider the degrees of freedom concept from a theoretical perspective, which starts with the random selection of observations to enter the sample. Suppose there was a sample containing five randomly selected observations (labelled C, H, M, P and T), which have the values 0.78, 0.26, 1.04, 0.81 and 0.76 for one of the measured variables (X). The mean of these values is 0.73. Now imagine that the process of random selection had not selected observation P but one labelled J instead and its value for the variable is 1.06 rather than 0.81. Now the five values produce a mean of 0.78. Looking at this example in another way, if the mean of the variable is truly 0.73, in the population as well as the first sample, then the total sum of five data values must equal 3.65. To achieve this total, four of the observations can take on any value, but once these are known then the fifth value is fixed. Hence there are four degrees of freedom. In this example, there is just one variable and the degrees of freedom equal the number of sampled observations minus one (df $= n - 1$), whereas if there are two or more variables the df will be calculated differently. Returning to the t distribution curves in Figure 6.2 for 30, 10 and 5 sampled observations in reality these correspond to the situations where there are 29, 9 and 4 degrees of freedom, respectively.

Box 6.2a   The t test.

t test statistic: $= \dfrac{|\overline{X} - \mu|}{s/\sqrt{n}}$

Box 6.2b   Application of the t test.

The t test requires knowledge of the population mean (μ) but substitutes the sample standard deviation (s) for the population parameter used in the Z test. The numerator of the t test statistic equation is identical to the one for the Z test. The denominator is very similar to the sample standard deviation substituting for the population parameter to estimate the standard error. The t distribution varies with sample size and degrees of freedom (df $= n - 1$) and probabilities associated with the value of t for different df enable a decision to be reached about acceptance or rejection of the Null and Alternative Hypotheses. The purpose of the t

(Continued)

Box 6.2b (Continued)

test is essentially the same as the *Z* test, namely, to decide the probability of having obtained the sample mean from the randomly selected sample and whether the difference between it and the known or hypothesised population mean has arisen through chance or sampling error.

The *t* test has been applied below to the same example data used for the *Z* test in Box 6.1a in order to allow comparison between the results of the two tests. The population mean score per county for Isle of Wight residents is again 0.283, but the sample standard deviation (0.456) was used to calculate the standard error of the mean. Therefore, the *t* test statistic (2.942) is different from the *Z* test statistic (2.163) in Box 6.1d. The degrees of freedom in this example are 29 (30−1): the number of counties chosen by any of the respondents minus 1. The *t* test results support those of the *Z* test, namely that the sampled respondents reveal Isle of Wight residents to have distinctly (non-random) preferences for living in certain English and Welsh counties.

The key stages in carrying out a *t* test are:

- *State Null Hypothesis and significance level*: the difference between the sample and population mean with respect to the preferences of Isle of Wight residents for living in the counties in England and Wales has arisen through sampling error and is not significant at the 0.05 level of significance.
- *Calculate test statistic (t)*: the calculations for the *t* statistic are given below.
- *Select whether to apply a one- or two-tailed test*: in this case, there is no prior reason to believe that the sample mean would be larger or smaller than the population mean, so a two-tailed test is required.
- *Determine the probability of the calculated t*: the probability of obtaining $t = 2.942$ is $p = 0.006$ with 29 degrees of freedom.
- *Accept or reject the Null Hypothesis*: the probability of *t* is <0.05, therefore reject the Null Hypothesis. By implication accept the Alternative Hypothesis, recognising that this might be an erroneous decision 5 times in 100.

Box 6.2c Assumptions of the *t* test.

There are five main assumptions:

- Random sampling should be applied.
- Data values should be independent of each other in the population.
- The variable should be normally distributed, although it may be difficult to determine and low to moderate skewness is permissible.
- A highly skewed distribution is likely to achieve a poor fit between the theoretical t distribution and the sampling distribution of the mean. A sufficiently large sample will help to overcome this difficulty.
- A negatively or positively skewed distribution affects the symmetry of the t distribution and the equal division of probabilities on either side of the mean.

Box 6.2d Calculation of the *t* test statistic.

Counties	x_n	$(x-\bar{x})^2$	Counties ...	x_n cont'd	$(x-\bar{x})^2$ cont'd
Avon	0.423	0.038	Kent	0.423	0.252
Bedfordshire			Lancashire		
Berkshire			Leicestershire		
Bucks	0.077	0.023	Lincolnshire		
Cambridgeshire			Merseyside		
Cheshire	0.154	0.006	Mid Glamorgan		
Cleveland			Norfolk		
Clwyd	0.192	0.001	North Yorkshire	0.154	0.006
Cornwall	1.462	1.520	Northamptonshire	0.038	0.036
Cumbria	0.308	0.006	Northumberland	0.038	0.036
Derbyshire			Nottinghamshire		
Devon	1.769	2.374	Oxfordshire	0.423	0.038
Dorset	1.269	1.083	Powys		
Durham			Shropshire	0.115	0.013
Dyfed	0.077	0.023	Somerset	0.385	0.024
East Sussex	0.115	0.013	South Glamorgan		
Essex	0.269	0.002	South Yorkshire		
Gloucestershire	0.538	0.096	Staffordshire	0.038	0.036
Greater London	0.808	0.335	Suffolk	0.077	0.023
Greater Manchester			Surrey	1.038	0.656
Gwent			Tyne & Wear		
Gwynedd	0.154	0.006	Warwickshire	0.192	0.001
Hampshire	0.731	0.252	West Glamorgan		
Hereford and Worcestershire	0.115	0.013	West Midlands		
Hertfordshire	0.115	0.013	West Sussex	0.115	0.013
Hertfordshire			West Yorkshire		
Humberside	0.731	0.038	Wiltshire	0.192	0.001
				$\sum x = 12.115$	$\sum (x-\bar{x})^2 = 6.017$

Mean	$\bar{x} = \dfrac{12.115}{30} = 0.404$
Standard deviation	$s = \sqrt{\dfrac{\sum (x-\bar{x})^2}{(n-1)}} = \sqrt{\dfrac{6.017}{(30-1)}} = 0.456$
Standard error of the mean	$s_{\bar{x}} = \dfrac{s}{\sqrt{n}} = \dfrac{0.456}{\sqrt{29}} = 0.085$
t statistic	$\dfrac{|\bar{x}-\mu|}{s/\sqrt{n}} = \dfrac{|0.404-0.283|}{0.085} = 2.942$
Probability	$p = 0.006$

6.2.2 Comparing Differences Between Pairs of Measurements for a Sample Divided into Two Parts

There are occasions when a single sample of observations can be divided into two parts or measured twice to produce a paired set of data values. In these circumstances a special version of the *t* test can be used, known as the paired sample *t* test, to investigate whether the difference between the means of the two sets of data values is significant. In Box 6.3a the temperature measurements for the melt water stream from Les Bossons Glacier in the French Alps where it melts from the glacier and where it exits the outwash plain are treated as a single sample of observations from one water course. These measurements can be paired because they were taken at the same time of day and it is reasonable to argue that they represent two sets of data values collected simultaneously for the same stream. The Null Hypothesis in the paired t test usually maintains that the difference between the sample means in the population of observations is zero, in other words, no difference exists, and any that has arisen in a particular set of sample data is the result of sampling error. The test seeks to discover whether this sampling error is of a magnitude that might be attributed to chance or large enough to indicate a significant difference between the two occasions or two parts of the observations when they were measured.

Identify other situations in which pairs of measurements might be made with respect to geographical, environmental or geo-scientific phenomena. Remember you are looking instances of where the sample of observations can be split into two parts or measured twice to obtain two paired sets of data values for one interval or ratio scale variable.

Box 6.3a   The paired sample *t* test.

t test statistic: $= \dfrac{|\,\bar{d} - 0\,|}{s_d / \sqrt{n}}$

Box 6.3b   Application of the paired *t* test.

The paired *t* test normally hypothesises that the population mean (μ) of the differences (*d*) between the paired data values equals zero and uses the standard deviation of the differences in the sample data (s_d) to estimate the population standard error. The degrees of freedom are calculated in the same manner as for the 'standard' *t* test (df = $n - 1$) and the fate of the Null and Alternative Hypotheses is decided by comparing the paired *t* test statistics with the level of significance, usually ≤ 0.05. The purpose of the paired *t* test is to discover the probability of having obtained the difference in means between the paired values and to reach a decision on whether this is significantly different from zero (no difference).

   The application of the paired t test below relates to the paired measurements of temperature in the melt water stream flowing from an alpine glacier at the snout (x_s) and the exit of the outwash plain (x_e). The test proceeds by calculating the difference between the two sets of temperature measurements, which have been made at the same time of day and treat the stream as a single water course. Not surprisingly the test results confirm the Alternative Hypothesis that the temperature of the water near the glacier snout is lower than near the exit from the outwash plain. Since the test examines the differences between the measurements irrespective of where they were made, the test results would be the same if the pairs of values were reversed (i.e. higher values near the snout and lower ones at the exit). It is only through the interpretation of the results by the researcher that meaning can be attached to the outcome of the test.

The key stages in carrying out a *t* test are:

- *State Null Hypothesis and significance level*: the mean of the differences between the pairs of water temperature measurements in Les Bossons Glacier stream and zero has arisen through sampling error and is not significant at the 0.05 level of significance.
- *Calculate test statistic (t)*: the calculations for the paired *t* statistic are given below.
- *Select whether to apply a one- or two-tailed test*: in this example, it is reasonable to argue that water temperature closer to the glacier snout will be lower than those recorded further away and therefore a one-tailed would be appropriate which involves halving the probability.
- *Determine the probability of the calculated t*: the probability of obtaining $t = 7.425$ is 0.000000992 with 17 degrees of freedom.
- *Accept or reject the Null Hypothesis*: the probability of *t* is <0.05, therefore reject the Null Hypothesis. By implication, the Alternative Hypothesis is accepted, recognising that this might be an erroneous decision 5 times in 100.

Box 6.3c   Assumptions of the paired *t* test.

There are four main assumptions that are the same as for the standard *t* test (see Box 6.2c).

Box 6.3d   Calculation of the paired *t* test statistic.

	x_s	x_e	$x_e - x_e = d$	$(d - \bar{d})^2$
09.00 hrs	4.10	4.30	−0.20	10.43
09.30 hrs	4.60	4.70	−0.10	11.09
10.00 hrs	4.60	4.80	−0.20	10.43
10.30 hrs	4.70	5.20	−0.50	8.58
11.30 hrs	4.80	6.70	−1.90	2.34
12.00 hrs	5.20	10.10	−4.90	2.16
12.30 hrs	5.40	10.50	−5.10	2.79
13.00 hrs	6.30	11.20	−4.90	2.16
13.30 hrs	6.90	11.40	−4.50	1.14
14.00 hrs	7.40	11.80	−4.40	0.94
14.30 hrs	7.10	12.30	−5.20	3.13
15.00 hrs	6.90	11.90	−5.00	2.46
15.30 hrs	6.30	11.40	−5.10	2.79
16.00 hrs	6.10	11.10	−5.00	2.46
16.30 hrs	5.50	10.20	−4.70	1.61
17.00 hrs	5.50	9.30	−3.80	0.14

(Continued)

Box 6.3d (Continued)

	x_s	x_e	$x_e - x_e = d$	$(d - \bar{d})^2$
17.30 hrs	5.30	8.80	−3.50	0.00
18.00 hrs	5.10	7.80	−2.70	0.53
			$\sum d = -61.70$	$\sum (d - \bar{d})^2 = 65.22$

Mean difference	$\bar{d} = \dfrac{-61.70}{18} = -3.43$
Standard deviation	$s = \sqrt{\dfrac{\sum (d - \bar{d})^2}{(n-1)}} = \sqrt{\dfrac{65.22}{(18-1)}} = 1.96$
Standard error of the mean	$s_{\bar{d}} = \dfrac{s}{\sqrt{n}} = \dfrac{1.96}{\sqrt{18}} = 0.462$
t statistic	$\dfrac{|\bar{d} - 0|}{s_d/\sqrt{n}} = \dfrac{|3.43 - 0|}{0.462} = 7.425$
Probability	$p = 0.000000992$

6.3 Two Samples and One Variable

Many projects focus on using samples of observations from two distinct populations in order to determine if they are different from or similar to each other. The next stage in our exploration of parametric statistical tests is to examine such situations where a researcher has identified two populations of observations, for example outcrops of sedimentary and igneous rocks, households in a developed and developing country, or areas of predominantly coniferous and broad-leafed woodland. Again, the focus of attention is on the data values for one or more continuous interval or ratio variables capable of being measured with respect to the individual entities in these populations. These variables are assumed to adhere to the normal probability distribution and sampled observations to have been selected randomly. The differences under scrutiny by the statistical tests are between measures of central tendency and dispersion in the two samples and by implication whether they are 'genuinely' from two populations or in reality from one.

> Reflect on how projects with two samples from distinctly different populations contrast with situations in which pairs of measurements are discussed at the end of the previous section. Identify other examples of projects in which there are two distinct populations geographical, environmental or geo-scientific phenomena. Remember you are looking for instances of sampled observations that can be measured on at least one interval or ratio scale variable from two distinct populations.

The hypothesised difference is often assumed to equal zero with the Null and Alternative Hypotheses phrased in such a way that they examine whether the two populations from which the sampled observations have been drawn are statistically different from each other. If the test suggested that Null Hypothesis should be accepted, the investigator should conclude that the populations are in fact the same with respect to the variable that has been examined, whereas acceptance of the Alternative Hypothesis suggests that they can be regarded as statistically different. In other words, continuing with the example of sedimentary and igneous outcrops, if differences between the mean quartz content of rock samples taken from random locations across these areas proved not to be statistically significant, then it might be possible to conclude that there was only one population rather than two. Obviously, other variables would need to be examined to determine whether they supported this conclusion.

6.3.1 Comparing Two Sample Means with Population Means

The theoretical starting point for examining the statistical tests available for answering this situation is to imagine there are two populations with known means (μ_1 and μ_2) and variances (σ_1^2 and σ_2^2) in respect of a certain variable (X) where subscripts 1 and 2 denote the two populations. We have already seen that according to the Central Limit Theorem random samples with N observations selected from each of the populations would produce a series of sample means that would follow the Normal Curve. Suppose that pairs of independent random samples were selected from the two populations, a process that could be carried on indefinitely, and the differences calculated between each pair of means in respect of X. These differences could be plotted as a frequency distribution and would form the **sampling distribution of the difference in means**. Descriptive statistics (e.g. mean, variance and standard deviation) could be determined for this set of differences between the sample means for populations 1 and 2. Theoretically, the difference between the population means ($\mu_1 - \mu_2$) equals the mean of these differences. However, some sampling error is likely to be present and the standard error of the differences can also be calculated.

Application of this theory for statistical testing requires knowledge of whether the sampling distribution of the difference in means follows the normal distribution. This is governed by two factors: the nature of the distributions of the parent populations and sample size. The sampling distribution of the difference in means will be normal in two situations:

- the frequency distributions of data values in both parent populations are normal **and** the sample size is finite;

  or

- the frequency distributions of data values in the parent populations are not normal **but** the sample size is infinite.

In these circumstances, the Z test can be used to examine whether the observed and hypothesised difference between the means of two samples selected from populations with known variances is significantly different. However, just as in the single sample case, the Z test requires knowledge of the population mean and variance, so the two-sample Z test demands an investigator to know the means and variances of two populations, although the former is usually assumed to be equal with a difference of zero. It may also be difficult to achieve or determine normality in two populations and sample size will invariably be finite. An estimate of Z ($\hat{Z}$) can be obtained by substituting the sample variances for the population ones if the sample sizes are sufficiently large, but this begs the question of how large is sufficiently large.

Fortunately, an alternative solution exists, namely, to apply the two sample t test, which although slightly less robust than the Z test has the advantage of using the sample variances. There are alternative versions of the two sample t test available depending upon whether the population variances are assumed equal (homoscedastic t test) or different (heteroscedastic t test). The decision on which version of the test should be applied may be made by reference to the results of an F test of the difference between variances (see Section 6.3.2). If the population variances are known to be equal or an F test indicates that they are likely to be so, the homoscedastic version t test should be applied which pools or combines the sample variances to estimate the standard error of the difference in means. If equality of population variances does not exist, then heteroscedastic t test should be used. Box 6.4a illustrates the application of the two sample t test, in this case in respect of whether residents on the Isle of Wight comprise two statistical populations in relation to their knowledge of the names of counties in England and Wales in which they might wish to live. Statistical software will usually offer both versions of the test and the choice over which to employ simply involves carrying out a preliminary F test of difference between variances.

Box 6.4a The two sample *t* test.

t test statistic: $= \dfrac{|\bar{x}_1 - \bar{x}_2|}{\sqrt{\dfrac{s_1^2}{n_1} + \dfrac{s_2^2}{n_2}}}$ unequal population variances

$= \dfrac{|\bar{x}_1 - \bar{x}_2|}{\sqrt{\dfrac{n_1 s_1^2 + n_2 s_2^2}{n_1 + n - 2}}\sqrt{\dfrac{n_1 + n_2}{n_1 n_2}}}$ equal population variances

1 Northumberland	17 Hereford and Worcester
2 Tyne & Wear	18 Warwickshire
3 Cleveland	19 Northamptonshire
4 Lancashire	20 Cambridgeshire
5 Merseyside	21 Gloucestershire
6 Greater Manchester	22 Oxfordshire
7 West Yorkshire	23 Buckinghamshire
8 South Yorkshire	24 Bedfordshire
9 Humberside	25 Hertfordshire
10 Derbyshire	26 Wiltshire
11 Nottinghamshire	27 Berkshire
12 Lincolnshire	28 Greater London
13 Shropshire	29 Cornwall
14 Staffordshire	30 Hampshire
15 West Midlands	31 Isle of Wight
16 Leicestershire	32 West Sussex
	33 East Sussex
	34 Gwynedd
	35 Powys
	36 West Glamorgan
	37 Mid Glamorgan
	38 South Glamorgan
	39 Gwent

Maps used with samples of residents on the Isle of Wight.

Box 6.4b   Application of the two-sample *t* test.

The two-sample *t* test is often used to test whether the observed difference between sample means selected from two populations has arisen through sampling error and, by implication, whether the true difference is zero, since the population means are hypothesised to have a difference of zero. Alternative formulae for the standard error of the difference in means ($s_{\bar{x}_1} - s_{\bar{x}_2}$) exist for use in the two sample t test equation depending upon whether the population means are known or believed because of statistical testing to be equal. The degrees of freedom are calculated differently from the 'standard' *t* test (df = $(n_1 - 1) + (n_2 - 1)$), since the samples need not contain the same number of observations. A decision on the Null and Alternative Hypotheses is reached by comparing the probability of having obtained the *t* test statistic with the chosen level of significance, usually whether ≤ 0.05.

The two-sample *t* test has been applied to two samples of residents on the Isle of Wight who were selected at random and completed an interview survey. The respondents in one sample were shown a map of counties in England and Wales that included county names and those in the other sample were given an unlabelled map to look at. Both sets of interviewees were asked to indicate and then rank the five counties where they would like to live if their choice was unrestricted in any way. The samples purport to divide the Isle of Wight residents into two statistical populations because of their knowledge of the administrative geography of England and Wales. The preliminary *F* test (see Box 6.5a) indicated that the variances of the parent populations were equal and therefore the homoscedastic version of the two sample tests should be applied.

The procedure for carrying out the two sample *t* test proceeds by calculating the means and variances of each sample. These calculations have been exemplified elsewhere and so Box 6.4d simply gives the raw data (mean aggregate rank scores per named (x_n) and unnamed county (x_u)) together with the values of these statistics. The standard error of the mean is calculated using a complicated-looking equation when applying the homoscedastic version of the test. The difference between the sample means is more straightforward to determine. These values are used to produce the *t* test statistic. The test results indicate that the difference between the sample means is so small that it could easily have arisen by chance, and there was no significant difference in respondents' preferences for living in other counties in England and Wales.

The key stages in carrying out a two-sample *t* test are:

- *State Null Hypothesis and significance level*: the means of the sampled populations are the same and any observed difference between the sampled has arisen through sampling error and is not significant at the 0.05 level of significance.
- *Determine whether population variances are equal or likely to be so*: apply *F* test (see Section 6.3.2)
- *Calculate test statistic (t)*: the calculations for the two sample t statistic with unequal population variances are given below.
- *Select whether to apply a one- or two-tailed test*: there is no prior reason for arguing that one of the samples should have a higher or lower mean than the other, therefore a two-tailed is appropriate.
- *Determine the probability of the calculated t*: the probability of obtaining $t = 0.613$ and $p = 0.470$ with 64 degrees of freedom.
- *Accept or reject the Null Hypothesis*: the probability of *t* is >0.05, therefore accept the Null Hypothesis, recognising that this might be an erroneous decision. By implication, the Alternative Hypothesis is rejected.

Box 6.4c   Assumptions of the two sample *t* test.

There are four main assumptions:

- Random sampling should be applied and observations should not form pairs.
- Data values should be independent of each other in the population.
- The variable under investigation should be normally distributed in both parent populations.
- Assessment of the homogeneity or heterogeneity (similarity/dissimilarity) of population variances guides the selection of the appropriate version of the two sample test.

Box 6.4d   Calculation of the two sample *t* test statistic.

Counties	x_n	x_u	Counties ...	x_n ...	x_u ...
Avon	0.423	0.154	Kent	0.423	0.115
Bedfordshire		0.077	Lancashire		0.154
Berkshire		0.077	Leicestershire		0.269
Bucks	0.077	0.154	Lincolnshire		0.192
Cambridgeshire			Merseyside		
Cheshire	0.154		Mid Glamorgan		
Cleveland			Norfolk		0.500
Clwyd	0.192		North Yorkshire	0.154	0.115
Cornwall	1.462	1.885	Northants	0.038	
Cumbria	0.308	0.231	Northumberland	0.038	0.077
Derbyshire		0.500	Nottinghamshire		0.038
Devon	1.769	1.115	Oxfordshire	0.423	0.115
Dorset	1.269	0.846	Powys		0.077
Durham			Shropshire	0.115	0.154
Dyfed	0.077	0.192	Somerset	0.385	0.346
East Sussex	0.115	0.385	South Glamorgan		
Essex	0.269	0.154	South Yorkshire		0.077
Gloucestershire	0.538		Staffordshire	0.038	0.077
Greater London	0.808	0.269	Suffolk	0.077	0.308
Greater Manchester		0.154	Surrey	1.038	0.192
Gwent			Tyne & Wear		
Gwynedd	0.154	0.038	Warwickshire	0.192	
Hampshire	0.731	0.692	West Glamorgan		
Hereford & Worcestershire	0.115		West Midlands		
Hertfordshire	0.115	0.192	West Sussex	0.115	0.923
Hertfordshire		0.192	West Yorkshire		

Humberside		0.731	0.154	Wiltshire	0.192	0.808
				$\sum x_n = 12.115$		$\sum x_u = 11.846$
Mean	Sample n (named counties)			$\bar{x}_n = \dfrac{12.115}{30} = 0.404$		
	Sample u (unnamed counties)			$\bar{x}_u = \dfrac{11.846}{36} = 0.329$		
Variance	Sample n (named counties)			$s_n^2 = \dfrac{\sum (x_n - \bar{x}_n)^2}{n_n - 1} = \dfrac{6.017}{29} = 0.207$		
	Sample u (unnamed counties)			$s_u^2 = \dfrac{\sum (x_u - \bar{x}_u)^2}{n_u - 1} = \dfrac{5.063}{35} = 0.145$		
Standard error of the difference in sample means				$\sqrt{\dfrac{n_n s_n^2 + n_u s_u^2}{n_n + n_u - 2}} \sqrt{\dfrac{n_n + n_u}{n_n n_u}} =$		
				$\sqrt{\dfrac{30(0.207) + 36(0.145)}{30 + 36 - 2}} \sqrt{\dfrac{30 + 36}{30(36)}} = 0.105$		
t statistic				$\dfrac{\lvert \bar{x}_{n1} - \bar{x}_u \rvert}{\sqrt{\dfrac{n_n s_n^2 + n_u s_u^2}{n_n + n_u - 2}} \sqrt{\dfrac{n_n + n_u}{n_n n_u}}} = \dfrac{0.075}{0.105} = 0.716$		
Degrees of freedom				$df = (n_n - 1) + (n_u - 1) = (30 - 1) + (36 - 1) = 64$		
Probability				$p = 0.470$		

6.3.2 Comparing Two-Sample Variances with Population Variances

The tests examined so far have focused on differences relating to the mean as a measure of central tendency. However, as outlined in Chapter 4, two sets of numbers can have the same mean value but quite different variances and standard deviations. A difference in variance between samples from two populations with respect to a variable signifies some variation in the relative dispersion of their data values. Putting this into context, the mean pH of two samples of 50 bottles each with 10 cl of sea water collected during one day from two different beaches might be the same, but the dispersion of individual measurements in each set might be different, one spread out and the other tightly packed around the mean. Therefore, the samples have different variances and standard deviations, but is this because of sampling error or the populations having distinct characteristics? The point of departure for understanding how to test for a significant difference between two sample variances involves starting with two populations of observations that can be measured with respect to a common variable that follows the normal distribution. If their means and variances were equal (i.e. $\mu_1 = \mu_2$ and $\sigma_1^2 = \sigma_2^2$), the populations would be regarded as statistically identical with respect to these parameters: the difference between the means is 0 and the ratio between their variances is 1. Such equality of variances is unlikely to arise for populations let alone samples of 'real world' phenomena and so one variance will always be larger than another. The ratio of the larger to the smaller variance provides the statistic, F, which quantifies the magnitude of the difference between them. Table 6.2 shows how the ratio between two pseudo samples with 'variances' ranging from 1 to 14: when both 'variances' are similar (e.g. 2 and 4, or 4 and 6) compared with when they are dissimilar (e.g. 2 and 12, or 4 and 14). A succession of samples each with a specific

Table 6.2 Differences in ratio between 'samples' with small and large variances whose values range from 2 to 14 (to 1 dec pl.).

Sample 1 variance values		Sample 2 variance values						
		2	4	6	8	10	12	14
	2	1.0	2.0	3.0	4.0	5.0	6.0	7.0
	4		1.0	1.5	2.0	2.5	3.0	3.5
	6			1.0	1.3	1.6	2.0	2.3
	8				1.0	1.3	1.5	1.8
	10					1.0	1.2	1.4
	12						1.0	1.2
	14							1.0

number of observations selected from two populations may include some overlap in respect of individual entities or each pair may be entirely separate; nevertheless, they all purport to measure the variance (and mean) of their respective parent population. However, sampling error is likely to result in the sample variances being different with the larger values coming from either of the populations. The selection of pairs of such samples and calculation of their F ratios could be carried on indefinitely to produce a frequency distribution of F values. Each combination of sample sizes from two populations (e.g. 25 from population A and 50 from population B) produces a distinct asymmetrical distribution with a high number of small F values and fewer large ones (see Figure 6.3).

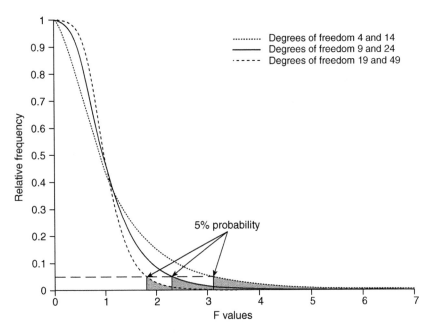

Figure 6.3 Examples of the F distribution with selected combinations of degrees of freedom (df).

The probabilities associated with these distributions form the basis for assessing a difference between sample variances using the F test. The test provides a way of determining whether the variances of the populations from which two samples have been selected may be considered equal. If the difference and therefore also the F ratio are so small that they can be regarded as having occurred through chance, then the population variances would be regarded as the same. As we have already seen, the application of the F test is usually a precursor to carrying out a two-sample t test, since alternative versions of this test are used depending on whether the population variances are equal. Calculation of the F test statistic (see Box 6.5a) is straightforward: it simply involves dividing the larger variance by the smaller. The probability associated with an F ratio for a particular combination of sample sizes, from which the degrees of freedom for each sample can be determined, enables an investigator to decide whether the F values are likely to have arisen through chance or are indicative of disparity in the dispersion of the data values in the two populations. If the probability associated with the calculated F statistic is less than or equal to the chosen level of significance (e.g. 0.05 or 0.01), the Null Hypothesis would be rejected in favour of the Alternative Hypothesis stating that the population variances are significantly different.

Box 6.5a   The two sample F test.

F test statistic: $= \dfrac{s_l^2}{s_s^2}$

Box 6.5b   Application of the F test.

The F test assesses the difference between the variances of two samples drawn from different populations in order to determine whether the population variances are equal or unequal. The Null Hypothesis argues that the population variances are the same ($\sigma_1^2 = \sigma_2^2$). The F test statistic is simply the ratio of the larger to the smaller sample variance (denoted by the subscripts l and s), a value that will always be greater than 1 unless the sample variances are equal, in which case an absence of sampling error is assumed. The samples do not need to contain the same number of observations and the degrees of freedom for each is $df_1 = n_1 - 1$ and $df_2 = n_2 - 1$. The fate of the Null Hypothesis is decided by reference to the probability of obtaining the F statistic or one greater in relation to the chosen level of significance, usually whether ≤ 0.05.

The F test has been applied to the samples of residents on the Isle of Wight who were shown different maps of counties in England and Wales, one with the names and one without. The samples purport to reflect a division in the Isle of Wight residents between those who have knowledge of the administrative geography of England and Wales and can identify counties in which they would like to live without the aid of county names and those whose knowledge is less. The calculations for the sample variances are given in Box 6.4d, therefore Box 6.5d simply illustrates the procedure for obtaining the F test statistic and the degrees of freedom. The F test results indicate that the difference in the variances is so small (insignificant) that it might arise through sampling error on 30.7% of occasions.

(Continued)

Box 6.5b (Continued)

The key stages in carrying out an *F* test are:

- *State Null Hypothesis and significance level*: the variances of the populations from which samples have been selected are the same: inequality of sample variances between the two groups of Isle of Wight residents is the result of sampling error and is not significant at the 0.05 level of significance.
- *Calculate test statistic (F)*: the calculations for the *F* statistic appear below.
- *Select whether to apply a one- or two-tailed test*: there is no prior reason for arguing that the variance of one population would be larger than the other therefore a two-tailed is appropriate.
- *Determine the probability of the calculated F*: the probability of obtaining *F* = 1.434 is 0.307 with 29 degrees of freedom for the sample with the larger variance and 35 from the one with the smaller variance.
- *Accept or reject the Null Hypothesis*: the probability of *F* is >0.05, therefore accept the Null Hypothesis, recognising that this might be an erroneous decision. By implication, the Alternative Hypothesis is rejected.

Box 6.5c Assumptions of the *F* test.

The *F* test has three main assumptions:

- Sampled observations should be chosen randomly with replacement unless the populations are of infinite size.
- Data values for the variable under investigation should be independent of each other in both populations.
- The variable should be normally distributed in both parent populations since minor skewness can potentially bias the outcome of the test.

Box 6.5d Calculation of the *F* test statistic.

Variance	Sample n (named counties)	From Box 6.5d	$\dfrac{6.017}{29} = 0.207$
	Sample u (unnamed counties)	From Box 6.5d	$\dfrac{5.063}{35} = 0.145$
F statistic			$\dfrac{s_n^2}{s_u^2} = \dfrac{0.207}{0.145} = 1.434$
Degrees of freedom	Sample n (named counties) larger variance		$\mathrm{df}_n = n_n - 1 = 30 - 1 = 29$
	Sample u (unnamed counties) smaller variance		$\mathrm{df}_u = n_u - 1 = 36 - 1 = 35$
Probability			$p = 0.307$

6.4   Three or More Samples and One Variable

The next logical extension to our examination of parametric statistical tests continuing from the previous sections is to explore techniques involving one variable with respect to three or more populations and samples. For example, an investigator might be interested in predominantly arable, dairy, beef and horticultural farms, in people travelling to work by car, train, bus/tram, bicycle/ motorcycle and by walking, or outcrops of sedimentary, igneous and metamorphic rocks. The intention is to compare measures of central tendency and dispersion calculated from the data values of the same variables measured on the interval or ratio scales for observations from each of the populations rather than to examine how two or more variables relate to each other. The observations in the populations are assumed to follow the normal distribution with respect to each variable under investigation. The statistical tests applied to such data are concerned with helping an investigator decide whether the populations are really (statistically) different from each other with regard to certain variables according to a specified level of significance. The key question posed by the Null and Alternative Hypotheses associated with this type of test is whether the populations can be regarded as statistically distinct. This assessment is usually made on the basis of samples of observations selected at random from their parent populations.

6.4.1   Comparing Three or More Sample Means with Population Means

An extension of the two-sample F test (Section 6.3.2) provides one of the most important statistical tests available allowing researchers to discover whether the means of samples selected from three or more different populations are statistically different from each other. The technique examines if the observed differences between the sample means are so large that they could have come from different populations or so small that they could be from the same population. The technique is known as Analysis of Variance (ANOVA), or more accurately One-way ANOVA since only one variable is being considered, whereas 'Analysis of Means' might seem more logical. Recall that the variance measures the spread of a set of data values about their mean and represents the sum of the squared differences between these numbers and their mean divided by n−1, where n equals the number of values. Now suppose there was a set of samples coming from separate populations that were measured in respect of a single identically defined variable. The variance for each sample and the overall pooled variance of all data values could be calculated and, regardless of whether the samples were in fact from different populations, there is no reason to assume that the dispersion of the variable would vary. In other words, the variances could be equal to the variance of all sample values, irrespective of whether the sample means are different or the same.

  This suggests that the overall variance can be divided into two parts: one variance is calculated for the dispersion of the data values within each sample and the other for the dispersion between each sample. Dividing the within-groups by the between-groups variance produces the F ratio. A large between-groups variance in comparison with the within-groups variance will produce a high F ratio and signify that inter-group differences account for a major part of the overall variance. In contrast, a small F ratio denotes the reverse, namely that there is a substantial overlap of the data values associated with the different samples. This separation of the total variance into two parts introduces the idea that some statistical explanation might account for the differences. The within-groups variance encapsulates some unexplained variation between the data values in each sample, whereas the between-groups variance relates to genuine differences between the samples. Figure 6.4 helps to explain these principles. The series of identical curves in Figure 6.4a represents samples with identical variances but different means: hence they are separated from each other

(a)

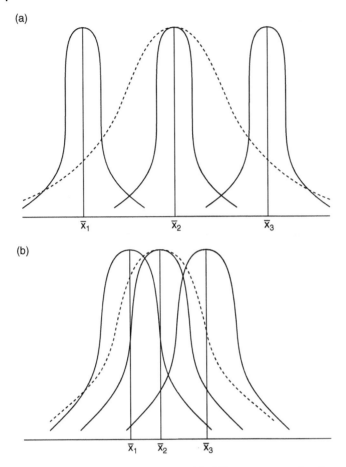

(b)

Figure 6.4 Comparison of samples with different means and variances. (a) samples with identical variances but different means produce a larger overall variance (broken line); (b) samples with identical variances and similar means produce a smaller overall variance (broken line).

along the horizontal axis. The curve representing the relative frequency distribution of data values pooled from all of the samples (broken line) is wider and has a larger variance, since it has been calculated from values throughout the range overall. In contrast, the identical curves in Figure 6.4b are for samples with means that are similar and closer together along the horizontal axis. The curve representing the complete set of data values covers a smaller range and its shape is closer to that of the individual samples. Furthermore, the **sum of the squares** of the differences between the sample means and the overall mean is smaller than the equivalent figure for the samples in Figure 6.4a. Is it possible that the samples might be from one single population rather than four separate ones?

The probabilities of the F distribution used with respect to the two sample F test are also employed in ANOVA and are used to help with answering this question with reference to the chosen level of significance. Box 6.6a illustrates One-way ANOVA with respect to four samples. The test is used to examine whether the difference between the within- and between-groups variances is so small that might well have arisen through chance or sampling error. If this is the case, then it is not only reasonable to argue that population and sample variances are very nearly the same but that their means are also very similar. The calculations for the within- and between-groups variances are laborious if attempted by hand even with the aid of a calculator and Box 6.6d summarises how these are carried out. Degrees of freedom were previously defined as the number of values that were 'free to vary' without affecting the outcome of calculating the mean, standard deviation, etc. In the case of One-way ANOVA there are two areas where such freedom needs to be considered. First, in relation to the overall mean of the entire set of data values, how many of these can be altered without changing the overall grand mean (mean of the means)? The answer is the number of values minus the number of samples. Second, with respect to the means of the individual samples, how many of these can vary without affecting the overall grand mean (mean of the means)? The answer is the number of samples minus one. These, respectively, are known as the within- and between-groups degrees of freedom. The within-groups variance will always be larger than or equal to the between-groups variance, so the F statistic is obtained by dividing the within-groups variance by the between-groups variance. The probability associated with an F statistic calculated for One-way ANOVA is the same as the F ratio for an F test for a particular combination of degrees of freedom. If the outcome from One-way ANOVA indicates that the Null Hypothesis should be rejected (i.e. the means are significantly different), this does not allow an investigator to conclude that one of the samples is 'more different' from the others, even if its mean is seemingly 'more separated' from the others.

> Why is the within-groups variance always larger than or equal to the between-groups variance? Hint: have another look at Figure 6.4.

Box 6.6a   One-way analysis of variance (ANOVA).

Within-groups variance $S_w^2 = \sum \frac{\sum (x_{ij} - \bar{x}_j)^2}{n-k}$

Between-group variance $S_b^2 = \frac{\sum n_j (\bar{x}_j - \bar{x}_t)^2}{k-1}$

ANOVA F statistic: $= \frac{S_w^2}{S_b^2}$

Box 6.6b   Application of One-way ANOVA.

One-way ANOVA assesses the relative contribution of the within-groups and the between-groups variances to the overall or total variance of a variable measured for samples selected from three or more populations. The Null Hypothesis states that population means are the same and only differ in the samples because of sampling error. The procedure involves the calculation of the within-groups and between-groups variances, respectively identified as S_w^2 and S_b^2. In the definitional formulae for these quantities above the subscripts refer to individual data values in each sample (i), the identity of the separate samples or groups (j) and the total data values (t). Adding together the numerators of the equations for the within- and between-groups variances produces a quantity known as the total sum of squares. The ratio of the within-groups variance to between-groups variance produces the F statistic, which will be greater than one unless they

(Continued)

Box 6.6b (Continued)

happen to be equal. The degrees of freedom for the within-groups variance are defined as $df_w = n - k$ and the between-groups as $(df_b = k - 1)$, where n represents the total number of data values and k is the number of groups or samples. The probability of having obtained the F statistic or a larger value helps an investigator decide between the Null and Alternative Hypotheses at the chosen level of significance, usually whether it is ≤ 0.05.

One-way ANOVA has been applied to unequal-sized samples of households in four villages (Kerry (13); Llanenddwyn (19); St Harmon (16); and Talgarth (16)) in mid-Wales. The face-to-face questionnaire survey asked respondents to say where they went to do their main household shopping for groceries before the era of online shopping. Using this information together with grid references for the households' addresses and the shopping towns, 'straight line' distances were calculated to indicate how far people travelled to do their shopping. These distances are tabulated in Box 6.6d with the settlements identified by the subscripts K, L, SH and T together with the sample means and variances. There is some variability in the sample means and variances with the survey suggesting that households in Talgarth travel further to shop. In contrast, households in Llanenddwyn show the least variability by having the smallest variance with most people travelling to one town, Barmouth, approximately 8 km away.

The detailed calculations of the within- and between-group variances would be tedious to reproduce. These have been summarised along with the computation of the F statistic. The results suggest there is a significant difference between the mean distances travelled by the populations of residents in the four settlements to do their shopping, although the test examines the interplay between means and variances, it does specify which sample, or samples is (are) different.

The key stages in carrying out One-way ANOVA are

- *State Null Hypothesis and significance level*: the mean distances travelled to shop by the populations of the four villages are the same and they are only different in sampled households because of sampling error. This variation in sample means is not significant at the 0.05 level of significance.
- *Estimate the within- and between-groups variances:* the results from these calculations are summarised below.
- *Calculate ANOVA statistic (F)*: the calculations for the F statistic appear below ($F = 3.36$).
- *Determine the probability of the calculated F:* the probability (p) of obtaining this F value is 0.025 with 60 degrees of freedom for the within-groups variance and three for the between-groups variance.
- *Accept or reject the Null Hypothesis*: the probability of F is <0.05 (0.025 < 0.05), therefore reject the Null Hypothesis, recognising that this might be an erroneous decision. By implication, the Alternative Hypothesis is accepted.

Box 6.6c   Assumptions of One-way ANOVA.

One-way ANOVA makes three main assumptions:

- Variables should be defined and measured on the interval or ratio scales identically for each of the samples.
- Observations should be chosen randomly with replacement unless the populations are of infinite size.
- Variables should be normally distributed in each of the populations.

Box 6.6d   Calculation of One-way ANOVA.

No.	Kerry (x_n)	Llanenddwyn (x_n)	St Harmon (x_n)	Talgarth (x_n)
1	12.41	14.89	55.05	13.21
2	11.96	8.07	1.30	13.63
3	11.96	8.07	0.45	13.96
4	11.96	0.58	4.62	13.96
5	11.55	8.07	3.67	13.96
6	0.36	7.68	1.92	19.68
7	0.36	7.74	1.92	19.68
8	0.36	7.74	3.67	19.68
9	0.36	7.75	9.11	21.61
10	0.42	15.33	4.62	19.68
11	0.42	7.74	3.67	19.68
12	0.42	0.99	3.67	21.80
13	0.36	7.75	4.43	8.42
14		7.75	4.43	5.41
15		8.45	41.10	8.42
16		7.41	4.43	5.36
17		8.80		
18		8.80		
19		8.60		
Sums	$\sum x_K = 69.20$	$\sum x_L = 152.19$	$\sum x_{SH} = 148.06$	$\sum x_T = 238.16$
Means	$\bar{x}_K = 4.84$	$\bar{x}_L = 8.01$	$\bar{x}_{SH} = 9.25$	$\bar{x}_T = 14.89$
Variances	$s_K^2 = 32.43$	$s_L^2 = 11.55$	$s_{SH}^2 = 239.87$	$s_T^2 = 31.96$

Variance	Within-groups	$S_w^2 = \sum \dfrac{\sum (x_{ij} - \bar{x}_j)^2}{n-k} = \dfrac{4697.809}{64-4} = 262.88$
	Between-groups	$S_b^2 = \dfrac{\sum n_j (\bar{x}_j - \bar{x}_t)^2}{k-1} = \dfrac{788.639}{4-1} = 78.29$
F statistic		$\dfrac{s_w^2}{s_b^2} = \dfrac{262.88}{78.29} = 3.36$
Degrees of freedom	Within-groups	$df_w = n - k - 64 - 4 = 60$
	Between-groups	$df_b = k - 1 = 4 - 1 = 3$
Probability		$p = 0.025$

The worked example in Box 6.6a uses straight line distance between the address and shopping town. Why might this give a misleading indication of the amount of travelling people undertake to do their shopping?

6.5 Confidence Intervals

A sample of observations is often our only route into discovering or estimating anything about the characteristics of its parent population. Even Z and t tests, which test sample data against a specified population mean and variance, or just the population mean, will often not know these details for the current population but rely on undertaking the comparison by reference to an 'older' or former population. An obvious example would be a sample of data collected in 2025 with the population parameters obtained from the British Population Census in 2021. Obtaining statistically significant Z and t test results in these circumstances would either indicate that the population had changed and the parameters relating to the 2021 population were no longer valid or if they are still valid the sample included observations that were untypical of the population. Chapter 2 outlined why investigating all members of a population is rarely a feasible option for researchers, irrespective of the scale of resources available to the study. If a sample is the only source of information about the population, then it would be useful to have some way of knowing how reliable the statistics are produced from its data values. How confident are we that the sample data are providing a reliable result? Thinking about the sampled measurements of glacial melt water stream temperatures taken at the entrance and exit points of the outwash plain, how confident are we that the means produced from these data values (5.7°C and 9.1°C, respectively) are a true reflection of the population parameters?

The answer to this question lies in calculating **confidence limits**, which represent a range within which we can be confident to a certain degree that the population parameter falls. We cannot be absolutely certain that it lies within this range, since sampling error may be present: we express our degree of confidence in percentage or proportionate probability terms. Such confidence limits are typically 95 or 99 per cent confidence, which respectively indicate that we are prepared to be wrong about the population parameter lying within the limits 5 or 1 per cent of times. Confidence limits can be computed from the probabilities associated with the Normal, t or F distributions. Confidence limits with respect to a statistical quantity, such as the mean, comprise two numbers, one larger and one smaller than the mean that defines a range within which we can be 95 or 99 per cent confident that the true population parameter falls. The wider this range, the less precisely are we able to state the population mean, a narrower range indicates that the sample provides a more constrained estimate of the parameter. There is an interaction between the size of the difference and the probability or confidence level: it is possible to be 95 or 99 per cent confident in respect of a wide or narrow confidence limit range.

Calculation of the confidence limits for the Z and t distributions (see Box 6.7a) are similar, although in the latter case, it is necessary to consider that the critical value of t associated with a particular level of probability varies according to the degrees of freedom. The values of Z or t inserted into the equations for calculating confidence limits represent the probability of the population parameter occurring in the tail of the distribution. Thus the 95 per cent confidence limits signify that there is a 5 per cent chance of the population parameter lying beyond this range which is normally split equally between the tails of the distribution, 2.5 per cent in each. In the case of the Z distribution, this point always occurs at a Z value of 1.96, whereas the t statistic corresponding to the same probability or confidence limit varies according to the degrees of freedom. Calculation of confidence limits based on the Z distribution requires the population mean and standard deviation to be known, but if this is the case there seems little point. Confidence limits based on the t distribution are more useful because they can be calculated from sample data. Confidence limits calculated with reference to the Z distribution will be narrower than those based on the t distribution for

---

Box 6.7a   Confidence limits.

Z distribution confidence limits: $-Z_p\sigma_e(\overline{x})$ to $+Z_p\sigma_e(\overline{x})$
t distribution confidence limits: $-t_{p,\,df}s_{\overline{x}}(\overline{x})$ to $+Z_{p,\,df}s_{\overline{x}}(\overline{x})$

Box 6.7b   Application of confidence limits.

The value of Z in the confidence limits equation depends simply on the chosen level of confidence required ±1.96 for 95% and ±2.57 99%. The value of t differs according to the level of confidence and the degrees of freedom: the t values corresponding to the 95 and 99% confidence levels with 10 and 25 degrees of freedom are respectively ±2.23/±2.06; and ±3.17/±2.79. Both types of confidence intervals entail multiplying the Z or t value just given by the standard error of the mean (the population standard error in the case of Z confidence limits). These results are multiplied by the sample mean ($\overline{x}$) to produce a value lower than and another higher than the population mean. The sample mean thus estimates that the true population mean lies within this range with the specified level of confidence.

In practice it is unlikely that the population standard error σ_e will be known without also knowing the population mean μ, thus confidence limits based on the Z distribution are rarely needed. The t distribution limits are much more useful since the population mean is frequently missing. However, both sets of calculations are given in Box 6.7d with respect to the sample of Isle of Wight residents who were given the maps showing county names. The mean score per county was 0.404 from the sample of respondents. The confidence limits based on the Z distribution are narrower than those from the t distribution. The 'known' population mean (0.283) lies outside of the interval produced from the Z confidence limits but within the t distribution limits. This result indicates the relative strength of these distributions.

The key stages in calculating confidence limits (intervals) are:

- *Decide on degree of confidence*: the usual levels are 95 and 99% corresponding to 0.05 and 0.01 probabilities.
- *Determine suitable probability distribution*: Z or t distributions are the main ones used with normally distributed variables measured on the interval or ratio scales.
- *Identify critical value of distribution*: this value will vary according to the degrees of freedom with certain distributions.
- *Insert critical value into confidence limits equation:* calculate limits using equation.

Box 6.7c   Assumptions of Z and t confidence limits.

The assumptions relating to the Z and t tests also apply to their confidence limits:

- Variables should be normally distributed in the population or the sample so large that the sampling distribution of the mean can be assumed normal (Z distribution).
- Observations should be independent of each other.
- Variables should be defined and measured on the interval or ratio scales.
- Observations should be chosen randomly with replacement unless the populations are of infinite size.

Box 6.7d   Calculation of confidence limits.

Z distribution confidence limits	$-Z_p\sigma_e(\bar{x})\,to + Z_p\sigma_e(\bar{x}) =$	$0.404 - (1.96(0.559))$ to $0.404 + (1.96(0.559)) = +0.294$ to $+0.513$
t distribution confidence limits	$-t_{p,\,df}s_{\bar{x}}(\bar{x})\,to + Z_{p,\,df}s_{\bar{x}}(\bar{x}) =$	$0.404 - (2.05(0.846))$ to $0.404 + (2.05(0.846)) = +0.231$ to $+0.576$

---

the same set of data values. Smaller sample sizes will usually produce wider confidence limits than larger ones and thus be less clear about the true value of the population parameter.

6.6   Closing Comments

This chapter has indicated that as investigators we should raise questions about the data values obtained and the descriptive statistics produced from sampled data. It has focused on interval or ratio scale variables and the parametric statistical tests that can be applied to help with reaching decisions about how much confidence can be placed in the information and whether results derived from a sample are significantly different from what might have occurred if random forces had been 'at work'. When applying statistical testing techniques, it is rarely if ever possible to be certain that we have made the right decision with respect to the Null and Alternative Hypotheses, even if we have diligently followed the rules and assumptions of the particular test, entered and checked our data entry into a computer and used reliable software to undertake the computations. There is always a risk of committing a type I or a type II error. These arise because there always remains a chance the results obtained from the sample data upon which the calculations have been based is one of those rare instances when:

- The true hypothesis is rejected (Type I); or
- The false hypothesis is accepted (Type II).

The risk is making a Type I error is given by the level of significance (e.g. 0.05 or 0.01). It might therefore seem prudent to choose a very low significance level in order to minimise the chance of making such errors. However, a low significance level increases the probability is committing a Type II error. We can never be certain which hypothesis is true and which is false, so it is not possible to know whether one of these two types of erroneous decisions has been made.

Further Reading

Agresti, A. (1996) *An Introduction to Categorical Data Analysis*, New York, John Wiley and Sons.

Davies, W.S. (2002) Quantitative methods: Bayesian inference, Bayesian thinking. *Progress in Human Geography*, **26**, 553–566.

Diamond, I. and Jeffries, J. (1999) *Introduction to Quantitative Methods*, London, Sage.

Ebdon, D. (1984) *Statistics in Geography*, 2nd edn, Oxford, Blackwell.

Harris, R. (2016) *Quantitative Geography: The Basics*, Sage Publications Ltd.

Perry, J.N., Liebhold, A.M., Rosenberg, M.S., Dungan, J., Meriti, M., Jakomulska, A. and Citron-Pousty, S. (2002) Illustrations and guidelines for selecting statistical methods for quantifying pattern in ecological data. *Ecography*, **25**, 578–600. DOI: 10.1034/j.1600-0587.2002.250507.x.

Rogerson, P.A. (2006) *Statistical Methods for Geographers: A Student's Guide*, Los Angeles, Sage.

Wrigley, N. (1985) *Categorical Data Analysis for Geographers and Environmental Scientists*, Harlow, Longman.

7

Non-parametric Tests

Chapter 7 has a similar overall structure to the previous chapter, but in this case, the focus is on non-parametric testing. These are examined in relation to the number of samples and/or attributes (variables) and the types of summary measures under scrutiny. It mainly deals with frequency counts of nominal and ordinal attributes as opposed to scalar (ratio/interval) variables. The statistical procedures covered in this chapter are some of the most used by students and researchers in Geography, Earth and Environmental Science and related disciplines.

Learning Outcomes

This chapter will enable readers to:

- Apply non-parametric tests to nominal and ordinal attributes;
- Examine techniques concerned with the median, ordinal and unordered frequency counts;
- Decide whether to accept or reject the Null and Alternative Hypotheses in the case of non-parametric tests;
- Continue planning an independent research investigation in Geography, Earth Science and related disciplines making use of non-parametric procedures.

7.1   Introduction to Non-parametric Tests

The assumptions and requirements of many of the parametric techniques examined in the previous chapter often mean that they are difficult to apply in Geography, Earth and Environmental Sciences. This is especially so when the aim is to investigate people and the connections between them and the groups, institutions and organisation in which they live their lives, and the physical and environmental context in which they act or perform daily. Interest in diverse aspects of this connectivity and the ways in which actions in the past and present will impact future generations and the Earth itself have become an increasingly important focus for our research. Sustainability and global climate change are now primary areas for research and have become central to policy at international, national and local scales. It is a 'classic' example of an issue relating to the interactions between human and physical environments to which geographers, Earth and environmental

Practical Statistics for Geographers and Earth Scientists, Second Edition. Nigel Walford.
© 2025 John Wiley & Sons Ltd. Published 2025 by John Wiley & Sons Ltd.
Companion website: www.wiley.com/go/PracticalStatistics2e

scientists are making significant contributions. The importance of these topics does not diminish the currency of challenges facing humankind such as geopolitics, international migration, inequality and development as well issues falling more in the domain of physical geography and environmental science including eco-systems services, re-wilding, flooding, natural hazards and wildfires. The difficulties of applying parametric procedures are particularly common where attention focuses on the actions and behaviour of humans in different geographical contexts. It is rare for some let alone all of the attributes and variables relevant in such investigations to conform to the assumptions of parametric statistical tests covered in Chapter 6. We saw that parametric tests require adherence to the normality assumption so that a valid comparison may be made between the test statistic calculated from the sample data and the corresponding value in the normal distribution or at least its 'close relative' the t distribution because the probabilities associated with known values of Z and t are intrinsic to their corresponding distributions. Parametric tests typically concern the mean, variance or standard deviation and are often applied to samples of data values in order to estimate a population parameter.

Non-parametric tests do not require that the variables under scrutiny in a sample's parent population should conform to the normal distribution and are sometimes referred to as 'distribution-free'. Most non-parametric procedures examine the median obtained from ordinal (ranked) data, frequency counts of data values produced when samples of observations are distributed between categories or classes, which may or may not be ordered and sequences of discrete events. The order of the classes in a frequency distribution is also important in some non-parametric tests when these represent a numerical sequence or ranking, such as when they are based on measurements of distance, time, elevation, currency, etc. The position in the sequence where each class occurs (its relative position) matters when investigating this type of ordered frequency distribution.

When the data values of scalar (interval or ratio) variables are distributed between classes, the mean can be estimated even when the individual values are unknown as explained in Chapter 5. Thus, although non-parametric techniques do not directly estimate the value of a population parameter with a known probability expressed as a confidence interval, they do help to decide whether the distribution of observed values is significantly different from a random one and whether the approximated mean or other measure of central tendency is in the correct range. The Null Hypothesis in non-parametric tests concerned with frequencies can be stated as examining whether the observed distribution of frequency counts is significantly different from what would have occurred if the entities had 'fallen' randomly into the finite number of available categories.

This chapter parallels the structure of Chapter 6 by grouping non-parametric tests according to the number of attributes (variables) under examination. Table 7.1 provides a summary of the types of comparison examined by the statistical tests in the subsections of this chapter. Most of the procedures are used with either nominal or ordinal frequency distributions and the following sections progress through those dealing with one, two or three or more samples. Pearson's Chi-square test features in several sections since it may be employed in different circumstances. In contrast, some tests, such as the enticingly named the Kolmogorov–Smirnov test, deal with a single sample of observations from which a frequency distribution of a ranked or sequenced attribute can be produced. Some of the non-parametric tests in the remainder of this chapter can be viewed as a non-parametric equivalent of one of the parametric tests discussed in Chapter 6. Wilcoxon's signed ranks test for a single sample of observations measured twice on a variable, which is the non-parametric equivalent of the paired sample t test with less stringent assumptions, is also included in this chapter in section 7.2.1.

Table 7.1 Subsections in Chapter 7 with non-parametric statistical tests for one, two and three or more samples.

	One attribute
One sample	Section 7.2 7.2.1 Comparison of sample and population mean of paired values 7.2.2 Comparison of sample and population nominal attribute counts 7.2.3 Comparison of sample and population ordinal attribute counts 7.2.4 Comparison of sample and population for an ordinal sequence of dichotomous event outcomes
Two samples	Section 7.3 7.3.1 Comparison of sample and population nominal attribute counts (also for one sample and population for two nominal attribute counts) 7.3.2 Comparison of sample (one or two) and population median
Three or more samples	Section 7.4 7.4.1 Comparison sample and population for three or more nominal attribute counts (or two or more nominal attributes for three or more samples) 7.4.2 Comparison of sample and population median (or one sample separated into three or more groups)

7.2 One Variable and One Sample

Section 6.2 introduced parametric tests by saying that student- or large-scale projects are very unlikely to focus on the analysis of just one variable, although attention may quite legitimately focus on a single sample of observations. In parallel with our examination of parametric tests, we begin looking at non-parametric procedures by considering the relatively simple question of whether a frequency distribution produced by assigning the data values of a single attribute for a sample of observations into a set of categories is significantly different from what would have occurred if they had been randomly allocated. The comparison made here, as in most non-parametric tests, does not relate to a single value (the mean or variance) but to the difference between the count of observed and expected frequencies in the classes. Just as with parametric tests, the aim is to discover the probability of a difference having arisen by chance or whether something else might be responsible. If the test produces a statistically significant outcome, the difference is too large to have occurred through chance or sampling error and the researcher needs to account for it in terms of some human or physical process.

This raises the now familiar question of how to determine the probabilities associated with what would occur if a random process had assigned the observations to the classes. The framework of rows and/or columns into which a particular set of entities are distributed provides three types of constraint – the numbers of categories, attributes (variables) and observations – that control the probabilities. The choice of *a posteriori* (after the event) and *a priori* (before the event) probabilities exist, although the frequency distribution derived from population data with which to compare a sample is often difficult to determine. There may be frequency distributions of the same type of observations categorised with respect to the same attributes that were produced from data collected previously or in respect of a different locality. For example, a sample-based frequency distribution may be compared with tables of statistics from an earlier national census or from research in a different region. These statistics can be used to give the empirical probabilities or

Table 7.2 Alternative ways of obtaining expected frequency counts.

Area size class (ha)	Count from earlier census	Proportion	Expected count for a sample of 100 farms based on: Empirical proportions	Equal proportions	Randomness
0.00–99.9	20,923	0.867	87	20	16
100.0–299.9	2389	0.099	10	20	24
300.0–499.9	523	0.217	2	20	25
500.0–699.9	155	0.006	1	20	17
700.0 and over	118	0.005	0	20	18
	24,108	1.000	100	100	100

Note: Expected counts rounded to integers.

proportions of the total number of observations that previously occurred in each of the classes. An alternative approach is to argue that the probability of an observation occurring in a particular category is in proportion to the number of categories or cells available. Table 7.2 illustrates three ways of obtaining the expected counts of a sample of 100 farms in five size categories. The first is based on empirical probabilities, where the proportions of farms in the size groups that occurred in a previous census dictate the expected number in the sample. The second approach simply assigns an equal probability (0.2) to each of the cells. Finally, a random number generator was used to obtain 100 integers between 1 and 5 corresponding to the five categories, which resulted in some clustering in the second and third categories and rather less in the first.

7.2.1 Comparing a Sample Mean with a Population Mean

The distinction between parametric and non-parametric statistical tests is not simply based on the former's focus on summary quantities and the latter's emphasis on frequency counts, since Chapter 5 shows that measures such as the mean and variance can be estimated for continuous (non-integer) data values expressed as frequencies. It is more of a question that parametric tests assume variables follow the normal distribution whereas non-parametric ones do not. There are some non-parametric tests that make less severe assumptions than their parametric counterparts when focusing on statistical measures rather than counts. The **Wilcoxon Signed Ranks** test illustrates this point since it is used to investigate whether the difference in mean for a set of paired measurements with respect to a scalar (interval or ratio) variable for one sample of observations is significant (Box 7.1a). It is the non-parametric equivalent of the paired sample *t* test.

Box 7.1a   The Wilcoxon Signed Ranks *W* test.

W test statistic: $= \dfrac{|\bar{d} - 0|}{s_d / \sqrt{n}}$

Box 7.1b   Application of the Wilcoxon Signed Ranks *W* test.

The Wilcoxon signed ranks test hypotheses that if there is no difference between the means of the paired sets of data values for a single variable in a sample of observations (i.e. subtracting one mean from the other equals zero), the sum of the ranks for the paired measurements for each observation also equals zero. Calculations for the test statistic *W* exclude pairs of measurements where the difference is zero. The probabilities associated with possible values of *W* vary according to the number of non-zero differences. The purpose of the *W* test is to discover the probability of obtaining the observed difference in the ranks as a result of chance and to decide whether the difference is significantly different from zero.

The calculations for the test statistic are simple from a computational point of view, although they do involve several steps. First, calculate the differences between the paired measurements and then, excluding any zero values, sort these from smallest to largest ignoring the positive and negative signs (i.e. +0.04 counts as smaller than −1.25). Now assign the rank score of 1 to the smallest difference, 2 to the next and so on, applying the same rank score to runs of equal differences. Now assign the + or − signs to the rank scores and separately sum the positive and negative ranks: the smaller of these totals is the required test statistic, *W*. Depending on the nature of differences under scrutiny a one- or two-tailed *W* test is needed, with the former involving the probability associated with the test statistic being halved. A decision on the fate of the Null and Alternative Hypotheses may be made according to the chosen level of significance either by reference to the probability of obtaining the *W* test statistic itself or by transforming *W* into *Z* and then determining its probability. Transformation of *W* to *Z* is preferable when there are more than 20 non-zero differences since the *W* distribution approximates the normal distribution.

Information obtained from the samples of residents on the Isle of Wight who were asked about their residential preferences using maps of counties in England and Wales with and without names is used to illustrate the application of the Wilcoxon signed ranks test. The first two data columns in Box 7.1d represent the mean score of each county (excluding those not favoured by any survey respondent) using the named (x_n) and unnamed maps (x_u). Having calculated the differences between these mean scores, the remaining columns represent the different steps involved in calculating *W*. The sums of the positive and negative signed ranks are, respectively, +439 and −465 leading to *W* equalling 439. Since there are more than 20 differences this has been transformed in *Z* to obtain the test statistic's probability. The probability of obtaining a *Z* (or *W*) test value of this size is high and thus at the 0.05 level of significance, the Null Hypothesis can be accepted. There does not appear to be a genuine difference between people's residential preferences when looking at maps with and without names.

The key stages in carrying out a Wilcoxon Signed Ranks test are

- *State Null Hypothesis and significance level*: the difference between the mean rank scores of the two sets of Isle of Wight residents and zero is the result of sampling error and is not significant at the 0.05 level of significance.
- *Calculate test statistic (W)*: the results of the calculations are given below.
- *Select whether to apply a one- or two-tailed test*: in this case, there is no sound reason to argue that the difference should be in one direction from zero or the other since people's residential preferences should not change simply because they have looked at a map with county names as opposed to one without.

(Continued)

Box 7.1b (Continued)

- *Determine the probability of the calculated W (Z):* the probability of obtaining $Z = 0.150$ is 0.881.
- *Accept or reject the Null Hypothesis:* the probability of Z is >0.05, therefore accept the Null Hypothesis and reject the Alternative Hypothesis, recognising that this might be an erroneous decision.

Box 7.1c   Assumptions of the Wilcoxon Signed Ranks *W* test.

There are three main assumptions of the *W* test:

- The differences are independent of each other.
- Zero value differences arise because rounding values to a certain number of decimal places creates equal values, thus their exclusion is arbitrary since they would disappear if more decimal places were used.
- The test is less powerful than the equivalent parametric paired *t*-test, although the decision in respect of the Null and Alternative Hypotheses may be the same.

Box 7.1d   Calculation of the Wilcoxon Signed Rank *W* test statistic.

County	x_n	x_u	$x_n - x_u = d$	Sorted counties	Sorted absolutes	Rank	Signed rank
Avon	0.154	0.423	0.269	Hants.	0.038	4.5	4.5
Beds	0.077	0.000	−0.077	North Yorks.	0.038	4.5	4.5
Berks.	0.077	0.000	−0.077	N'hants.	0.038	4.5	4.5
Bucks.	0.154	0.077	−0.077	Northum.	0.038	4.5	−4.5
Ches.	0.000	0.154	0.154	Notts.	0.038	4.5	−4.5
Clwyd	0.000	0.192	0.192	Shrops.	0.038	4.5	−4.5
Cornwall	1.885	1.462	−0.423	Staffs.	0.038	4.5	−4.5
Cumbria	0.231	0.308	0.077	Somerset	0.038	4.5	4.5
Derbs.	0.500	0.000	−0.500	Beds.	0.077	12	−12
Devon	1.115	1.769	0.654	Berks.	0.077	12	−12
Dorset	0.846	1.269	0.423	Bucks.	0.077	12	−12
Dyfed	0.192	0.077	−0.115	Cumbria	0.077	12	12
East Sussex	0.385	0.115	−0.269	Herts.	0.077	12	−12
Essex	0.154	0.269	0.115	Powys	0.077	12	−12
Glos.	0.000	0.538	0.538	South Yorks.	0.077	12	−12
Greater London	0.269	0.808	0.538	Essex	0.115	17.5	17.5
Greater Manch.	0.154	0.000	−0.154	Dyfed	0.115	17.5	−17.5

County	x_n	x_u	$x_n - x_u = d$	Sorted counties	Sorted absolutes	Rank	Signed rank
Gwynedd	0.038	0.154	0.115	Gwynedd	0.115	17.5	17.5
Hants.	0.692	0.731	0.038	Heref & Worcs.	0.115	17.5	17.5
Heref. & Worcs.	0.000	0.115	0.115	Ches.	0.154	21	21
Herts.	0.192	0.115	−0.077	Greater Manch.	0.154	21	−21
Hum'side	0.192	0.000	−0.192	Lancs.	0.154	21	−21
Kent	0.115	0.731	0.615	Clwyd	0.192	24.5	24.5
Lancs.	0.154	0.000	−0.154	Hum'side	0.192	24.5	−24.5
Leics.	0.269	0.000	−0.269	Lincs.	0.192	24.5	−24.5
Lincs.	0.192	0.000	−0.192	Warks.	0.192	24.5	24.5
Norfolk	0.500	0.000	−0.500	Suffolk	0.231	27	−27
North Yorks.	0.115	0.154	0.038	Avon	0.269	29	29
N'hants.	0.000	0.038	0.038	Leics.	0.269	29	−29
Northum.	0.077	0.038	−0.038	East Sussex	0.269	29	−29
Notts.	0.038	0.000	−0.038	Oxfords.	0.308	31	31
Oxfords.	0.115	0.423	0.308	Dorset	0.423	32.5	32.5
Powys	0.077	0.000	−0.077	Cornwall	0.423	32.5	−32.5
Shrops.	0.154	0.115	−0.038	Derbs.	0.5	34.5	−34.5
Somerset	0.346	0.385	0.038	Norfolk	0.5	34.5	−34.5
South Yorks.	0.077	0.000	−0.077	Glos.	0.538	36.5	36.5
Staffs.	0.077	0.038	−0.038	Greater London	0.538	36.5	36.5
Suffolk	0.308	0.077	−0.231	Kent	0.615	38.5	38.5
Surrey	0.192	1.038	0.846	Wilts.	0.615	38.5	−38.5
Warks.	0.000	0.192	0.192	Devon	0.654	40	40
West Sussex	0.923	0.115	−0.808	West Sussex	0.808	41	−41
Wilts.	0.808	0.192	−0.615	Surrey	0.846	42	42
Sum of the negative ranks			−465				
Sum of the negative ranks			439				
W statistic			439				
Z statistic			$\dfrac{\lvert W - \mu_W \rvert}{\sigma_W} = \left\lvert 439 - \dfrac{42(42+1)}{4} \right\rvert - 0.5 \Big/ \sqrt{\dfrac{42(42+1)((2 \times 42)+1)}{24}}$				
			$Z = 0.150$				
Probability			$p = 0.881$				

7.2.2 Comparing a Sample's Nominal Counts with a Population

The Chi-square test, or more correctly **Pearson's Chi-square test**, is one of the most useful and versatile non-parametric tests. It is used in different guises with nominal (categorical) counts, although the chi-square (χ^2) probability distribution which provides the basis for assessing probabilities relates to continuous scalar measurements rather than frequencies. We have seen that the data values of a normally distributed variable can be transformed or standardised into Z scores (see Chapter 5). A series of n such Z scores (Z_1, Z_2, Z_3, $\cdots$, Z_n) can be squared and summed to produce a value known as Chi-square (χ^2), thus:

$$\chi^2 = (Z_1)^2 + (Z_2)^2 + (Z_3)^2 + \cdots + (Z_n)^2$$

The value of Chi-square depends on the amount of variation in the series of Z scores, if they are all identical the statistic would equal zero with an infinite amount of variation producing a maximum value of plus infinity. The number of data values (the sample size) provides the degrees of freedom, which influences the form and shape of the χ^2 distribution. Small sample sizes (degrees of freedom) produce a distribution curve that is highly skewed to the left (negative), whereas once the number of cases approaches 40 or more the shape and form is like that produced by the normal distribution (Figure 7.1). Given that the form and shape of the curve vary with sample size, but that individual values of χ^2 can be obtained from different sets of data values, it follows that the probabilities associated

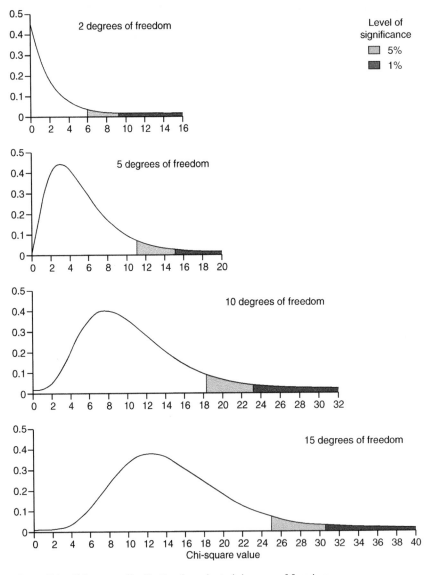

Figure 7.1 Chi-square distribution for selected degrees of freedom.

with the test statistic are related to the degrees of freedom. The probability of obtaining a χ^2 value of 11.34 is 0.50 when there are 12 degrees of freedom (non-significant at 0.05), but 0.01 if there are 3 degrees of freedom (significant at 0.05). The implication is that it is more difficult to obtain a larger chi-square with a relatively small number of observations. Since the minimum value of χ^2 is always zero and its curve extends towards plus infinity at the extreme right of the chart, the probabilities relate to the chance of obtaining the particular χ^2 value or a larger one. Thus, if 100 samples with 3 observations were selected randomly from the same population, five samples should produce χ^2 equal to at least 11.34; similarly, 100 samples with 12 observations in each should result in 50 with a χ^2 of at least 11.34.

The Chi-square statistic can be used to test for a difference between variances, but much more often the closely related Pearson's chi-squared distribution is used with counts of (n) observations distributed across (j) classes or categories. The purpose of Pearson's Chi-square test is to examine whether the difference between the observed and expected frequency distributions of counts is more or less than would be expected to occur by chance. Pearson's Chi-square test statistic, also commonly but incorrectly referred to as a χ^2, is also calculated from the summation of a series of discrete elements:

$$\text{Pearson's chi-square } \chi^2 = \frac{(O_1 - E_1)^2}{E_1} + \frac{(O_2 - E_2)^2}{E_2} + \frac{(O_3 - E_3)^2}{E_3} + \cdots + \frac{(O_j - E_j)^2}{E_j}$$

where O and E, respectively, refer to the observed and expected frequency counts in each of j classes or categories. The result of summing the elements in this series is to produce a statistic that very closely approximates the χ^2 distribution and its associated probabilities with ($j-1$) degrees of freedom.

Box 7.2a   Pearson's Chi-square (χ^2) test.

Pearson's Chi-square (χ^2) test statistic: $= \sum_j^1 \frac{(O_j - E_j)^2}{E_j}$

Box 7.2b   Application of Pearson's Chi-square (χ^2) test.

Pearson's Chi-square test adopts the now familiar cautious stance (hypothesis) that any difference between the observed (O) and expected (E) frequency counts has arisen because of sampling error or chance. It works on the difference between the observed and expected counts for the complete set of observations in aggregate across the groups. Calculations for the test statistic do not involve complex mathematics, but, as with most tests, it can readily be produced using statistics software. The degrees of freedom are defined as the number of groups or cells in the frequency distribution minus 1 ($j-1$). The probabilities of the test statistic depend on the degrees of freedom. The calculations involve dividing the squared difference between the observed and expected frequency count in each cell by the expected frequency and then summing these values to produce the test statistic. Comparison of the probability of having obtained this test statistic value in relation to the chosen level of significance enables a decision to be made regarding whether to accept or reject the Null Hypothesis, and by implication the Alternative Hypothesis.

Pearson's Chi-square test has been applied to households in four villages in mid-Wales that have been pooled into a single sample with respect to the frequency distribution of the mode

(Continued)

Box 7.2b (Continued)

of transport used when they went to do their main household shopping for groceries. This comprises the univariate frequency distribution reproduced in Box 7.2d below. The expected frequencies have been calculated in two ways: on the basis that households would be distributed in equal proportions between the four modes of transport; and alternatively, that they would occur in the unequal proportions obtained from the results of a national survey of households' shopping behaviour. The test statistics for these two applications of the procedure are respectively 9.12 and 14.80, whose corresponding probabilities are 0.023 and 0.002. These are both lower than the 0.05 significance level and so the Alternative Hypothesis is accepted.

The key stages in carrying out a Pearson's Chi-square test are:

- *State Null Hypothesis and significance level*: the difference between the observed and expected frequencies has arisen because of sampling error and is not significant at the 0.05 level of significance. The sample frequency distribution is not significantly different from either an equal allocation between the categories or what would be expected according to a previous national survey.
- *Calculate test statistic (χ^2)*: the results of the calculations are given below.
- *Determine the degrees of freedom*: in this example, there are 4 − 1 = 3 degrees of freedom.
- *Determine the probability of the calculated χ^2*: the probability of obtaining $\chi^2 = 9.12$ and $\chi^2 = 14.80$ each with three degrees of freedom are respectively 0.023 and 0.002.
- *Accept or reject the Null Hypothesis*: the probability is <0.05 for both χ^2 values, therefore reject the Null Hypotheses and accept the Alternative Hypothesis.

Box 7.2c Assumptions of Pearson's Chi-square (χ^2) test.

There are five main assumptions of Pearson's Chi-square test:

- Simple or stratified random sampling should be used to select observations that should be independent of each other.
- The number of categories or classes has a major effect on the size of the test statistic. In the case of nominal categories (as in this transport to shop example) these may be pre-determined, although could be collapsed but not expanded (family car and other cars could be merged) if required. However, there are many possibilities for classifying continuous interval or ratio scale variables (see Chapter 5).
- Frequencies expressed as percentages (relative frequencies) will produce an inaccurate probability for Pearson's Chi-square statistic: if the sample size is <100 it will be too low and if >100 to high. If, as in the worked example, the sample size happens to equal 100 then absolute and relative frequencies are the same and the probability is accurate.
- The minimum expected frequency count per cell is five and overall, no more than 20% of the cells should be at or below this level. The number of cells with low expected frequencies may be adjusted by redefining or merging classes or categories.
- The categories or classes should not be ordered or ranked (e.g. a sequence of distance groups or ordered social classes) because the same χ^2 will be obtained irrespective of the order in which categories are tabulated.

Box 7.2d   Calculation of Pearson's Chi-square (χ^2) test.

Transport mode	Observed sample frequencies	Exp freq (equal props)	$\frac{(O-E)^2}{E}$	Exp freq (props from national survey)	$\frac{(O-E)^2}{E}$
Family car	36	25	4.84	40	0.40
Other car	18	25	1.96	10	6.40
Public transport	18	25	1.96	30	4.80
Walk or cycle	28	25	0.36	20	3.20
Total	100	100	9.12	100	14.80
Pearson's Chi-square statistic	Equal proportions	$\chi^2 = 9.12$		Proportions from national survey	$\chi^2 = 14.80$
Probability	Equal proportions	$p = 0.023$		Proportions from national survey	$p = 0.002$

7.2.3   Comparing a Sample's Ordinal Counts with a Population

The requirement that Pearson's Chi-square test should only be used with attributes or variables whose categories are simple nominal groups and not ordered or ranked in any way might initially seem unproblematic. After all, there must be lots of attributes and variables that simply differentiate between observations and do not put them in order. However, the more you think about it the more classified variables measuring such things as distance, area, temperature, acidity/alkalinity, volume and population as well as various financial quantities (land value, income, expenditure, rental value, etc.) do in fact summarise an ordered sequence of data values. For example, parcels of land classified according to different types of use and assigned to zones from a city centre constitute an ordered sequence of distance zones. Perhaps there are only relatively few truly nominal categories such as gender, household tenure, mode of transport, ethnicity, rock type and energy source. Nevertheless, some form of statistical test is required that enables investigators to get around the assumption that Pearson's Chi-square test should not be used with ordered categories. One such test is the **Kolmogorov–Smirnov D test**.

List another five nominal attributes, more if you want, that do **not** have some implied order in the categories or classes. Reflect on why ordered classes are a problem with Pearson's Chi-square test.

Hints: what do ordered classes suggest when interpreting a frequency counts? What happens to the Pearson's Chi-square statistic if you re-order the classes (e.g. change the sequence from lowest income through to highest-income class to highest-income class through to lowest income class)?

One of the most typical applications of the Kolmogorov–Smirnov D test involves comparing an observed set of frequency counts for a ranked attribute or variable with the frequencies expected to arise by applying a probability distribution, such as the Poisson distribution. This test focuses on the cumulative observed and expected probabilities of the number of observations in each of the series of ordered classes. The D statistic is simply the maximum absolute difference between the cumulative observed and expected probabilities. Because the observations are distributed between an

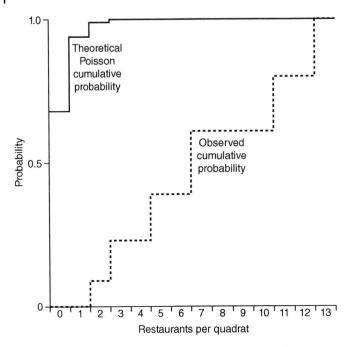

Figure 7.2 Comparison of Poisson and observed cumulative probability distributions.

ordered series of classes, the cumulative probabilities for each successive class represent the proportion of observations that have occurred or would be expected to occur according to the Poisson distribution up to that point. Figure 7.2 illustrates how the two cumulative probabilities increment in a stepwise fashion and that the maximum difference between them indicates the goodness of fit between the two distributions. The Poisson distribution predicts nearly all (99 per cent) of the restaurants in the dataset falling in the first, second and third categories (labelled 0, 1 and 2 in Figure 7.2 denoting grid squares (quadrats) with none, one and two). The observed cumulative frequencies show that only about 10 per cent of all restaurants were in squares with 0, 1 or 2 occurrences. Application of the Kolmogorov–Smirnov is these circumstances is not universally accepted as legitimate by statisticians, since calculation of the Poisson expected frequencies relies on using the mean or average density of the sampled observations, which contravenes the expectation that they are obtained independently of the sample data. Despite these reservations, provided that a sufficiently rigorous level of significance is applied and caution is exercised in marginal cases, then the *D* test can prove to be a useful way of testing a ranked frequency distribution.

Box 7.3a illustrates use of the test to examine whether the distance travelled by households in four villages in mid-Wales to the settlement where they do their main shopping is similar to what might occur in the process of choosing where to go was random or otherwise. The distances have been classified into seven groups with the first containing the count of households travelling less than 0.5 km, regarded for this purpose as zero, the next class 0.5–10.4 km and so on up to the seventh with just one household that journeyed in the range 50.5–60.4 km. The test results in this case indicate there is a moderate probability of obtaining *D* equal to 0.77 with 64 observations and so, despite some statisticians' reservations about the procedure, it may be considered reasonably reliable in this case.

Box 7.3a   The Kolmogorov–Smirnov *D* test.

The Kolmogorov–Smirnov *D* test statistic: = MAX Abs(Cum P_{obs} – Cum P_{Exp})

Box 7.3b   Application of the Kolmogorov–Smirnov *D* test.

The starting point for the Kolmogorov–Smirnov test is that any difference between the cumulative observed and expected frequency distributions of an ordered (ranked) attribute or classified variable for a single sample of observations is a consequence of sampling error. Application of the test proceeds by calculating the difference between the cumulative observed (Cum P_{Obs}) and expected (Cum P_{Exp}) probabilities across the ordered sequence of classes and the largest difference provides the *D* test statistic. The probability of having obtained the particular value of *D*, which varies according to sample size, n, helps with deciding on whether to accept or reject the Null Hypothesis at a chosen level of significance.

The Kolmogorov–Smirnov test has been applied to the distance travelled to the main settlement for shopping by households in the group of mid-Wales villages, with the original continuous variable classified into distance bands (classes). Having tabulated the survey data into the observed frequency counts (f_{Obs}), the observed and expected probabilities are required. Box 5.7 (Chapter 5) shows how the Poisson probability distribution could be applied to produce such probabilities and has been used in this case. The observed probabilities are obtained by dividing the number of households in each class by the total (e.g. $64/35$ = 0.55). There are seven classes and 64 households producing an overall mean density of households per ranked class (λ) of 0.11 which has been inserted into the Poisson distribution equation to produce the expected probabilities. Each set of probabilities is cumulated across the classes and taking the absolute difference between these cumulative probabilities yields a column of differences. The *D* statistic can be identified as the maximum difference, which is 0.77 in this case. Since the probability of having obtained a maximum difference of this size is 0.17, which is higher than the standard significance level (0.05), the Null Hypothesis is accepted.

The key stages in carrying out a Kolmogorov–Smirnov test are:

- *State Null Hypothesis and significance level*: sampling error accounts for the difference between the cumulative observed and expected frequencies and its magnitude is likely to have arisen more often than 95 times in 100. The observed ordered frequency distribution from the sample of observations is not significantly different from what would be expected according to the Poisson probability distribution
- *Calculate test statistic (D)*: the tabulated ordered frequency distributions and calculations are given below showing the maximum difference is 0.77.
- *Determine the probability of the calculated D*: the probability of the test statistic varies according to sample size and in this case with a sample of 64 observations (households) it is 0.17.
- *Accept or reject the Null Hypothesis*: the probability of *D* is <0.05, so the Null Hypothesis should be accepted with the consequence that the Alternative Hypothesis is rejected, recognising that this might be an erroneous decision.

Box 7.3c   Assumptions of the Kolmogorov–Smirnov (*D*) test.

There is one main assumption of the Kolmogorov–Smirnov test:

- Observations should be selected by means of simple random sampling and be independent of each other.

Box 7.3d   Calculation of the Kolmogorov–Smirnov *D* statistic.

Distance (km)	f_{Obs}	P_{Obs}	P_{Exp}	Cum P_{Obs}	Cum P_{Exp}	Abs(Cum P_{obs} – Cum P_{Exp})
0	8	0.13	0.90	0.13	0.90	**0.77**
0.5–10.4	35	0.56	0.10	0.67	0.99	0.32
10.5–20.4	17	0.27	0.01	0.94	1.00	0.06
20.5–30.4	2	0.03	0.00	0.97	1.00	0.03
30.5–40.4	0	0.00	0.00	0.97	1.00	0.03
40.5–50.4	1	0.02	0.00	0.98	1.00	0.02
50.5–60.4	1	0.02	0.00	1.00	1.00	0.00
	64	1.00	1.00			

Kolmogorov–Smirnov statistic: $D = \text{MAX Abs}(\text{Cum } P_{obs} - \text{Cum } P_{Exp}) = 0.77$

Probability: $p = 0.17$

7.2.4   Comparing the Ordinal Sequence of Dichotomous Outcomes for a Sample with a Population

The Kolmogorov–Smirnov test is used with an ordered series of classes in a frequency distribution, in other words, the observations have been allocated to a set of integer classes, but the test pays no attention to the order in which the events or outcomes themselves occurred. This characteristic can be examined by the **Runs (or Wald-Walfowitz) test**, which focuses on the sequence of individual dichotomous (two option) outcomes through time and across space and, as the name suggests, whether 'runs' of the same outcome occur or if they are jumbled up. The test can be used to examine whether the outcomes for a sample of observations can be regarded as independent of each other. The test is founded on the notion that the outcomes of a certain set of events can only be ordered in a finite series of combinations, since the number of events is not open-ended. If the number of runs observed in a particular sample exceeds or falls short of what might be expected, then the independence of the observations is brought into question. A run is defined as either one or an uninterrupted series of identical outcomes. Suppose an event involves tossing a coin four times in succession to find out how many combinations or sequences of outcomes there are that include a total of two heads and two tails: the answer is that there are six:

HHTT; HTHT; HTTH; TTHH; THTH; and THHT

> Give two examples of tossing a coin four times and not getting a total of two heads and two tails. How many runs of the same outcome are there in each of these sequences?
> Hint: how many runs with TH?

Box 7.4a applies the Runs test to a more geographical problem. Box 7.4b shows a 250 m stretch of road in a residential area along which there is a series of 50 segments and if there was at least one car or another type of vehicle parked at the kerb in a 5 m segment it was counted as occupied otherwise it was unoccupied. There were 19 and 31 segments respectively occupied and unoccupied on the day when the data were recorded, indicating that 38 per cent had vehicles parked in 25 runs. Driving along the road from one segment to the next would mean that a new run started when going from a segment with parked vehicles to one without or *vice versa*, whereas adjacent segments with the same outcome would be part of the same run. The Runs test is applied in this case to determine whether there is sufficient evidence to conclude that the numbers of runs along the stretch of road on this occasion is random. The test seeks to establish whether the observed number of runs is more or less than the expected mean given that the overall sequence of dichotomous outcomes is finite. The comparison is made using the Z statistic and its probability distribution. Figure 7.3 relates to the example in Box 7.4a where there are a total of 50 outcomes in the sequence and illustrates the distinctive shape of probability distribution. There are two areas where the combinations of dichotomous outcomes (e.g. total segments with and without parked vehicles) fall below the 0.05 significance level (9 and 41; 10 and 40; 11 and 39; 12 and 38; and 13 and 37), where the Null Hypothesis would be rejected. The example of 19 and 31 segments in the two outcomes is not one of these and

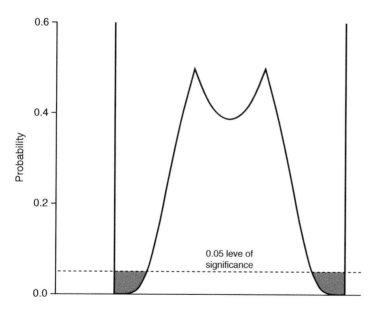

Figure 7.3 Probability distribution associated with the Runs test.

so the Null Hypothesis is accepted. It is worth noting before moving on from the Runs test that the sequence of occupied and unoccupied road segments in this example may display **spatial autocorrelation**, a tendency for similar data values to occur close together in space, a topic that we will look at in Chapter 8 and the Explicitly Spatial Section. However, for the time being, it is useful to consider the idea that in a sequence of 50 dichotomous outcomes a combination of 19 and 31 runs of the two types could produce a number of different spatial patterns in addition to the one shown in Box 7.4d.

Box 7.4a   The Runs (Wald-Walfowitz) test.

R = number of changes of dichotomous outcomes in a complete sequence

Conversion of R to Z test statistic: $= \dfrac{R - \mu_R}{\sqrt{\sigma_R^2}}$

Box 7.4b   Application of the Runs (Wald-Walfowitz) test.

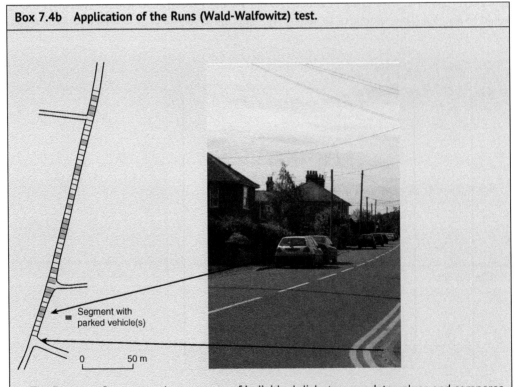

Segment with
■ parked vehicle(s)

0   50 m

The Runs test focuses on the sequence of individual dichotomous data values and compares the observed number of runs (R) with the average (μ_R) that be expected if changes in the sequence of outcomes had occurred randomly. The expected mean (μ_R) and variance (σ_R^2)

are calculated empirically from the sample data and are used to convert the observed number of runs into a Z statistic. The Null Hypothesis is evaluated by comparing the probability of Z in relation to the chosen level of significance, typically 0.05.

Cars and other vehicles parked at kerbs along roads going through residential areas are commonplace: from the motorists' perspective they represent an obstacle impeding and slowing progress, for residents they constitute a convenient parking place, for cyclists they are a potential hazard if vehicle occupants open roadside doors and for pedestrians, they present a potential hazard when attempting to cross a road. If we consider any given stretch of road capable of being divided into segments of equal length (say 5 m), the Runs test has been used to examine the hypothesis that the presence or absence of one or more vehicles in a sequence of segments is mutually independent of each other. The Runs test has been used in respect of the stretch of road shown above together with a photo to illustrate kerbside parking. The sequence of segments occupied and unoccupied by cars or other vehicles is tabulated in Box 7.4d with the runs highlighted. The 25 runs were of various lengths ranging from single segments up to a run of five segments without any parked vehicles. The expected number of runs was 24.56 and the variance 10.85, which produces a Z statistic with high probability and so the Null Hypothesis of a random number of runs is upheld. It would seem that the outcome in each segment of the road is independent of the others.

The key stages in performing a runs test and calculating the associated Z statistic are:

- *State Null Hypothesis and significance level*: the number of runs is no more or less than would be expected to have occurred randomly and the sequence of outcomes indicates the observations are independent of each other at the 0.05 level of significance.
- *Determine the number of runs, R*: the sequence of road segments with and without parked vehicles is given below and the alternated shaded and unshaded sections indicate there are 25 runs.
- *Calculate Z statistic*: the Z statistic requires that the mean and variance of the sequence be obtained before calculating the test statistic itself, which equals 0.134 in this case.
- *Determine the probability of the Z statistic*: this probability is 0.897.
- *Accept or reject the Null Hypothesis*: the probability of Z is much greater than 0.05 and so the Null Hypothesis should be accepted, recognising that this might be an erroneous decision.

Box 7.4c   Assumptions of the Runs test.

There is one assumption of the Runs test:

- Observations are assumed independent of each other (the test examines whether this is likely to be the case).

Box 7.4d Calculation of the Runs test.

Segment	1	2	3	4	5	6	7	8	9	10	11	12
Car	NC	NC	NC	NC	C	C	NC	NC	C	C	NC	C
Runs	1	1	1	1	2	2	3	3	4	4	5	6
Seg. (cont.)	13	14	15	16	17	18	19	20	21	22	23	24
Car (cont.)	NC	C	NC	NC	C	NC	NC	NC	C	C	C	C
Runs (cont.)	7	8	9	9	10	11	11	11	12	12	12	12
Seg. (cont.)	25	26	27	28	29	30	31	32	33	34	35	36
Car (cont.)	NC	NC	C	NC	C	NC	NC	NC	NC	C	C	NC
Runs (cont.)	13	13	14	15	16	17	17	17	17	18	18	19
Seg. (cont.)	37	38	39	40	41	42	43	44	45	46	47	48
Car (cont.)	NC	NC	NC	C	NC	NC	NC	NC	NC	C	C	NC
Runs (cont.)	19	19	19	20	21	21	21	21	21	22	22	23
Seg. (cont.)	49	50										
Car (cont.)	C	NC		Segs. with cars $N_C = 19$					Segs. without cars $N_{NC} = 31$			
Runs (cont.)	24	25		Total runs $R = 25$								

Mean

$$\mu_R = \left(\frac{(2(N_C \times N_{NC}))}{N}\right) + 1 = \left(\frac{2(19 \times 31)}{50}\right) + 1 = 24.56$$

Variance

$$\sigma_R^2 = \frac{(2(N_C N_{NC})(2(N_C N_{NC} - N_C - N_{NC}))}{(N)^2(N-1)} = \frac{2(1178(1178 - 19 - 31))}{250(49)} = 10.85$$

Z

$$Z = \frac{R - \mu_R}{\sqrt{\sigma_R^2}} = \frac{25 - 24.56}{\sqrt{10.85}} = 0.134$$

Probability

$$p = 0.897$$

7.3 Two Samples and One (or More) Variable(s)

Many investigations are concerned with examining the differences and similarities between two populations through separate random or stratified random samples rather than the more limited case of focusing on a single set of observations. Our exploration of non-parametric tests now moves on to such instances as samples drawn from populations of arable and dairy farms, acidic and alkaline soils, sedimentary and igneous rocks, and coastal and inland cities. The majority of questions to be examined in these situations involve nominal or ordinal frequency counts rather than measurements for continuous variables and are concerned with deciding the fate of Null Hypotheses that claim the observations from the two samples are randomly distributed amongst the categories of an attribute or classified variable. Such tests seek to establish if the samples are genuinely from two significantly different populations or whether they should really be viewed as just one population. The tests work by comparing the observed frequency counts with those that would be expected according to a random probability distribution.

Some of the procedures examined in this section can also be applied in situations where there is only one sample. In these circumstances, rather than concentrating on the differences between two samples, the objective is to examine the frequencies of observations for two nominal attributes or categorised variables. For example, the data values in a set of water samples collected at separate locations along a river might be cross-tabulated with respect to an attribute recording the land use at the sample point and the acidity/alkalinity of the water with the aim of investigating any association between these characteristics. Most investigations using the non-parametric procedures examined here will carry out a series of such analyses to explore or mine the data values in the sample(s) with the aim of reaching conclusions about the research questions.

7.3.1 Comparing Two Attributes for One Sample (or One Attribute for Two or More Samples)

Section 7.2.2 outlined the theoretical background to Pearson's Chi-square test and illustrated how it could be applied to a univariate frequency distribution. The same test can be used in two further ways:

- to examine whether a sample of observations cross-tabulated on two (or more) attributes or classified variables are independent of each other.
- to discover whether the frequency distributions of observations from two (or more) samples in respect of the same attribute or categorised variable are significantly different from each other.

When used in the second situation Pearson's Chi-square test is the non-parametric equivalent of One-way Analysis of Variance (see Chapter 6). Although conceptually different, the observations in each case are usually shown for the purpose of statistical analysis in the form of a cross-tabulation comprising rows (r) and columns (c), which define the size of the table in terms of total cells. The simplest such table is a two-by-two cross-tabulation where the total observations have been distributed across four cells. There is no reason why the two dimensions of a table must be equal and it is entirely feasible for the observations in a sample to be tabulated in respect of one attribute with three categories and another with four to produce a table with 12 cells.

> How many cells would there be in a cross-tabulation of five samples by an attribute with three categories?

The count of observations in the cells of a table is finite in any particular dataset and so the more cells there are the greater the chance that some will have few or no observations. In many cases the number of cells is also finite, being determined by the number of attribute/variable categories and/or samples. However, there are occasions when it is justifiable to introduce some flexibility by combining or collapsing the categories of an attribute or variable to increase counts in at least some of the cells. Farms in the parishes stretching across the South Downs in East and West Sussex have been allocated to two sets of size classes in Table 7.3 according to their land areas as recorded in the 1941 Agricultural Census. The first classification has 13 groups with eight of these referring to farms under 100 ha, whereas the second is more general with only two classes below this size threshold. The total number of farms in each classification is the same and low counts in some cells have been overcome by collapsing some of them together. The interaction between the number of rows and columns and the total count of observations not only defines the cell size of a table but also constrains how the bivariate distribution is free to vary, in other words, they control the degrees of freedom. Referring to Table 7.3 only one cell is free to vary within each column, because total farms in the South Downs parishes in the two counties are fixed. However, when distributing the observations in a single sample across the intersecting rows and columns defined by two categorised attributes/variables, the degrees of freedom are defined as $(r-1)(c-1)$.

Table 7.3 The effect of adjusting attribute or variable categories on frequency distribution counts.

Detailed classification	East Sussex	West Sussex	Total	General classification	East Sussex	West Sussex	Total
0.0–1.9	3	1	4	0.00–49.9	174	177	351
2.0–4.9	54	54	108				
5.0–9.9	39	38	77				
10.0–19.9	30	27	57				
20.0–29.9	23	23	46				
30.0–39.9	17	18	35				
40.0–49.9	8	16	24				
50.0–99.9	40	42	82	50–99.9	40	42	82
100.0–199.9	48	49	97	100.0–299.9	61	68	129
200.0–299.9	13	19	32				
300.0–499.9	4	17	21	300.0 and over	8	20	28
500.0–699.9	3	1	4				
700.0 and over	1	2	3				
Total	283	307	590	Total	283	307	590

Source: MAFF Agricultural Statistics, 1941.

How many degrees of freedom are there in this cross-tabulation?	A	B	C	D	E
1					
2					
3					

Pearson's Chi-square test offers a flexible and conceptually straightforward means of addressing research questions in Geography, Earth and environmental sciences together with many of the social sciences. It is concerned with testing a Null Hypothesis with respect to observations that are independently allocated to the cells in the cross-tabulation.

In the case of a two-by-two table, the fact that an observation is assigned to the top row has no influence on whether it is assigned to the left or right column. Pearson's Chi-square test is robust, although since the probabilities associated with test statistic are influenced by sample size the minimum number of observations is usually set at 30. However, if the dimensions of the cross-tabulation result in more than 20 per cent of cells having an expected count of five or less then the total sample size should be larger. This issue should be addressed when planning the data collection stage to have a large enough sample and to think in detail about cross-tabulations you intend to produce. It is usually impossible to make the necessary changes after you have collected your data. Box 7.5a illustrates how to apply Pearson's Chi-square test when there are two or more samples and one attribute using data from the survey of households in four settlements in mid-Wales. The application examines whether people in the four settlements use different modes of transport to carry out their major shopping trip, or equivalently whether distinct types of transport are favoured by residents in the sampled settlements.

Box 7.5a   The Bivariate Pearson's Chi-square (χ^2) test.

Pearson's Chi-square (χ^2) test statistic: $= \sum\limits_{j}^{1} \dfrac{\left(O_j - E_j\right)^2}{E_j}$

Box 7.5b   Application of the Bivariate Pearson's Chi-square (χ^2) test.

Pearson's Chi-square test focuses on the overall difference between the observed (f_O) and expected (f_E) frequencies. The test statistic formula calculates the sum of the squared differences between the observed and expected counts divided by the expected for each cell. The first step in carrying out the test is to distribute the observations into a cross-tabulation and then calculate the expected frequencies. Calculating the expected frequencies for any individual cell involves multiplying the corresponding marginal row and column totals and dividing the result in each cell by the total number of observations (see the upper part of Box 7.5d). There is an observed count of eight households in the case of the top left cell (travel by family car and Barmouth sample) and the expected value is $7.2 \left(\dfrac{36 \times 20}{100} \right)$: the difference is very small (0.8) in this cell, which contrasts with the lower right cell where the difference is 4.8. The second step involves squaring the difference in each cell and dividing by the expected count (lower part of Box 7.5d) and then summing these to obtain χ^2. The probability of obtaining the χ^2 value varies according to the degrees of freedom. A decision on the Null Hypothesis is reached by comparing this probability with the chosen level of significance. If the probability of having obtained the test value is equal to or less than 0.05 (or another selected significance level), the Null Hypothesis should be rejected.

The samples of households in the four settlements in mid-Wales have been tested with Pearson's Chi-square to see if there is a significant difference in the mode of transport used when doing their main household shopping. The χ^2 statistic is 10.440, which has a probability of 0.316 with nine degrees of freedom. This implies that the Null Hypothesis should be accepted since for a four-by-four cross-tabulation (i.e. nine degrees of freedom) the test statistic value would have had to reach at least 16.619. The test points to the conclusion that households in the four settlements tend to divide between the different modes of transport in the same fashion.

Application of Pearson's Chi-square test has six main stages:

- *Declare the Null Hypothesis and significance level*: any differences between the observed and expected frequencies are a consequence of sampling error and they are not significant at the 0.05 level.
- *Computer the expected frequencies* (f_E): these come from dividing the product of the marginal totals corresponding to each cell by the overall total number of observations.
- *Calculate test statistic* (χ^2): this is the sum of the squared differences between f_O and f_E divided by f_E.
- *Determine the degrees of freedom*: in this example df $= (4-1)(4-1)$ and so there are nine degrees of freedom.
- *Determine the probability of the calculated χ^2*: the test statistic equals 10.440, which has a probability of 0.316 with nine degrees of freedom.
- *Accept or reject the Null Hypothesis*: Null Hypothesis should be accepted since 0.316 is >0.05 and would be obtained in approximately 32 samples out of a 100, if households in the four settlements were sampled on that number of occasions, recognising that this might be an erroneous decision.

Box 7.5c   Assumptions of Pearson's Chi-square (χ^2) test.

There are five main assumptions of Mann-Whitney test:

- Observations selected for inclusion in the sample(s) should be selected by simple or stratified random sample and be independent of each other.
- Adjustment of attribute/variable classes may be carried out, although care should be taken to ensure the categories are still meaningful.
- The total count of observations in the four samples from the mid-Wales settlements happens to equal 100 and the cell counts are percentages of the overall total. However, the test should not be applied to data recorded as relative frequencies (percentages or proportions) and should be converted back to absolute counts before carrying out a Pearson's Chi-square test.
- Low expected frequencies, usually taken to mean ≤ 5 counts in no more than 20% of cells, are likely to mean that the χ^2 test statistic is unreliable. The problem may be overcome by re-defining attribute/variable classes.
- The test should not be applied if any of the attributes or grouped variables are ordered or ranked at all, since the calculated test statistic will be the same regardless of the ranking.

Box 7.5d   Calculation of Pearson's Chi-square (χ^2) test.

		Sample settlement			Row totals
Transport mode	**Barmouth**	**Brecon**	**Newtown**	**Talgarth**	
Family car f_O	8	6	16	6	36
f_E	$\dfrac{(36 \times 20)}{100} = 7.2$	$\dfrac{(36 \times 26)}{100} = 9.4$	$\dfrac{(36 \times 35)}{100} = 12.6$	$\dfrac{(36 \times 19)}{100} = 6.8$	36
Other car f_O	4	3	6	5	18
f_E	$\dfrac{(18 \times 20)}{100} = 3.6$	$\dfrac{(18 \times 26)}{100} = 4.7$	$\dfrac{(18 \times 35)}{100} = 6.3$	$\dfrac{(18 \times 19)}{100} = 3.4$	18
Public transport f_O	3	5	8	2	18
f_E	$\dfrac{(18 \times 20)}{100} = 3.6$	$\dfrac{(18 \times 26)}{100} = 4.7$	$\dfrac{(18 \times 35)}{100} = 6.3$	$\dfrac{(18 \times 19)}{100} = 3.4$	18
Walk or cycle f_O	5	12	5	6	28
f_E	$\dfrac{(28 \times 20)}{100} = 5.6$	$\dfrac{(28 \times 26)}{100} = 7.3$	$\dfrac{(28 \times 35)}{100} = 9.8$	$\dfrac{(28 \times 19)}{100} = 5.3$	28
Column totals	20	26	35	19	100

Family car	$\dfrac{(8-7.2)^2}{7.2} = 0.09$	$\dfrac{(6-9.4)^2}{9.4} = 1.21$	$\dfrac{(16-12.6)^2}{12.6} = 0.92$	$\dfrac{(6-6.8)^2}{6.8} = 0.10$
Other car	$\dfrac{(4-3.6)^2}{3.6} = 0.04$	$\dfrac{(3-4.7)^2}{4.7} = 0.60$	$\dfrac{(6-6.3)^2}{6.3} = 0.01$	$\dfrac{(5-3.4)^2}{3.4} = 0.73$

| | Sample settlement | | | | Row |
Transport mode	Barmouth	Brecon	Newtown	Talgarth	totals
Public transport	$\frac{(3-3.6)^2}{3.6}=0.10$	$\frac{(5-4.7)^2}{4.7}=0.02$	$\frac{(8-6.3)^2}{6.3}=0.46$	$\frac{(2-3.4)^2}{3.4}=0.59$	
Walk or cycle	$\frac{(5-5.6)^2}{5.6}=0.06$	$\frac{(8-7.2)^2}{7.2}=0.09$	$\frac{(5-9.8)^2}{9.8}=02.35$	$\frac{(6-5.3)^2}{5.3}=0.09$	
Pearson's chi-square statistic			$\chi^2=10.440$		
Degrees of freedom			$df=(r-1)(c-1)=(4-1)(4-1)=9$		
Probability			$p=0.316$		

7.3.2 Comparing the Medians of an Ordinal Variable for One or Two Samples

The basis of the **Mann–Whitney U test** is that a set of ordered data values for a variable can be partitioned into two groups. This can arise in two ways

- there are two samples that have been measured in respect of the same ordinal variable.
- there is one sample containing a dichotomous attribute, which may be spatial or temporal, enabling the observations to be split into two groups across an ordinal variable.

The test works by pooling the ordinal measurements across the complete set of observations and then compares the medians for each group or sample. The test establishes whether the quantitative difference between the medians is small enough simply to be the result of sampling error. If the test is applied to variables that were originally collected on the interval or ratio scales of measurement, then some loss of information is implied by sorting into ascending order and then assigning rank scores. However, where the assumption of normality of the variables cannot realistically be upheld, the Mann–Whitney test provides a robust alternative to the (unpaired) two sample t test. Box 7.6a applies the test to the sample of road segments with or without vehicles analysed previously using the Runs test.

Box 7.6a   Mann–Whitney U test.

Mann–Whitney U test statistic: $=\sum r-\dfrac{n_r(n_r+1)}{2}$

Conversion of U to Z test statistic: $=\dfrac{|U-\mu_U|}{\sigma_U}$

Box 7.6b   Application of Mann–Whitney U test.

The Mann–Whitney U test is commonly used to examine the difference between the medians of an ordinal variable for two samples or two groups of observations from one sample. The objective in both instances is to investigate whether the difference is of such a size that it is likely to have arisen from sampling error, in other words the true difference is zero. In general terms, the Null Hypothesis is that the rank scores and therefore the medians of observations in the two samples or groups from a sample are such that they balance each other and so they really originate from one undifferentiated population with respect to the ordinal variable being examined. The calculations for the test statistic are straightforward. First, the observations are sorted in ascending order with

(Continued)

Box 7.6b (Continued)

respect to the variable and their ranks (R) are assigned (smallest value assigned rank 1, second rank 2 and so on). The second step involves sorting the observations' rank scores according to the sample or group to which they belong. The rank scores are summed for each group of observations and the smaller total is designated $\sum r$. The value of this quantity is inserted into the equation for the test statistic together with n_r – the number of observations in the group that provides $\sum r$.

The test statistic's probability is determined in different ways depending on the number of observations in each group, although using statistical software to apply the test usually removes the need to choose between the methods. The probability distribution of U becomes approximately Normal when there are more than 20 observations in each sample or group and the probability of the test statistic is obtained by transforming U into Z having calculated a population mean and standard deviation of U given the known, fixed number of observations in each sample or group (see calculations below). The fate of the Null and Alternative Hypotheses is decided by reference to the specified significance level, typically whether ≤ 0.05.

The Mann–Whitney test has been applied to the 250 m stretch of road divided into 5 m segments used in Box 7.4a. The distance from the start of the sequence of segments increments by 5 m when progressing from one segment to the next and they are categorised into two groups by means of the dichotomous attribute denoting whether there are parked vehicles on at least one side of the road: these are denoted by the subscripts N and NC respectively. The raw data (distance measurements and rank scores) are given below. The Null Hypothesis is that the median distance of segments in the two groups from the start of the stretch of road is the same and that sampling error accounts for the observed difference. Although strictly speaking there are just too few observations in the group with parked vehicles ($n_C = 19$), the U test statistic has been converted to Z to illustrate the procedure. The Z value 0.290 has a probability of 0.772, which clearly results in the Null Hypothesis being accepted at the 0.05 significance level.

Application of the Mann–Whitney test has seven main stages:

- *State the Null Hypothesis and significance level*: any difference between two groups of road segments is the result of sampling error and it is not significant at the 0.05 level.
- *Sort and assign ranks (R)*: since the segments are numbered from the start of the length of the road and the distances increment in 5 m units from this point, the sorted rank scores are the same as the segment identification numbers. Re-sort the segments according to dichotomous grouping attribute or sample membership.
- *Sum the rank scores of each group to Identify $\sum r$*: the results of these additions are shown below.
- *Calculate test statistic (U)*: insert $\sum r$ into the equation and calculate the test statistic, which equals 470.
- *Determine the probability of the calculated U*: in this example, since the number of observations in each group is very nearly >20, U has been converted into Z (0.290), which has a probability of 0.772.
- *Select one- or two-tailed probability*: if there is any prior, legitimate reason for arguing that one group of segments be nearer to the start of the road than the other, apply a one-tailed test by dividing the probability in half.
- *Accept or reject the Null Hypothesis*: Null Hypothesis should be accepted since there is an extremely high probability of having obtained the test statistic by chance; the division of the road segments according to the presence or absence of parked vehicles seems to be random along the entire length of the road (0.772 > 0.05), recognising that this might be an erroneous decision.

<table>
<tr><td colspan="2">**Box 7.6c   Assumptions of the Mann–Whitney test.**</td></tr>
</table>

There are three main assumptions of Mann-Whitney test:

- Observations should be selected by means of simple random sampling.
- The observations should be independent of each other in respect of the ordinal, ranking variable. (Note: it is arguable whether this is the case in this example).
- The test is robust with small and large sample sizes and is almost as powerful as the t test with large numbers of observations.

Box 7.6d   Calculation of the Mann–Whitney test.

Segment	Vehicle parked	Distance (x) in metres	Overall rank (R)	Segment ID sorted	Sorted rank
1	N	5	1	1	1
2	N	10	2	2	2
3	N	15	3	3	3
4	N	20	4	4	4
5	Y	25	5	7	7
6	Y	30	6	8	8
7	N	35	7	11	11
8	N	40	8	13	13
9	Y	45	9	15	15
10	Y	50	10	16	16
11	N	55	11	18	18
12	Y	60	12	19	19
13	N	65	13	20	20
14	Y	70	14	25	25
15	N	75	15	26	26
16	N	80	16	28	28
17	Y	85	17	30	30
18	N	90	18	31	31
19	N	95	19	32	32
20	N	100	20	33	33
21	Y	105	21	36	36
22	Y	110	22	37	37
23	Y	115	23	38	38
24	Y	120	24	39	39
25	N	125	25	41	41

(Continued)

Box 7.6d (Continued)

Segment	Vehicle parked	Distance (x) in metres	Overall rank (R)	Segment ID sorted	Sorted rank
26	N	130	26	42	42
27	Y	135	27	43	43
28	N	140	28	44	44
29	Y	145	29	45	45
30	N	150	30	48	48
31	N	155	31	50	50
32	N	160	32	5	5
33	N	165	33	6	6
34	Y	170	34	9	9
35	Y	175	35	10	10
36	N	180	36	12	12
37	N	185	37	14	14
38	N	190	38	17	17
39	N	195	39	21	21
40	Y	200	40	22	22
41	N	205	41	23	23
42	N	210	42	24	24
43	N	215	43	27	27
44	N	220	44	29	29
45	N	225	45	34	34
46	Y	230	46	35	35
47	Y	235	47	40	40
48	N	240	48	46	46
49	Y	245	49	47	47
50	N	250	50	49	49

Sum of the ranks of segments with parked vehicles	$\sum R_C = 470$
Sum of the ranks of segments without parked vehicles	$\sum R_{NC} = 805$

Mann–Whitney U statistic

$$U = \sum r - \frac{n_r(n_r + 1)}{2} = 470 - \frac{19(19 + 1)}{2} = 280$$

Population mean U

$$\mu_U = \frac{n_C n_{NC}}{2} = \frac{19(31)}{2} = 294.50$$

Standard deviation U

$$\sigma_U = \sqrt{\frac{n_C n_{NC}(n_C + n_{NC} + 1)}{12}} = \sqrt{\frac{19 \times 31(19 + 31 + 1)}{12}} = 50.03$$

Z test statistic

$$Z = \frac{|U - \mu_U|}{\sigma_U} = \frac{|280 - 294.50|}{50.03} = 0.290$$

Probability

$$p = 0.772$$

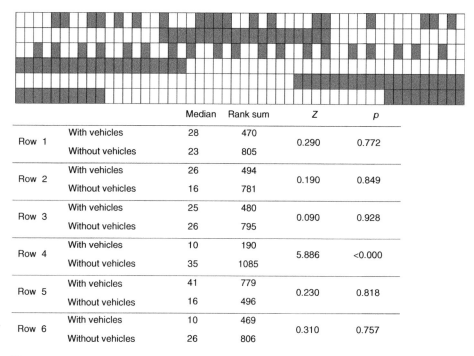

		Median	Rank sum	Z	p
Row 1	With vehicles	28	470	0.290	0.772
	Without vehicles	23	805		
Row 2	With vehicles	26	494	0.190	0.849
	Without vehicles	16	781		
Row 3	With vehicles	25	480	0.090	0.928
	Without vehicles	26	795		
Row 4	With vehicles	10	190	5.886	<0.000
	Without vehicles	35	1085		
Row 5	With vehicles	41	779	0.230	0.818
	Without vehicles	16	496		
Row 6	With vehicles	10	469	0.310	0.757
	Without vehicles	26	806		

Figure 7.4 Combinations of road segments in groups with 19 and 31 observations.

The results of the Mann–Whitney in Box 7.6a support the previous conclusion that the occurrence of the two types of segments is not significantly different from what might be expected if they were to have occurred randomly along the length of the road. However, there are many different ways in which 50 observations, in this case road segments, could be allocated to two groups containing 19 and 31 cases. The top row of Figure 7.4 reproduces the situation shown in Box 7.4a, while the others present five more distinctive patterns. The clearest difference in medians occurs in the fourth row where all the segments with parked vehicles are at the start and the probability of such extreme differentiation occurring by chance is very small. Even the arrangement in rows 2, 5 and 6 with all the occupied segments clumped in one or two groups have relatively high probabilities of occurring randomly (>0.750). The regular, almost systematic spacing of vehicles along the stretch of road represented in row 3 is the one most likely to occur through a random process ($p = 0.928$). These examples help to reveal that the results of the test are dependent not only on the degree to which the ordinal variable separates observations between the two groups but also on the size of the groups. Rows 4 and 5 as just as differentiated as each other, but in one case the occupied segments are at the start and in the other at the end.

7.4 Multiple Samples and/or Multiple Variables

This section focuses on those situations where there are two or more samples and two or more attributes, or one sample and three or more attributes or classified variables. These non-parametric techniques are less demanding in terms of the assumptions made about the underlying distribution of the data values than their parametric counterparts, but nevertheless they can provide important insights into whether the observed data are significantly different from what might have occurred if random processes were operating. The data values are either measurements recorded on the ordinal scale (i.e. placing the observations in rank order, but not quantifying the size of the differences between them) or categorical attributes or classified variables that allow the observations to be tabulated as counts. The cross-tabulations of frequency counts considered here have at least three dimensions. For example, a single sample of households might be categorised in respect of their housing tenure, the economic status of the oldest member of the household and whether anyone in the household had an overseas holiday during the previous 12 months. Supposing there were four tenure categories, six types of economic status and the holiday attribute is dichotomous, then there would be 48 cells in the cross-tabulation ($4 \times 6 \times 2$). Such a table is difficult to visualise and would become even more so with additional dimensions. Nevertheless, some statistical tests exist to cope with these situations. Pearson's Chi-square test examined in Section 7.3.1 can be extended in this way and the Kruskal–Wallis H test is an extension of the Mann–Whitney test.

7.4.1 Comparing Three or More Attributes for One Sample (or Two or More Attributes for Three or More Samples)

Extension of Pearson's Chi-square test to those situations where the cross-tabulation of counts has three or more dimensions can occur in two situations where there is (are):

- one sample and three (or more) categorised attributes.
- three or more samples and two (or more) categorical variables.

Both situations are represented in the three dimensional cross-tabulation shown in Table 7.4. Attributes A and B are categorised into five (I, II, III, IV and V) and three (1, 2 and 3) classes respectively. The third dimension is shown as jointly representing either three classes of attribute C (a, b and c) or three samples X, Y or Z. Overall Table 7.4 represents a five-by-three-by-three cross-tabulation. The counts in the cross-tabulation are for illustrative purposes only and each sample (or group for attribute C) conveniently contains 100 observations, thus the individual cells and the marginal totals in each part of the table sum to 100. The calculation of Pearson's Chi-square test statistic proceeds in the same way in this situation as with a two-dimensional cross-tabulation. Once the expected frequencies have been computed, the squared difference between the observed and expected count divided by the expected $\left(\dfrac{(O - E)^2}{E} \right)$ is calculated for each cell. The sum of these gives the test statistic. The degrees of freedom are defined as $(X_1 - 1)(X_2 - 1) \cdots (X_j - 1)$ where X represents the number of categories in attributes 1 to j. In the case of the cross-tabulation in Table 7.4 there are 16 degrees of freedom $(5 - 1)(3 - 1)(3 - 1)$. The probability associated with the test statistic with these degrees of freedom enables a decision to be reached regarding the Null Hypothesis. Since most multi-dimensional cross-tabulations will have more cells than the simpler two-dimensional ones, there is a greater likelihood that low expected frequencies might occur. It is therefore important to ensure that sample sizes are sufficiently large to reduce the chance of this problem occurring during the planning stage of an investigation.

7.4.2 Comparing the Medians of an Ordinal Variable for Three or More Samples (or One Sample Separated Into Three or More Groups)

The **Kruskal–Wallis H** test represents an extension or generalisation of the Mann–Whitney test that enables more complex types of question to be investigated. The test can be used in two situations where the focus of attention is on differences in the ranks of an ordinal variable in respect of a sample subdivided into three or more groups, or there are three or more samples.

The H test is the non-parametric equivalent of the One-way ANOVA procedure examined in Section 6.3.1. One-way ANOVA explores whether the differences in the means of a continuous

Table 7.4 Three-dimensional cross-tabulation with three categorised attributes or three samples with two categorised attributes.

			Attribute A					
Attribute C, group a OR Sample X	Attribute B		I	II	III	IV	V	Total
		1	15	8	4	3	5	35
		2	7	6	2	5	0	18
		3	6	10	4	11	13	44
	Total		28	24	10	19	18	100
			Attribute A					
Attribute C, group b OR Sample Y	Attribute B		I	II	III	IV	V	Total
		1	6	5	3	10	4	28
		2	5	7	11	9	4	36
		3	9	2	1	14	10	36
	Total		20	14	15	35	18	100
			Attribute A					
Attribute C, group c OR Sample Z	Attribute B		I	II	III	IV	V	Total
		1	11	6	10	15	0	42
		2	7	7	3	5	6	28
		3	7	3	0	4	16	30
	Total		25	16	13	24	22	100

variable for three or more samples are statistically significant, the Kruskal–Wallis test performs an equivalent task with respect to differences in the medians. Are differences in the median of a continuous non-normal variable for three or more samples statistically significant? The ordinal (or sorted continuous) variable provides a unifying thread to each of these situations. The data values for this variable are pooled, just as with the Mann–Whitney test, across the three or more samples, or sub-groups from a single sample and rank scores are assigned with average ranks used for any tied observations. The general form of the Alternative and Null Hypotheses for the Kruskal–Wallis test, respectively, assert that the difference in medians between the three or more samples or sub-groups from a sample in respect of an ordinal variable either are significant (Alternative) or not significant (Null). Just as with the Mann–Whitney test, the Kruskal–Wallis explores the extent to which the samples or sub-groups comprising the complete set of observations are interleaved or interspersed with regard to their data values on an ordinal variable.

The Kruskal–Wallis has been applied to the samples of households in mid-Wales and the length of their major shopping journeys to investigate whether there are differences between the populations of households in the four sampled settlements. Box 7.7a examines this application of the test procedure and reaches the conclusion that there is a significant difference between the median length of the journeys (in km) made by households in the four settlements where the sample surveys were carried out. Examination of the sample data reveals that the median distance for Talgarth households is greater than the other three settlements. However, from the perspective of the test results, this does not necessarily indicate that households in this settlement stand out from the rest. Just as the One-way ANOVA technique focused on the overall difference in the means of three or more samples with respect to a ratio or interval scale variable, so the Kruskal–Wallis test examines the overall, not the specific, difference in medians between the samples.

Box 7.7a   Kruskal–Wallis (*H*) test.

Kruskal–Wallis *H* test statistic: $= \dfrac{12 \sum N \left(\overline{R}_i - \overline{R} \right)^2}{N(N+1)}$

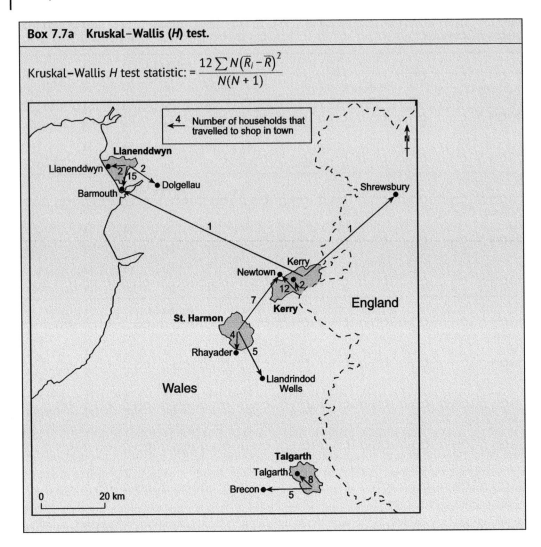

Box 7.7b   Application of Kruskal–Wallis (*H*) test.

The Kruskal–Wallis test is used in two ways, either where one sample has been subdivided into three or more groups or where there are three or more samples. In both cases, there is an ordinal variable that has been measured across the sub-groups or samples. The calculations for computing the Kruskal–Wallis test statistic (*H*) are simple summation and division, although performing these by hand can be tedious, especially if the number of observations in the samples or groups is large, because of the amount of data sorting required. The complete set of observations is pooled and sorted with respect to the ordinal variable, and then ranks are determined. Ties are dealt with by inserting the average rank, thus four observations with the same value for the ordinal variable in the 4th, 5th, 6th and 7th positions would each be assigned a rank score of 5.5. The sums of the rank scores for each sample or groups are calculated and designated scores $R_1 R_2, R_3, \cdots, R_i$ where *i* denotes the number of samples. The means of the rank sum for each sample or group are calculated as $\bar{R}_1, \bar{R}_2, \bar{R}_3, \cdots, \bar{R}_i$ with the number of observations in each represented as $n_1, n_2, n_3, \cdots, n_i$: both of these sets of values are inserted in the equation used to calculate H.

Examination of the probability associated with the calculated H test statistic in comparison with the chosen level of significance indicates whether the Null Hypothesis should be accepted or rejected. Provided that there are more than five observations in each of the samples or sub-groups, the probability distribution of the test statistic approximates Chi-square with $k - 1$, where k corresponds to the number of samples or groups. The Kruskal–Wallis test has been applied to the four samples of households in mid-Wales settlements whether any differences in the median distance travelled to carry out their main shopping are significantly greater than might have occurred randomly or by chance.

Application of the Kruskal–Wallis test can be separated into six main steps:

- *State the Null Hypothesis and significance level*: any differences between the sampled observations in the samples (or groups from one sample) have arisen as a result of sampling error and are no greater than might be expected to occur randomly at least 95 times in 100 (i.e. using a 0.05 significance level).
- *Sort and assign ranks (R)*: pool all the observations, sort in respect of the ordinal variable (distance to 'shopping town'), assign rank scores to each, averaging any tied values.
- *Sum the rank scores of each sample (group)*: the sum of the ranks for each sample or group are calculated and identified as R, where the subscript distinguishes the groups.
- *Calculate test statistic (H) and apply correction factor if >25% of observations is tied*: calculations using the equation for the test statistic are given below, which equals 21.90 in this example. Corrected H equals 31.43.
- *Determine the probability of the calculated H*: the number of observations in each sample is >5 and so the probability of having obtained the test statistic by chance can be determined from the Chi-square distribution with three degrees of freedom and is <0.001.
- *Accept or reject the Null Hypothesis*: the test statistic's probability is lower than the chosen significance level (0.05) and therefore the Null Hypothesis should be rejected. It is very unlikely that the difference in medians between the samples (groups) has arisen by chance, recognising that this might be an erroneous decision.

Box 7.7c   Assumptions of the Kruskal–Wallis (*H*) test.

There are three main assumptions of the Kruskal–Wallis test:

- Observations should be selected by means of simple random sampling.
- Sampled entities should be independent of each other and selected from populations of observations that have the same shape with respect to the ordinal variable.
- If the number of ties is large (more than 25% of the ordinal values), then the calculated H statistic should be divided by a correction factor $C = 1 - \dfrac{(T^3 - T)}{n_s^3 - n_s}$, where T and n_s are respectively the number of tied and the total number of observations.

Box 7.7d   Calculation of the Kruskal–Wallis test.

Settlement	Distance (km)	Rank (R)	Settlement (sorted)	s
Kerry	0.36	3	Kerry	3
Kerry	0.36	3	Kerry	3
Kerry	0.36	3	Kerry	3
Kerry	0.36	3	Kerry	3
Kerry	0.36	3	Kerry	3
Kerry	0.42	7	Kerry	7
Kerry	0.42	7	Kerry	7
Kerry	0.42	7	Kerry	7
St Harmon	0.45	9	Kerry	44
Llanenddwyn	0.58	10	Kerry	46
Llanenddwyn	0.99	11	Kerry	46
St Harmon	1.30	12	Kerry	46
St Harmon	1.92	13.5	Kerry	48
St Harmon	1.92	13.5	Llanenddwyn	10
St Harmon	3.67	16.5	Llanenddwyn	11
St Harmon	3.67	16.5	Llanenddwyn	26
St Harmon	3.67	16.5	Llanenddwyn	27
St Harmon	3.67	16.5	Llanenddwyn	29
St Harmon	4.43	20	Llanenddwyn	29
St Harmon	4.43	20	Llanenddwyn	29
St Harmon	4.43	20	Llanenddwyn	32
St Harmon	4.62	22.5	Llanenddwyn	32
St Harmon	4.62	22.5	Llanenddwyn	32
Talgarth	5.36	24	Llanenddwyn	35

Talgarth	5.41	25	Llanenddwyn	35
Llanenddwyn	7.41	26	Llanenddwyn	35
Llanenddwyn	7.68	27	Llanenddwyn	39
Llanenddwyn	7.74	29	Llanenddwyn	40
Llanenddwyn	7.74	29	Llanenddwyn	41.5
Llanenddwyn	7.74	29	Llanenddwyn	41.5
Llanenddwyn	7.75	32	Llanenddwyn	54
Llanenddwyn	7.75	32	Llanenddwyn	55
Llanenddwyn	7.75	32	St Harmon	9
Llanenddwyn	8.07	35	St Harmon	12
Llanenddwyn	8.07	35	St Harmon	13.5
Llanenddwyn	8.07	35	St Harmon	13.5
Talgarth	8.42	37.5	St Harmon	16.5
Talgarth	8.42	37.5	St Harmon	16.5
Llanenddwyn	8.45	39	St Harmon	16.5
Llanenddwyn	8.60	40	St Harmon	16.5
Llanenddwyn	8.80	41.5	St Harmon	20
Llanenddwyn	8.80	41.5	St Harmon	20
St Harmon	9.11	43	St Harmon	20
Kerry	11.55	44	St Harmon	22.5
Kerry	11.96	46	St Harmon	22.5
Kerry	11.96	46	St Harmon	43
Kerry	11.96	46	St Harmon	63
Kerry	12.41	48	St Harmon	64
Talgarth	13.21	49	Talgarth	24
Talgarth	13.63	50	Talgarth	25
Talgarth	13.96	52	Talgarth	37.5
Talgarth	13.96	52	Talgarth	37.5
Talgarth	13.96	52	Talgarth	49

(Continued)

Box 7.7d (Continued)

Settlement	Distance (km)	Rank (R)	Settlement (sorted)	s
Llanenddwyn	14.89	54	Talgarth	50
Llanenddwyn	15.33	55	Talgarth	52
Talgarth	19.68	58	Talgarth	52
Talgarth	19.68	58	Talgarth	52
Talgarth	19.68	58	Talgarth	58
Talgarth	19.68	58	Talgarth	58
Talgarth	19.68	58	Talgarth	58
Talgarth	21.61	61	Talgarth	58
Talgarth	21.80	62	Talgarth	58
St Harmon	41.10	63	Talgarth	61
St Harmon	55.05	64	Talgarth	62

	Kerry	Llanenddwyn	St Harmon	Talgarth
Sum of the ranks $\sum R_K, \sum R_L, \sum R_H, \sum R_T$	266	633	389	792
Number of households n_K, n_L, n_H, n_T	13	19	16	16
Mean of the ranks $\overline{R}_K, \overline{R}_L, \overline{R}_H, \overline{R}_T$	20.5	33.3	24.3	49.5

Kruskal–Wallis H statistic

$$H = \frac{12 \sum n_i (\overline{R}_i - \overline{R})^2}{N(N+1)} = \frac{12 (13(20.5 - 32.5)^2 + 13(33.3 - 32.5)^2 + 13(24.3 - 32.5)^2 + 13(49.5 - 32.5)^2}{64(64+1)} = 21.90$$

Correction factor

$$C = 1 - \frac{(T^3 - T)}{n_s^3 - n_s} \quad 1 - \frac{(75907 - 43)}{262144 - 64} = 0.697$$

Corrected H

$$H_{Cor} = \frac{H}{C} = \frac{21.90}{0.697} = 31.43$$

Degrees of freedom $\quad df = k - 1 \quad 4 - 1 = 3$

Probability $\quad p > 0.001$

7.5   Closing Comments

This chapter has examined a set of non-parametric statistical tests that parallel the parametric ones covered in Chapter 6. The main differences between the two groups of techniques relate to whether the data values for the variable(s) support the assumption of normality and whether they are continuous measurements on the ratio or interval scales in the case of parametric procedures, which test numerical quantities (the mean, variance and standard deviation) with respect to a population or between samples. The equivalent non-parametric tests covered in this chapter have either focused on frequency counts in ordered or unordered classes or ordinal ranks and measurements, in which case the main concern is with the median. Techniques concerned with frequencies look at whether the difference between observed and expected counts is more than might have occurred by chance through sampling error. Sometimes these tests are used to compare samples and populations, although the role of the population data is to calculate the expected frequencies.

The next chapter moves our exploration of statistical analysis in respect of non-spatial data into the realm of examining relationships between variables or attributes. However, this does not mean that we have finished with statistical testing, since the techniques used to calculate quantities to summarise the relationships between variables or attributes may well need to be tested to discover whether the value might have arisen by chance through sampling error. Given that such quantities are often only produced using sampled data, it may be necessary to generate confidence limits within which the population parameter lies.

A new dataset was introduced in this chapter depicting a sequence of 5 m road segments with or without parked vehicles along a 250 m stretch of road. Reflect on whether the values held by each segment (i.e. vehicle or no vehicle) are truly independent of each other.

Is it not possible that some of the vehicles are parked outside residential properties?

Is it likely that these are distributed with a certain regularity along the road and have narrower or wider frontages onto the road?

Are there less likely to be vehicles parked where there are parking restrictions, such as near a junction, school or pedestrian crossing?

What do these factors say about the possible lack of independence in the data?

Further Reading

Agresti, A. (1996) *An Introduction to Categorical Data Analysis*, New York, John Wiley and Sons.

Diamond, I. and Jeffries, J. (1999) *Introduction to Quantitative Methods*, London, Sage.

Ebdon, D. (1984) *Statistics in Geography*, 2nd edn, Oxford, Blackwell.

Griffith, D.A. (1978) A spatially adjusted ANOVA model. *Geographical Analysis*, **10**, 296–301.

Harris, R. (2016) *Quantitative Geography: The Basics*, Sage Publications Ltd.

Rogerson, P.A. (2006) *Statistical Methods for Geographers: A Student's Guide*, Los Angeles, Sage.

Scheffé, H. (1959) *The Analysis of Variance*, New York, John Wiley and Sons.

Wrigley, N. (1985) *Categorical Data Analysis for Geographers and Environmental Scientists*, Harlow, Longman.

Section IV

Forming an Association or Relationship

8

Correlation

Correlation is an overarching term used to describe those statistical techniques that explore quantitatively the strength and direction of relationships between attributes or variables. There are various types of correlation analysis that can be applied to variables and attributes measured on the ratio/interval, ordinal and nominal scales. The chapter also covers tests that can be used to find out the significance of correlation statistics. Correlation is often used by students and researchers in Geography, Earth and Environmental Science and related disciplines to help with understanding how variables relate to each other.

Learning Outcomes

This chapter will enable readers to:

- Carry out correlation analysis techniques with different types of variables and attributes;
- Apply statistical tests to find out the significance of correlation measures calculated for sampled data;
- Consider the application of correlation techniques when planning the analyses to be carried out in an independent research investigation in Geography, Earth Science and related disciplines.

8.1 Nature of Relationships Between Variables

Most of the statistical techniques explored in the previous two chapters have concentrated on one variable at a time and treated this in isolation from others that may have been collected at the same time for samples(s) of observations. These techniques may be applied to each variable in turn and the results assembled to provide answers to your research questions. However, many researchers find it more interesting to answer questions that demand that we explore how different attributes and variables are connected with (related to) each other. In statistics, such connections are called relationships. Most of us are probably familiar, at least in simple terms, with the idea that relationships between people can be detected by means of analysing DNA. Thus, if the DNA of each person in a randomly selected group of people was obtained and analysed, it would be possible to discover if any of them were related to each other. If such a relationship were to be found, it would imply that the two people had something in common, perhaps sharing a common ancestor. In everyday terms, we might expect to see some similarity between those pairs of people having a relationship, in terms

Practical Statistics for Geographers and Earth Scientists, Second Edition. Nigel Walford.
© 2025 John Wiley & Sons Ltd. Published 2025 by John Wiley & Sons Ltd.
Companion website: www.wiley.com/go/PracticalStatistics2e

of such things as eye, skin or hair colour, facial features and other physiological characteristics. If these characteristics were to be quantified as attributes or variables, then we would expect people who were related to each other to possess similar values and those who are unrelated to have contrasting ones.

It is invariably a mistake to push an analogy too far; nevertheless, this illustration of what it means for people to be related in their DNA and thereby possess similar values for certain physiological attributes does start to help us to understand what is meant by a relationship in statistics. Techniques for statistically analysing relationships are broadly divided into **bivariate** and **multivariate** procedures. The former deals with situations where two attributes or variables (or one attribute and one variable) are the focus of attention: correlation and regression are complementary techniques for analysing such relationships, which are explored in this and the following chapter. Multivariate correlation and regression involve the analysis of relationships between three or more attributes or variables at the same time.

Correlation is often defined as a way of measuring the direction and strength of the relationship between two attributes or variables. But what is meant by the phrase 'direction and strength of the relationship?' The simplest way of explaining these terms is by means of a hypothetical example. Suppose there is a single sample of observations that have been measured in respect of two variables designated as X and Y. The data values for these have been recorded as numbers according to either the interval or ratio measurement scales, which imply a difference of magnitude. In theory, these data values could range from minus to plus infinity, although the measurements for most of the variables familiar to geographers, Earth and environmental scientists will occur within a more limited section of the full numerical range. We will assume the sample of observations has produced values for X in the range 12.5 to 67.3 and for Y between 20.4 and 47.6. The direction of the relationship between X and Y refers to the relation between one set of data values and the other. Examine the three sets (A, B and C) of paired X and Y data values in Table 8.1 and consider for which set your answer would be 'Yes' for each of the following questions:

1) Do the values of X and Y for the observations increase or decrease in parallel with each other, so that values towards the lower end of the X range are matched with Y values at the lower end of its range, and *vice versa*?
2) Do the values of X and Y for the observations increase or decrease in an irregular way, so that values towards the upper end of the X range stand just as much chance of being matched with Y values at the upper, middle or lower part of its range, and *vice versa*?
3) Do the values of X and Y for the observations increase or decrease in opposite directions to each other, so that values towards the lower end of the X range are matched with Y values at the upper end of its range, and *vice versa*?

If the answer to question 1 is 'Yes', this indicates a positive relationship, whereas 'Yes' in answer to question 3 implies a negative relationship is present. An affirmative answer to question 2 denotes the absence of direction in the relationship or at least one that is unclear. In Table 8.1, it is fairly clear how 'Yes' answers to the three questions match with the three sets of data values A, B and C. However, in many cases, it might not be so easy. Most of the low X values are matched with low Y ones, but some are not, or the majority of low X values are paired with high Y ones, but a few are not. These less definite situations indicate positive and negative relationships, respectively, but there remains some doubt. This conundrum may be resolved by recognising that positive and negative relationships can be either strong or weak: if they are very weak, then this suggests the data values for X and Y increase or decrease in an irregular way (see question 2).

Table 8.1 Paired data values for observations with respect to variables X and Y.

Observation	Set A		Set B		Set C	
	X	Y	X	Y	X	Y
A	21.8	36.7	21.8	44.1	15.4	27.5
B	32.2	47.2	15.4	47.2	21.8	23.3
C	47.3	44.1	31.0	43.5	32.2	31.2
D	15.4	34.6	36.2	36.7	31.0	32.2
E	63.2	38.0	32.2	38.0	36.2	26.7
F	52.0	31.2	44.1	32.2	40.5	34.6
G	31.0	43.5	40.5	34.6	47.3	38.0
H	36.2	23.3	52.0	27.5	44.1	44.1
I	44.1	32.3	47.3	31.2	52.0	43.5
J	40.5	27.5	63.2	23.3	63.2	47.2
	Answer 'Yes' to question 1, 2 or 3?		Answer 'Yes' to question 1, 2 or 3?		Answer 'Yes' to question 1, 2 or 3?	

The various types of correlation analysis share one important characteristic: they all produce a correlation coefficient that acts as a statistical measure of the direction and strength of the relationship between variables. Like the mean and variance are descriptive measures of central tendency and dispersion, a correlation coefficient describes the relationship between two attributes or variables as revealed by their data values. A correlation coefficient is a standardised measure whose value will always fall within the range -1.0 to $+1.0$. These, respectively, correspond to very strong perfect negative and positive relationships. Standardisation of the coefficient enables different sets of paired X and Y data values to be compared with each other irrespective of differences in the numerical range or the units of measurement.

Questions 1, 2 and 3 all included the phrase 'and *vice versa*': in other words, question 1 could have been stated as:

1) Do the values of Y and X for the observations increase or decrease in parallel with each other, so that values towards the lower end of the Y range are matched with X values at the lower end of its range, and *vice versa*?

Referring again to Table 8.1 to answer this new version of the question should result in an affirmative 'Yes' answer being given for the same set of data values as previously. What does this tell us about the nature of the relationship between X and Y? The answer is that it is reciprocal. It does not matter which set of values is labelled X and which Y, the relationship is the same: if the answer to question 1 is 'Yes', they are connected in a positive way. The correlation coefficient calculated for the paired data values would be the same irrespective of which were labelled X and Y. Correlation allows such flexibility because neither X nor Y is thought to exert a controlling influence over the other. Chapter 9 will reveal that such flexibility in labelling variables does not exist with regression analysis where one variable is a 'cause' that produces an 'effect' on the other. No such distinction exists in correlation analysis between which variable is the cause and which is the effect, whereas this is part of the basic rationale for carrying out regression.

The application of correlation analysis to pairs of X and Y variables in research investigations will rarely, if ever, result in a correlation coefficient exactly equal to -1.0 or $+1.0$, although figures close to 0.0 are not unusual. The values of -1.0 and $+1.0$ represent the ideal or perfect form of the relationship in statistical terms. Far more common are correlation coefficient values somewhere within the range between these extremes. The sign denotes the direction of the relationship and its numerical value, the strength of the **covariation** between the data values (the extent to which they vary together). Interpretation of the strength of a correlation coefficient is a matter of relativity. It is clear that a coefficient of $+0.51$ is stronger than $+0.34$ and weaker than $+0.62$, but what about $+0.74$, $+0.78$ and $+0.82$? Should all three of these be treated as indicating the presence of strong relationships, although some are stronger than others? Suppose the correlation coefficients for three pairs of variables are calculated as -0.22, -0.39 and -0.46. In comparison with the previous set of values, these all indicate weak negative relationships, but -0.46 rounded up to -0.50 seems a little stronger. These examples are intended to demonstrate that the interpretation of correlation coefficients is not a simple matter, especially when trying to decide whether one is stronger than another. A certain amount of judgement is required when interpreting correlation coefficients. However, as we shall see later, carrying out a statistical test on the significance of a correlation helps with making this decision. Suppose you knew that the correlation coefficient -0.39 was significant at 0.05 level (i.e. obtaining this value by chance would only occur 5 times in 100), whereas -0.46 was not significant at 0.05 level (i.e. occurred more than 5 times in 100). You could justifiably reach the conclusion that the moderately weak negative relationship (-0.39) suggested the two variables are related, whereas those in the second instance are not.

A correlation coefficient is a statistical quantity measuring the direction and strength of a relationship, but before looking at details of correlation techniques it is useful to visualise the scatter of observations across the two-dimensional space defined by the X and Y attributes or variables. The 'unlikely to occur' instances of a perfect negative or positive relationship with observations along diagonal line from top left and bottom left of the chart, respectively, are not shown in Figure 8.1. More typical outcomes are shown with the observations forming a scatter of points distributed across the chart in such a way that they indicate the presence of a weak or strong negative or positive relationship. Figure 8.1a and 8.1b illustrates these situations using hypothetical data values with respect to two variables X and Y. The general direction or trend of the points in Figure 8.1a is from upper left to lower right, although they are clearly not perfectly aligned along the diagonal. Conversely, the scatter of points in Figure 8.1b follows the lower left to upper right direction but deviates from a perfect positive relationship. Figures 8.1c and 8.1d show a similar pair of strong negative and positive relationships, but in this case, for two attributes measured on the nominal scale labelled M and N. The obvious difference between the two sets of figures is that the categories of the attributes are discrete rather than continuous measurements. The observations are depicted in columns and rows with some data points being replaced by numbers to signify how many observations possessed a particular combination of categorical nominal values (codes). For example, there were three observations classed as C for attribute M and two for attribute N. The correlation coefficients given in each section of Figure 8.1 are indicative of the value that might be associated with the scatter of points shown, since we have yet to explore how to calculate coefficient values.

There are two main things to consider when deciding which of the distinct types of correlation analysis you should use: first, the type of sampling used; and the measurement scale(s) of the attributes and variables being investigated. Simple random sampling with interval or ratio scale

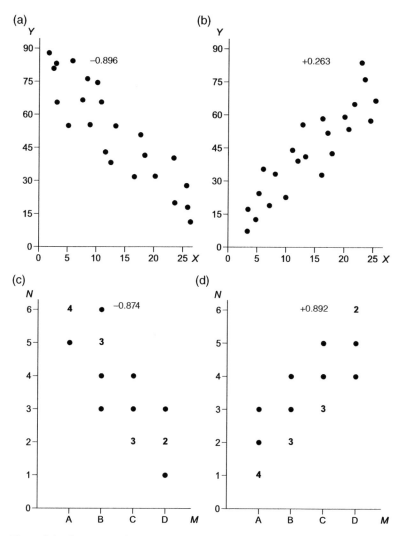

Figure 8.1 Scatter graphs showing different forms of relationship between two scalar variables (X and Y); and two nominal attributes M and N. (a) indicative strong negative scatter of continuous data values; (b) indicative strong positive scatter of continuous data values; (c) indicative strong negative scatter of integer data values; (d) indicative strong positive scatter of integer data values.

variables allows more robust correlation techniques to be applied, whereas stratified random sampling and nominal scale attributes would only allow weaker forms of correlation. We have already seen that as you progress through the different measurement scales from interval/ratio (scalar), ordinal and nominal, there is a reduction of information. Scalar measurements record differences of magnitude between individual data values, ordinal measurements show differences in their sequence or rank order and nominal attributes where the data values are simply labelled categories.

Figure 8.2 illustrates the effect of this reduction in the level of information recorded for two variables. The scatter of points in Figure 8.2a indicates the presence of a strong positive correlation

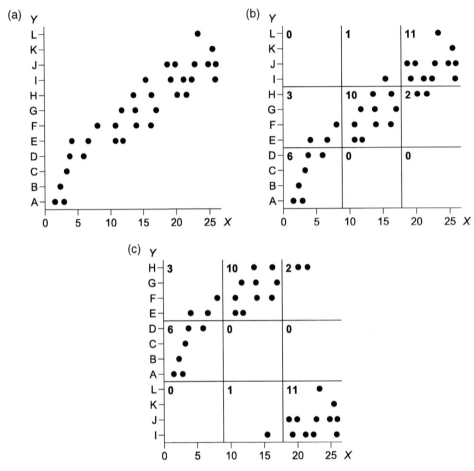

Figure 8.2 Connections between scatter graphs and cross-tabulation of data values and form of relationship between two variables (X and Y). (a) strong positive relationship between variables; (b) regular square grid overlain on scatter of data points; (c) re-ordering of rows in the grid with the data points overlain in the squares.

between the data values for variables X and Y. Now, let us reveal that one of the variables (Y) has not been measured on the ratio/interval scale, but is in fact an attribute with a large number of categories, for example representing nominal codes for 12 industrial sectors. The correlation analysis technique to be used in the circumstances should reflect the fact that there are different scales of measurement present in the two variables. Such techniques usually work on data shown as a cross-tabulation, and there are many ways in which the raw data values could be allocated to classes. One example is shown in Figure 8.2b by means of a grid overlain over the scatter of points and annotated with their frequency count. Testing for association (as opposed to correlation) between the classified variables by means of Pearson's Chi-square test produces a χ^2 statistic equalling 18.846 with four degrees of freedom and a probably of 0.00084, which indicates the Null Hypothesis should be rejected at the 0.05 level of significance. Given that Y represents a nominal attribute there is no reason the labels for each industrial sector should be shown in alphabetical order. Figure 8.2c shows the scatter of points with a similar overlain grid, but this time the sequence of industrial sector categories has been altered. The consequence of this change is that the scatter of points does not indicate the presence of such a strong correlation but the outcome of applying Pearson's Chi-square test again is to reject the Null Hypothesis ($\chi^2 = 18.846$, $p = 0.0084$). The change has altered the relative location of the cells in each row of the cross-tabulation, for example those originally in the bottom row have moved to the centre. However, this does not alter the observed and expected frequencies of each cell since the row and column totals stay the same.

Look back at the self-assessment question in Section 7.2.3 that invited you to think about why ordered classes are problematic with Pearson's Chi-square test. Figure 8.2 should help to clarify the issue.

What does Figure 8.2 reveal about the connection between analyses of association as examined in a cross-tabulation and of correlation as displayed in a scatter graph?

Another reason for examining the scatter of X and Y data values for a set of observations before carrying out a particular type of correlation analysis is that it might indicate that the variables are related in a non-linear fashion (i.e. not along a straight line). Examine the two sets of paired X and Y data values in Table 8.2 and decide which set would answer the following questions:

1) Do low and high values of variable X tend to be related to low values of variable Y, whereas mid-range X values are related to higher Y data values, or *vice versa*?
2) Do low and above mid-range values of variable X tend to be related to low values of variable Y, whereas below mid-range and high values of X relate to higher Y data values?

Figures 8.3a and 8.3b show the scatters of points for the pairs of data values tabulated in Table 8.2 and should help you answer these questions. Both examples clearly suggest the relationship between the variables is not a simple linear one. The scatter of points in Figure 8.3a displays a curved form whereas those in Figure 8.3b appear wavy or cyclical, in other words, there are peaks and troughs moving left to right across the scatter graph. The situations represented in Figure 8.3 require the application of more advanced correlation techniques than can be covered in an introductory text.

Table 8.3 provides a summary of the types of correlation examined in the subsections of this chapter. Pearson's Product Moment Correlation Coefficient sits on a similar pinnacle with respect to the different correlation techniques used to examine bivariate linear relationships as the Z test. Next in line are Spearman's Rank Correlation and Kendall's Tau Rank Correlation Coefficients, which, given their name, unsurprisingly deal with pairs of ranked or ordinal data values. Lastly, we come to three correlation techniques, the Phi Coefficient, Kappa Index and Cramer's V that can be used to correlate nominal attributes. All of these techniques have a common purpose, namely, to quantify the direction and strength of bivariate relationships.

8.2 Correlation of Normally Distributed Scalar Variables

Application of correlation analysis and the computation of a correlation coefficient between a pair, indeed several pairs, of variables are not restricted to ratio or interval scale data values. Just as the Z test provided a 'gold standard' starting point for examining parametric univariate statistical tests,

Table 8.2 Paired data values for observations in respect of variables *X* and *Y* in Figure 8.3.

Observation	Data values		Data values	
	X	*Y*	*X*	*Y*
A	3	20	1	24
B	4	17	2	29
C	4	30	3	34
D	5	42	4	35
E	6	38	5	45
F	8	42	6	42
G	8	45	7	47
H	9	46	8	48
I	10	42	9	44
J	10	50	10	33
K	12	45	11	36
L	14	42	13	35
M	14	51	14	30
N	16	46	14	31
O	16	42	15	24
P	17	43	16	24
Q	17	39	17	28
R	17	30	18	24
S	18	35	20	23
T	21	30	21	27
U	22	19	22	31
V	22	18	23	35
W	23	14	23	40
X			24	47
Y			25	48
Z			26	46
	Answer 'Yes' to question 1 or 2?		Answer 'Yes' to question 1 or 2?	

we will begin our coverage of correlation with a similarly robust but demanding procedure with respect to sampling method and normality of data values. The subsections in this chapter explore the different correlation techniques with respect to the relationship between two variables. However, in most projects, there will be groups of variables and researchers will be interested in the relationships between a whole series of pairs: variables 1 and 2, variables 1 and 3; variables 1 and 4; up to variables 3 and 4. The correlation coefficients produced for each pair from such analysis can be summarised in the form of a correlation matrix (see Table 8.4), which will be symmetrical about the diagonal representing the correlation of one variable with itself. The correlation coefficients shown in Table 8.4 are fictitious and illustrate a range of weak and strong, positive and

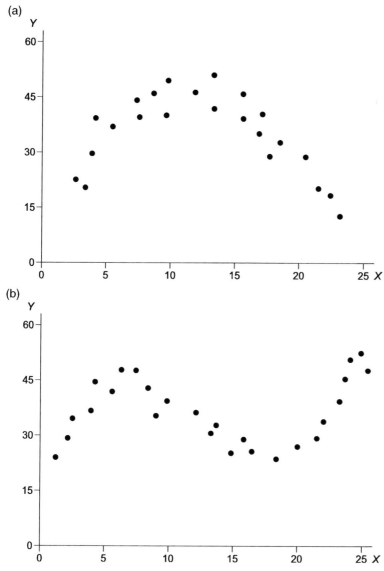

Figure 8.3 Scatter graphs showing curvilinear and cyclical forms of relationship between two scalar variables (X and Y). (a) curvilinear relationship with one curve; (b) curvilinear relationship with two curves.

Table 8.3 Subsections in Chapter 8 with correlation analysis procedures according to sampling method and measurement scale of variables or attributes.

	Data values for two or more pairs of variables
Sample selected randomly; normally distributed data values	Section 8.2 8.2.1 Relationship between variables measured on ratio or interval scale
Sample selected randomly or nearly randomly; non-normally distributed data values	Section 8.3 8.3.1 Relationship between variables measured on ratio/interval or ordinal scales 8.3.2 Relationship between variables measured on ordinal scale (distribution-free technique)
Sample selected randomly or non-randomly; non-normally distributed data values	Section 8.4 8.4.1 Relationship between attributes on nominal scale with two classes or categories on each (2 × 2 cross-tabulation) 8.4.2 Relationship between attributes on nominal scale with more than two classes or categories

Table 8.4 Example of simple correlation matrix with four variables (attributes).

Variable	1	2	3	4
1	1.0	0.67	−0.05	0.71
2	0.67	1.0	0.54	−0.43
3	−0.05	0.54	1.0	0.23
4	0.71	−0.43	0.23	1.0

negative relationships. A table such as this provides a succinct summary of results from correlation analysis and can easily be augmented by the inclusion of annotation (e.g. bold and/or italic font) to denote which coefficients are statistically significant at various levels.

8.2.1 Pearson's Product Moment Correlation Coefficient

Chapter 4 explained that the variance is a numerical quantity that measures the dispersion or spread of the data values for an individual ratio or interval scale variable. It quantifies the extent to which the data values of a set of observations are spread out or concentrated together in relation to the mean. An extension of the variance called **covariance** is an underpinning concept in Pearson's Product Moment Correlation. It can be thought of as a 'two dimensional' variance quantifying the extent to which data values of two variables are jointly dispersed or spread out. More formally, the means of the two variables locate a point known as the **centroid** that lies within the two dimensional statistical space of the variables: covariance represents their mean (average) deviation from this point. There are slightly different equations for calculating covariance depending on whether the observations are a population or a sample, the main difference being the denominator is N (the number of observations) in the case of a population and $n - 1$ for a sample.

$$\text{Population} = \frac{\sum (X - \mu_X)(Y - \mu_Y)}{N} \quad \text{sample} = \frac{\sum (x - \bar{x})(y - \bar{y})}{n - 1}$$

The individual deviations of each X and Y value from their respective means are not squared before summation since the multiplication of one deviation by the other means that the numerator does not sum to zero (compare with variance equation in Chapter 4). Figure 8.4 shows several combinations of dispersion for a sample of 10 observations measured in respect of variables X and Y. These combinations depict the following situations:

- X and Y values are widely dispersed along their observed numerical ranges and the points are distant from the centroid (Figure 8.4a).
- X values are widely dispersed and Y values are close together and the points are horizontally balanced on either side of the centroid (Figure 8.4b).
- X values are close together and Y values are spread out and the points are vertically balanced on either side of the centroid (Figure 8.4c).
- X and Y values are close together along their numerical ranges, and the points are clustered around the centroid (Figure 8.4d).

The direction of the relationship is broadly positive (i.e. trending lower left to upper right), although clearly its strength is different, in each case. The covariance of each scatter of points

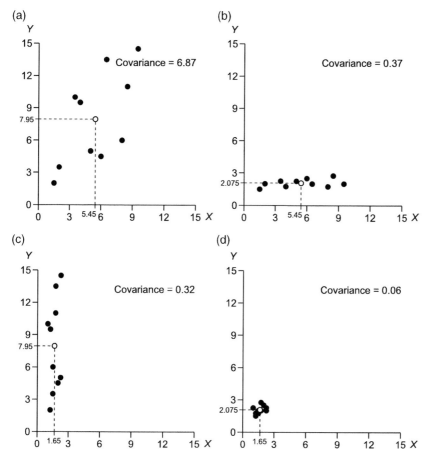

Figure 8.4 Scatter graphs showing the effect on covariance of different dispersion of X and/or Y variable data values. (a) wide dispersion of data values on X and Y axes; (b) wide dispersion of data values on X axis and small dispersion of data values on Y axis; (c) small dispersion of data values on X axis and wide dispersion of data values on Y axis; (d) small dispersion of data values on X and Y axes.

has been included and reveals the most dispersed pattern to have the highest value and the least dispersed the lowest one.

Sampling error can affect the values of a correlation coefficient for a sample of observations, just as it can influence any other statistic. In other words, the value obtained is not identical with or even close to the equivalent population parameter and so we need to apply a suitable statistical test to discover whether the difference is more or less than might be expected to have arisen through random influences. Turning this problem on its head, if the true value of the population parameter is unknown, it is useful to know how much confidence can be placed in the sample statistic. This requires the calculation of confidence limits for the sample correlation coefficient. These issues inevitably raise questions about probability. How probable is it that the difference between the correlation coefficient derived from a sample and its parent population is purely random? What is the probability that the population correlation coefficient lies within five per cent of the sample statistic? The normal distribution discussed previously is a univariate probability distribution dealing with one variable. Pearson's correlation coefficient is a bivariate technique so a probability distribution encompassing two variables simultaneously is needed. This is known as the **bivariate normal distribution**.

The normal distribution is often referred to as being bell-shaped, although its graphical representation is really a cross section or slice through the 'bell' (see Chapter 5). The bivariate normal distribution may be thought of as representing a three-dimensional bell shape, although not necessarily with a perfectly circular form around the central point when viewed from above. This central point corresponds to the centroid of the distribution, at the intersection of the means of the X and Y data values. This three-dimensional shape is achieved by the **conditional** and **marginal frequency** distributions of the X and Y variables in the population following the normal distribution. This confusing terminology needs further explanation. The conditional distributions of X and Y refer to the distribution of Y values for each cross section drawn through the bell shape, which will be normally distributed and symmetrical about their own means. The marginal distributions of X and Y are respectively the frequency distributions for **all** values of X and Y in the population. Horizontal slices through the population of data values will not necessarily be circular but may be ellipsoidal since the bell shape may be elongated in one direction. This occurs because although marginal distributions of are required to be normal it is not necessary for each to have an identical variance and standard deviation (e.g. one may be more dispersed than the other, see Figure 8.4). The variance and standard deviation of X may be larger than Y because its data values in relative terms are more spread out around their mean. This difference in dispersion results in the shape of the frequency distribution being stretched along the numerical scale of the X axis in comparison with the Y axis.

In Chapter 5, we saw that according to the Central Limit Theorem, the sampling distribution of a normally distributed ratio or interval scale variable will also be normal regardless of sample size (i.e. the means of many samples from the population will themselves follow the normal distribution). The sampling distribution of the bivariate normal distribution is affected by sample size and the population correlation coefficient. The sampling distribution becomes closer to being normal and symmetrical with increasing sample size provided the population correlation coefficient does not equal zero. There are two ways of testing whether a sampled product-moment correlation coefficient is statistically significant. The first hypothesis is that the population correlation coefficient does equal zero and therefore that the variables are unrelated. It tests whether the difference between zero and the sample coefficient is the result of sampling error by calculating a t statistic. However, sometimes it is argued that the population coefficient does not equal zero but some specific value, which might have been obtained from some previous analysis of the same population of observations. An alternative test can be used in this situation to investigate whether the difference between a known population correlation coefficient and the sample coefficient has arisen through sampling error. The sample product moment coefficient is transformed into a quantity known as **Fisher's Z**, which is not itself the same as Z discussed earlier, but it can be converted into a Z score so that its probability can be determined from the Z distribution.

Box 8.1a illustrates the application of Pearson's Product Moment correlation technique to two variables that have been measured with respect to the Burger King restaurants in Pittsburgh. The grid coordinates of the city centre have been determined (24.5, 26.0) and the distance of each restaurant has been calculated to this point using the calculations associated with Pythagoras' theorem outlined elsewhere. The second variable, although recording hypothetical values, represents the daily footfall of customers at the 22 restaurants. This example explores the idea that outlets closer to the city centre might have a higher number of customers, on account of the population being inflated by workers in the downtown area during the day. This idea connects with the broader notion of retail and catering outlets competing for space and prime locations within urban centres and of consumers' behaviour in journeying to fast food outlets. Customers to a complete set of

---

Box 8.1a   Pearson's Product Moment Correlation Coefficient.

Pearson's Correlation Coefficient

Population: ρ or $R = \dfrac{\dfrac{\sum(X-\mu_X)(Y-\mu_Y)}{N}}{\sqrt{\dfrac{\sum(X-\mu_X)^2}{N}}\sqrt{\dfrac{\sum(Y-\mu_Y)^2}{N}}}$ Sample: $r = \dfrac{\dfrac{\sum(x-\bar{x})(y-\bar{y})}{(n-1)}}{\sqrt{\dfrac{\sum(x-\bar{x})^2}{n-1}}\sqrt{\dfrac{\sum(y-\bar{y})^2}{n-1}}}$

Alternative, simpler equations:

Population: ρ or $R = \dfrac{N\sum XY - (\sum X)(\sum Y)}{\sqrt{(N\sum X^2)-(\sum X)^2}\sqrt{(N\sum Y^2)-(\sum Y)^2}}$

Sample: $r = \dfrac{n\sum xy - (\sum x)(\sum y)}{\sqrt{(n\sum x^2)-(\sum x)^2}\sqrt{(n\sum y^2)-(\sum y)^2}}$

Box 8.1b   Application of Pearson's Product Moment Correlation Coefficient.

The symbols for a population and sample Pearson's Product Moment Correlation Coefficient for are, respectively, ρ (or R) and r. Two versions of the definitional equations for these quantities are given above: the upper part (nominator) of the first pair of equations represents the covariance and the lower (denominator) is the product of the standard deviations. Manual calculation of the coefficient is of course unnecessary when using statistical software for your analysis, nevertheless in the application of the technique shown below only the simpler version involving the data values for X and Y variables is used. There are two ways of testing the significance of Pearson's Product Moment Correlation Coefficient obtained from a sample of data depending on whether the population parameter is hypothesised as zero or some other value. The Null Hypothesis states that the difference between the sample and population coefficient has arisen

(Continued)

Box 8.1b (Continued)

through sampling error. In the first case, the difference is converted into a *t* statistic, which is assessed with $n-2$ degrees of freedom, in the latter *r* is transformed into Fisher's *Z*, which is itself turned into the more familiar *Z* statistic. The probability of having obtained this difference is assessed in relation to the chosen level of significance in the normal way to reach a decision on the fate of the Null Hypothesis.

Pearson's Product Moment coefficient has been calculated for the Burger King restaurants in Pittsburgh. The two variables are the distance of each restaurant from the city centre and the number of daily customers (hypothetical data values). The scatter graph suggests that the number of customers declines as the distance from the city centre increases. This indicates the presence of a distance decay function. The result of calculating Pearson's correlation coefficient seems to confirm this impression with a moderately strong negative relationship between the two variables. Testing this result by converting the coefficient into a *t* statistic reveals that there is an extremely low probability that the coefficient value has arisen through chance, so although the relationship is not very strong it is possible to be confident that it is reliable. However, the value of r^2 (the coefficient of determination) is only 33.2%, which suggests that other factors also contribute to the variation in customer numbers.

The key stages in applying and testing the significance of Pearson's Product Moment Correlation Coefficient are

- *Calculate correlation coefficient, r:* use this quantity to indicate the direction and strength of the relationship between the variables;
- *State Null Hypothesis and significance level:* the difference between the sample *r* (−0.5766) and population (0.0) correlation coefficients in respect of distance of Burger King restaurants from the centre of Pittsburgh and the number daily number of customers has arisen through sampling error and is not significant at the 0.05 level of significance.
- *Calculate either the t or Fisher's Z test statistic:* the calculation of the *t* test statistic is given below.
- *Select whether to apply a one- or two-tailed test:* in this case, there is no prior reason to believe that the sample correlation coefficient would be larger or smaller than the population one.
- *Determine the probability of the calculated t:* the probability is 0.00000537
- *Accept or reject the Null Hypothesis:* the probability of *t* is <0.05, therefore reject the Null Hypothesis, recognising that this might be an erroneous decision.

Box 8.1c Assumptions of the Pearson's Product Moment Correlation Coefficient.

There are three main assumptions:

- Random sampling with replacement should be employed or without replacement from an infinite population provided that the sample is estimated at less than 10% of the population.
- In principle the data values for the variables *X* and *Y* should conform to the bivariate normal distribution: however, such an assumption may be difficult, if not impossible, to state in some circumstances. It is accepted the provided sample size is sufficiently large, which is itself often difficult to judge, the technique is sufficiently robust to accept some unknown deviation from this assumption.
- The observations in the population and the sample should be independent of each other with respect to variables *X* and *Y*, which signifies an absence of autocorrelation.

Box 8.1d Calculation of the Pearson's Product Moment Correlation Coefficient.

No.	x	y	xy	x^2	y^2
1	13.54	1200.00	16,244.38	183.25	1,440,000.00
2	19.14	7040.00	134,729.12	366.25	49,561,600.00
3	11.67	1550.00	18,092.56	136.25	2,402,500.00
4	6.26	4670.00	29,257.47	39.25	21,808,900.00
5	9.66	2340.00	22,596.45	93.25	5,475,600.00
6	10.31	8755.00	90,244.44	106.25	76,650,025.00
7	2.75	9430.00	25,932.50	7.56	88,924,900.00
8	7.16	3465.00	24,805.62	51.25	12,006,225.00
9	2.50	8660.00	21,650.00	6.25	74,995,600.00
10	6.18	7255.00	44,869.70	38.25	52,635,025.00
11	2.40	7430.00	17,832.00	5.76	55204900.00
12	1.80	5555.00	10,014.42	3.25	30,858,025.00
13	10.00	4330.00	43,300.00	100.00	18,748,900.00
14	2.69	6755.00	18,188.39	7.25	45,630,025.00
15	6.50	1005.00	6532.50	42.25	1,010,025.00

(Continued)

Box 8.1d (Continued)

No.	x	y	xy	x^2	y^2
16	7.50	4760.00	35,700.00	56.25	22,657,600.00
17	15.91	1675.00	26,655.67	253.25	2,805,625.00
18	14.60	1090.00	15,917.36	213.25	1,188,100.00
19	13.87	1450.00	20,104.87	192.25	2,102,500.00
20	13.12	2500.00	32,811.01	172.25	6,250,000.00
21	12.66	1070.00	13,545.12	160.25	1,144,900.00
22	13.65	1040.00	14,193.24	186.25	1,081,600.00
	$\sum x = 203.88$	$\sum y = 93025.00$	$\sum xy = 683216.81$	$\sum x^2 = 2420.07$	$\sum y^2 = 574582575.0$

r (product moment correlation coefficient)

$$r = \frac{n\sum xy - (\sum x)(\sum y)}{\sqrt{(n\sum x^2) - (\sum x^2)}\sqrt{(n\sum y^2) - (\sum y^2)}}$$

$$\frac{22(683216.81) - (203.88)(93025.00)}{\sqrt{(22(2420.07) - 203.88^2}\sqrt{(22(57458255.00) - 93025.00^2}}$$

$$\frac{-3933880.3}{(108.04)(63144.01)} = -0.5766$$

t statistic

$$t = \frac{(|r|\sqrt{(n-2)})}{\sqrt{(1-r^2)}} = t = \frac{(0.5767\sqrt{(22-2)})}{\sqrt{(1-(-0.5767^2))}} = 5.11$$

Probability $\qquad p = <0.0001$

Coefficient of determination $\qquad (r^2) = 0.3347$ or 33.47%

Burger King (or any other type of fast-food restaurant) in a large city such as Pittsburgh will be influenced by a number of factors in determining which one to visit. Therefore, our analysis is unlikely to reveal the whole story: there are likely to be a number of other variables that should also be considered and, at present, the analysis treats each restaurant as having the same floor space and capacity to serve its clientele.

One further important aspect of Pearson's Product Moment Correlation Coefficient r should be considered before exploring alternative correlation analysis techniques. The square of r, which will inevitably be a smaller positive value than the coefficient itself, provides a quantity known as the **coefficient of determination** (r^2). This acts as a shorthand indication of the amount of the total variance that is explained by the correlation between the two variables. The coefficient of determination for the example in Box 8.1a equals 0.3347 indicating that 33.47 per cent of the total variance between the two variables has been explained. The unexplained percentage of the variance is attributed to other variables that were not included in the analysis. Such explanation (or lack of it) is of course statistically defined in terms of the particular numerical values from which the statistic r has been calculated. It would thus be entirely feasible to perform a correlation analysis between two variables, calculate a strong positive or negative r, obtain a very high (r^2) and reach potentially rather ludicrous conclusions about their relationship.

8.2.2 Correlating Ordinal Variables

Some investigators take a relaxed view of the assumptions associated with Pearson's Product Moment Correlation Coefficient and apply the technique in situations where they are known to be violated or simply disregarded. Such a cavalier approach is prone to criticism not only from statisticians but also from others working in the same subject area in the course of the process of peer review for publication. Perhaps more relevant in the present context is that such disregard of the assumptions is likely to affect the assessment of students' dissertations. Fortunately, there are alternative non-parametric techniques that are rather less demanding and may be used with ranked ratio/interval scale variables or simply with ordinal data. The following sections examine two of these correlation analysis techniques.

8.3 Correlation of Non-normally Distributed or Ordinal Variables

Ordinal correlation techniques start by sorting the data values of one of the variables, X or Y, into numerical rank order from the smallest to the largest when one or both is (are) not normally distributed. Alternatively, the original data values of one or both variables might form a numerical sequence, for example field size in ha ranked from smallest to largest for a sample of farms in a case study area being correlated with accessibility to farm buildings. These situations are quite common, and **Spearman's Rank Correlation** is a very convenient and acceptable alternative to Pearson's Product Moment Correlation.

8.3.1 Spearman's Rank Correlation

Spearman's Rank Correlation is extensively used in geographical investigations and produces a correlation coefficient referred to by the symbol r_s. The underlying idea behind the technique is that if two variables, X and Y, for a sample have a relationship with each other (i.e. are correlated), their rank scores in respect of these variables will be similar (positive correlation) or opposite (negative correlation). The questions asked with respect to the data values shown in Table 8.1 can be re-stated to reflect the ranks of the variables:

1) Do the ranks of paired X and Y for the observations differ in parallel with each other, so that lower ranks for X are matched with lower Y ranks, and *vice versa*?
2) Do the ranks of paired X and Y for the observations differ in an irregular way, so that those ranked high on X stand just as much chance of being matched with upper, middle or lower ranks for Y, and *vice versa*?
3) Do the ranks of paired X and Y for the observations differ in opposite directions to each other, so that lower ranks for X are matched with upper ranks for Y, and *vice versa*?

Some of the pre-processing of data values involved in calculating Spearman's rank correlation coefficient resembles the preparation of data to apply univariate statistical tests to ordinal data values. The rank scores for a sample of observations in respect of X and Y are determined taking into account tied cases by assigning the mean (average) rank in these instances. The correlation coefficient is calculated from the squared differences between the rank scores (see Box 8.2a) rather than the raw data values.

One of the problems associated with using Spearman's rank correlation is that converting the raw data values into rank scores can mask an underlying non-linear relationship between the variables

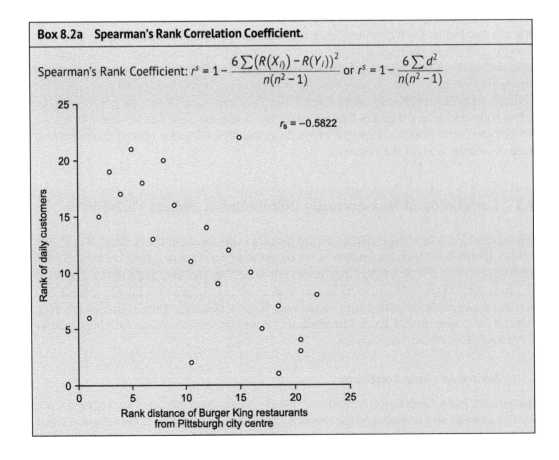

Box 8.2a   Spearman's Rank Correlation Coefficient.

Spearman's Rank Coefficient: $r^s = 1 - \dfrac{6\sum(R(X_i) - R(Y_i))^2}{n(n^2 - 1)}$ or $r^s = 1 - \dfrac{6\sum d^2}{n(n^2 - 1)}$

$r_s = -0.5822$

Rank of daily customers (y-axis)

Rank distance of Burger King restaurants from Pittsburgh city centre (x-axis)

Box 8.2b Application of Spearman's Rank Correlation Coefficient.

The basis of Spearman's rank correlation coefficient is that it examines the difference in rank scores for a set of observations with respect to two variables. These ranks are obtained by a process of sorting the data values for X and Y into ascending order and assigning a rank according to their position in the sequence taking due account of ties. The observations are then re-sorted back to their original order so the difference (*d*) in ranks can be determined. The subscripts *i* in the first of the definitional equations refers to an individual observation and R to its rank scores in respect of X and Y. The term *n* refers to the number of observations in the usual way in both equations. Testing the significance of Spearman's rank correlation coefficient helps to determine whether the value of r_s obtained in respect of a sample of observations is significantly different from what might have been expected to arise through random sampling error.

Spearman's rank coefficient has been applied to the rank scores of the distance of each Burger King restaurant from the centre of Pittsburgh and the number of daily customers (hypothetical data values). The distribution of the rank scores in the scatter graph above shows a similar general trend to the raw data examined in Box 8.1a. The rank scores associated with greater distance from the city centre seem to be associated with lower ranks for the daily number of customers. Spearman's rank correlation coefficient is −0.5734, which is remarkably similar to the value for Pearson's coefficient. Converting Spearman's coefficient into a t statistic reveals that there is a small chance of having obtained this value through sampling error, which provides support for rejecting the Null Hypothesis.

The key steps in applying and testing Spearman's rank correlation are

- *Obtain and assign rank scores:* determine the rank of X and Y data values by independently sorting each variable into ascending order and assigning corresponding rank to each observation.
- *Match correct rank pairs for each observation and calculate difference:* re-sort observations by their sequence or identifier number to match up correct rank score pairs and calculate difference in ranks (*d*).
- *Calculate r_s:* insert relevant values into the definitional equation for Spearman's rank correlation coefficient.
- *State Null Hypothesis and significance level:* the correlation coefficient has been produced from data values relating to a sample of observations and is subject to sampling error. The population parameter is 0.0 or some other specified value. The difference between these values has arisen through sampling error and is not significant at the 0.05 level of significance.
- *Calculate test statistic (t) and its probability or refer to standard r_s significance graph/table if n <* 100: the t test statistic equals 15.900, and the sample size is 22; plotting the sample r_s on Figure 8.6 with *n* − 2 (20) degrees of freedom reveals its probability is <0.05 but >0.01.
- *Select whether to apply a one- or two-tailed test:* there is typically no prior reason to believe that the sample correlation coefficient would be larger or smaller than the hypothesised population coefficient.
- *Accept or reject the Null Hypothesis:* the probability of *t* is <0.05, therefore reject the Null Hypothesis. This decision implied that the Alternative Hypothesis is accepted, recognising this might be an erroneous decision.

Box 8.2c   Assumptions of Spearman's Rank Correlation Coefficient.

There are two main assumptions:

- Ideally random sampling with replacement should be used with respect to a finite population, although random sampling without replacement is allowed from an infinite population. This restriction can be relaxed if the sample constitutes no more than 10% of the population and is sufficiently large.
- The technique ignores the size of differences between the X and Y data values for the observations and focuses on the simpler issue of their rank position.

Box 8.2d   Calculation of Spearman's Rank Correlation Coefficient.

No.	x	y	$R(x_i)$	$R(y_i)$	d^2
1	13.54	1200.00	6	18	144
2	19.14	7040.00	1	6	25
3	11.67	1550.00	9	16	49
4	6.26	4670.00	16	10	36
5	9.66	2340.00	12	14	4
6	10.31	8755.00	11	2	81
7	2.75	9430.00	19	1	324
8	7.16	3465.00	14	12	4
9	2.50	8660.00	20	3	289
10	6.18	7255.00	17	5	144
11	2.40	7430.00	21	4	289
12	1.80	5555.00	22	8	196
13	10.00	4330.00	10	11	1
14	2.69	6755.00	18	7	121
15	6.50	1005.00	15	22	49
16	7.50	4760.00	13	9	16
17	15.91	1675.00	2	15	169
18	14.60	1090.00	3	19	256
19	13.87	1450.00	4	17	169
20	13.12	2500.00	7	13	36
21	12.66	1070.00	8	20	144
22	13.65	1040.00	5	21	256

$$\sum d^2 = 2802$$

r_s (Spearman's rank correlation coefficient) $\quad r^s = 1 - \dfrac{6\sum d^2}{n(n^2-1)} \qquad 1 - \dfrac{6(2802)}{22(484-1)} = -0.5822$

t statistic $\qquad t = \dfrac{(n-2)}{\sqrt{(1-(r_s))}} \qquad \dfrac{(22-2)}{\sqrt{(1-(-0.5822))}} = 15.900$

Probability $\qquad\qquad p = <0.0001$

Coefficient of determination $\qquad (r^2) = 0.3390$ or 33.90%

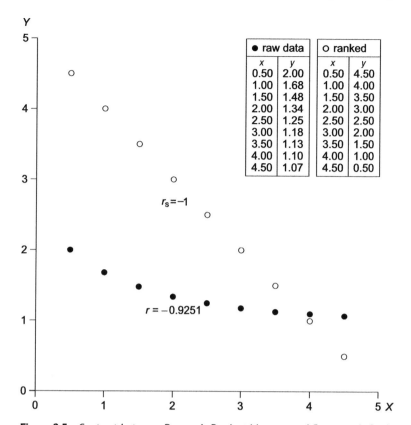

Figure 8.5 Contrast between Pearson's Product Moment and Spearman's Rank correlation analyses.

that might more appropriately be explored by means of correlation techniques more suited to this type of relationship. Figure 8.5 illustrates this by a series of data values that have a curved form, which has Pearson's correlation coefficient of −0.9251. In order that the rank scores can be shown on a similar numerical scale to the raw data values, they have been recorded as the ascending ranks 0.5, 1.0, 1.5, etc. for X and the corresponding descending ones 4.5, 4.0, 3.5, etc. for Y. These reveal a second issue associated with the Spearman's rank correlation coefficient, namely that they display perfect negative correlation ($r_s = -1.000$) because the ascending and descending rank scores are mirror images of each other. Such an extreme case is unusual and has been included to emphasise the need for caution when applying correlation analysis. If there is genuine a linear relationship between the two variables, Spearman's rank correlation will usually produce a slightly less robust outcome than Pearson's product-moment correlation.

There are two approaches to testing the significance of Spearman's rank correlation coefficient depending on the number of sampled observations. If there are less than 100 observations in the sample, the significance of Spearman's correlation coefficient can be obtained by consulting Figure 8.6. This shows the critical values of r_s at selected levels of significance (0.05 and 0.01) for different degrees of freedom. The degrees of freedom are defined as $n - 2$, where n is the number of observations in the sample. If the sample coefficient when plotted on the chart falls below the 0.05 significance line for the known degrees of freedom, then it is possible that the correlation has arisen by chance and the Null Hypothesis should be accepted. Whereas if Spearman's coefficient falls on or above the line, you can be 95 per cent confident that the result is statistically significant and you can accept the Alternative Hypothesis. If there is a larger sample, r_s can be converted into a t statistic and its probability be determined with $n - 2$ degrees of freedom.

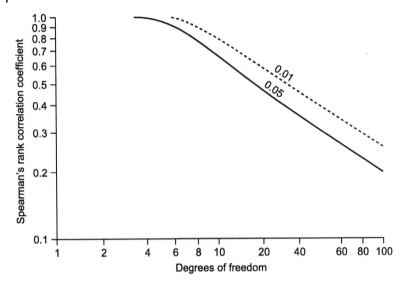

Figure 8.6 Selected Spearman's Rank correlation coefficients' significance and degrees of freedom.

Analysis of the distance of the Pittsburgh Burger King restaurants from the city centre and their daily customer footfall in Box 8.1a, tacitly assumed that these variables complied with the assumptions of Pearson's correlation procedure, that they followed the bivariate normal distribution. Box 8.2a presents the alternative Spearman's rank correlation analysis of these variables. The analysis is based on the idea that the rank scores of the raw data values will display some form of regular pattern if there is a strong relationship between them. In other words, if the restaurants were ranked in order of size in terms of their daily customer footfall (one being the first or most visited restaurant, two the second, three the third, and so on), then smaller ranks would be for outlets close to the city centre and larger ranks for those further away. The coefficient of determination, the ratio of the explained variance to the total variance, indicates Spearman's Rank Correlation achieved a very similar amount of explanation to that obtained with Pearson's Product Moment Correlation (33.9%).

> How could this pattern be expressed in the reverse way with respect to the ranked distance variable?

8.3.2 Kendall's Tau Correlation Coefficient

One of the advantages of **Kendall's τ (tau) rank correlation coefficient** is that its significance can be determined without reference to a specific probability distribution (i.e. it is distribution free). Another advantage is that compared with Spearman's rank correlation, it measures the direction and strength of the relationship between two variables in a more direct and simpler fashion. Both work with the rank scores of the observations rather than the raw data values, but Spearman's Rank Coefficient is Pearson's Product Moment calculated using these ranks. Kendall's τ works out how jumbled up the two sets of ranks are in relation to the ideals of perfect concordance (agreement) or total discordance (disagreement) between the sets of ranks produced by the two variables. However, one disadvantage of Kendall's τ technique is that the calculation of the coefficient is more complicated if there are ties in the rank scores of the data values. Mathematically simple but computationally tedious, the technique proceeds by independently ordering the pairs of data values for two variables, X and Y, and assigning their ranks. Each pair is examined to determine whether the rank for X is greater than the rank for Y, in which case a pair is labelled as concordant otherwise they are referred to as discordant. The quantity S comes from the difference between the number of concordant n_c and discordant n_d pairs. This quantity is referenced in the alternative definitional equation, which comes into play if there are any ties in rank scores (see Box 8.3a).

The technique can also be thought of as identifying the number of neighbour swaps required to convert one set of rank scores into another. Suppose there is a set of four observations ranked as 1, 2, 3 and 4 in respect of variable X and 4, 2, 1, 3 for variable Y (see Table 8.5). How many swaps are required to turn the first sequence into the second? We could start by swapping Y ranks 2 and 1 to produce 4, 1, 2, 3; then swap 4 and 1 to obtain 1, 4, 2, 3; and then 4 and 2 to give 1, 2, 4, 3 and finally 4 and 3 to yield the desired 1, 2, 3, 4: a total of four swaps is required to convert from one rank sequence into another. This constitutes a measure of how 'mixed up' or different are the two sets of rank scores. Recalling that two sets of perfectly in-sequence or out-of-sequence would respectively produce correlation coefficients of +1.000 and −1.000, Kendall's tau may be thought of as

---

Box 8.3a   Kendall's Tau Rank Correlation Coefficient.

Kendal's rank correlation coefficient: $\tau = \dfrac{n_c - n_d}{n(n-1)/2}$

Alternative Kendall's rank correlation coefficient if there are ties:

$$\tau_b = \frac{S}{\sqrt{\left[n(n-1)/2 - \sum_{i=1}^{t} t_i(t_i-1)/2\right]\left[n(n-1)/2 - \sum_{i=1}^{u} u_i(u_i-1)/2\right]}}$$

	1	2	3	4	5	6	7	8	9	10	11	12	13	14	15	16	17	18	19	20	21
2	C																				
3	C	C																			
4	C	C	D																		
5	C	C	C	C																	
6	C	C	D	C	D																
7	C	D	D	D	D	D															
8	C	C	C	C	D	C	C														
9	C	C	D	D	D	D	C	D													
10	C	D	D	D	D	D	D	D	D												
11	D	D	D	D	D	D	D	D	D	D											
12	C	D	D	D	D	D	C	D	D	C	C										
13	C	D	D	D	D	D	D	D	D	D	C	D									
14	C	D	D	D	D	D	D	D	D	C	C	D	C								
15	C	C	C	C	C	C	C	C	C	C	C	C	C	C							
16	C	D	D	D	D	D	D	D	D	D	C	D	C	D	D						
17	D	D	D	D	D	D	D	D	D	D	C	D	D	D	D	D					
18	C	D	D	D	D	D	D	D	D	D	C	D	D	D	D	D	C				
19	D	D	D	D	D	D	D	D	D	D	D	D	D	D	D	D	D	D			
20	D	D	D	D	D	D	D	D	D	D	C	D	D	D	D	D	D	D	C		
21	D	D	D	D	D	D	D	D	D	D	C	D	D	D	D	D	D	D	C	C	
22	C	D	D	D	D	D	D	D	D	D	C	D	D	D	D	D	C	C	C	C	C
Y																					
X	1	2	3	4	5	6	7	8	9	10	11	12	13	14	15	16	17	18	19	20	21

Box 8.3b Application of Kendall's Tau Correlation Coefficient.

The simpler version of Kendall's τ coefficient, used when none of the observations are tied with the same rank score for the two variables, is calculated from the number of concordant n_c and discordant n_d pairs of ranks. The denominator of this equation $(n(n-1)/2)$ represents the total number of possible pairings of X with Y ranks when there are n observations. In the equation that is used to calculate τ_b if there are ties, the subscript i refers to the number of tied observations at a certain rank score for X and u to the number of tied observations at a certain rank score for Y.

Kendall's rank correlation coefficient has been applied to variables X and Y denoting the distances of the Burger King restaurants from the centre of Pittsburgh and their daily customer footfall in order to provide some consistency with the worked examples for Pearson's and Spearman's correlation techniques. Kendall's rank correlation untangles the complexity of two sets of paired rank scores for a set of observations. The Cs and Ds in the grid above represent the concordant and discordant connections between the 22 Burger King restaurants in Pittsburgh (numbered 1 to 22) with respect to the ranks for X and Y. Reading up the columns of the grid, where each number identifies a different restaurant, if a cell contains a C then reading along the row there is a concordant link with the numbered restaurant shown on the Y axis indicating that the Y rank is larger than the X rank. In contrast, a D indicates that Y rank is lower than the X one. There are no tied ranks (see Box 8.3d) and the simpler, standard equation τ can be used. The ranks for the observations are shown in ascending order for variable X (distance from city centre). Kendall's τ coefficient is −0.429, which suggests a much weaker relationship between the two variables than either Pearson's or Spearman's correlation analyses. As usual, the probability of having obtained this value by chance should be determined to discover whether the sample data can be regarded as providing a statistically significant result.

The key stages in carrying out Kendall's rank correlation are:

- *Calculate correlation coefficient τ:* this quantity indicates the extent to which two sets of ranks are different and thus the direction and strength of the relationship between the ranked variables. Calculations for the quantity are given below.
- *State Null Hypothesis and significance level:* the two sets of ranks are mutually independent of each other and are not different from what would have occurred if they had been randomly assigned; the correlation coefficient is not significant at the 0.05 level.
- *Select whether to apply a one- or two-tailed test:* in most cases, a two-tailed test is used since there is rarely a prior reason for assuming a concordant or discordant relationship should exist.
- *Determine the probability of the calculated τ:* the probability of obtaining $\tau = -0.429$ is $p < 0.005$.
- *Accept or reject the Null Hypothesis:* the probability associated with this value of τ is <0.05, therefore reject the Null Hypothesis. By implication the Alternative Hypothesis is accepted, recognising that this might be an erroneous decision.

Box 8.3c Assumptions of Kendall's Tau Correlation Coefficient.

There is only one main assumption:

- Random sampling should be applied.

Box 8.3d   Calculation of Kendall's Tau Correlation Coefficient.

No.	$R(x_i)$	$R(y_i)$	Concordant (n_c)	Discordant (n_d)
2	1	6	16	5
17	2	15	7	13
18	3	19	3	16
19	4	17	4	14
22	5	21	1	16
1	6	18	2	14
20	7	13	4	11
21	8	20	1	13
3	9	16	1	12
13	10	11	3	9
6	11	2	10	1
5	12	14	1	9
16	13	9	3	6
8	14	12	1	7
15	15	22	0	7
4	16	10	0	6
10	17	5	2	3
14	18	7	1	3
7	19	1	3	0
9	20	3	2	0
11	21	4	1	0
12	22	8	0	0
			$\sum n_d = 6$	$\sum n_d = 165$

τ (Kendall's tau rank correlation coefficient)	$\tau = \dfrac{n_c - n_d}{n(n-1)/2}$	$\tau = \dfrac{66 - 165}{22(22-1)/2} = -0.429$
Z test	$Z = \dfrac{3(\tau)\sqrt{n(n-1)}}{\sqrt{2(2n+5)}}$	$\dfrac{3(-0.429)\sqrt{22(22-1)}}{\sqrt{2(44)+5}} = -2.792$
Probability	$p < 0.005$	

measuring how the rank scores in respect of two variables for a set of observations approach either of these extreme forms of relationship.

The probability of the value of Kendall's τ obtained for a sample of observations may not be made directly by reference to one of the standard probability distributions such as Z or t, but nevertheless it needs to be determined to test whether the coefficient is statistically significant. There are two approaches depending upon whether there are ties in the rank scores for some of the observations and therefore on the version of the coefficient that is calculated (see Box 8.3a). The τ coefficient may be viewed as representing the probability that ordering a set of observations by means of two

Table 8.5 Swapping ranks of variable *Y* to reproduce ranks of *X* and identification of concordant and discordant ranks.

X	*Y*	*Y* 1st swap 1 and 2	*Y* 2nd swap 4 and 1	*Y* 3rd swap 4 and 2	*Y* 4th swap 3 and 4
1	4	4	1	1	1
2	2	1	4	2	2
3	1	2	2	4	3
4	3	3	3	3	4
		Concordant	Discordant		
1	4	0	3		
2	2	1	1		
3	1	1	0		
4	3				
Totals		2	4		
Kendall's τ	$\dfrac{(2-4)}{(4-2)} = -2$				

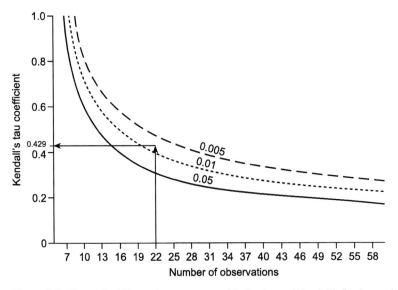

Figure 8.7 Two-tailed five and one per cent critical values of Kendall's Rank correlation coefficient according to sample size.

variables will produce the same or opposite outcome. A random ranking of the observations would have a 0.5 or 50 per cent probability of occurrence. The curves in Figure 8.7 trace the critical values of the τ coefficient that are associated with the 0.05, 0.01 and 0.005 (5, 1 and 0.5%) levels of significance. If the value of the coefficient is equal to or greater than the graphed value for a sample with n observations, the result is significant. The lines on the chart in Figure 8.7 also represent the critical values for a one-tailed test of significance for Kendall's rank correlation coefficient at 0.025, 0.005

and 0.0025 levels of significance since the probabilities are halved. Box 8.3a provides a worked example of Kendall's τ rank correlation coefficient with respect to the Burger King restaurants in Pittsburgh. The coefficient in this case has been plotted on Figure 8.7 with 22 observations and shows that the result is significant, since it falls above the curve representing 0.05 and 0.01 probability levels.

8.4 Correlation of Nominal Scale Attributes

The links between the different scales of measurement (interval/ratio, ordinal and nominal) and the likelihood that research investigations will involve variables from each of these scales has been explored in other chapters. There is a group of neglected correlation techniques that enable the direction and strength of the relationship between nominal attributes to be examined. Several of these correlation techniques use the χ^2 distribution to assess the probability of the coefficient when testing for significance. These correlation techniques may not be as powerful as Pearson's, Spearman's and Kendall's coefficients nevertheless they do allow the direction and strength of the relationship between nominal attributes to be quantified.

8.4.1 Phi Correlation Coefficient

The **Phi ϕ Coefficient** measures the degree of association between two nominal attributes or ratio/interval scale variables both of which are classified into dichotomous (binary) categories. The coefficient is therefore used with data expressed as a frequency distribution and measures the extent to which the observations are distributed into the cells along one of the diagonals of the cross-tabulation or are 'off the diagonal' and allocated among the four cells. The 2×2 cross-tabulations on the left and right of Table 8.6 represent the extremes of perfect negative and perfect positive relationships with all the observations falling along the diagonal cells. The central cross-tabulation depicts the distribution of observations across all four cells in an uneven fashion. The ϕ coefficient can in principle be applied to larger contingency tables than 2×2, but this is rare since its value can exceed +1.000 in these circumstances and its interpretation is less straightforward.

> How many observations would there need to be in each of the cells in Table 8.2 for the calculated correlation coefficient to indicate the complete absence of a relationship (i.e. 0.0000)?

The calculation of the phi coefficient refers to the four cells in a 2×2 cross-tabulation and the marginal totals of the rows and columns by letters. The cells are labelled A, B, C and D, and the marginal totals E, F, G and H as shown in Table 8.7 together with N for the total number of

Table 8.6 Two-by-two cross-tabulations illustrating diverse types of relationship between two attributes.

Strong positive relationship			Weak relationship			Strong negative relationship		
A	**B**		**A**	**B**		**A**	**B**	
C	0	47	47	C	21	24	45	C	53	0	53
D	53	0	53	D	33	22	55	D	0	47	47
	53	47	100		54	46	100		53	47	100

Table 8.7 Standard labelling of cells and marginal totals of a two-by-two cross-tabulation for the phi correlation coefficient calculations.

	Attribute Y class 1	Attribute Y class 2	Marginal row totals
Attribute X class 1	A	B	E
Attribute X class 2	C	D	F
Marginal column totals	G	H	N

observations. The ϕ coefficient shares the common feature of all correlation analysis techniques that its value is constrained to fall within the range of -1.000 to $+1.000$, thus making its interpretation comparable with other forms of correlation coefficient. Values towards the extremes beyond -0.700 and $+0.700$ are accepted as indicating a strong relationship (negative or positive), -0.300 (to -0.700) and $+0.300$ (to $+0.700$) a moderate one (negative or positive) and -0.300 to $+0.300$ little discernible relationship. The definitional equation for the ϕ coefficient given in Box 8.4a is really a simplified version of Pearson's Chi-square equation reflecting the dichotomous nature of the two variables, although the extreme values (-1.000 and $+1.000$) will only be achieved if both the row totals and the column totals are equal. The examples of strong negative and positive relationships in the contingency tables in Table 8.6 will not result in ϕ coefficient of -1.000 and $+1.000$ since the pairs of row and column totals are very nearly but not exactly equal.

The ϕ coefficient is closely related to χ^2 (Chi-square) and another way of defining the coefficient that is equivalent to the one given in Box 8.4a is to say that it is the number of observations divided by the square root of χ^2. This gives a clue as to how the statistical significance of the phi coefficient can be assessed. The coefficient is converted into χ^2 by multiplying the squared coefficient by the total number of observations. The number of degrees of freedom used when assessing the probability of χ^2 is always the same, since the correlation technique is normally only applied to a 2×2 cross-tabulation and so $(c-1)(r-1)$ invariably equals one.

Box 8.4a   Phi Correlation Coefficient.

Phi correlation coefficient: $\phi = \dfrac{AD - BC}{\sqrt{((A+B)(B+C)(A+C)(B+D))}}$ or $\phi = \dfrac{AD - BC}{\sqrt{(EFGH)}}$

Box 8.4b   Application of the Phi Correlation Coefficient.

Most applications of the Phi Correlation Coefficient require that the observations are cross-tabulated into 2×2 contingency data using two dichotomous attributes (or binomially classified) variables. The calculations for the coefficient are relatively straightforward requiring that the difference between the products resulting from multiplying the frequency counts in specific cells in the table are divided by the square root of the product of the marginal totals. Testing the significance of the ϕ coefficient is achieved by converting it to χ^2 and determining the probability of obtaining this value for comparison with the chosen level of significance, typically 0.05. This probability guides the investigator towards deciding whether to accept or reject the Null Hypothesis.

The worked application of the coefficient relates to the sample survey of households in Mid-Wales, whose shopping habits have been explored previously. The sample survey of randomly selected households in the communities was also asked about whether a laptop PC was available in the household and the gender of the individuals present, including the person who was designated as the 'head of household'. The 188 households have been distributed across the four cells of a 2 × 2 cross-tabulation of these two nominal attributes in Box 8.4d below. The majority of household heads were male and about a third of households had a laptop PC. Does the gender of the head of household and the presence or absence of a PC correlated? The Null Hypothesis states that there is no relationship between the attributes and any indication to the contrary in the sample data has occurred because of sampling error. The ϕ coefficient equals +0.089 and converting this into χ^2 to test its significance with one degree of freedom indicates that the Null Hypothesis should be retained. This result suggests that at the time the survey was carried out, there was no relationship between the gender of the head of household and the presence or absence of a laptop PC. Clearly, the results could be different now and/or in another locality.

The key stages in applying and testing the significance of the ϕ coefficient are

- *Prepare the 2 × 2 cross-tabulation and calculate the ϕ coefficient:*
- *State Null Hypothesis and significance level:* the direction and strength of the relationship between the gender of the head of household and the presence or absence of a PC from the sample survey has arisen by chance and is not significant at the 0.05 level of significance.
- *Calculate test statistic (χ^2):* the calculations for the test statistic are given below.
- *Determine the probability of the calculated χ^2:* the probability of obtaining $\chi^2 = 1.497$ and $p = 0.221$.
- *Accept or reject the Null Hypothesis:* the probability of χ^2 is >0.05, therefore accept the Null Hypothesis. By implication, the Alternative Hypothesis is rejected.

Box 8.4c   Assumptions of the Phi Correlation Coefficient.

There are three main assumptions:

- Observations should be chosen at random from the parent population.
- The value of the coefficient will fall within a limited range of the marginal totals of the contingency table are very dissimilar.
- Ordering the nominal categories for each attribute affects the sign of the coefficient and thus the direction of any relationship. By convention, dichotomous categories such as Yes and No, Present and Absent, etc. are usually given in this order. If the opposite order is used for one or both, there will be an effect on the sign of the coefficient.

Box 8.4d   Calculation of the Phi Correlation Coefficient.

			PC in household					PC in household		
			Yes	No				Yes	No	
	Gender	Male	A	B	E	Gender	Male	58	114	172
		Female	C	D	F		Female	3	13	16
			G	H	N			61	127	188

ϕ coefficient	$\phi = \dfrac{AD - BC}{\sqrt{(EFGH)}}$	$\phi = \dfrac{(58)(13) - (114)(3)}{\sqrt{(172)(16)(61)(127)}} = +0.089$
χ^2 statistic	$\chi^2 = \phi^2 n =$	$+0.089^2 (188) = 1.497$
Probability	$p =$	0.221
Coefficient of determination	$(r^2) =$	0.0488 or 4.88%

8.4.2   Cramer's *V* Correlation Coefficient and the Kappa Index of Agreement

The **Cramer's *V* Coefficient** provides a further tool for analysing nominal attributes since it produces a quantity that normally lies within the range 0.000 and +1.000 and unlike the Phi Coefficient is not restricted to 2 × 2 contingency tables. Cramer's *V* is one of a number of measures that are linked to χ^2 and it is defined as the square root of χ^2 divided by *nm*, where *n* denotes sample size and *m* is the smaller of the number of rows minus one or columns minus one. The application of Cramer's *V* therefore simply entails calculating χ^2 for a contingency table and then applying this adjustment to produce the required statistic. One way of interpreting Cramer's *V* is to view it as measuring the maximum possible variation of two variables. The value of the coefficient is influenced by the degree of equality between the marginal totals, which was noted as one of the assumptions of the Phi coefficient. If the marginal totals are unequal, then the extreme value of 1.00 is unattainable. The Cramer's *V* is often used with grid-based spatial datasets, particularly where two classification schemes with an unequal number of classes in each are being compared. The coefficient provides a measure of agreement between the two schemes.

The **Kappa Index** is another technique for comparing two classifications of attribute data that have been assembled as a cross-tabulation. The counts in the cells of the contingency table are the number of units allocated to particular categories according to classification schemes A and B. If both classification schemes allocated units to the same categories, then the observations would all occur in the cells forming the diagonal of the table. The Kappa Index is calculated from the difference between the observed and expected proportions of units thus matched along the diagonal ($O - E$). This provides the numerator for the definitional Kappa Index equation which is divided by $1 - E$ the denominator. Both the Kappa Index and Cramer's *V* can have negative values (i.e. between −1.00 and 0.00), but this is unusual since it indicates an unusually low degree of agreement between the classification schemes.

8.5   Closing Comments

This chapter has examined some of the most important and useful techniques for measuring the correlation between variables and the association between attributes. One critical issue that has been reserved for consideration at a later point is autocorrelation and in particular spatial autocorrelation. The worked example for three of the correlation techniques referred to the location of Burger King's restaurants in Pittsburgh in respect of their distance from the city centre and the typical daily footfall of customers. As a conclusion to this chapter and foretaste of our later exploration of spatial autocorrelation, consider the following question. Is it not likely that the restaurants located close together near the centre of the city will have similar numbers of customers every day partly because they are located near to each other and convenient for the city centre workforce and shoppers, whereas those that are further away will be drawing on a different population of potential customers from the residential and suburban areas of the city who are mostly travelling to the restaurants by private transport?

Further Reading

Diamond, I. and Jeffries, J. (1999) *Introduction to Quantitative Methods*, London, Sage.

Ebdon, D. (1984) *Statistics in Geography*, 2nd edn, Oxford, Blackwell.

Harris, R. (2016) *Quantitative Geography: The Basics*, Sage Publications Ltd.

Rogerson, P.A. (2006) *Statistical Methods for Geographers: A Student's Guide*, Los Angeles, Sage.

9

Regression

Chapter 9 explores regression analysis, which contrasts with correlation in two ways: At least one of the variables is treated as exerting a controlling influence on the other, and it may be able to predict unknown values of the dependent variable(s). The starting point is ordinary least square regression which focuses on linear relationships between two variables with more advanced multivariate regression techniques introduced as a basis for exploring correlation and regression of spatial data in subsequent chapters. The chapter also explores the two main ways of testing the significance of regression models.

Simple regression techniques, such as those examined here, are rarely capable of fully explaining the complexities of geographical, geological and environmental processes and should be used with some caution by students undertaking independent projects in these areas, although an understanding of the principles of these techniques provides a fuller appreciation of the research literature.

Learning Outcomes

This chapter will enable readers to:

- Apply simple linear regression analysis to ratio/interval scale variables;
- Specify Null Hypotheses in a format appropriate to regression analysis and undertake significance testing;
- Appreciate the ways of extending these simpler regression techniques to examine relationships where there is more than one independent variable;
- Recognise situations in which regression analyses could be applied in an independent research investigation in Geography, Earth Science and related disciplines.

9.1 Specification of Linear Relationships

Chapter 8 explains that the main purpose of correlation analysis is to summarise the direction and strength of the relationship between two variables: it describes how their data values are connected. However, the data values for the X and Y variables are interchangeable, and when

Practical Statistics for Geographers and Earth Scientists, Second Edition. Nigel Walford.
© 2025 John Wiley & Sons Ltd. Published 2025 by John Wiley & Sons Ltd.
Companion website: www.wiley.com/go/PracticalStatistics2e

computing correlation coefficients, it does not matter whether the values for the variable X are 'plugged into' the equation for the terms X or Y the outcome (i.e. the value of the coefficient) would be the same. Correlation techniques do not attempt to specify the form of the relationship in the sense of identifying which of the two variables is in control. This is the purpose of regression that designates each of the two variables as having a specific role in the relationship. The **independent variable** (X) is treated as exerting a controlling effect on the other, which is known as the **dependent variable** (Y). What this means is that the variable X at least partially determines the value of Y. Putting it another way, if we know the value of X then we can estimate or predict the value of Y with a certain level of confidence. The predicted value of Y may not be absolutely correct, but statistically testing regression analysis helps you to decide if it is acceptable at a specified level of significance This raises the question, why should we want to estimate the value of Y if we already have a set of paired X and Y measurements. There are three main ways to answer this question:

- The X and Y data values may relate to a sample and we want to estimate the Y values of other observations in the same population for which only the X data values are known.
- The data values recorded for the independent X variable by the sampled observations may not be the ones for which estimated Y values are required.
- The form of the relationship revealed by regression analysis for one sample of observations from a specific area or time period may be applicable to comparable populations of the same type of entity in other areas or at different times.

The labels dependent and independent apply to two groups of variables when analysing more complex relationships: there will be more than one independent variable and possibly more than one dependent variable as well. Whether the focus is on complex multivariate or simpler bivariate relationships, regression analysis seeks to express the form of the relationship mathematically by means of an equation. There are standard regression equations representing different forms of relationship that link together the dependent and independent variables in a particular way (e.g. linear, curvilinear and exponential). These equations are sometimes described as modelling the relationship. The form of regression analysis or model that an investigator chooses to apply depends on how the dependent and independent variables appear to be related when examined as a scatter graph. Some of the distinct types of scatter graphs are examined in Chapter 8 (see Figure 8.3). From a computational perspective, it would be feasible to apply simple bivariate linear regression to the X and Y data values that produced either of the scatters of points shown in Figure 8.3. However, a simple straight line is clearly not the most suitable way of summarising the two forms of the relationship since one is curvilinear and the other cyclical.

The separation of variables into dependent and independent groups in regression analysis requires that an investigator thinks conceptually about how to model the process or relationship. What variables might cause people to emigrate from one country to another? What variables might cause water quality in rivers to decline over a period? What factors might increase the speed at which shoppers can evacuate a shopping centre in an emergency? These examples illustrate that regression analysis is often used to explore relationships where there is a cause and an effect, one thing resulting in another. There is often an interest in the Geographical Sciences, particularly Physical Geography and the Earth/Environmental Sciences but also those involving human interaction with the environment, in exploring how natural processes or systems work, for example how soil erosion occurs or how the market value of land changes.

However, perhaps one of the simplest examples is to imagine farmers growing crops in an area where there is limited precipitation during the summer growing season (e.g. Mediterranean areas).

A farmer choosing to irrigate his/her crops is likely to increase the yield in comparison with a farmer who opts to rely on precipitation. Furthermore, irrigation at one rate (e.g. 25 l/hr for 10 hrs/day) may have a different effect on yield compared with another (e.g. 75 l/hr for 4 hrs/day). The same total quantity of water is applied (250 l), but the duration of irrigation is different. A lower rate of application over a longer period may be more beneficial towards increasing yield compared with a higher rate over a shorter period. The former might help to counteract the effect of the irrigated water evaporating in high temperatures, whereas the latter might increase erosion of soil through overland flow. These two rates of irrigation consume the same quantity of water and therefore the cost of irrigation is the same. However, suppose two neighbouring farmers both irrigate their crops at a rate of 25 l/hr, but one does so for 4 hours and the other for 8 hrs/day: the water charges for the first farmer would be half of those of the second. The second farmer might achieve a higher increase in crop yield than the first, but unless this is at least double the yield achieved by the first farmer, some of the money paid out to the water supplier will have been wasted. We could carry on adding other factors or variables into the analysis or adopt the strategy of some investigators and discuss the results with 'other things being equal' or *ceteris paribus* to use the term employed by economists. What this means is that quite possibly there are some other variables influencing crop yield that have not been measured and analysed, but these are assumed to have an equal impact (on crop yield) for both farmers irrespective of the value of the independent variable(s). No matter what the irrigation rate, the effect of these 'other things' on crop yield does not vary.

This description of the relationship between crop yield (dependent variable), irrigation rate and payment for water (independent variables) has made certain unspecified or tacit assumptions. First, it has been assumed that agricultural crop plants need water in order to produce a yield (e.g. seeds in the case of cereal crops). Second, water supplied to crops by means of irrigation and precipitation is equivalent; in other words, there is nothing particularly beneficial or detrimental to crop growth associated with either of these types of water supply. Third, water is charged per litre and the rate does not vary with the quantity used, whereas there might be a financial penalty or incentive for high or low usage. Fourth, all farmers are regarded as equal and it is assumed that there are no subsidies for certain types of farm/farmer (e.g. small or large farms) or for those growing certain crops. Fifth, the agricultural crops grown in the area are assumed to have equal demand for water to achieve the same growth rate. Some leafier plants such as maize or corn and potatoes tend to have higher water requirements in the hot summer period having a larger surface area available for evapotranspiration in comparison with cereals (wheat, barley, oats, etc.), which are ripening in this period, and demand less water.

Some of the issues raised by these assumptions could be investigated by undertaking a series of regression analyses. For example, the water supply and growth rate relationship could be explored for each of several types of crops and even for different varieties of each crop type. The charge for irrigation water could be varied according to the amount used or farmers growing certain low-value staple crops could pay less for irrigating these rather than more exotic high-value ones. In this fashion a complex system, the relationship between yield, irrigation rate and water charges are broken down into its simpler components. Recognition that even complex multivariate relationships can be subdivided into their constituent parts provides a suitable point for introducing the next section which examines bivariate regression techniques; in other words, simple linear or ordinary least squares (OLS) regression where there are just two variables in the mathematical model or equation. We will then progress to non-linear (curvilinear) bivariate relationships and finally introduce multivariate (multiple) regression where there is more than one independent variable.

9.2 Bivariate Regression

The application of regression techniques is sometimes described as fitting a regression line to a set of data points. The purpose of this exercise is to determine how closely the points fit a particular line and to specify the relationship between the dependent and independent variables as an equation. The form or shape of the line produced by the distinct types of regression equation is fixed. The righthand side of Figure 9.1 shows the equations and form of the lines for three ideal types of

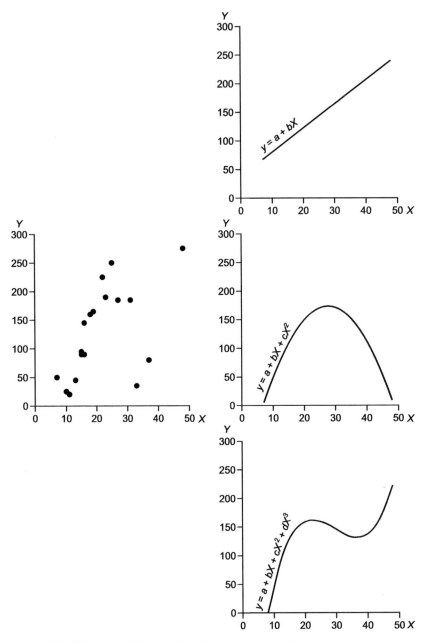

Figure 9.1 Fitting standard regression lines to a scatter of points by eye.

regression (simple linear, quadratic and cubic) where two variables are analysed. Each of these represents a relationship between one dependent variable (Y) and one independent variable (X). The left side of Figure 9.1 depicts a scatter graph of data points (observations) plotted in positions according to their paired data values on X and Y. It is clear that the point cloud does exactly match any of the three ideal regression equations, although it is clear that it is not just a random scatter. It approximates the shape of one of the standard regression lines.

Figure 9.1 illustrates that the process of fitting a regression line (equation) to a set of data points is about discovering the form of the relationship between the variables. Simply looking at the three regression lines and scatter plot is a hit-and-miss way of applying regression techniques. Even if you can decide which of the three standard regression lines best fits the scatter of points, they do not fall perfectly along any of the lines. Most points measured along the X-axis (the independent variable) deviate from where they should be if they conform exactly to the regression equation that defines the form of the line. The equations alongside the lines in Figure 9.1 define the form of the regression and are known as the linear, quadratic and cubic forms of bivariate regression. The following sections will examine how to fit a scatter of data points to these distinct types of bivariate regression model.

9.2.1 Simple Linear (Ordinary Least Squares) Regression

The previous description of fitting data points to a regression line has approached the question from a graphical perspective and works from the principle of looking for the line that most closely matches the points. If the known values of X perfectly determine the corresponding values of Y, they fall along a regression line with a particular form. Each form of regression line can be represented as an equation and in the case of **simple linear** regression the general form of the equation is:

$$Y = a + bX$$

where a and b are constants, and X denotes the known values of the independent variable and Y the values of the dependent variable. If the values of the constants are given, then inserting each known X data value into the equation will inevitably result in a pre-determined series of Y values. This series of Y values lies along the regression line and equal its height above or below the horizontal axis. These values may be viewed as the amount that the independent variable (X) can shift the dependent variable (Y) in a positive direction above or a negative direction below the horizontal axis (where Y equals zero). In some applications when $X = 0.0$ Y will also equal zero, whereas as in other cases, there will always be some 'background' value of Y irrespective of the value of X. For example, there are background values of the different radioactive isotopes in the atmosphere regardless of whether there is any source in an area potentially raising this level.

This description helps to explain the role of the constants a and b in the simple linear regression equation. The constant a equals the value of Y where the regression line passes through the vertical axis and $X = 0.0$; it is known as the **intercept**. Suppose the equation omitted the constant b and appeared as:

$$Y = a + X$$

This would mean that the value of Y would always equal the X value added to the intercept: so, with an intercept of 10.0 the series of Y values associated with $X = 1, 2, 3, 4 \ldots 10$ would equal 11, 12, 13, 14 ... 20.

> Visualise the graph of this relationship between X and Y for this series of X and Y values. What is the direction of the line in relation to the X-axis? Change the value of the intercept, perhaps to 15 and recalculate the X values. What would be the direction of this line plot on the graph?

These questions illustrate the importance of the b constant in the regression equation: its value and + or – sign control the **gradient** or **slope** of the regression line. Without the b constant the angle between the line and the horizontal axis would always be 45° and this effectively means that the slope of the line is 1.0 in all cases. The b constant is the amount that Y increases per unit of X: a large value produces a steep slope and a small value produces a gentle one. A positive or negative sign for the slope coefficient has the same significance as for a correlation coefficient: a positive b means that the regression line ascends from lower left to upper right whereas a negative one denotes the reverse (it descends from upper left to lower right).

The perfect relationship defined by the simple linear regression equation is unlikely to occur when the pairs of data values forming a scatter of points on a graph have been obtained from a population or sample of observations. Figure 9.2a shows a selection of the infinite number of straight lines that could be drawn through a given scatter of data points. Which, if any, of these is the 'best fit' line? How do we know which is the best-fit line? What does 'best' mean in this context? What criteria might be useful in deciding whether one line is a better fit than another? Obviously, the line cannot zigzag across the graph connecting all the data, since it would not be a straight line. The means of the X and Y data values measure the central tendency of the two variables, which can be plotted as the middle or centroid of the point distribution (see Chapter 8). Given this role, it is reasonable to argue that the regression line should pass through the centroid, which acts as an 'anchor point' through which pass an infinite number of straight lines like the spokes of a wheel. One of these is shown in Figure 9.2b, but even though this passes through the centroid it may not be the 'best fit' line. To discover whether it is, we need to remember that the values of X control the values of Y (at least to some extent) and not the other way round. We are seeking to discover the **linear regression of Y on X** and not the reverse. The Y values do not fall along straight for some unknown reason, which represents the extent to which the independent variable does not fully control the dependent one.

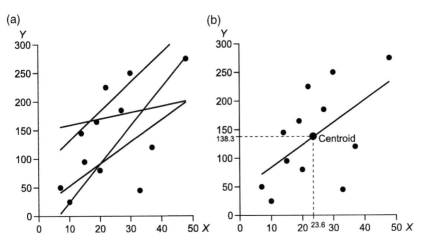

Figure 9.2 Selection of possible regression lines. (a) Selection of straight lines drawn through a scatter of points; (b) potential regression line through passing through centroid of data points.

The dependent variable will imperfectly predict the value of Y in most applications and there will be differences between observed and predicted Y values for known values of X. This difference or deviation is known as the **residual** or **error.** A positive residual occurs when the regression equation overestimates the observed value of Y and a negative one when it underestimates it. Smaller positive or negative residuals indicate that the regression line and its equation provide a better prediction of the Y values in comparison with larger differences. The size of the residuals is therefore relative. If we could quantify the overall size of the total difference represented by the residuals, then this might help in determining the line that best fits a given scatter of data points. The solution is to select the line that minimises the total sum of the positive and negative residuals. However, if we simply add up the residuals for any line that passes through the centroid of the data points, the result will always equal zero. This is the same problem that arises when summing the positive and negative differences of a set of data values from their mean. This difficulty with the absolute positive and negative residuals always adding up to zero can be overcome by squaring the values and then adding them up. Recall the need to square the differences between a set of data values and their mean before dividing by the number of observations to obtain a variance (see Chapter 4).

The 'best fit' regression line can now be defined as the line that passes through the centroid of the data points and minimises the sum of the squared deviations, differences or residuals in Y units measured vertically from each data point to the line. It is unrealistic to use the trial-and-error method of drawing various lines and calculating the sum of the squared deviations to find the best fit one. We need a more straightforward approach to calculate the coefficients (constants) a and b in the regression equation. Once these are known the equation can be solved for any value of X and the line drawn through the predicted Y values. The simple linear regression equation given previously should be slightly rewritten to indicate that the Y values are predicted and differ from the observed ones:

$$\hat{Y} = a + bX$$

The overscript ^ ('hat') symbol denotes that values of the dependent variable along the regression may differ from the observed values in a sample. The equivalent linear regression equation for a population of observations is

$$\mu = \alpha + \beta X$$

The use of μ, which normally symbolises a population mean, in place of $\hat{Y}$ makes the key point that the regression line and the Y values lying along it are the mean or average of the relationship between the two variables in a population of observations for each value of X. These mean values for the dependent variable at each known value of X and the conditional distributions associated with them lie parallel to the vertical axis at each point along the line and follow the normal probability distribution. The population mean of Y is predicted by any known value of X.

Figure 9.3 shows how the same regression line and equation can apply to more than one scatter of data points. However, the residuals in Figure 9.3a are smaller than those in Figure 9.3b: the regression equation seems to act as a more accurate way of predicting the dependent variable from the independent one in the former case. However, given that the equation relates to both distributions, it would be useful to have some way of quantifying the overall variability in the residuals or errors and the extent to which the lines 'best fit' the data points. We need a statistic that measures their overall dispersion about the regression line in a comparable way to a variance or standard deviation that quantifies the dispersion of data values about their mean. The **standard error of the estimate**

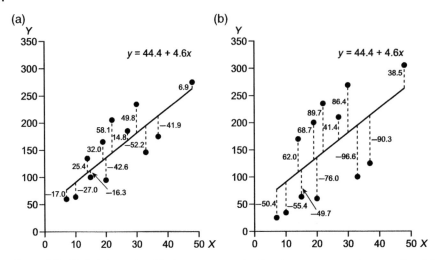

Figure 9.3 Varied sizes of residuals on a regression line for same regression equation. (a) small residual values; (b) large residual values.

(σ_{YX} for a population and s_{yx} for a sample) performs this function and is similar to the standard error of the mean examined in Chapter 6. It is effectively the standard deviation of the residuals from the best-fit regression line, which are usually shown as ε (epsilon) for a population and e for a sample and indicates that they represent the error in the estimation of the true Y values. The standard error of the estimate acts as a summary measure, like the mean and standard deviation and allows different samples to be compared regarding the amount that their data points are dispersed or scattered about the regression line. It measures the overall accuracy of the predicted Y values and is the standard deviation of the sampling error.

Box 9.1a applies simple linear regression with respect to the distance of Burger King and McDonald's restaurants from the centre of Pittsburgh and illustrates how to calculate the constants a and b to define the specific straight-line equation that fits the two sets of data points. However, we shall later examine some of the difficulties associated with using a spatial variable, such as distance from the city centre, in simple linear regression. The application explores whether distance from the city centre has a controlling influence over the daily footfall of customers at the two types of restaurants and whether there is a difference between the two brands. The distances have been coded to the nearest 2.5 km, which represents an attempt to address the assumption that the data values of X should be chosen by the researcher rather than be allowed to arise from the sample data. The regression lines and equations for the two distributions are reasonably similar, trending at a moderately steep angle from higher numbers of customers near the centre to lower values in the suburban areas. The dispersion of data points for the Burger King restaurants appears to be slightly greater with two particularly well-patronised outlets at the 10 and 20 km points, the latter untypical of other suburban sites. Box 9.1a includes calculations for the standard error of the estimate for the two sets (samples) of restaurants. In comparison with data values for distance from city centre used previously, the adjusted values result in the data points in the scatter plots appearing in vertical lines above the X-axis corresponding to 2.5, 5.0, 7.5, etc. km. Recoding of the original distances to the nearest 1 km or even 500 m would have produced different scatters of points that would still have been arranged in vertical lines.

Box 9.1a   Ordinary Least Squares (OLS) Regression.

Simple Linear Regression equation:
Population: $Y = \alpha + \beta X$ Sample: $y = a + bx$

α and β or a and b coefficients: $b = \dfrac{n\sum xy - \sum x \sum y}{n\sum x^2 - (\sum x)^2}$ $a = \dfrac{\sum y}{n} - \dfrac{b\sum x}{n}$

Estimation of Y:
Population: $\mu = \alpha + \beta X$ Sample: $\hat{y} = a + bx$
Standard error of the estimate:

Population: $\sigma_{Y.X} = \sqrt{\dfrac{\sum \left(Y - \hat{Y}\right)^2}{N}}$

Sample: $s_{y.x} = \sqrt{\dfrac{\sum (y - \hat{y})^2}{n - 2}}$

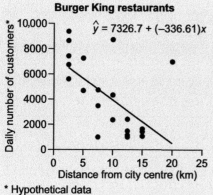

Burger King restaurants

$\hat{y} = 7326.7 + (-336.61)x$

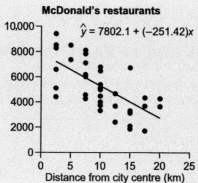

McDonald's restaurants

$\hat{y} = 7802.1 + (-251.42)x$

Daily number of customers*

Distance from city centre (km)

* Hypothetical data

Box 9.1b   Application of OLS Regression.

Simple linear or OLS regression defines a straight line by means of a mathematical equation that connects two sets of paired data values, one measuring a dependent variable (Y) and the other an independent variable (X). The most straightforward way of obtaining the regression equation and line for a particular sample of observations is by calculating the constants a and b and inserting the resulting values into the standard equation $\hat{Y} = a + bX$. These calculations are normally carried out by statistical or spreadsheet software, although a worked example helps to clarify the technique. The series of $\hat{Y}$ values for the dependent variable lie along the regression line and are estimates or predictions of its value in the statistical population for the known values of X. The standard error estimate statistic (s_{YX})

(Continued)

Box 9.1b   (Continued)

summarises the overall accuracy of the regression equation in predicting Y and provides a means of comparing different samples.

Simple linear or OLS regression has been applied to investigate whether distance from the centre of Pittsburgh acts as a good predictor of daily customer footfall for two types of fast-food restaurants. The values of the independent variable (X) should not be discovered empirically but be pre-determined by an investigator. This requirement has been at least partially satisfied by categorising the distances of the Burger King and McDonald's restaurants at intervals of 2.5 km. In the interests of brevity, only the calculations involved in obtaining the a and b constants in respect of the data values for the Burger King restaurants are shown in Box 9.1d, although the standard errors for both are shown. The scatter plots of the two sets of data values and the nest fit regression lines are similar, although the gradient of the McDonalds line is slightly less and the set of Burger King restaurants includes a few that have rather more customers than would be predicted by the regression equation. This visual interpretation of the two distributions is supported by the standard errors of the estimate for the two sets of data (2447.30 for Burger King and 1529.99 for McDonalds). There is more dispersion in the residuals for the Burger King restaurants compared with the McDonald's ones.

The key stages in applying Simple Linear Regression are:

- *Calculate constants a and b:* these denote where the regression line intercepts the vertical Y axis and the gradient or slope of the line;
- *Calculate the standard error(s) of the estimate:* this statistic provides a summary of the overall fit of the data points to the regression line/equation.

Box 9.1c   Assumptions of OLS Regression.

There are five main assumptions:

- The values of the independent variable X should be determined by the investigator and are NOT sample values.
- The dependent variable Y should be normally distributed.
- The dependent variable should have the same variance for all values of the independent variable X.
- The complete set of residuals or errors follows the normal distribution.
- There should be no systematic pattern in the values of the residuals (i.e. no autocorrelation is present).

Box 9.1d Calculation of OLS Regression (Burger King).

No.	x	y	xy	x^2	y^2	$(y-\hat{y})^2$
1	12.5	1200.0	15,000.0	156.3	1,440,000.0	3,683,002.4
2	20.0	7040.0	140,800.0	400.0	49,561,600.0	41,543,954.6
3	12.5	1550.0	19,375.0	156.3	2,402,500.0	2,462,121.9
4	5.0	4670.0	23,350.0	25.0	21,808,900.0	948,072.2
5	10.0	2340.0	23,400.0	100.0	5,475,600.0	2,626,474.0
6	10.0	8755.0	87,550.0	100.0	76,650,025.0	22,985,887.8
7	2.5	9430.0	23,575.0	6.3	88,924,900.0	8,671,758.7
8	7.5	3465.0	25,987.50	56.3	12,006,225.0	1,788,010.2
9	2.5	8660.0	21,650.0	6.3	74,995,600.0	4,729,689.8
10	5.0	7255.0	36,275.0	25.0	52,635,025.0	2,596,319.9
11	2.5	7430.0	18,575.0	6.3	55,204,900.0	892,618.7
12	2.5	5555.0	13,887.50	6.3	30,858,025.0	865,299.9
13	10.0	4330.0	43,300.0	100.0	18,748,900.0	136,426.8
14	2.5	6755.0	16,887.50	6.3	45,630,025.0	72,783.9
15	7.5	1005.0	7537.50	56.3	1,010,025.0	14,418,462.0
16	7.5	4760.0	35,700.0	56.3	22,657,600.0	1777.9
17	15.0	1675.0	25,125.0	225.0	2,805,625.0	363,114.7
18	15.0	1090.0	16,350.0	225.0	1,188,100.0	1,410,370.0
19	15.0	1450.0	21,750.0	225.0	2,102,500.0	684,905.2
20	12.5	2500.0	31,250.0	156.3	6,250,000.0	383,303.4

(Continued)

Box 9.1d (Continued)

No.	x	y	xy	x^2	y^2	$(y-\hat{y})^2$
21	12.5	1070.0	13,375.0	156.3	1,144,900.0	4,198,872.3
22	12.5	1040.0	13,000.0	156.3	1,081,600.0	4,322,719.2
	$\sum x = 202.5$	$\sum y = 93025.0$	$\sum xy = 673700.0$	$\sum x^2 = 2406.3$	$\sum y^2 = 574582575.0$	$\sum(y-\hat{y})^2 = 11985945.6$

Slope constant

$$\text{Burger King } b = \frac{n\sum xy - \sum x \sum y}{n\sum x^2 - (\sum x)^2} = \frac{22(673700.00) - 202.50(93025.00)}{22(2406.25) - 202.50^2} = -336.61$$

Intercept constant

$$\text{Burger King } a = \frac{\sum y}{n} - \frac{b\sum x}{n} = \frac{93025.00}{22} - \frac{-336.61(202.50)}{22} = 7326.74$$

Regression equation

$$\text{Burger King } \hat{y} = a + bx \qquad \hat{y} = 7326.74 + -336.61x$$

Standard error of the estimate

$$\text{Burger King } s_{y.x} = \sqrt{\frac{\sum(y-\hat{y})^2}{n-2}} \qquad \sqrt{\frac{11978945.66}{22-2}} = 2447.30$$

McDonald's $\qquad \sqrt{\frac{91294159.81}{40-2}} = 1529.99$

What does the standard error of the estimate statistic given in Box 9.1a tell us about the dispersion of the types of restaurants about their respective regression lines?

9.2.2 Testing the Significance of Simple Linear Regression

The differences between the predicted and observed Y values can occur for three main reasons when applying regression analysis to populations or samples of observations:

- there is an **alternative** independent variable that exerts stronger control over the dependent variable.
- there are **additional** independent variables not yet considered that also exert stronger control over the dependent variable in combination with the currently selected independent variable.
- there is sampling error in the data.

The first two reasons suggest that further analysis is required that considers other independent variables, whereas the third calls for testing the significance of the regression statistics. There are two main approaches to this task: one involves applying analysis of variance (see Chapter 6) to test the whole regression model or equation and the other uses a series of t tests to investigate the significance of its component parts $(a, b \text{ and } \hat{Y})$. Confidence intervals around the regression line and the intercept, which is really the predicted value of Y when $X = 0$, are similar to such limits around a univariate mean. They define a range within which an investigator can be 95 or 99 per cent confident that the true population regression line falls.

9.2.2.1 Testing the Whole OLS Regression Model

We have seen that Analysis of Variance (ANOVA) can be used as a statistical test to examine whether the means of a group of independent samples are significantly different from each other and that the procedure divides the total variance into the between groups and within groups parts. When applied to simple linear regression, the total variance of the dependent variable (Y) is similarly separated into two parts: one connected with the predicted $\hat{y}$ values along the regression line and the other with the residuals of the observed values from the line. These are respectively known as the regression and residual variances. The purpose of dividing the total variance into these two parts is to indicate how much of the variability of the Y values is explained by the linear regression model and how much by the residuals. If a sizable proportion of the total variance comes from the regression equation, then it is reasonable to conclude that this model and the independent variable (X) together act as a good predictor of the dependent variable (Y). If the residuals' variance accounts for a considerable proportion of the total, this suggests that other excluded independent variables explain more of the variability in the dependent variable.

Consideration of extreme or unlikely situations sometimes helps to explain how a statistical technique works. In the case of simple linear regression, it would be very unusual for the regression line to be horizontal, which would indicate that the value of Y was constant for all values of the independent variable. This would mean that there was no slope $(b = 0.0)$ and $\bar{y} = \hat{y}$. The regression sum of squares and variance would both equal zero and the total variance would equal the residual variance. The regression line invariably slopes upwards or downwards through a scatter of points from the left of the scatter graph, which means that some of the total sum of squares and variance are attributed to the regression model. If the Y values all lie along a sloping regression line, this produces another extremely unlikely situation in which the total sum of squares and variance of Y are only associated with the regression model (i.e. there is no variance from any residuals). The size of F

test statistic (see Box 9.2a) indicates whether the amount of variability in the Y values attributed to the regression model is significant in relation to that connected with the residuals or prediction errors.

9.2.2.2 Testing the Constants and Predicted Y Values in OLS Regression

An alternative approach to testing the significance of regression analysis is to examine the sampling error that might be present in the intercept and slope constants as estimates of the corresponding population parameters α and β, and in the predicted values of Y. The starting point for testing the significance of these quantities is to recognise that the calculated values are obtained from a particular sample of observations and that this is just one of an infinite number of such samples that could have been selected from the population. This represents the standard starting point for most statistical tests, namely that we do not know for certain whether the statistics obtained from our sample are typical or unusual in relation to the entire population of observations. Theoretically, if regression analysis in respect of X and Y variables was applied to a series of samples selected from a population, each slope coefficient and the mean of all the slope coefficients would be an estimate of the population parameter, β. This set of coefficients would have a degree of dispersion which can be measured by their standard error (s_b), which can be estimated from a single sample (see Box 9.2a). This quantity can be used in two ways:

- to calculate a t test statistic that examines the Null Hypothesis that the differences between the sample slope coefficient and zero are not significant
- to obtain confidence limits within which the population parameter might be expected to fall

---

Box 9.2a Significance Testing of OLS Regression.

Whole model or equation using Analysis of Variance:

Population: $F = \dfrac{\sigma_\mu^2}{\sigma_\varepsilon^2}$ Sample: $F = \dfrac{s_{\hat{y}}^2}{s_e^2}$

Regression variance: $s_{\hat{y}}^2 = \dfrac{\sum(\hat{y}-\bar{y})^2}{k}$ Residual variance: $s_e^2 = \dfrac{\sum(y-\hat{y})^2}{n-2}$

Separate components of equation:

Standard errors:

Slope: $s_b = \dfrac{s_{y.x}}{\sqrt{\left(\sum x^s - \dfrac{(\sum x)^2}{n}\right)}}$ Intercept: $s_a = s_{y.x}\sqrt{\dfrac{\sum x^2}{n\sum x^2-(\sum x)^2}}$

Predicted Y: $s_{\hat{y}} = s_{y.x}\sqrt{\dfrac{1}{n}+\left(\dfrac{n(x-\bar{x})^2}{n\sum x^2-(\sum x)^2}\right)}$

Slope: $t = \dfrac{|b-\beta|}{s_b}$ Intercept: $t = \dfrac{|a-\alpha|}{s_a}$ Predicted Y

Confidence limits of regression equation or line:

Slope: $b - t_{0.05}s_b$ to $b + t_{0.05}s_b$

Intercept: $a - t_{0.05}s_a$ to $a + t_{0.05}s_a$

Predicted Y: $\hat{y} - t_{0.05}s_{\hat{y}}$ to $\hat{y} + t_{0.05}s_{\hat{y}}$

Box 9.2b   Application of Significance Testing to OLS Regression.

Using Analysis of Variance to test the significance of the whole regression model involves calculating the regression variance $\left(s_{\hat{y}}^2\right)$ and the residual variance $\left(s_e^2\right)$. The F test statistic is obtained by dividing the regression variance by the residual variance and its probability determined with k degrees of freedom for the regression variance and $n-k-1$ for the residual variance. n and k .are, respectively, the number of observations and the number of independent variables, which is always 1 in the case of simple linear regression. In general terms, the Null Hypothesis states that the variability in the dependent variable (Y) is not explained by differences in the value of the independent variable (X). The fate of the Null Hypothesis is judged by specifying a level of significance (e.g. 0.05) and determining the probability of having obtained the value of the F test statistic. If the probability is less than or equal to the chosen significance threshold, the Null Hypothesis would be rejected.

The alternative method of testing the significance of regression analysis is to focus on the individual elements in the regression equation, the slope and intercept constants and the predicted Y values, which estimate the mean of the dependent variable for each known value of X. The procedure is similar in each case and recognises that if an infinite series of samples were to be selected from a population there would be a certain amount of dispersion in the slope, intercept and predicted Y values. This can be quantified by means of calculating the standard error, which is used to produce t test statistics and confidence limits in each case. The general form of the Null Hypothesis for each of these tests is that any difference between the sample-derived statistics (b, a and $\hat{y}$) and the known or more likely assumed population parameters (β, α and μ) has arisen through sampling error. There are $n-2$ degrees of freedom when determining the probability associated with the t test statistic calculated from the sample data values. As usual, the decision on whether to accept the Null Hypothesis or the Alternative is based on the probability of the t test statistic in relation to the stated level of significance (e.g. 0.05).

The calculation of confidence limits for the slope, intercept and predicted Y values provides a range of values within which the corresponding population parameter is likely to lie with a certain degree of confidence (e.g. 95 or 99%). In each case the standard error is multiplied by the t statistic value for the chosen level of confidence with $n-2$ degrees of freedom and the result is added to and subtracted from the corresponding sample-derived statistic in order to define upper and lower limits. In the case of confidence limits for the Y values predicted, these can be calculated for several known values of X and shown on the scatter plot to indicate a zone within which the corresponding population parameter is expected to fall.

Several types of significance testing have been applied to the simple linear regression equations in Box 9.1a relating to the Burger King and McDonald's restaurants in Pittsburgh. The Null Hypothesis relating to the regression model states that distance from the centre of the city does not explain the daily customer footfall in respect of either type of restaurant. Each of the Null Hypotheses in the second group of tests asserts that the slope and intercept coefficients and predicted dependent variable values are not significantly different from would occur randomly. The probabilities associated with $F = 10.26$ (Burger King) and $F = 27.84$ (McDonald's) respectively with 1/20 and 1/39 degrees of freedom in the ANOVA tests clearly suggest that the regression models for both types of restaurants provide evidence that there is a small probability that distance from city centre does not partly explain daily customer footfall. The individual t tests

(Continued)

Box 9.2b (Continued)

produce statistically significant results for both types of restaurants in respect of the slope and intercept coefficients. The probability of the t test statistic for b ($t = 3.20$) is 0.004 and for a ($t = 6.67$) is <0.000 for Burger King restaurants and the equivalent figures for McDonald's outlets are $t = 5.28$ $p < 0.000$ and $t = 14.71$ and $p < 0.0$. The critical t distribution values at the 0.05 significance level with 20 and 39 degrees of freedom respectively for the Burger King and McDonald's restaurants are 2.086 and 2.021. These have been inserted into the calculations for the confidence limits in Box 9.2c.

The key stages in testing the significance of Simple Linear Regression are:

- *Choose type of testing procedures:* it is possible to test the whole regression model using analysis or the slope, intercept and predicted Y values, although in practice statistical software allows both to be carried easily.
- *State Null Hypothesis and significance level:* the regression model does not indicate that the independent variable (X) explains any variation in the dependent variable (Y) beyond that which might be expected to occur through sampling error at the 0.05 level of significance OR the slope, intercept and predicted Y values are not significantly different from what occurs by chance at the same level of significance.
- *Calculate the appropriate test statistic (F or t):* the calculations for the test statistics are given below.
- *Select whether to apply a one- or two-tailed test:* in most cases, there is no reason to believe the difference between the sample statistics would be smaller or larger than the hypothesised population values.
- *Determine the probability of the calculated test statistics:* the probabilities of obtaining the individual test statistics by chance using the appropriate degrees of freedom.
- *Accept or reject the Null Hypothesis:* these probabilities are <0.05 and therefore the Null Hypothesis should be rejected and the Alternative Hypothesis is accepted, recognising that this might be an erroneous decision.

Given earlier comments about the improbability that the regression line will be horizontal, it might seem strange to assign the value zero to β to test whether the difference between b and β is significant. However, if the slope coefficient was zero, this would imply that the relationship between the dependent and independent variables was fixed irrespective of the value of X. Testing whether the sample slope coefficient is significantly different from zero seeks to discover whether the independent variable exerts a controlling influence. If the value of Y was constant for all values of X, the latter would clearly not be exerting any control. There are some occasions when a value other than zero might be assigned to β in the Null Hypothesis, for example when a complete statistical population of observations had been analysed previously and the current investigations seek to determine whether the regression of Y on X has changed over time.

The theory behind testing the significance of the intercept follows a similar line of argument. If an infinite number of samples was selected from a given population and their X values equalled zero, then the mean of all the intercepts would be expected to equal α, the population parameter. It would not be necessary for the intercepts of all the samples to equal α and so there would be some degree of dispersion which could be measured by their standard error (s_a). This can again be estimated from

Box 9.2c   Calculation of Significance Testing Statistics for OLS Regression.

Analysis of variance

Regression variance

Burger King $s_{\hat{y}}^2 = \dfrac{\sum(\hat{y}-\bar{y})^2}{k}$ $\dfrac{61448873.67}{1} = 61448873.67$

McDonald's $= 65179528.15$

Residual variance

Burger King $s_e^2 = \dfrac{\sum(y-\hat{y})^2}{n-2}$ $\dfrac{119785945.66}{22-2} = 5989597.28$

McDonald's 2340875.89

Analysis of variance and probability

Burger King $F = \dfrac{s_{\hat{y}}^2}{s_e^2}$ $\dfrac{61448873.67}{5989597.28} = 10.26$

$p =$ $= 0.004$

McDonald's $F =$ $= 27.84$

$p =$ < 0.000

t tests and confidence limits

Standard error of the slope

Burger King

$s_b = \dfrac{s_{y.x}}{\sqrt{\left(\sum x^s - \dfrac{(\sum x)^2}{n}\right)}}$ $\dfrac{2447.30}{\sqrt{\left(2406.25 - \dfrac{202.50^2}{22}\right)}} = 105.09$

McDonald's $= 47.65$

t test statistic for slope and probability

Burger King

$t = \dfrac{|b-\beta|}{s_b} =$ $\dfrac{|-336.61-0|}{105.09} = 3.20$

$p =$ $= 0.004$

McDonald's $t =$ $= 5.28$

$p =$ < 0.000

Confidence limits of slope coefficient

Burger King $-336.61 - 2.086(105.09)$ to $-336.61 + 2.086$ (105.09)

$b - t_{0.05}\,s_b$ to $b + t_{0.05}\,s_b$ -555.82 to -117.39

McDonald's $-251.42 - 2.021(47.65)$ to $-251.42 + 2.021$ (47.65)

    -347.72 to -155.12

Standard error of the intercept

Burger King

$s_a = s_{y.x}\sqrt{\dfrac{\sum x^2}{n\sum x^2 - (\sum x)^2}}$ $2447.3\sqrt{\dfrac{2406.25}{22(2406.25) - 202.5^2}} = 1099.45$

McDonald's $= 530.46$

(Continued)

Box 9.2c (Continued)

t test statistic for intercept and probability	Burger King $$t = \frac{|a - \alpha|}{s_a} =$$	$$\frac{|7326.74 - 0|}{1099.45} = 6.67$$
	$p =$	<0.000
	McDonald's $t =$	$= 14.71$
	$p =$	<0.000
Confidence limits	Burger King	$7326.74 - 2.086(1099.45)$ to $73\,26.74 + 2.086$ (1099.45)
	$a - t_{0.05}\, s_a$ to $a + t_{0.05}\, s_a$	5033.29 to 9620.19
	McDonald's	$7802.10 - 2.021(530.46)$ *to* $7802.10 + 2.021$ (530.46) 6730.04 to 8874.16
Standard error of the predicted Y	Burger King $$s_{\hat{y}} = s_{y.x}\sqrt{\frac{1}{n} + \left(\frac{n(x - \bar{x})^2}{n\sum x^2 - (\sum x)^2}\right)}$$	Burger King example with $x = 15.0$ $$2447.3\sqrt{\frac{1}{22} + \left(\frac{22(15.0 - 9.20)^2}{22(2406.25) - (202.5^2)}\right)} = 12.32$$

	x	$\hat{y}$	$s_{\hat{y}}$	$\hat{y} - t_{0.05}\, s_{\hat{y}}$ (lower)	$\hat{y} + t_{0.05}\, s_{\hat{y}}$ (upper)
Confidence limits of the predicted Y	2.5	14.23	6484.77	6456.01	6513.54
	5.0	8.92	5643.25	5625.21	5661.29
	7.5	3.62	4801.72	4794.41	4809.04
	10.0	1.71	3960.20	3956.74	3963.66
	12.5	7.01	3118.67	3104.50	3132.85
	15.0	12.32	2277.15	2252.25	2302.05
	17.5	17.63	1435.62	1399.99	1471.26
	20.0	22.94	594.10	547.73	640.47

	x	$\hat{y}$	$s_{\hat{y}}$	$\hat{y} - t_{0.05}\, s_{\hat{y}}$ (lower)	$\hat{y} + t_{0.05}\, s_{\hat{y}}$ (upper)
Confidence limits of the predicted Y	2.5	9.06	7173.55	7155.23	7191.87
	5.0	6.02	6545.00	6532.83	6557.17
	7.5	2.98	5916.45	5910.43	5922.47
	10.0	0.17	5287.90	5287.55	5288.25
	12.5	3.12	4659.35	4653.04	4665.66
	15.0	6.17	4030.80	4018.34	4043.26
	17.5	9.21	3402.25	3383.64	3420.86
	20.0	12.26	2773.70	2748.93	2798.47

any single sample of the observations (see Box 9.2a). The standard error can be used to test the significance of the intercept in a particular sample of observations and to determine confidence intervals. In most cases, the Null Hypothesis assumes that population parameter (α) equals zero, since the only evidence available to suggest that the regression does not pass through the origin of the scatter plot (i.e. $X = 0.0$ and $Y = 0.0$) comes from the sampled set of observations and these might be untypical of the population. *A priori* reasoning often leads to the conclusion that Y must equal zero when X equals zero. For example, an investigation into the relationship between wind speed (independent) and energy generated (dependent) by wind turbines might assume that when there is no wind there will also be no energy output. Equally, there are other occasions when *a priori* reasoning might lead to a Null Hypothesis that assumes Y equals some value more or less than zero. For instance, an examination of the relationship between an increase (or decrease) in tax per litre of fuel (independent) and people's monthly expenditure on this item (dependent) might assume that when $X = 0.0$ Y will be > or < 0.0 because other factors affect people's consumption patterns. Simple linear regression focuses on one independent variable and there may be others that lead to a positive or negative value of Y when X equals zero; therefore, the best option is to assume α is zero.

The predicted values of the dependent variable $(\hat{Y})$ are estimates of the population means (μ) for each of the normally distributed conditional distributions that are associated with the different values of X. If a series of samples was drawn from a population, the means of their Y values for each value of X would be expected to equal μ or $\hat{Y}$, the predicted value of the dependent variable in the population. It is certain the sampled Y values and quite possibly those with any individual sample would display some degree of dispersion that could be measured by the standard error $(s_{\hat{y}})$. For example, the scatter plots in Box 9.1a show that some fast food restaurants at the same distance from the centre of Pittsburgh had different average daily customer footfalls. It is possible to carry out *t*-tests with respect to Y for each of the known values of X, although it is more useful to gain an overall impression of the predictive strength of the regression equation by calculating the confidence limits around the regression line. Chapter 6 examined how a sample mean $(\bar{x})$ can be used to determine confidence limits within which a population mean (μ) is likely to fall in a certain percentage of randomly selected samples (e.g. 95 or 99 per cent) from the population with reference to the probabilities of the t distribution. A wide confidence interval indicates that the specific value of the sample mean $(\bar{x})$ is a poor estimate of the population mean (μ), whereas a narrower interval suggests the sample statistic is a good estimate. These principles can be applied to produce confidence intervals (or limits) for the predicted values of the dependent variable. The procedure involves calculating the standard error of Y for each value of X and inserting this into the formula shown in Box 9.2a. The confidence limits for each value of X appear as curves around the regression line because more extreme values of the independent variable are further from its mean and produce higher values of $s_{\hat{y}}$.

9.2.3 Explanatory Power of OLS Regression and Effect of Limits in Range of *X* Values

There are two further points to note before moving on from simple bivariate linear regression analysis into the more complex realms of non-linear relationships. The **coefficient of determination** (r^2) was explored in relation to Pearson's product-moment correlation and can be calculated for simple linear regression as the ratio between the variances of the predicted and observed Y values, $(s_{\hat{y}}^2)$ divided by (s_y^2). It quantifies the extent to which X explains Y with values in the range 0.0–1.0: when expressed as a percentage this value indicates the amount of the variance of the dependent variable that is accounted for by the independent or explanatory variable.

Most applications of simple linear regression generate the regression equation from sample data: as a result, the predictions for Y are based on the known values of X that have been included. Although the research may have been planned so that the full range of known X values has been covered, the current range can expand or contract over time. In both cases, the analysis could give rise to misleading results and an incorrectly specified regression equation connecting the two variables might lead to riskier predictions. If a narrower range of X values is required, then some of the original more extreme values where the confidence limits are wider may have produced a weaker relationship than is the case. In contrast, if a wider range of X values extending beyond the original set is needed, then simple linear regression may produce a stronger relationship than genuinely exists. Figure 9.4a includes an original

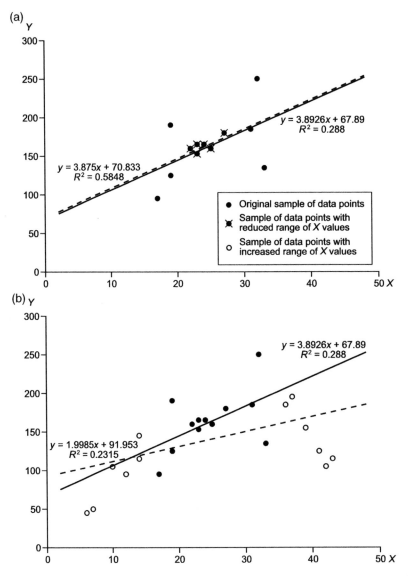

Figure 9.4 Differences in the 'goodness of fit' in relation to a range of values of independent variable. (a) Narrower range of X data values than original set; (b) wider range of X data values than original set.

set of data points (solid circles) with X values in the range 17–33, which produces $r^2 = 0.288$ and a regression of $y = 67.89 + 3.89x$. However, suppose the range of X values changes over time and narrows to 23–27: now the regression equation and line are virtually identical to the previous one ($y = 70.83 + 3.88x$) but with a higher coefficient of determination ($r^2 = 0.585$) indicating the independent variable explains a higher percent of the variance (58.5 per cent). Figure 9.4b shows the case where the range of X values increases (6–43) with the original set of data points retained. The coefficient of determination now reduces to 23.1 per cent and the regression equation becomes $y = 91.95 + 2.00x$. Examination of the extra data points in Figure 9.4b (open circles) suggests that the additional X values beyond the original minimum and maximum are associated with lower Y values. Thus, there is some indication that the original regression equation provides less accurate predictions Y at the upper and to some extent lower ends of the wider range of X values. Additionally, the new regression equation for the full range of X values has a lower coefficient of determination ($r^2 = 0.231$) and indicates a weaker relationship between the variables.

9.3 Non-linear Bivariate Relationships

The previous example reveals there is no guarantee that the linear relationship measured by a regression equation applies beyond the values of the independent variable recorded by a particular sample of observations: a different linear or a **curvilinear regression** equation better fits the data points. Visual inspection of the scatter plot should be sufficient to indicate whether a non-linear relationship might exist and that curvilinear bivariate regression might be applied. The purpose of applying curvilinear regression is the same as OLS regression, namely, to measure and specify the relationship between a pair of ratio or interval scale variables and, when based on sample data, to test hypotheses about the explanatory and predictive power of the independent variable over the dependent one. Extending our exploration of bivariate regression reminds us that what we are attempting to do is discover the equation for the line that most closely fits the observed data. Shifting from the simple linear to a curvilinear equation implies that we are willing to accept a model in which the controlling influence of X over Y differs across the range of its data values. Perhaps two areas or zones of data values of the independent X variable produce high values for Y and in a range in between these the variable X has a weaker effect on Y and thus produces lower dependent variable (Y) values. However, weaker and stronger are perhaps misleading terms, since what is really happening is that two sections in the range of X data values predict higher Y values and one section predicts lower ones.

The main difference is that the curvilinear regression equation defines a curved rather than a straight line. What this means is that the way in which the value of the dependent variable relates to the independent variable differs across the range of X values. The slope of the regression line is constant throughout its length in simple linear regression, whereas the slope of a curved line varies along its length. One approach to dealing with curved trends in a set of data points is to transform the original data values of either X and Y or both variables, for example turning them into logarithms and then applying simple linear regression to the transformed values. A major drawback of this approach is that it may be difficult to interpret the results in terms of the real measurement scale of the variables.

Table 9.1 Specific and generic form of linear and curvilinear bivariate regression equations.

Polynomial equation	Alternative form of polynomial equation	Name	Number of curves
$\hat{y} = a + b_1 x_1$	$\hat{y} = a + bx$	Linear	Straight line – no curve
$\hat{y} = a + b_1 x_1 + b_2 x_1^2$	$\hat{y} = a + bx + cx^2$	Quadratic	One curve
$\hat{y} = a + b_1 x_1 + b_2 x_1^2 + b_3 x_1^3$	$\hat{y} = a + bx + cx^2 + dx^3$	Cubic	Two curves
$\hat{y} = a + b_1 x_1 + b_2 x_1^2 + b_3 x_1^3 + b_4 x_1^4$	$\hat{y} = a + bx + cx^2 + dx^3 + ex^4$	Quartic	Three curves
......	......	......	......
$\hat{y} = a + b_1 x_1 + b_2 x_1^2 + b_3 x_1^3 + b_n x_n^4$	$\hat{y} = a + bx + cx^2 + dx^3 + mx^n$	Generic equation	$n-1$ curves

9.3.1 Forms of Non-linear and Curvilinear Bivariate Relationship

Once we admit the possibility that a bivariate relationship varies along the range of values of the independent variable, it is realistic to imagine that the line may have not one but a series of curves (see lower right part of Figure 9.1). We will explore curvilinear regression using **polynomial** equations of the general form shown in Table 9.1. Just like polygon means a geometrical shape with many sides, polynomial means an equation has many terms, although it is evident that a simple linear regression equation is the starting point for these. Each equation seeks to predict the value of Y and includes an intercept and a series of one, two, three, four or more (up to n) terms where the X values are multiplied by a different slope coefficient. Moving through the equations in Table 9.1 from linear through quadratic, cubic and quartic x refers to the same independent variable throughout and the terms $b_1, b_2, b_3, b_4, \cdots, b_n$ or $b, c, d, e, \cdots, m$ represent alternative ways of representing the slope coefficients. The distinct types of polynomial equations are differentiated by their degree, which refers to the largest exponential power (e.g. cubic degree denotes the cube of X (raised to the power 3)). These power terms introduce curves into the regression line as indicated in Table 9.1.

Curvilinear regression represents a potentially more successful way of determining the bivariate regression equation and line that best fits a scatter of data points for a sample of observations. It explicitly recognises the possibility that the best-fit line may not be straight but curved. The process of applying curvilinear regression analysis is essentially the same as that used with simple linear regression, namely, to determine the slope coefficients and the intercept. Box 9.3a illustrates the application of curvilinear regression to a subgroup of 20 farms that have been randomly selected from a larger set to keep the dataset to a reasonable size. The dataset relates to these farms in 1941 and the analysis explores whether farm size measured in terms of hectares acts as an explanatory variable in respect of the total horsepower of wheeled tractors on the farms. Such motive power was a comparatively new feature on British farms at this time and one aspect of the investigation focuses on whether larger farms showed evidence of having invested in this type of machinery more in comparison with smaller ones. The first four orders or degrees of bivariate polynomial regression analysis are illustrated using this example in Box 9.3a.

Box 9.3a   Curvilinear (Polynomial) Bivariate Regression.

Bivariate Polynomial Regression equation:

Population: $Y = \alpha + \beta_1 X_1 + \beta_2 X_1^2 + \beta_3 X_1^3 + \beta_n X_1^n$

Sample: $y = a + b_1 x_1 + b_2 x_1^2 + b_3 x_1^3 + b_n x_1^n$

Estimation of Y:

Population: $\mu = \alpha + \beta_1 X_1 + \beta_2 X_1^2 + \beta_3 X_1^3 + \beta_n X_1^n$

Sample: $\hat{y} = a + b_1 x_1 + b_2 x_1^2 + b_3 x_1^3 + b_n x_1^n$

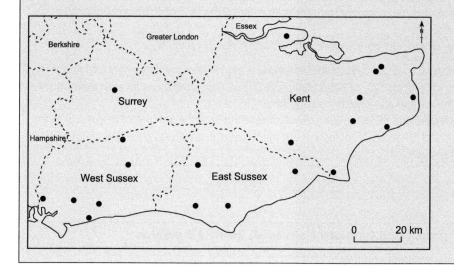

Box 9.3b   Application of curvilinear (polynomial) bivariate regression.

The bivariate polynomial regression equation is an extension of OLS regression and the latter is usually carried out first. Further terms are added that incrementally increase the power by which the values of the independent variable (X) are raised. It is usually only necessary to progress to the cubic or quartic level in order to discover the curvilinear regression line that best fits the data values of the dependent variable (Y). Polynomial regression is typically found by calculating the constants a and $b_1, b_2, \cdots, b_n$, and inserting these into the appropriate definitional equation given above. The predicted values of the dependent variable ($\hat{y}$) lie along the regression line, which has zero, one, two, three, etc. curves according to the order of the polynomial equation. These are estimates or predictions of the dependent variable's value in the statistical population for known values of X. The overall accuracy of these predictions can be assessed by means of the standard error of the estimate (s_{YX}), which is calculated for each polynomial regression equation in a series (i.e. first order, second order, third order, etc.). This provides a way of comparing their explanatory power.

Bivariate polynomial regression has been applied to a randomly selected sub-sample of farms in the South East of England that were being farmed in the early years of the Second World War. Ideally, the values of the independent variable should not have been discovered empirically (i.e.

(Continued)

Box 9.3b (Continued)

---

simply from the sample data) but should have been determined by randomly selecting farms of a particular size. The independent variable is the total farm area measured in hectares and the dependent variable is the total horsepower of wheeled farm tractors, which were relatively new in British farming at the time. The slope and intercept constants together with the standard error of the estimates for the first four orders of the polynomial regression series have been included in Box 9.3d, although details of the calculations are omitted. The standard error of the estimate statistics suggests that there is least dispersion in the first order or OLS regression equation, although it might appear from the scatter plots that second or third polynomial equations are a better fit.

The key stages in applying bivariate polynomial regression are:

- *Calculate constants a and b coefficients:* the former is the value of the dependent variable at the point where the regression line cuts through its axis and the latter together with their sign are the gradient or slope and direction of the line;
- *Calculate the standard error(s) of the estimate*: these indicate the amount of dispersion in the residuals around the series of regression lines and therefore suggest which line/equation best fits the data values.

Box 9.3c Assumptions of Curvilinear (Polynomial) Bivariate Regression.

There are three main assumptions:

- Random sampling with replacement should be employed or without replacement from an infinite population provided that the sample is estimated at less than 10% of the population.
- In principle the data values for the variables X and Y should conform to the bivariate normal distribution: however, such an assumption may be difficult, if not impossible, to state in some circumstances. It is accepted the provided the sample size is sufficiently large, which is itself often difficult to judge, the technique is sufficiently robust to accept some unknown deviation from this assumption.
- The observations in the population and the sample should be independent of each other with respect to variables X and Y, which signifies an absence of autocorrelation.

9.3.2 Testing the Significance of Bivariate Polynomial Regression

Polynomial regression adds successive terms to the simple linear regression in order to determine whether a line with one, two, three or more curves provides a better fit to the data points. Unless visual inspection of the scatter plot clearly suggests the form of the curved line (i.e. how many curves), the application of polynomial regression proceeds by adding successive terms to the equation each one increasing the degree or power to which the independent variable is raised. Each term that is added to a polynomial regression equation will increase the r^2 value and the series of Null

Box 9.3d   Calculation of Curvilinear (Polynomial) Regression.

No.	x	y	No. contd.	x contd.	y contd.
1	292.41	34	11	341.82	80
2	442.26	45	12	104.49	48
3	178.00	48	13	512.63	146
4	768.20	50	14	230.89	15
5	114.41	0	15	301.32	48
6	67.84	27	16	97.20	20
7	1549.73	50	17	343.44	22
8	53.46	30	18	196.59	86
9	129.60	0	19	50.90	20
10	130.41	26	20	65.41	0

Polynomial regression scatter plots and equations

First order $\hat{y} = a + b_1 x_1$

Second order $\hat{y} = a + b_1 x_1 + b_2 x_1^2$

	First order	Second order	
Slope constant	$b_1 = 0.033$	Slope constants	$b_1 = 0.17, b_2 = -0.000095$
Intercept constant	$a = 29.77$	Intercept constant	$a = 7.55$
Equation	$\hat{y} = 29.77 + 0.033x$	Equation	$\hat{y} = 7.55 + 0.17x - 0.000095x^2$
Standard error of the estimate	$s_{y.x} = 33.22$	Standard error of the estimate	$s_{y.x} = 62.07$

(Continued)

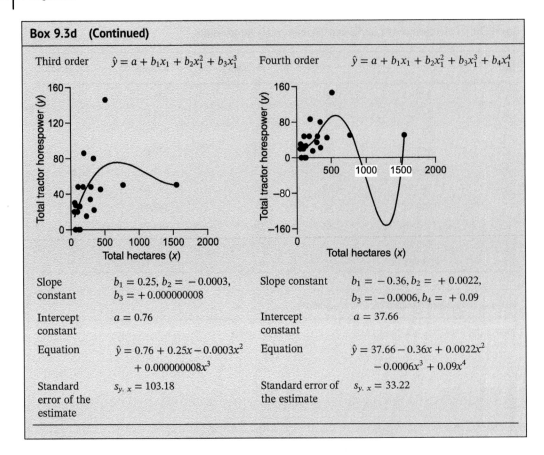

Box 9.3d (Continued)

Third order $\hat{y} = a + b_1 x_1 + b_2 x_1^2 + b_3 x_1^3$ Fourth order $\hat{y} = a + b_1 x_1 + b_2 x_1^2 + b_3 x_1^3 + b_4 x_1^4$

	Third order		Fourth order	
Slope constant	$b_1 = 0.25, b_2 = -0.0003,$ $b_3 = +0.000000008$		Slope constant	$b_1 = -0.36, b_2 = +0.0022,$ $b_3 = -0.0006, b_4 = +0.09$
Intercept constant	$a = 0.76$		Intercept constant	$a = 37.66$
Equation	$\hat{y} = 0.76 + 0.25x - 0.0003x^2$ $+ 0.000000008x^3$		Equation	$\hat{y} = 37.66 - 0.36x + 0.0022x^2$ $- 0.0006x^3 + 0.09x^4$
Standard error of the estimate	$s_{y \cdot x} = 103.18$		Standard error of the estimate	$s_{y \cdot x} = 33.22$

Hypotheses seek to determine whether the increase in r^2 values between one polynomial regression equation and the next is more than might be expected to occur by chance. The series of r^2 values allow the explanatory power of the independent variable in the different equations to be assessed. Each curvilinear regression equation can also be tested to determine whether it fits the data points better than a horizontal line, which is like applying a t test to the slope coefficient in simple linear regression. Box 9.4a tests the results of applying polynomial regression with respect to the dependent and independent variables examined in Box 9.3a, namely the sub-sample of farms in South East England.

Box 9.4a Significance testing of Bivariate Polynomial Regression.

Coefficient of determination: $r^2 = \dfrac{s_{\hat{y}}^2}{s^2}$

Confidence limits of predicted Y: $\hat{y} - t_{0.05}\, s_{\hat{y}}$ to $\hat{y} + t_{0.05}\, s_{\hat{y}}$

Box 9.4b **Application of significance testing to Bivariate Polynomial Regression.**

The most straightforward way of testing significance in polynomial regression is to examine the incremental increase in the value of r^2 (the coefficient of determination) that occurs with additional term in the equation. The starting point is that the simple linear regression equation is entirely adequate and the series of Null Hypotheses examine whether the difference in r^2 values between one equation and the next in the sequence is more than might be expected to occur by chance. The degrees of freedom for the numerator are the order polynomial regression (i.e. 1, 2, 3, etc.) and for the denominator they are $n - 2 - i$, where n is the number of observations and i is the power to which the independent variable is raised in the polynomial regression equation (i.e. $1, 2, 3, \cdots k$). A decision on whether to accept or reject each Null Hypothesis with respect to the F statistic is made by reference to the probability of the difference having occurred by chance in relation to the level of significance, typically 0.05 or 0.01. If the probability is less than or equal to the stated level, the Null Hypothesis should be rejected.

This form of significance testing with bivariate polynomial regression has been applied to farm survey data introduced in Box 9.3a. The separate r^2 statistics associated with the four polynomial regression equations are distinguished from each other by their subscripts 1, 2, 3 and 4. The numerator and denominator degrees of freedom for the first order (the simple linear equation) are 1 and 20-1; 2 and 20-2-1 for the second-order polynomial equation; 3 and 20-2-2 for the third and 4 and 20-2-3 for the fourth. The F statistics corresponding to these combinations of degrees of freedom for the numerator and denominator and their probabilities are given below and indicate that the second-order polynomial regression equation provides a statistically significant increase in the value of r^2 compared with the linear, cubic and quartic versions. Whether there is any logical reasoning that can be adduced for **why** the total horsepower of tractors on farms in the South East of England in 1941 should be related to and at least partly explained by $\hat{y} = 7.55 + 0.17x - 0.000095x^2$ is another matter.

The confidence limits around curvilinear regression can also be calculated and plotted. These indicate a zone or range around the regression line within which the population parameter (μ) of the predicted dependent variable ($\hat{y}$) will fall with a certain degree of confidence for known values of the independent variable (x). The standard error of the estimate is multiplied by the t statistic value at the specified confidence level (e.g. 95%) with $n - 2$ degrees of freedom and the resultant value is added to and subtracted from $\hat{y}$ for any known value of x. In this application, the 95% confidence limits are wide, which suggests that the polynomial regression equation provides only a moderately reliable prediction of the population parameter for any given value x.

The key stages in testing the significance of Bivariate Polynomial Regression are:

- *Calculate the F test statistics*: the F statistics are usually calculated for the individual r^2 values and the difference between these for successive polynomial regression equations. The sequence of F statistics for these differences in this application is 5.399, 0.224 and 2.057.
- *State Null Hypotheses and significance level*: the general form of the Null Hypothesis is that the difference between the pair of r^2 values is not significantly greater than would occur by chance at the 0.05 level of significance.
- *Determine the probability of the calculated test statistics:* the probabilities associated with the F statistics with 2 and 17 degrees of freedom is 0.015, with 3 and 16 degrees of freedom is 0.879 and with 4 and 15 degrees of freedom is 0.138.

(Continued)

Box 9.4b (Continued)

Accept or reject the Null Hypothesis: these probabilities indicate that the increase in r^2 achieved between the first- and second-order polynomial regression equation is more than might be expected at the 0.05 level of significance and therefore the Null Hypothesis associated with this difference is rejected and the other two Null Hypotheses are accepted, recognising that this might be an erroneous decision.

Box 9.4c Calculation of significance testing statistics for Bivariate Polynomial Regression.

First-order Coefficient of determination

$$r^2 = \frac{s_{\hat{y}}^2}{s^2}$$

$$\frac{134.09}{1179.88} = .1138$$

F test statistic for r_1^2 and its probability

$$F = \frac{n-2(r_1^2)}{1-r_1^2}$$

$$\frac{20-2(0.1138)}{1-0.1138} = 1.935$$

$$p = \qquad\qquad = 0.146$$

Second-order Coefficient of determination

$$r_2^2 = \frac{s_{\hat{y}}^2}{s^2}$$

$$\frac{134.09}{384.79} = .3274$$

F test statistic and its probability

$$F = \frac{n-2-1(r_2^2)}{1-r_2^2}$$

$$\frac{20-2-1(0.3274)}{1-0.3274} = 5.239$$

$$p = \qquad\qquad = 0.034$$

F test statistic for difference between r_1^2 and r_2^2 and its probability

$$F = \frac{n-2-1(r_1^2 - r_1^3)}{1-r_2^2}$$

$$\frac{20-2-2 10.3274 - 0.1138)}{1-0.3274} = 5.399$$

$$p = \qquad\qquad = 0.015$$

Third-order Coefficient of determination

$$r^2 = \frac{s_{\hat{y}}^2}{s^2}$$

$$\frac{134.09}{397.25} = .3367$$

F test statistic and its probability

$$F = \frac{n-2-2(r_3^2)}{1-r_3^2}$$

$$\frac{20-2-2(0.3367)}{1-0.3367} = 5.050$$

$$p = \qquad\qquad = 0.080$$

F test statistic for difference between r_2^2 and r_3^2 and its probability

$$F = \frac{n-2-2(r_3^2 - r_2^3)}{1-r_3^2}$$

$$\frac{20-2-2(0.3367 - 0.3274)}{1-0.3367} = 0.224$$

$$p = \qquad\qquad = 0.879$$

Fourth-order Coefficient of determination

$$r^2 = \frac{s_{\hat{y}}^2}{s^2}$$

$$\frac{134.09}{491.657} = .4167$$

F test statistic and its probability

$$F = \frac{n-2-3(r_4^2)}{1-r_4^2}$$

$$\frac{20-2-3(0.4167)}{1-0.4167} = 6.251$$

$$p = \qquad\qquad = 0.031$$

F test statistic for difference between r_3^2 and r_4^2 and its probability.

$$F = \frac{n-2-3(r_4^2 - r_3^3)}{1-r_4^2}$$

$$\frac{20-2-3(0.4167 - 0.3367)}{1-0.4167} = 2.057$$

$$p = \qquad\qquad = 0.138$$

	x	$\hat{y}$	$s_{\hat{y}}$	$\hat{y} - t_{0.05}\, s_{\hat{y}}$ (lower)	$\hat{y} + t_{0.05}\, s_{\hat{y}}$ (upper)
Confidence limits of the predicted Y	0	7.55	12.10	−17.87	35.56
	200	38.23	22.74	−9.54	37.54
	400	61.31	30.91	−3.63	39.15
	600	76.79	36.48	0.15	40.31
	800	84.67	39.34	2.01	40.93
	1000	84.95	39.45	2.07	40.95
	1200	77.63	36.78	0.35	40.38
	1400	62.71	31.67	−3.82	39.27

9.4 Complex Relationships

This text has followed the accepted norm by introducing bivariate linear and curvilinear regression analysis before progressing to multivariate or multiple regression. However, it soon becomes clear when explaining bivariate curvilinear models between one dependent and one independent variable that they are insufficient to understand relationships and 'why things happen'. We appreciate that our 'real world' research questions are more complex since they include more than one independent variable to explain differences in the value of the dependent variable. This means that the dependent and independent variables are connected in such a way that members of the latter group can affect the values of the dependent variable in diverse ways. For example, suppose there are three independent variables, A, B and C:

- Variable A is positively correlated with the dependent variable so that low values of A are associated with low values of Y and *vice versa*;
- Variable B is negatively correlated with the dependent variable so that low values of B are linked to high values of Y and *vice versa*;
- Mid-range values of Variable C are associated with high values of Y, and low and high values of C with low Y values.

Imagine a version of the sport known as 'tug of war' (tug o' war) where rather than two teams pitted against each other pulling in opposite directions there are three, four or more teams all pulling on ropes that are knotted together in the middle. The variable strength of the teams results in the knot (representing the predicted value of the dependent variable Y) moving its location as the competing teams pull their rope in different directions. This potentially offers a more satisfying way of investigating 'real' world events, since they rarely have only one cause, and often different causes will produce an outcome in diverse ways. Multiple regression is the means of modelling such complex relationships in statistical analysis.

9.4.1 Multivariate (Multiple) Regression

Multiple regression still refers to the dependent variable as Y, while the separate independent variables are usually labelled X, but with different subscripts to distinguish them from each other $x_1, x_2, x_3, \cdots, x_n$. Computationally multivariate regression is like polynomial regression in so far as different variables are added into the regression equation to discover if they increase the r^2 value significantly. The overall purpose is to discover which combination of independent variables provides the equation with the highest explanatory power, which is normally assessed by examining the r^2 values and testing the significance of different combinations. A bivariate scatter plot provides a relatively straightforward way of visually assessing whether a simple linear or curvilinear model is needed. However, it is more difficult to see a pattern in a single multivariate scatter plot where there are two independent variables and one dependent variable, and impossible if there are more than two independent variables. Thus, the decision on which independent variables to include in the regression equation and the order in which they should be entered is a little more problematic. There are two main ways of achieving this objective: either to introduce all of the independent variables into the analysis together or to enter them one after the other. The inclusion of more than one independent variable in regression analysis is only appropriate if this improves the predictive power of the regression equation or to test theoretical models relating to complex concepts, such as a landscape's attractiveness.

Visualisation of such multivariate relationships by means of a scatter plot is not only exceedingly difficult but also would not easily enable the predictions to be made. We usually define a multivariate relationship by means of an equation, which has the generic form:

$$\hat{y} = a + b_1 x_1 + b_2 x_2 + b_3 x_3 + \cdots b_n x_n$$

Each of the independent variables $(x_1, x_2, x_3, \cdots, x_n)$ has its own slope coefficient $(b_1, b_2, b_3, \cdots b_n)$ as a multiplier, although there remains only one intercept (a). Calculation of the slope coefficients and intercept for multivariate regression is more complex than is the case for simple bivariate linear regression and the details are not examined here. However, it is worth considering why these computations are more complex. Simple linear regression assumes that all of the variance of the dependent variable is accounted for by one independent variable and so two or more independent variables might be expected to account for this variance jointly. However, some or all these independent variables may be correlated with each other, and so they exert an overlapping or duplicate effect on the values of the dependent variable. Figure 9.5 illustrates the impact of this statement in a hypothetical situation where there are two independent variables $(x_1$ and $x_2)$ and one dependent variable (y). The upper part of Figure 9.5 separately shows the bivariate scatter plots and linear regression equations between each independent variable and the dependent variable, which has the same values throughout, where x_1 and x_2 are uncorrelated with each other (Pearson's $r = +0.056$). The results of bivariate regression analysis indicate that x_1 has a strong explanatory relationship with y, since the slope coefficient equals 0.72 and $r^2 = 0.85$ indicating that it accounts for some 85 per cent of the variability in the independent variable. In contrast, x_2 has much weaker explanatory power with $r^2 = 0.05$. The lower pair of bivariate scatter plots and regression analyses in Figure 9.5 show a comparable situation when the two independent variables are perfectly positively correlated with each other (Pearson's $r = +1.000$). In this instance, the visual appearance of the scatter plots seems identical at first sight, although close inspection reveals that the x_2 values are each 10 units less than the x_1 ones, hence their perfect positive correlation. The slope coefficients and the r^2 values for the two regression equations are identical.

> Why are the values of the intercept different in the bivariate regression equations in the lower part of Figure 9.5? Describe what has happened to the regression line.

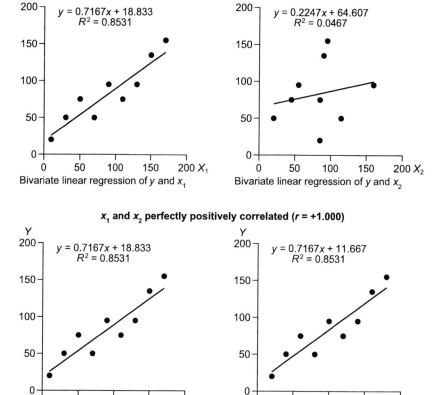

Figure 9.5 Bivariate scatter plots between one dependent and two independent variables contrasting the effect of different strengths of correlation between x_1 and x_2.

A perfect positive or negative correlation between any pair of independent variables means that one of them is unnecessary in the regression analysis and could be omitted without any loss of explanation for or capacity to predict the values of the dependent variable. There is an equivalent statistic to the coefficient of determination that is used in multivariate regression analysis, known as the **squared multiple correlation coefficient** (r^2), that indicates the percentage or proportion of the dependent variable's total variance that is accounted for by the independent variables. The r^2 value for the multiple regression of x_1 and x_2 with y when applied to the data values in Figure 9.5 equals 0.88 (88%). This indicates that the addition of the second independent variable marginally improves on accounting for the variance of y compared with using the first independent variable on its own. One consequence of these issues is that when applying multivariate regression investigators should select independent variables that are only weakly or uncorrelated with each other but that are strongly correlated with the dependent variable. The first stage in your regression analysis should be to carry out bivariate correlation on each pair of independent variables and between the dependent variable and each independent ones to identify those that can be eliminated from multiple regression.

Multivariate regression potentially offers an investigator a more comprehensive statistical explanation of the causes of variability in a dependent variable, since it explicitly recognises that outcomes can have multiple causes. The calculations involved with applying multivariate regression analysis are not examined in detail, since statistical and spreadsheet software is readily available for the task. Box 9.5a illustrates the application of multiple regression to the same sub-sample of farms in South East England. However, in this case, we have the total areas of the 20 farms from three separate surveys carried out in 1941, 1978 and 1998. The underlying research question aims to determine whether farm size as measured by land area in the first two years provides a statistical explanation of where these farms were in the overall spectrum of farm size in the latest survey year.

Box 9.5a   Multivariate (Multiple) Regression.

Multiple Regression equation:

Population: $Y = \alpha + \beta_1 X_1 + \beta_2 X_2 + \beta_3 X_3 + \cdots \beta_n X_n$

Sample: $y = a + b_1 x_1 + b_2 x_2 + b_3 x_3 \cdots b_n x_n$

Population: $\mu = \alpha + \beta_1 X_1 + \beta_2 X_2 + \beta_3 X_3 + \cdots \beta_n X_n$ Sample: $\hat{y} = a + b_1 x_1 + b_2 x_2 + b_3 x_3 \cdots b_n x_n$

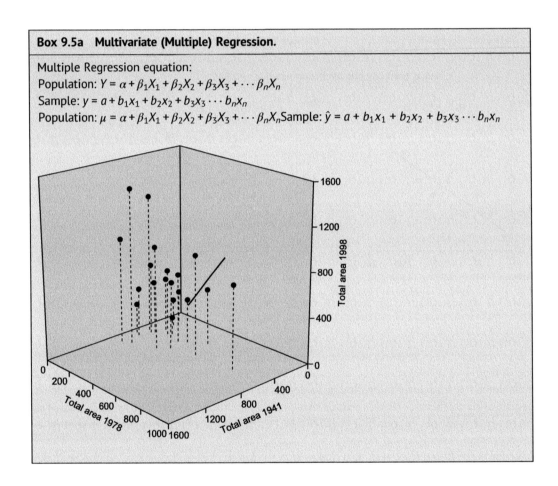

Box 9.5b   Application of Multivariate Regression.

Multivariate linear regression represents an extension of OLS regression to include two or more independent variables labelled $x_1, x_2, x_3, \cdots, x_n$. Each of these has a slope coefficient ($b_1, b_2, b_3, \cdots, b_n$) that denotes the gradient of its relationship with the dependent variable: in other words, the amount that the value of y (the dependent variable) increases per unit of the corresponding x.

The multivariate linear regression equation includes the intercept (a), which represents the point where the regression plane passes through the axis of the dependent variable. The squared multiple coefficient of determination (r^2), which can easily be converted into a percentage by moving the decimal point two spaces to the right, indicates the explanatory power of all or some of the independent variables.

Multivariate linear regression has been applied to a sample of farms in the South East of England (Kent, Surrey, East and West Sussex). Box 9.5d below includes a sub-sample of 20 of these farms in the interest of limiting the amount of data presented. The total area of the sampled farms has been obtained by means of a questionnaire survey on three separate occasions: the agricultural census on 4 June 1941 and author-conducted interview surveys in 1978 and 1998. The regression model seeks to explore whether the independent variables x_1 and x_2, respectively, the 1941 and 1978 areas accurately predict farm size in 1998 (y). It is difficult to determine from the 3D scatter plot how well the independent variables account for the variability in the dependent variable, which can be assessed by means of the standard error of the estimate, which equals 272.46 in this case.

The key stages in applying Multivariate Linear Regression are

- *Calculate a and $b_1, b_2, b_3, \cdots, b_n$:* these are constants in the regression equation and the slope coefficients act as multipliers for known values of the independent variables that define the dimensional regression plane, which intercepts the vertical Y axis and the gradient or slope of the line;
- *Calculate the standard error(s) of the estimate:* a small standard error of the estimate indicates that the data points have a close fit with the linear multivariate regression equation.

Box 9.5c   Assumptions of Multivariate Regression.

There are three main assumptions:

- Sampled observations should be selected randomly with replacement or without replacement from an infinite population if the sample is considered to be less than 10% of the population.
- The deviations or residuals from the regression line (plane) follow a normal distribution.
- The residuals from the regression line have a uniform variance

---

Box 9.5d   Calculation of Multivariate Regression.

No.	x_1	x_2	y	No. cont.	x_1 cont.	x_2 cont.	y cont.
1	292.41	725.00	930.00	11	341.82	617.10	607.04
2	442.26	329.70	512.00	12	104.49	530.10	647.51

(Continued)

Box 9.5d **(Continued)**

No.	x_1	x_2	y	No. cont.	x_1 cont.	x_2 cont.	y cont.
3	178.00	423.40	600.00	13	512.63	572.00	910.57
4	768.20	532.10	748.68	14	230.89	369.80	404.69
5	114.41	259.20	250.91	15	301.32	475.90	627.28
6	67.84	817.60	930.79	16	97.20	661.80	444.00
7	1549.73	749.00	748.68	17	343.44	422.80	404.69
8	53.46	411.50	424.93	18	196.59	396.20	141.64
9	129.60	379.80	485.63	19	50.90	625.80	283.19
10	130.41	789.10	1401.05	20	65.41	486.00	1254.17

Slope constants	$b_1 =$		$b_1 = -0.034$	
	$b_2 =$		$b_2 = 1.228$	
Intercept constant	$a =$		$a = -1.098$	
Regression equation	$y = a + b_1 x_1 + b_2 x_2 + b_3 x_3 \cdots b_n x_n$		$y = -1.098 - 0.34 x_1 + 1.228 x_2$	
Standard error of the estimate	$s_{y.x_1 x_2} =$		$= 272.46$	
Squared multiple correlation coefficient	$r^2 = 1 - \dfrac{\sum (y - \hat{y})^2}{\sum (y - \bar{y})^2}$		$1 - \dfrac{1261975.73}{3457465.03} = 0.365$	

9.4.2 Testing the Significance of Multivariate Regression

The application of multivariate regression analysis to a particular research problem produces a multiple regression equation that best fits the sample data and that predicts the values of the dependent variable in the statistical population. The elements in this equation, the intercept and slope coefficients, which are multipliers of the independent variables, predict $\hat{y}$ values that lie on a sloping plane within the multi-dimensional space defined by the total number of variables. However, just as with bivariate regression, the observed values y in the sample data are unlikely to fall exactly on this regression plane. In other words, there will be an error or residual difference between the predicted $\hat{y}$ value and the observed y value. The regression plane summarises the relationship between the variables so that the total sum of the residuals between the data points and the plane is minimised. Figure 9.6 shows the data values associated with a 3-dimensional scatter plot.

There are again two main approaches to testing the significance of multivariate regression with sample data: one involves testing the whole regression model and the other focuses on the individual parts or elements in the equation. The partitioning of the total variability of the dependent variable between the regression and residual effects that we came across with respect to bivariate regression also works with multivariate regression. The total variability or sum of squares of the dependent variable is defined as $\sum (y - \bar{y})^2$, which is divided between the regression sum of squares $\left(\sum (\hat{y} - \bar{y})^2 \right)$ and the residual sum of squares $\left(\sum (y - \hat{y})^2 \right)$. Significance testing involves using

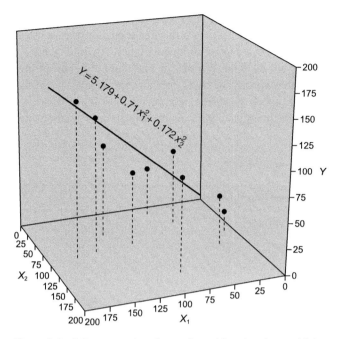

Figure 9.6   3-D scatter plot of x_1 and x_2 with y showing multiple regression equation and line.

analysis of variance to examine whether the regression sum of squares is significantly different from zero. However, the application of ANOVA to test the significance of the multivariate regression equation is not as simple as in bivariate regression where there is only one independent variable and a significant difference between the regression sum of squares and zero also indicates that the slope coefficient (b) and correlation between the variables ($r_{y,\,x}$) are significant at the chosen level. The situation is slightly more complicated with multivariate regression because there is more than one independent variable. A significant difference between the regression sum of squares and zero in these circumstances does not indicate which of the independent variables is significantly related to a dependent variable.

This issue can be resolved by testing the significance of the slope coefficients with Null Hypotheses that $b = 0.0$ using a series of t tests. Each of the t values is calculated by dividing the slope coefficient by the corresponding standard error, for example in respect of the first independent variable:

$$t = \frac{b_1}{s_{b_1}}$$

The t values are assessed by reference to the critical value in the t distribution corresponding to the chosen level of significance with the degrees of freedom equal to the total number of observations minus the number of independent variables minus one. An investigator examines each t test result to determine which of the independent variables plays a statistically significant role in the multivariate relationship. This examination may indicate that all, some or none are significant and each outcome can arise with both significant and non-significant ANOVA results. This can occur because the common variance (i.e. that associated with all the independent variables) does not form part of the individual *t* tests for the slope coefficients. The surprising outcome of a significant ANOVA result without any significant independent variables occurs when the latter share a large amount of the common variance but they each have a small unique variance. Any independent variable with a non-significant slope coefficient does not provide an accurate prediction of the dependent variable and so can realistically be ignored. Box 9.6a provides the results of testing the significance of the multiple regression equation produced by applying it to the sub-sample of farms in South East England in order to investigate the possible explanatory relationship between farm areas at contrasting times.

Box 9.6a Significance testing of Multivariate (Multiple) Regression.

Whole model or equation using Analysis of Variance:

Population: $F = \dfrac{\sigma_\mu^2}{\sigma_\varepsilon^2}$ Sample: $F = \dfrac{s_{\hat{y}}^2}{s_e^2}$

Regression variance: $s_{\hat{y}}^2 = \dfrac{\sum(\hat{y}-\bar{y})^2}{n}$ Residual variance: $s_e^2 = \dfrac{\sum(y-\hat{y})^2}{n}$

Separate components of equation:

Slope: $t = \dfrac{|b_{1\cdots k}-\beta_{1\cdots k}|}{s_b}$ Intercept: $t = \dfrac{|a-\alpha|}{s_a}$ Predicted Y

Confidence limits of regression equation or line:

Predicted Y: $\hat{y}-t_{0.05}\,s_{\hat{y}}$ to $\hat{y}+t_{0.05}\,s_{\hat{y}}$

Box 9.6b Application of Significance Testing to Multivariate (Multiple) Regression.

Testing the significance of multiple regression proceeds in a similar way to simple linear regression by using Analysis of Variance to focus on the whole regression model by calculating an *F* statistic by dividing the regression variance $\left(s_{\hat{y}}^2\right)$ by the residual variance $\left(s_e^2\right)$ and/or by computing *t* test statistics for the individual coefficients in the regression equation ($a, b_2, b_2, \cdots b_k$), where *k* is the number of independent variables and the predicted values of the dependent variable ($\hat{y}$). Calculation of the *t* test statistics and the confidence limits requires that the relevant standard error is computed first since this is used as the divisor in the *t* statistic equation. These calculations are not shown below, since although they follow the principle outlined previously for similar linear regression, they need to consider the covariance between the independent variables.

When using ANOVA there are degrees of freedom relating to the regression variance (*k*) and the residual variance ($n-k-1$), which are used to determine the probability of *F*. The degrees of freedom for the different *t* tests on the coefficients are $n-2$. The ANOVA Null Hypothesis states that the set of independent variables included in the analysis do not provide a statistically significant explanation for the variability in the dependent variable at a particular level of significance (e.g. 0.05 or 95%). The Null Hypotheses relating to each of the slope coefficient assert that they are not significantly different from zero (i.e. constant for all values of the corresponding independent variable. The Null Hypothesis for the intercept (*a*) also states that the 'true' value is zero and the difference between the observed value and zero has only arisen through sampling error. If the probability associated with these test statistics is less than or equal to the chosen significance (e.g. 0.05 or 0.01), the Null Hypothesis would be rejected.

The confidence limits for the slope, intercept and predicted *Y* values are also usually calculated in order to define a range of values within which the corresponding population parameters ($\beta_1, \beta_2, \cdots \beta_k, \alpha$ and μ) are likely to lie with a certain degree of confidence (e.g. 95 or 99%). In each case the standard error is multiplied by the t statistic value for the chosen level of confidence with $n - k - 1$ degrees of freedom and the result is added to and subtracted the corresponding sample-derived value to define upper and lower limits.

Both types of significance testing have been applied to farm survey data introduced in Box 9.3a, where the dependent variable is total farm area in 1998 and the two independent variables are farm area in 1941 and 1978. It is difficult to determine from the 3D scatter plot given above whether the independent variables provided a strong explanation for farm size in 1998. The significance test results calculated below indicate that historical farm size does provide some explanation for subsequent total area, but that the area 20 years previously has a stronger relationship than this combined with the area nearly 60 years before.

The key stages in testing the significance in Multiple Regression are:

- *Choose type of testing procedures:* most statistical software will carry out both types of significance testing and the investigator's task is to interpret the information provided correctly.
- *State Null Hypothesis and significance level:* the H_o with the ANOVA procedure states the combination of independent variables does not explain any variability in the dependent variable beyond what might be expected through sampling error at the 0.05 level of significance OR the slope and intercept coefficients are not significantly different from zero at the same level of significance.
- *Calculate the appropriate test statistic (F or t):* the calculations for the test statistic are given below.
- *Determine the probability of the calculated test statistics:* the probability of obtaining $F = 4.89$ is 0.021 when there are 2(*k*) and 17($n - k - 1$) degrees of freedom respectively for the regression and residual variances; the probability of the *t* test statistic for b_1 ($t = 0.183$) is 0.857, or b_2 ($t = 3.079$) is 0.007 and for the intercept (*a*) ($t = 0.005$) is 0.996.
- *Accept or reject the Null Hypothesis:* these test results and probabilities the Null Hypothesis for the ANOVA procedure would be rejected, whereas only the b_2 slope coefficient is significantly different from zero at the 0.05 level, recognising that this might be an erroneous decision. These results suggest that it was 1978 total farm area that had greater explanatory power with respect to influencing the 1998 farm area.

Box 9.6c   Calculation of significance testing statistics for Multivariate (Multiple) Regression.

Regression variance	$s_{\hat{y}}^2 = \dfrac{\sum(\hat{y}-\bar{y})^2}{k}$	$\dfrac{725423.711}{2} = 362711.86$
Residual variance	$s_e^2 = \dfrac{\sum(y-\hat{y})^2}{n-k-1}$	$\dfrac{1261976.09}{20-2-1} = 74233.87$
Analysis of variance and probability of F	$F = \dfrac{s_{\hat{y}}^2}{s_e^2}$	$\dfrac{362711.89}{74233.89} = 4.89$
	$p =$	0.021
Test statistics for slope coefficients b_1 and b_2, and probabilities of their t values	$t = \dfrac{|b_1 - \beta_1|}{s_{b_1}} =$	$\dfrac{|-0.034-0|}{0.19} = 0.183$
	$p =$	$= 0.857$
	$t = \dfrac{|b_2 - \beta_2|}{s_{b_2}} =$	$\dfrac{|1.228-0|}{0.40} = 3.079$
	$p =$	$= 0.007$
Test statistic for intercept and probability of t	$t = \dfrac{|a - \alpha|}{s_a} =$	$\dfrac{|-1.08-0|}{219.60} = 0.005$
	$p =$	$= 0.996$

	x_1	x_2	$\hat{y}$	$s_{\hat{y}}$	$\hat{y} - t_{0.05}\, s_{\hat{y}}$ (lower)	$\hat{y} + t_{0.05}\, s_{\hat{y}}$ (upper)
Confidence limits of the predicted Y	0	0	−1.10	215.01	−452.83	450.63
	200	200	237.70	142.43	−61.60	536.95
	400	400	476.50	84.90	298.13	654.76
	600	600	715.30	83.50	539.86	890.56
	800	800	954.10	139.98	660.00	1247.96
	1000	1000	1192.90	212.30	746.86	1638.65
	1200	1200	1431.70	288.80	824.94	2038.11
	1400	1400	1670.50	366.86	899.72	2440.87

9.5   Closing Comments

This chapter has examined simple linear regression and the OLS model in some detail before intro-ducing the more complex techniques of polynomial and multiple or multivariate regression. How-ever, in some respects, it has only scratched the surface of the latter techniques and a full account is beyond the scope of an introductory quantitative and statistical analysis text such as this. We could for example have moved on to consider polynomial multiple regression where the plane on which the predicted value of the dependent variable is warped and curved in different directions. This chapter has mentioned that there are separate ways of entering independent variables into a mul-tivariate regression analysis, but detailed examples have not been provided. The intention has been to provide sufficient information to enable students to apply regression analysis in their project in a sensible and meaningful way, even if more complex techniques could have been tried. Recall that the basic principle underlying regression is that we are attempting to model 'how the world works' using a discrete set of variables. We are unlikely to be able to achieve this completely, but we can successfully investigate part of such relationships and by doing so gain new insights.

There is one further crucial point to make before closing this chapter and moving on from the non-spatial statistical techniques that we have used with geographical, environmental and geological data. Our discussion has glossed over the difficulties that often arise with applying standard statistical tests, correlation and regression to spatially distributed phenomena. These are examined in the next, Explicitly Spatial section of this text. By way of introduction, it is worth recalling Tobler's first law of geography that is stated in Chapter 2: 'Everything is related to everything else, but near things are more related than distant things' (Tobler, 1970: 236). The implication is that phenomena of the same type (i.e. from the same population) that are located spatially close to each other are likely to be more like each other than they are to occurrences of the same phenomena that are found further away. This contravenes the assumption of independence in the data values of the variables for a set of observations that is the fundamental basis of most of the correlation and all the regression techniques examined in this and the previous chapter, and the statistical tests in the earlier chapters. The following chapter will explore this problem of **spatial autocorrelation** in more detail and examine geospatial analysis techniques that can be applied to harness the benefits of Tobler's first law of geography.

Multiple regression analysis explicitly recognises that the set of independent variables included in the final model after all the candidate variables have been tried will not predict the dependent variable perfectly – something will be missing. This is acknowledged by the inclusion of an error term in the definitional equation for multiple regression analysis (see Box 9.4a). The presence of this term in the population and sample equations (respectively ε) and (e) represents the combined residuals of the independent variables. It is tempting to suggest that sometimes this missing term represents the geographical or spatial location of phenomena.

Reference

Tobler, W. (1970) A computer movie simulating urban growth in the Detroit region. *Economic Geography*, **46**(2), 234–240.

Further Reading

Diamond, I. and Jeffries, J. (1999) *Introduction to Quantitative Methods*, London, Sage.

Draper, N.R. and Smith, H. (2003) *Applied Regression Analysis*, 2nd edn, Chichester, John Wiley and Sons.

Ebdon, D. (1984) *Statistics in Geography*, 2nd edn, Oxford, Blackwell.

Everitt, B.S. and Dunn, G. (2001) *Applied Multivariate Data Analysis*, London, Arnold.

Harris, R. (2016) *Quantitative Geography: The Basics*, Sage Publications Ltd.

Kleinbaum, D., Kupper, L.L. and Muller, K.E. (1988) *Applied Regression Analysis and Other Multivariate Methods*, 2nd edn, Boston, PWS-Kent Publishing Company.

Lane, S.N. (2003) Numerical modelling in physical geography: understanding explanation and prediction, in *Key Methods in Geography* (eds N. J. Clifford and G. Valentine), London, Sage.

Rogerson, P.A. (2006) *Statistical Methods for Geographers: A Student's Guide*, Los Angeles, Sage.

Section V

Explicitly Spatial

10

Exploring Spatial Aspects of Geographical Data

This chapter focuses on the issues that can arise when attempting to apply inferential statistical tests, correlation and regression analysis techniques to data recording the specific location of observations. The main issues arise from the likely presence of spatial and/or temporal autocorrelation in geographical data. Exploratory spatial data analysis comprises a set of tools for exploring the patterns produced by spatial entities and gain further understanding of the processes that produced them. This chapter, after re-enforcing the underlying principles accounting for the location of spatial entities, shows how relatively simple measures can be calculated and, in some cases, tested statistically to avoid the pitfalls of unwittingly ignoring the lack of independence in spatial data by students and researchers in Geography, Earth and Environmental Science and related disciplines.

Learning Outcomes

This chapter will enable readers to:

- Describe the characteristics and implications of spatial autocorrelation.
- Calculate and apply suitable measures to describe and analyse spatial patterns.
- Consider how to incorporate these measures when analysing geographical datasets in an independent research investigation in Geography, Earth Science and related disciplines.

10.1 Location of Spatial Entities

The enhanced capability of Geographical Information Systems (GIS) to generate spatial statistics for users should not conceal the fact that the beginnings of such techniques amongst statisticians and geographers interested in the spatial arrangement of geographical features go back over 100 years. Certainly, by the 1950s and 1960s, early formulations of some techniques now in regular use had already seen the light of day (for a fuller account, see Getis and Ord, 1996). The previous four chapters covered non-spatial statistical techniques whose focus is on whether descriptive statistical quantities, summary coefficients measuring the strength and direction relationships and models specifying their form, when computed from sample data, are significantly different from what might have occurred through chance. The important difference between those techniques and spatial statistics is that when investigating the spatial distribution of phenomena (geographical features), we are also concerned with whether their location is random or otherwise. Has their location in space occurred through chance? For example, in the case of a group of points, we test whether their observed location within

Practical Statistics for Geographers and Earth Scientists, Second Edition. Nigel Walford.
© 2025 John Wiley & Sons Ltd. Published 2025 by John Wiley & Sons Ltd.
Companion website: www.wiley.com/go/PracticalStatistics2e

a given area of space is significantly different from what would have occurred if they were randomly distributed. An area of space may be regarded as an infinite collection of points, each of which may be occupied by a discrete occurrence of a type of spatial entity (e.g. oil wells, churches, telephone masts, trees, etc.) or ordered sets of points forming linear or areal features (e.g. roads, rivers, lakes, land covers, soil types, etc.). However, from both computational and visualisation perspectives a particular set of such entities cannot occur everywhere, since their location is dictated by the level of precision or detail in the grid coordinates that ties them to a georeferenced location.

There are certain types of geographical, geophysical and environmental phenomena and variables, such as air temperature, elevation, depth, water and atmospheric pressure that do exist everywhere over or underneath the Earth's surface, or in the atmosphere above. These variables create spatial entities by virtue of variations in the intensity or magnitude of their data values over the extent of the Earth's surface. Thus, anticyclones and cyclones are meteorological features with spatial extent within the Earth's atmosphere as a result of high- and low-pressure values near each other, respectively. Sources of minerals such as bauxite are concentrations of the geochemical compounds constituting a particular type of rock under the surface. These types of phenomena and variable will usually only be capable of measurement with any degree of certainty with respect to discrete spatial units, such as pixels (or voxels (pixels in a three-dimensional grid)), obtained from remotely sensed data with intermediate values being interpolated by modelling. Even some quantifiable socio-economic variables (e.g. population density and monetary land value) also exist over most of the Earth's terrestrial surface and some extend over marine areas as well. Arguably, even some qualitative attributes such as political affiliation or development status are spread across discrete portions of the Earth's surface so that the political party in government extends its authority over the whole country lived in by the electorate.

First, we will consider how geographical space comes to be occupied by entities or phenomena and by implication how the patterns produced by these occupants of space can change over time. Figure 10.1 represents a 1000-m^2 portion of the Earth's surface superimposed with a regular grid of 2500 squares each representing 0.4 m^2. Some of the individual squares in the left version of the grid (Figure 10.1a) have been shaded to indicate they are locations where a species of ground-nesting

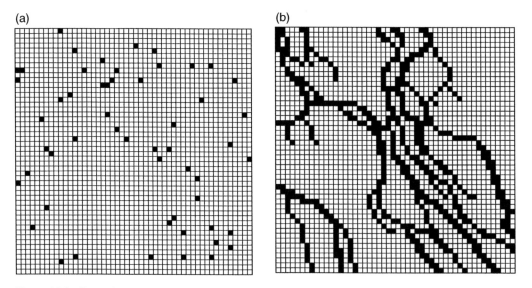

Figure 10.1 Examples of random points and linear spatial processes. (a) Squares in the grid occupied by migratory ground nesting birds; (b) squares in the grid occupied by flowing water.

migrant bird has chosen to nest and lay its eggs. When selecting a location each pair of birds has three options: select a location at random, opt for regular equidistant spacing or cluster together to provide some protection against predators. Spatial statistics usually analyse a complete set or population of spatial entities; nevertheless, the occupied squares effectively represent a sample of the potentially infinite, but often in reality limited, locations where the entities could be found, in this case where the birds could have nested. Suppose there had been 2500 pairs of birds flying over this portion of the Earth's surface in search of a nesting site they could each have selected to occupy their own location. Obviously, the 'real' land surface is not covered by a grid of 2500 squares each 0.4 m^2, but each pair is aware that it needs at least this amount of space to raise its offspring. However, nesting in such proximity to each other might increase competition for food and encourage conflict between birds, and some over-flying birds might decide to carry on to another part of the Earth's surface where more space was available. Such rationalisation of the birds' selection of a nest site relates to a behavioural process operating in the species and 'learnt' over successive generations.

The right version of the grid (Figure 10.1b) represents a different scenario. The shaded squares indicate the presence of a high quantity of surface water on a flat, unvegetated portion of the Earth's surface adjacent to a major river channel, which experiences a diurnal change in flow because of rising and falling tides that leads to overtopping of the banks to produce a network of braided streams. Again, the shaded squares may be considered a sample of the population of squares that might have sufficient surface water to form channels. Just as all the squares could have become occupied by nesting birds, it is reasonable to imagine that all of them could have been inundated with water as it flowed over the bank of the main river channel. However, small-scale differences in the surface and in the flow of water are likely to mean that the water bifurcates to produce a braided network. Testing the occupied versus the non-occupied locations in both these examples (i.e. location of nesting bird points and intermittent water channel lines) involves deciding whether the observed pattern in either case differs significantly from what would have arisen if those entities were located randomly. Statistical testing in these cases is underpinned by an assumption that the entities are located where they are because of a process and the analysis aims to establish whether the process is random or deterministic. Can the spatial pattern produced by the process be explained or understood by reference to some, unidentified, controlling factors, or is it just random or haphazard?

The spatial statistical techniques used to address this question are concerned with **pattern in location**. They seek to discover whether the observed pattern formed by the location of a set of spatial features is significantly different from what would have occurred if they were distributed randomly. Although not the same, it is similar to testing whether chance or sampling error accounts for a difference between a population and sample mean or variance. Many techniques coming under the broad heading spatial statistics deal with point rather than line or area patterns, and these fall into two main groups: distance statistics and grid-based techniques for point sets. Both sets of procedures refer to probability distributions, some of which are used in non-spatial statistics such as the Z and t distributions, to test whether the pattern produced by a particular set of points is significantly different from a random one.

10.2 Introduction to Spatial Autocorrelation

The correlation and regression techniques examined in Chapters 8 and 9 are often used to investigate research questions in the geographical, environmental, and geological sciences and if the specific geographical locations of the individual observations (entities) are not themselves part of the correlation or modelling analysis this is likely to be unproblematic. For example, applications of correlation and regression may be carried out in a geographical context, such as with respect to manufacturing businesses operating in a certain city region or to the concentration of pollutants along the channels in a drainage basin. Provided that the spatial distribution and location of the population and sample of observations are only of background interest and not part of the statistical analysis, both correlation and regression can be carried out with relative ease. However, once the spatial location of the entities starts to be regarded as relevant to the investigation, for example the distribution of businesses in relation to each other or to some other place, such as the centre of the city or the sites along the river channels where water is sampled in relation to potential sources of pollution, issues emerge that overshadow the application of correlation and regression as described in the previous chapters.

The origin of these problems arises from the fact the individual entities that make up a given collection of spatial units (points, lines and areas) are rarely, if ever, entirely independent of each other. Yet, a fundamental assumption of correlation and regression is that the data values possessed by each observation in respect of the variables and attributes being analysed should be independent. If this dependence between the entities is ignored, then its effect on the results of the correlation and regression will be undetected. For example, it might have artificially increased or decreased the value of the correlation coefficient so indicating a stronger or weaker relationship than is present. Similarly, it might have affected the form of the regression equation and could lead to unreliable predicted values for the dependent variable. The potential problems that might arise from a lack of independence between spatial features are illustrated in Box 10.1 with respect to some of the moraine that has emerged from Les Bossons Glacier near Chamonix in France and been deposited on the sandur or outwash plain. There is mixture of sizes of material in the area of moraine shown

---

Box 10.1   Spatial autocorrelation.

Sub-sample of debris shown with dark shading.

A lack of independence in the data values for a collection of *n* observations is likely to mean that there is some systematic pattern in the size of the residuals along the regression line. It could be that the lower and upper ends of the range of values for the variable x produce larger residuals and so a poorer prediction of the dependent variable in regression, or there is a repeating pattern of large and small residuals along the range x values: either way, these patterns indicate the presence of autocorrelation in the data.

The image of part of the moraine of Les Bossons Glacier suggests that the size (area) and long axis length of debris material is not distributed randomly. For example, there seems to be a group of large boulders towards the upper left and a larger number of small items in the upper and central right areas. The 20 debris items shaded black were randomly selected as a sub-sample of the full 100 boulders, stones and pebbles in the full sample. Their surface area has been measured, and simple linear regression analysis has been applied to examine the relationship that hypothesises area as an explanatory variable in respect of axis length. The calculations for the regression analysis have not been included since the standard procedures discussed in Chapter 9 have been followed. The regression equation is $\hat{y} = 7.873 + 0.001x$, and with $r^2 = 0.905$, there is a strong indication that surface area has significant explanatory power in respect of long axis length. The residuals from this regression analysis seem to display some systematic pattern along the regression line with smaller residuals at the lower and larger residuals at the upper ends of the range of x values. The residuals become progressively larger towards the upper end. The confidence limits are further from the regression line at the upper compared with the lower end, suggesting that we should be less confident in the ability of the regression model (equation) to predict long axis length for debris with larger surface areas.

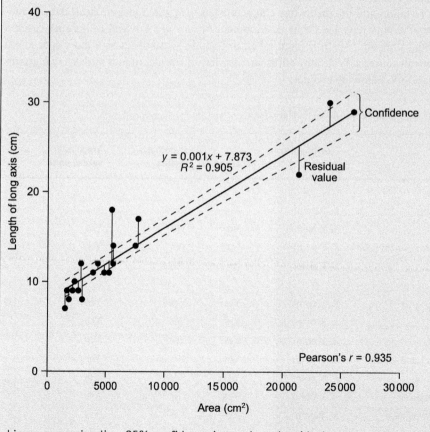

Linear regression line, 95% confidence intervals and residuals.

and a superficial examination clearly shows that different sized materials are not randomly distributed. There are patches of individual boulders, stones and pebbles together with finer material not visible in the image. A random sample of these boulders, stones and pebbles was selected, and their upward facing surfaces were digitised. These polygons are shown in a 'map' superimposed on the image. The sampled items have been measured in respect of their surface area and their long axis.

Regression and correlation analyses have been carried out on a sub-sample of these items (in the interests of clarity) with surface area as the independent and axis length as the dependent variables. The results (Pearson's correlation coefficient and linear regression equation) are shown on the scatter plot in Box 10.1. These suggest a very strong positive relationship between the variables ($+0.951$) and with r^2 equal to 0.905: there is an indication that surface area explains 90.5 per cent of the variability in axis length. The scatter plot also shows the residuals of the sampled data points as vertical lines connected to the regression line, which represents the predicted value of the dependent variable for the known values of surface area (dependent). The results suggest that the residuals for stones with smaller surface areas were themselves smaller compared with the larger stones and their residuals. The confidence intervals support this notion since they curve away from the regression line towards the upper end of the independent variable axis. This suggests that the residuals vary in value in a systematic way along the regression line. This might not be a problem if the varied sizes of moraine material were randomly distributed across the area, but the image clearly shows this is not the case. Separate sub-samples of material from the different parts of the moraine could potentially produce contrasting and even contradictory results from their respective correlation and regression analyses.

Spatial autocorrelation and the distinctive characteristics of spatial (geographical) data in contrast with non-spatial data have led to the development of several analytical techniques covered by the term spatial statistics, which is often taken to include geostatistics as a more specialised branch of the overall concept. Table 10.1 offers an overview of the techniques in these two groups covered in this and the following chapter.

Table 10.1 Principal methods for the statistical analysis of spatial (geographical) data.

	Type of entity/ feature	Attributes	Hypothesis testing	Irregularly spaced units
Spatial statistics				
Nearest neighbour	X, Y point	No	Yes	Yes
Ripley's K (L)	X, Y point	No	Yes	Yes
Quadrat – variance–mean ratio	X, Y point	No	Yes	Yes
Kappa Index	X, Y point	No	Yes	Yes
Geostatistics				
Correlogram	X, Y point	Yes	Yes	Yes
Kriging	X, Y point	Yes	Yes	Yes
Variogram	X, Y point	Yes	Yes	Yes
Moran's I (global and local)	X, Y point	Yes	Yes	Yes
Geary's C	X, Y point	Yes	Yes	Yes
Trend surface analysis	X, Y point	Yes	No	Yes
Geographically weighted regression	X, Y point	Yes	Yes	Yes

Note: Some of the X, Y point-based techniques can be adapted for use with line or polygon features.
Source: Adapted from Perry *et al.* (2002).

10.3 Analysis of Spatial Patterns of Points

10.3.1 Statistics Based on Distance

10.3.1.1 Nearest Neighbour Analysis

Techniques based on distance focus on the space between points and the extent to which they are separated from each other across space, since this potentially sheds some light on the underlying process. **Neighbourhood statistics** constitute the most important group of techniques of this type and the most intuitive and useful of these focuses on nearest or first-order neighbours (i.e. the pairs of entities that are closest together), although others such as second-order or furthest (Nth order) neighbour may be appropriate. The mean distance for each order in this series can be calculated up to the $K-1$ order, where K represents the total number of points, which constitutes the 'global' or overall mean distance. There are three 'perfect' or 'ideal' types of point spatial distribution: the points all occur at the same location (i.e. on top of each other); they are equidistant from each other; or a random mixture of short, medium and long distances apart. These three situations translate into extreme values of the neighbour statistic R, namely 0.0000 completely clustered; 2.1491 an evenly spaced triangular lattice; and 1.0000 randomly distributed. The **Nearest Neighbour** (NN) Index (R) may be used as a descriptive measure to compare the patterns of various categories of phenomena within the same study area. For example, it may be used to investigate the patterns of different species of trees in a woodland. Each tree of the separate species is treated as being located at a specific point, and the NN index provides a way of describing whether some species are more or less clustered than others. The statistic R acts as a quantitative summary of the spacing or layout of a particular set of entities, trees in this example, although comparison between two groups of either the same or different sets of entities in separate study areas is problematic, for example coniferous trees in different areas of woodland. If the same set of tree species is found in another area of woodland that is larger (or smaller) than the first, then it would not be legitimate to compare their R values. These situations create problems because the size of the area is important when calculating the R index and the standard error, which is required for the Z statistic (see Box 10.2a).

A particular set of points is usually not a sample but the population of spatial entities where they all happen to be located (see Section 10.1). Nevertheless, because there are an infinite number of locations where the entities could potentially have been located, it may be necessary to discover whether the difference between the NN index value obtained from the observed set of points and that which would have been obtained if the points were randomly distributed could have arisen by chance. This difference can be tested using the Z distribution with a Null Hypothesis that R does not equal 1.0000 (randomly spaced) because of sampling error, provided that the number of points is large (e.g. >100). The R index is the ratio between the observed and expected mean nearest neighbour distance, and the Null Hypothesis therefore examines whether chance or sampling error accounts for these values being unequal.

Application of the NN index can be problematic in two situations: if the boundary of the study area or the minimum distance is ill-suited to the particular distribution of points. The boundary of the area is likely to be problematic either when the points are within an area that includes a large amount of empty space or when the area's boundary is so tightly drawn around the entities that they are artificially inhibited from displaying a more dispersed pattern. The sensible solution to these issues is to delimit the area regarding the nature of the entities, for example defined with reference to continuous built-up area in the case of features distributed in 'urban space' rather than some arbitrary rectangular or circular enclosing area. Nevertheless, the boundary may still be relatively arbitrary and the nearest neighbour of an entity might lie outside the defined study area. If

Box 10.2a   Nearest Neighbour Index and Z test statistic for McDonald's and Burger King restaurants in Pittsburgh.

R Nearest Neighbour Index: $= \dfrac{r_o}{r_e}$

Z test statistic for R: $= \dfrac{|r_o - r_e|}{s_d}$

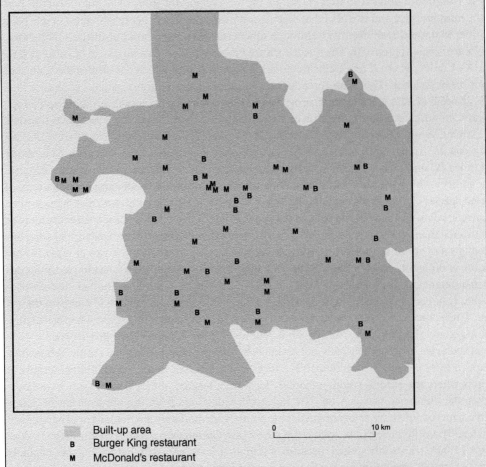

Built-up area

B   Burger King restaurant

M   McDonald's restaurant

0   10 km

Box 10.2b   Application of the Nearest Neighbour Index (R) and Z test statistic.

The Nearest Neighbour Index (R) is a summary measure describing the pattern in location of points representing a set of entities. The nearest distance (d) of each point to its closest neighbour is obtained by producing the distance matrix between all points in the set and sorting or searching the columns to determine the minimum value. The sum of these minimum values is divided by the number of points to produce the observed mean nearest neighbour distance

(r_o). The expected mean nearest neighbour distance (r_e) is obtained from a standard equation using the number of points N and the size of the study area A; thus, $1/(2\sqrt{N/A})$ where the term $\frac{N}{A}$ represents the density of points per unit area. The index value is the ratio between the observed and expected mean nearest distance between the points within a defined area of space.

A given set of points under investigation is usually regarded as a population, since there is little reason for sampling spatial entities to analyse their pattern in location. The Z test can therefore be applied, which requires the standard error (s_d) that is obtained from the following equation $\frac{0.26136}{\sqrt{N(N/A)}}$. The value of Z (see Box 10.2a) and its associated probability are used to examine whether the observed mean nearest distance is different from the expected value as a result of chance or sampling error. The decision on whether to accept the Null or Alternative Hypothesis is made by reference to the probability of Z in relation to the chosen level of significance.

The mean Nearest Neighbour Index has been applied to the combined set of 64 McDonald's and Burger King restaurants in Pittsburgh with the distances and area measured in km. The distance matrix between points is tedious to calculate by hand or using standard spreadsheet or statistical software and, since there was an example in Box 4.6c relating to the points representing the location one company's restaurants, details for the full set are not provided here. The R index value indicates the points are dispersed within the continuous built-up area of the city and this value is highly significant at the 0.05 level. It would occur by chance less than 0.1% of occasions. Visual scrutiny of the pattern reveals a certain amount of clustering of restaurants in parts of the urban area, but overall, they are dispersed. This is reflected in an R index value of 1.69, which is towards the middle rather than the upper end of the dispersed section of the range of values (0.0000–2.1491).

The key stages in calculating the Nearest Neighbour Index are as follows:

- *Calculate the distance matrix:* this requires the distance between each point and all of the others in the set.
- *Calculate the observed mean nearest distance:* determine the 1st nearest distance of each point; sum these and divide the result by the total number of points.
- *Calculate the expected mean nearest distance:* this is obtained from the standard equation above.
- *Calculate the Nearest Neighbour Index R:* this is the ratio between observed and expected mean nearest neighbour distances.
- *State Null Hypothesis and significance level:* the distribution of points is completely random and any appearance of clustering or regular spacing is the result chance and is not significant at the 0.05 level of significance.
- *Calculate test statistic (Z):* the calculations for the Z statistic are given below and $Z = 10.277$.
- *Determine the probability of the calculated Z:* the probability of obtaining this Z value is less than 0.001.
- *Accept or reject the Null Hypothesis:* the probability of Z is considerably <0.05, therefore reject the Null Hypothesis and by implication accept the Alternative Hypothesis that the dispersed pattern is not likely to be the result of chance, recognising that this might be an erroneous decision.

Box 10.2c   Nearest neighbour distances (*d*).

Point No.	*d*	Point No.	*d*	Point No.	*d*	Point No.	*d*
1	1.3	17	25.2	33	6.1	49	1.2
2	1.3	18	1.2	34	1.7	50	1.9
3	6.2	19	8.0	35	3.9	51	0.8
4	0.9	20	1.1	36	3.2	52	1.4
5	2.9	21	9.2	37	5.8	53	1.0
6	2.2	22	1.0	38	0.9	54	2.0
7	3.0	23	2.3	39	0.9	55	3.1
8	1.3	24	1.0	40	1.2	56	4.9
9	4.4	25	3.6	41	0.8	57	4.4
10	6.6	26	2.3	42	3.0	58	3.2
11	1.1	27	5.0	43	4.1	59	5.9
12	2.5	28	1.9	44	2.1	60	3.5
13	1.0	29	2.2	45	1.0	61	3.2
14	4.6	30	1.2	46	1.0	62	2.1
15	1.1	31	2.1	47	0.7	63	1.0
16	4.0	32	2.2	48	0.6	64	1.2
$\sum d=$							191.7

Box 10.2d   Calculation of the Nearest Neighbour Index, *R* and *Z* test statistic.

Observed mean nearest distance	$r_o = \dfrac{\sum d}{N} =$	$\dfrac{191.7}{64} = 3.02$
Expected mean nearest distance	$r_e = 1/(2\sqrt{N/A}) =$	$1/(2\sqrt{64/816.7}) = 1.79$
Nearest Neighbour *R* index	$R = \dfrac{r_o}{r_e} =$	$\dfrac{3.02}{1.79} = 1.69$
Standard error of mean nearest neighbour distance	$s_d = \dfrac{0.26136}{\sqrt{A(A/N)}} =$	$\dfrac{0.26136}{\sqrt{64(64/8.167)}} = 0.117$
Z statistic	$Z = \dfrac{\lvert r_o - r_e \rvert}{s_d}$	$\dfrac{\lvert 3.02 - 1.79 \rvert}{0.117} = 10.277$
probability	$p=$	<0.001

---

the point locations represent dynamic features of the landscape, such as parking spaces that become occupied and unoccupied or bird nesting sites that empty when the young have fledged, then the NN index may be subject to different values over a brief period. The minimum or nearest distance may be inappropriate if there is a series of discrete but regularly spaced groups containing a similar number of entities. This could occur in the countryside where there are clusters of buildings in dispersed settlements, in which case a higher order neighbour (e.g. the furthest neighbour distance) would be a more suitable way of analysing the overall pattern. The choice of order would be made with reference to the maximum number of points per cluster so that none of the distances used in calculating the *R* index came from the same group of points. Another common problem associated with using a single set of distances to summarise the pattern in location of a set of points is that pairs of points are often nearest neighbours of each other, they are reflexive (i.e. the nearest neighbour of point A is point B, and *vice versa*), which arises from the location of one being dependent upon that of another.

10.3.1.2 Ripley's *K* Function

An alternative technique for examining patterns of points in location that addresses some of the issues associated with the NN index is **Ripley's *K*** function (Ripley, 1979). Rather than focusing on **one** set of distances in respect of each point, for example its 1st-, 2nd- or k−1th-order neighbour, this technique embraces all point-to-point distances. Conceptually, circles of fixed radius are drawn around each point, the number of points in each circle is counted, and these counts are summed across all points. This is repeated for successive small increments of distance moving outwards from the first circle around each point in the set (see Figure 10.2a). The cumulative aggregate frequencies for each distance increment provide values for the function $K(d)$, where d is the radius around each point. The values of $K(d)$ can be plotted against distance to provide information on localised clustering in the set of points. The size and shape of the study area have an impact on whether the K statistic indicates point dispersion or clustering for identical sets of points. There may also be boundary effects in respect of points located towards the study area boundary (see Figure 10.2b), which results in parts of some concentric circles falling outside the study area. Most software used for calculating the K statistic includes an option to compensate for such boundary effects. Statistical testing can be applied to the $K(d)$ function, since the expected aggregate count of points for each distance zone is the overall point density multiplied by the area of each circle in turn. Subtracting the expected from the observed count in each circle transforms the $K(d)$ into $L(d)$, which would equal zero for a spatially random pattern, no difference between the two counts. Thus, $L(d)$ would be approximately zero for each circle radius (d); however, the sampling distribution of $L(d)$ is unknown, and in other words, there is no standard probability distribution against which to test the difference between the observed and expected values.

Several spatial statistics encounter this problem, and the typical solution is to resort to **Monte Carlo simulation** to obtain a cumulative frequency distribution based on random occurrence of events. This involves simulating a set of *N* entirely random events (points) within a given area (*A*), where *N* and *A* have the same values as for the observed number of points in the region of interest and are defined by reference coordinate space (i.e. only events possessing coordinates within the limits of region's boundary are permitted). This simulation is repeated many times (e.g. 99,999 or 9999) to generate probabilities by determining the maximum absolute difference between the simulated and theoretical cumulative distribution. This process may be compared with discovering the probability of events empirically (see Section 5.4). If the number of times the maximum absolute difference was greater than that between the observed and expected pattern is 499 and the number of simulations or trials is 9999, then:

$$p = \frac{1 + 499}{1 + 9999} = 0.05$$

This is equivalent to a 5 per cent probability of the difference having occurred through chance. Figure 10.3 shows an example of the upper and lower confidence intervals (see Section 6.5)

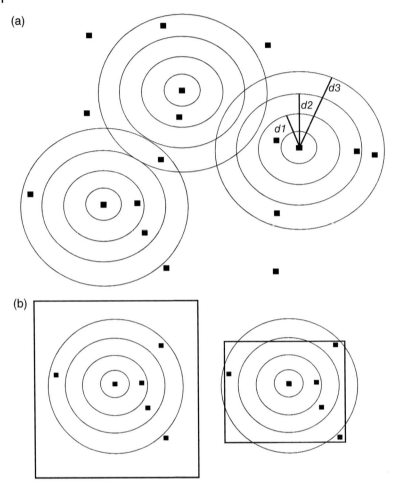

Figure 10.2 Ripley's *K* function showing three distance bands (*d1, d2* and *d3*) and the effect of study area size and boundary proximity. (a) Fixed radius concentric circles centred on each point used to obtain frequency count in each distance band; (b) frequency count in distance bands restricted for points located towards edge of study area (edge effect).

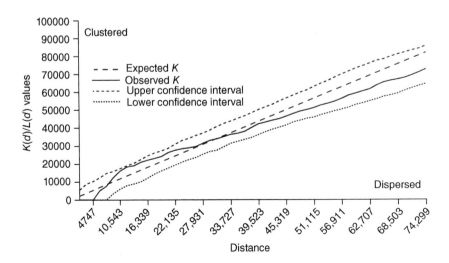

Figure 10.3 Confidence intervals resulting from Monte Carlo simulation.

produced with 999 simulation trials together the observed and expected K values. The observed plot line lying between these limits indicates the overall pattern of the points' location is random rather than clustered or dispersed. Significant local clustering of points is indicated at distances where the observed K 'bulges' above the upper confidence interval, whereas 'bulging' below the lower confidence interval reveals dispersion at that distance. Ideally, the Ripley's K technique should be applied when there are at least 100 points; nevertheless, Box 10.3a shows its use with respect to the 64 fast-food restaurants in Pittsburgh previously examined.

Box 10.3a   Ripley's K function statistic and confidence limits for McDonald's and Burger King restaurants in Pittsburgh.

Ripley's K function: $K(d) = \dfrac{A}{N}\sum_i\sum_j\dfrac{I(d_{ij})}{N}, i \neq j(d)$

Difference between observed and expected under complete spatial randomness:

$L(d) = \sqrt{\dfrac{K(d)}{n}} - d$

Box 10.3b   Application of Ripley's K function.

Ripley's K (or L) function makes use of the distances between all occurrences of an entity, not only those between nearest neighbours. The technique starts by creating a circle with radius d around each point, and then, the number of other events falling within the circle is counted, labelled j. Repeat this for each of the i points. This is equivalent to the definitional equation given above where $I(d_{ij}) = 1$ if the distance, d_{ij}, from i to j is less than d; otherwise, $I(d_{ij}) = 0$. These calculations are repeated for the required number of circles with increasing radii of a small, fixed distance, which is usually determined in relation to the size of the study area. The slight increase in distance d is often equal to the radius of a circle with the same area as the study area divided by 100. The set of distances is $K(d)$. The graph of the function $K(d)$ gives information about localised clustering of points.

The expected number of points in a circle with a radius d under conditions of complete spatial randomness is the point density, m (mean points per unit area), multiplied by the area of the circle, where m is estimated as the total number of points divided by A, the area of the study area (often a rectangular bounding box) $\left(E(\# < d) = \dfrac{N}{A}\pi d^2\right)$. The difference between the observed $K(d)$ and expected number ($E\#<d$) of points within a given circle with a radius d has an expected value of $L(d) = 0$ for a given d, thus measuring the difference between the observed pattern of points and what would be expected under complete spatial randomness. Software capable of calculating Ripley's K function will usually allow confidence limits to be computed by recalculating the $K(d)$ for 99, 999 or 9999 randomly distributed sets of points for each iteration. The $K(d)$ values that deviate most above and below the expected $K(d)$ are the upper and lower confident limits, respectively.

Ripley's K function has been applied to the combined set of 64 McDonald's and Burger King restaurants in Pittsburgh (see Box 10.2a) with the distances and area measured in km and km^2, respectively. Ten distance bands were specified (maximum distance from centre divided by number of iterations $\left(\dfrac{210}{10} = 21\right)$ and the minimum bounding rectangle was used to define the study area with Ripley's edge correction applied to reduce boundary edge effects. This

(Continued)

Box 10.3b (Continued)

correction examines the distance of each point from the edge of the study area and to all of its neighbours. Any points further away than the edge are given extra weighting, which is appropriate when the study is defined as the minimum bounding rectangle. The results provide some evidence of statistically significant clustering of restaurants in distance band 2 (circled area in Box 10.3d), but that at other distances the observed value is within the confidence intervals and close to what would be expected under complete spatial randomness.

The key stages in computing Ripley's K function are as follows:

- *Create a template of a circle with a fixed radius:* move the centre of the template to each individual point in turn and count the number of other points falling within it. Sum the results to produce $K(d)$.
- *Increase the radius of the template circle by the same fixed distance:* repeat the process of moving the (second) template to each individual point in turn and counting the number of other points falling within its boundary. Sum the results to produce $K(d)$ distance band two. Continue creating circle templates by incrementing the radius until desired number of distance bands is reached with $K(d)$ obtained for each one.
- *Calculate the expected K(d) for each distance band:* this is obtained from the standard equation above. Subtract the observed from the expect $K(d)$ for each distance band to obtain of $L(d)$. Under conditions of complete spatial randomness $L(d) = 0$.
- *State Null Hypothesis and significance level:* the overall distribution of points is completely random and any appearance of clustering or regular spacing is the result chance and is not significant at the chosen level of significance.
- *Calculate the upper and lower confidence limits:* these limits are calculated by undertaking a series of 99 or 999 simulations and define the range outside which you can be confident of the value of $K(d)$ at 0.05, 0.01 or another level of significance.
- *Accept or reject the Null Hypothesis:* if $K(d)$ is larger than the upper confidence limit, it suggests clustering is present at that distance, whereas if $K(d)$ is smaller than the lower confidence limit, there appears to be dispersion at that distance.

Box 10.3c Ripley's *K* statistic, observed and expected together with 95% confidence intervals, for 10 distance bands.

Distance band *d*	Expected *K(d)*	Observed *K(d)*	Lower 95% confidence interval	Lower 95% confidence interval
1 (0–21.9 m)	8819.32	7938.11	0.00	12,551.26
2 (21.0–41.9 m)	17,638.63	23,258.89	13,172.42	21,797.75
3 (42.0–62.9 m)	26,457.95	31,831.01	21,921.46	31,694.60
4 (63.0–83.9 m)	35,277.27	39,688.26	30,792.19	43,067.88
5 (84.0–104.9 m)	44,096.58	47,059.43	38,332.10	50,599.44
6 (105.0–125.9 m)	52,915.90	56,271.77	47,448.14	58,437.63
7 (126.0–146.9 m)	61,735.22	63,603.74	57,059.16	67,529.42
8 (147.0–167.9 m)	70,554.53	72,291.25	65,596.29	76,066.96
9 (168.0–187.9 m)	79,373.85	80,917.80	73,743.51	85,028.18
10 (188.0–209.9 m)	88,193.17	91,397.96	80,432.59	93,205.68

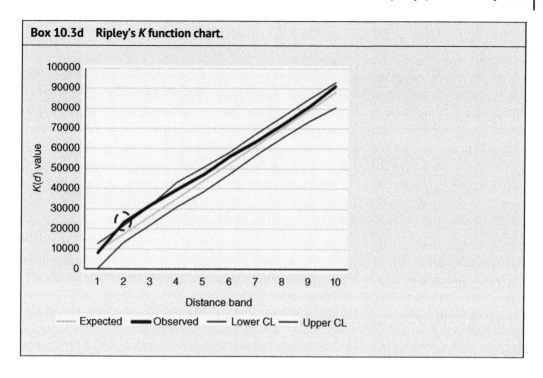

Box 10.3d   Ripley's *K* function chart.

Expected ▬▬ Observed ── Lower CL ── Upper CL

10.3.2   Statistics Based on Quadrats

Grid-based techniques provide another way of examining the pattern in location of spatial phenomena, especially those that can be represented as points. Section 10.3.1 explores neighbourhood techniques that use measurements of the distances between observations to investigate spatial pattern of points: quadrat or grid-based techniques offer another approach. These involve overlaying the complete set of points with a regular, but not necessarily rectangular (square), grid and counting the number of entities falling within each of its cells. In some ways, this is like cross-tabulating a set of observations in respect of two categorical variables and assigning them to a set of tabular cells. The main difference is that the grid lines forming the cells in the 'table' represent spatial boundaries that summarise the overall pattern produced by the points along ordered series of X and Y coordinates. The location of each grid square within this geographical framework and its count of observations are jointly of crucial importance in spatial statistics since these define the spatial pattern and density of the entities. The frequency distribution of the number of point entities per grid cell can be assessed in relation to what would be expected according to the Poisson probability distribution (see Chapter 5). The entities may be regarded as randomly distributed, if the observed and expected frequency distributions are sufficiently similar.

10.3.2.1   Variance–Mean Ratio

A useful feature of the Poisson distribution is that its mean and variance are equal, which enables us to formulate a Null Hypothesis stating that if the observed pattern of points is random, then according to the Poisson distribution, the **variance–mean ratio (VMR)** will equal 1.0, whereas a statistically significant difference from 1.0 suggests some non-random processes at least partially accounts for the observed pattern, which must necessarily tend towards either clustering or dispersion. We have seen that any set of points, even if it is the population of entities, should be considered

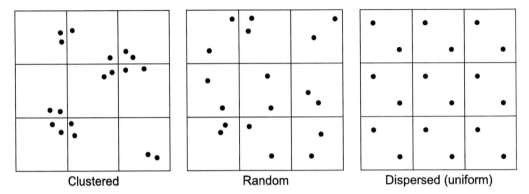

Figure 10.4 Clustered, random and dispersed spatial distributions of point features.

as but one of a potentially infinite number of such sets of the phenomena in question and should be regarded as a sample. A finite number of points distributed across a fixed number of grid cells will have a mean that does not vary (λ) irrespective of whether they are clustered, random or dispersed. The three grids in Figure 10.4 show three spatial patterns for 18 points, clustered, random and dispersed, subject to the condition that each cell in the overlain grid contains two points. The nature of the pattern is therefore indicated by differences in the variances of the point patterns. A variance to mean ratio significantly greater than 1.0 suggests clustering and conversely less than 1.0 uniformity or dispersion. Since any one set of points constitutes a sample, the VMR should be tested for its significance, which can be achieved by conversion into a t or χ^2 test statistic.

We have already come across the illustration in Box 10.4b, which shows a nine cell square grid overlain on the pattern of Burger King's restaurants in Pittsburgh. The numbers in the quadrats represent the observed frequency count of these restaurants, which seem to display some evidence of clustering in the pattern of their locations. The question is whether such visual evidence is supported by spatial statistics. The outcome of applying the Kolmogorov–Smirnov test in Box 10.4a suggests that the pattern is not significantly different from what might be expected to occur if a random process was operating. However, there remains an issue over whether this result is an artefact of the number of quadrats and where the grid is located. The size of the quadrats, 10 km, has been determined in an arbitrary fashion as have the orientation and origin of the grid. A grid with smaller or larger quadrats and/or rotated clockwise 45° is likely to have produced entirely different observed and therefore expected frequency distributions, although the underlying location of the points (restaurants) is unaffected. Help with deciding quadrat size is provided by the rule-of-thumb that these should be in the range D–$2D$, where D is the study area divided by the number of points. In this example, there are 22 points in a 90-km^2 study area ($D = 4.09$), which suggests the quadrats should have been in the range 4.09–8.18 km: the cells in the grid in Box 10.4b are slightly larger than this upper guideline at 10 km.

Box 10.4a The Variance to Mean Ratio.

$$\text{VMR} = \frac{s^2}{\lambda} = s^2 = \frac{1}{n-1}\left(\sum f_i x_i^2 - \frac{(\sum f_i x_i)^2}{n}\right) : \lambda = \frac{n}{n_q}$$

$$\text{Conversion of VMR to } t \text{ test statistic} := \frac{\text{VMR} - 1}{s_{\text{VMR}}}$$

Box 10.4b   Application and Testing of the Variance to Mean Ratio.

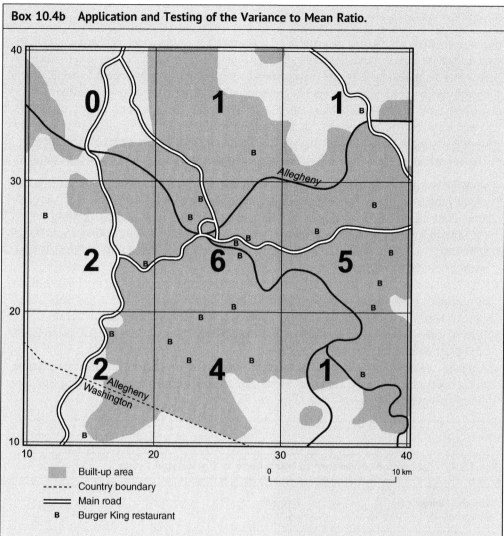

Built-up area
Country boundary
Main road
B   Burger King restaurant

A regular grid of quadrats is overlain on a map showing the location of the point features and a count of points per quadrat is made. However, before the variance to mean ratio (VMR) can be computed, it is necessary to calculate the variance and mean of the points in the quadrats. Calculating the mean or average (λ) number of points per quadrat is relatively easy, it is the number of points (n) divided by the total quadrats (k). The equation for calculating the variance is not unlike the one used to estimate the variance for a non-spatial frequency distribution in Chapter 5. The number of points per quadrat in the variance formula is denoted by x in each of i frequency classes going from 0, 1, 2, 3 ..., i, and the frequency count of quadrats in each class is f. There are two alternative test statistics that can be calculated to examine the significance of the VMR – the t statistic and the χ^2 continuous chi-square (not Pearson's chi-square) in both cases with $k - 1$ degrees of freedom. The probabilities associated with these can then be used in the usual way to decide whether to accept or reject the Null Hypothesis.

A regular grid containing nine cells or quadrats has been superimposed over the map of Pittsburgh, showing the location of the Burger King restaurants and a count of restaurants per quadrat has been made. The observed counts of restaurants in the centre, centre-right and centre-bottom quadrats are high (6 and 5, respectively), whereas the remainder are less well served. But is this

(Continued)

Box 10.4b   (Continued)

any different from what would be expected if the restaurants were randomly distributed? Or are they, as it might appear, mostly located in the main built-up area of the city? The mean number of restaurants per quadrat is 2.44 and the variance 1.630 resulting in a VMR equal to 0.667, which suggests the Burger King restaurants tend towards being dispersed in a uniform pattern. Calculating both types of test statistic with respect to this VMR value and their associated probabilities and with eight degrees of freedom indicates that we really cannot conclude that this dispersed pattern is any more than would be expected in a random locational process was operating.

The key stages in applying the variance to mean ratio and associated *t* test are as follows:

- *State Null Hypothesis and significance level*: the restaurants are distributed at random in the city and any suggestion from the VMR ratio index that they are either clustered or dispersed (uniform) is the result of sampling error and its value is not significant at the 0.05 level.
- *Calculate VMR ratio*: the tabulated ordered frequency distribution of Burger King's restaurants per quadrat and calculations to obtain the variance and mean, prerequisites of the VMR are given below (VMR = 0.667).
- *Calculate t and/or χ^2 test statistic(s)*: the *t* statistic requires that the standard error of the mean be obtained before calculating the test statistic itself, which equals −1.333 in this case. The χ^2 statistic is simply VMR multiplied by the degrees of freedom and here equals 5.336.
- *Determine the probability of t and/or χ^2 test statistic(s)*: these probabilities are 0.220 and 0.721 with eight degrees of freedom, respectively.
- *Accept or reject the Null Hypothesis*: both probabilities are such as to clearly indicate that the Null Hypothesis should be accepted at the 0.05 level of significance, recognising that this decision might be erroneous.

Box 10.4c   Calculation of Variance to Mean Ratio and associated *t* statistic.

Number of Burger King restaurants (x)	Frequency count (f)	fx	fx²
0	1	0	0
1	3	3	3
2	2	4	8
3	0	0	0
4	1	4	16
5	1	5	25
6	1	6	36
	9	22	88

Sample variance $\quad s^2 = \dfrac{1}{n-1}\left(\sum f_i x_i^2 - \dfrac{(\sum f_i x_i)^2}{k}\right) = \quad \dfrac{1}{22-1}\left(88 - \dfrac{(22)^2}{9}\right) = 1.630$

Sample mean $\quad \lambda = \dfrac{n}{k} = \quad \dfrac{22}{9} = 2.444$

Variance to mean ratio $\quad \text{VMR} = \dfrac{s^2}{\lambda} = \quad \dfrac{1.630}{2.444} = 0.667$

Number of Burger King restaurants (x)		Frequency count (f)	fx	fx²
Standard error of the mean	$VMR_s = \dfrac{2^2}{(k-1)} =$			$\dfrac{2^2}{(9-1)} = 0.50$
t statistic	$t = \dfrac{VMR - 1}{s_{VMR}} =$			$\dfrac{0.667 - 1}{0.5} = -1.33$
Chi-square statistic	$\chi^2 = (k-1)VMR$			$(9-1)0.667 = 5.336$
Probability	t statistic	$p = 0.220$	Chi-square statistic	$p = 0.721$

10.3.2.2 Pearson's Chi-square Test

An alternative grid-based way of examining the spatial pattern of a set of point observations to see whether they are randomly distributed across a set of quadrats is to carry out a Pearson's chi-square test using the Poisson distribution to generate the expected frequencies. Thus, the expected frequencies reflect what would have occurred if a random Poisson process was at work. The Pearson's chi-square test for a univariate frequency distribution has already been examined, and Box 10.5a simply illustrates the calculations involved with applying the procedure to the counts of Burger King's restaurants in Pittsburgh. The observed and expected frequency counts relate to the quadrats with different numbers of restaurants (i.e. 0, 1, 2, 3, 4, 5 and 6) not to the numbers of restaurants themselves. So, with a total of 22 restaurants and nine quadrats, the average number per cell is 2.44. Although this example deals with fewer points than would normally be the case, it is realistic to assume that the probability of an observation occurring at a specific location is small. There are two factors reducing the degrees of freedom in this type of application of the test, not only is the total count known but also calculating the expected frequencies using the Poisson distribution employs the sample mean. Thus, the degrees of freedom are the number of classes minus two and equal five in this case. With a Null Hypothesis that the observed distribution of points is random and referring to the results shown in Box 10.5a, we can see that there is an extremely low probability (>0.000) of having

Box 10.5a Calculation of Pearson's chi-square statistic.

Number of Burger King restaurants (x)	Frequency count O	Expected frequency count E	(O − E)	$\dfrac{(O-E^2)}{E}$
0	1	1.91	−1.91	1.91
1	3	4.67	−1.67	0.60
2	2	5.70	−1.70	0.51
3	0	4.65	−4.65	4.65
4	1	2.84	1.16	0.47
5	1	1.39	3.61	9.39
6	1	0.57	5.43	52.21
	9	21.73		69.74
Pearson's chi-square statistic		$\chi^2 = 69.74$		
Probability		$p < 0.000$		

obtained a χ^2 test statistic as high as 69.74 with five degrees of freedom, and so the Alternative Hypothesis can be accepted that restaurants are not randomly distributed.

10.4   Analysis of Spatial Patterns of Lines

Lines, sometimes referred to as second-order spatial entities requiring a minimum of two pairs of X, Y coordinates are used as spatial data to locate the position of inherently linear geographical features, such as rivers, railways and pipelines. Individual lines often combine to form closed or open networks, a national road transport network exemplifies the former and a drainage basin the latter. Analysis of linear networks, known as network analysis and embedded with our familiar GPS-based navigation systems involves navigation around networks considering barriers, connectivity and other temporary or permanent restrictions, lies beyond the scope of a text focused primarily on quantitative statistical techniques. However, some examination of techniques to describe and analyse the spatial patterns produced by collections of linear entities is called for.

Simple lines with two pairs of X, Y coordinates comprising start (origin) and end (destination) points are necessarily straight but may be of different lengths and orientations in terms of compass directions. Such straight lines traverse several other points along their length even if these points have not been digitally captured to visualise the feature in mapping software. Real-world linear features often display a degree of sinuosity; for example, even high-speed roads (motorways in the UK) are rarely designed as straight connections between cities to avoid intervening features (e.g. archaeological sites) and to limit the repetitiveness of driving on a straight sections of dual carriageway roads. Channels in a drainage basin are rarely straight, and if they are, it is usually because the meandering course of a river has been artificially straightened by human intervention. The separate parts of curved linear features are often referred to as arcs or segments, and the length of these composite sinuous lines is best measured as the sum of the individual lengths of their arcs rather than the straight line distance between the start and end points. In some types of investigation, it is not physical distance along linear features that is relevant but concepts such as journey time and travel cost that are more useful as they indicate how much time it takes or the cost involved to traverse a particular route.

The previous section introduced techniques to test whether the points in a set are distributed in a random, clustered or dispersed spatial pattern, but are there equivalent methods available for testing spatial patterns of sets of lines? They are less common, but one technique used with points, Ripley's K function, can be adapted for application to line (and polygon – see 10.4) features by using their centroids weighted by line length (area in the case of polygons) as points on which to base the statistic. There is some debate about what clustered, dispersed and random spatial patterns of straight lines might look like, and Figure 10.5 suggests one option with sets of straight lines all with the same length and orientation (horizontal). The midpoint (centroid) of the lines is at the same location as the points shown in the spatial patterns in Figure 10.4.

The lines forming the spatial patterns in Figure 10.5 are all straight with the same length and orientation (horizontal). Suppose the lines did not have these characteristics, they were curved and straight, had different lengths and were oriented at various angles. How would this influence a decision on whether the overall pattern was clustered, random or dispersed?

Sketch out some simple examples and decide on your instinctive reaction to the different patterns you produce.

Box 10.6a shows an application Ripley's K function to a set of lines comprising a sample of the slit trenches dug in one of the military training areas on the South Downs, England, that were

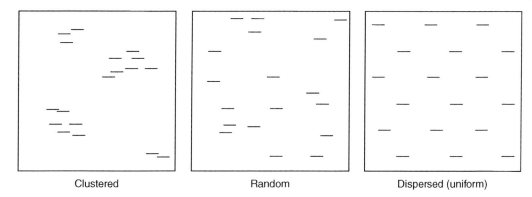

| Clustered | Random | Dispersed (uniform) |

Figure 10.5 Clustered, random and dispersed spatial distributions of linear features.

Box 10.6a   Ripley's K function statistic and confidence limits for a subset of slit trenches in World War II on South Downs, England.

Ripley's K function: $K(d) = \dfrac{A}{N} \sum_i \sum_j \dfrac{I(d_{ij})}{N}, i \neq j$

Difference between observed and expected under complete spatial randomness:

$$L(d) = \sqrt{\dfrac{K(d)}{n}} - d$$

Box 10.6b   Application of Ripley's K function.

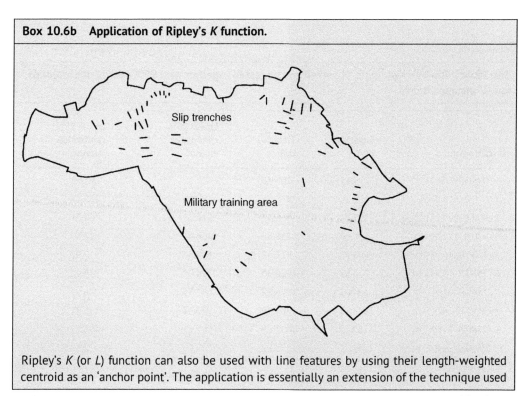

Ripley's K (or L) function can also be used with line features by using their length-weighted centroid as an 'anchor point'. The application is essentially an extension of the technique used

(Continued)

Box 10.6b   (Continued)

with point feature data outlined previously (see Box 10.3a). Here, we focus on the results of the application with line features rather than the computational procedure. Parts of the South Downs in England, especially those with relatively poor land and rough grass vegetation, were taken over for military training purposes during World War II, and in some places, slit trenches were dug. This application focuses on a sampled subset of these trenches south east of Lewes, county town of East Sussex.

The study area around the sample of lines is the minimum bounding rectangle, and ten distance bands were specified (maximum distance from centre divided by number of iterations 3100/10 = 310) with Ripley's edge correction applied to reduce boundary edge effects. The results provide statistically significant evidence of clustering of the sampled slit trenches along the range of distance bands (highlighted area in Box 10.6d), and by reference to the confidence limits, they do not appear to display a spatially random pattern.

The key stages in computing Ripley's K function are those specified in Box 10.3b with the addition of a weighting factor being applied based on the length of the line.

- *Calculate the upper and lower confidence limits:* here, these limits are calculated by undertaking a series of 99 simulations and define the range outside which you can be confident of the value of $K(d)$ at 0.05 level of significance (95%). Additional simulations (e.g. 999 or 9999) might have widened the confidence limits such that they enclosed the observed value of $K(d)$ at some distances, but examination of the chart in Box 10.6d suggests this would be unlikely.

Box 10.6c   Ripley's K statistic, observed and expected together with 95% confidence intervals, for 10 distance bands.

Distance band d	Expected $K(d)$	Observed $K(d)$	Lower 95% confidence interval	Lower 95% confidence interval
1 (0.0–309.9 m)	129.15	205.72	59.39	193.69
2 (310.0–619.9 m)	258.29	473.51	187.80	353.15
3 (620.0–929.9 m)	387.44	608.13	314.58	468.31
4 (930.0–1239.9 m)	516.59	758.82	456.05	617.57
5 (1240.0–1549.9 m)	645.73	871.63	599.56	710.93
6 (1550.0–1859.9 m)	774.88	1020.30	714.08	855.98
7 (1860.0–2169.9 m)	904.03	1176.82	842.75	1006.97
8 (2170.0–2479.9 m)	1033.17	1304.49	975.38	1155.01
9 (2480.0–2789.9 m)	1162.32	1431.69	1087.21	1298.71
10 (2890.0–3109.9 m)	1291.47	1522.82	1223.03	1422.38

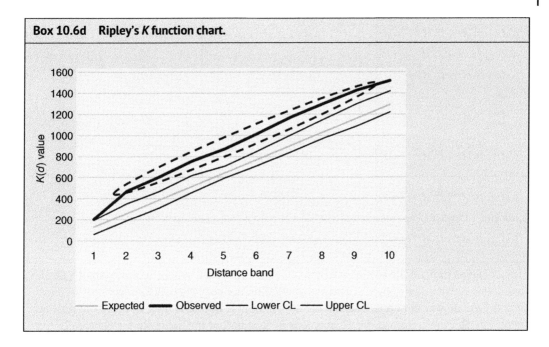

Box 10.6d   Ripley's *K* function chart.

requisitioned from landowners during World War II. The trenches were of different lengths, and the analysis has been applied to a sample using Ripley's *K* function weighted according to their length. The results suggest a non-random spatial pattern with evidence of clustering despite there being example lone individual trenches in the sample.

10.5   Analysis of Spatial Patterns of Polygons

Polygons are formed from ordered rings of points with each successive pair connected by straight lines commonly referred to as arcs or segments. Polygons can be classified in different ways, for example as **regular** circles, squares, rectangles or hexagons, or as **irregular** (i.e. not a standard geometrical shape). Polygons often represent diverse types of geographical area feature formed by human or physical processes, such as administrative regions, electoral constituencies, lakes, fields, building plots or land use covers. Some of these features cover the entirety of a study area, whereas others are discrete: in both cases, the individual entities of a single feature type (e.g. lakes) occupy a unique portion of the surface without overlapping. Some area feature types form nested hierarchies of different sizes in terms of physical area in combination with a quantifiable or qualitative attribute, such as total population or number of electors. Hierarchical governance structures in many countries are associated with such geographical hierarchies.

The introduction to section 10.3 indicated that Ripley's *K* function could also be adapted to test spatial patterns formed by sets of polygons. Again, the centroids of the polygons are used as an anchor point, but this time they are weighted by the physical size (area) of the feature. This means that their relative size is considered when computing *K(d)* using their centroids. Figure 10.6 offers an example of clustered, dispersed and random spatial patterns produced by regular sets of 18 parallelograms with each centred on the points as shown in Figure 10.4.

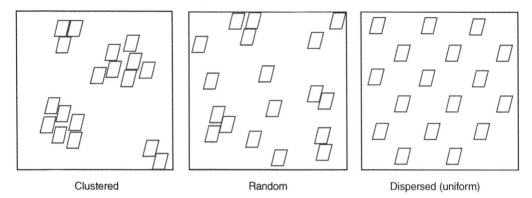

| Clustered | Random | Dispersed (uniform) |

Figure 10.6 Clustered, random and dispersed spatial distributions of polygonal features.

> The polygons forming the spatial patterns in Figure 10.6 are all simple non-overlapping parallelograms with the same physical area and orientation. Suppose the polygons were not identical in this way but were of different sizes and were oriented at various angles. What if there were no gaps between the areas and the 18 polygons in each part of the figure were expanded by the same or different amounts so that the whole portion of space enclosed in the square outer boundary was filled without any overlap? How would such changes influence a decision on whether the overall pattern was clustered, random or dispersed?
>
> It might help if you sketch out some simple examples and decide on your instinctive reaction to the different patterns you produce.

The application of Ripley's K function is outlined in Box 10.7a in relation to a set of non-overlapping, discrete irregular polygons. These are fields that were identified during 1940 and 1941 as suitable for being ploughed-up and re-sown or planted to arable crops on part of the South Downs, England. The process of identifying such under-productive fields was part of a nationwide campaign in Britain in the early years of World War II that had the aim of increasing the quantity of home-grown agricultural output. The analysis presented in Box 10.7a is concerned with whether there is evidence of statistically significant spatial clustering or dispersion of the sample of 'plough-up' fields or whether they were distributed at random across the study area. Clustering might indicate fields identified for ploughing-up were located on certain types land or only on farms where the farmer needed some assistance. Other issues, such as elevation or with what crops the fields would be re-sown, are not considered here.

Box 10.7a Ripley's K function statistic and confidence limits for a subset of fields ploughed-up for agricultural production in World War II on South Downs, England.

Ripley's K function: $K(d) = \dfrac{A}{N} \sum i \sum j \dfrac{I(d_{ij})}{N}, i \neq j$

Difference between observed and expected under complete spatial randomness:

$L(d) = \sqrt{\dfrac{K(d)}{n}} - d$

Box 10.7b   Application of Ripley's *K* function.

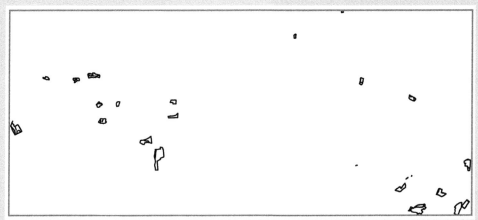

Fields identified to be ploughed-up in 1940 or 1941

When Ripley's *K* (or *L*) function is used with polygons, their centroids, weighted by physical area, are used as the 'points' with which to compute the statistic. Apart from this adaptation, the procedure is essentially the same as for point features (see Box 10.3a). This Box concentrates on the results obtained by applying this technique to a sample of fields in a section of the South Downs in England that were identified as suitable for being ploughed-up because they were largely unproductive on relatively poor land and then re-sown or planted with crops, such as wheat, barley, oats, linseed and potatoes to increase agricultural output during World War II. The sample of plough-up fields analysed here were north of Brighton in a rectangle 36.3 km (W-E) and 16.1 km (N-S).

Ripley's *K* function has been applied using a minimum bounding rectangle around the sample of polygons and ten distance bands with Ripley's edge correction reducing boundary edge effects. The sampled plough-up fields do not appear to display a spatially random pattern, and there is statistically significant evidence of clustering along the range of distance bands (highlighted area in Box 10.7d).

The key stages in computing Ripley's *K* function are those specified in Box 10.3b with the addition of a weighting factor being applied based on the area of the polygon.

- *Calculate the upper and lower confidence limits:* 99 simulations or permutations were used to compute these confidence limits, which define the range outside which you can be confident of the value of $K(d)$ at 0.05 level of significance. These limits might widen if 999 or 9999 simulations had been carried out such that they at least partially enclosed the observed value of $K(d)$. Examination of the chart in Box 10.7d suggests this would be unlikely.

Box 10.7c   Ripley's *K* statistic, observed and expected together with 95% confidence intervals, for 10 distance bands.

Distance band *d*	Expected $K(d)$	Observed $K(d)$	Lower 95% confidence interval	Lower 95% confidence interval
1 (0.0–1959.9 m)	910.35	2475.89	377.56	1289.15
2 (1960.0–3919.9 m)	1820.71	3372.19	1412.36	2288.77
3 (3920.0–5879.9 m)	2731.07	4623.21	2263.89	3413.11

(Continued)

Box 10.7c (Continued)

Distance band *d*	Expected K(*d*)	Observed K(*d*)	Lower 95% confidence interval	Lower 95% confidence interval
4 (5880.0–7839.9 m)	3641.42	5879.28	3144.10	4290.26
5 (7840.0–9799.9 m)	4551.77	6840.55	4125.73	5046.36
6 (9800.0–11,759.9 m)	5462.13	8127.61	5090.69	5996.25
7 (11,760.0–13,719.9 m)	6372.48	8893.72	5833.67	7037.75
8 (13,720.0–15,679.9 m)	7282.84	9609.18	6850.08	8082.19
9 (15,680.0–17,639.9 m)	8193.19	10,026.21	7644.26	9068.43
10 (17,640.0–19,599.9 m)	9103.55	10,594.42	8483.39	9905.19

Box 10.7d Ripley's *K* function chart.

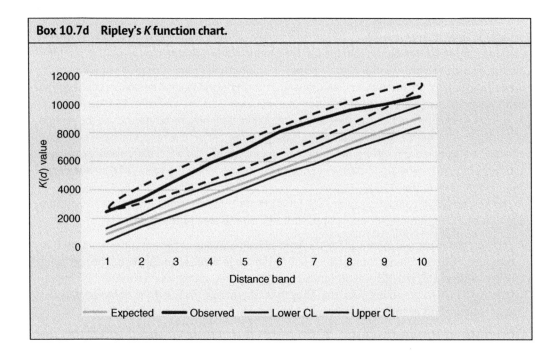

10.6 Closing Comments

When carrying out analyses on features distributed across the Earth's surface, apart from examining the values of their variables and attributes, we might be interested in the patterns formed by the locations of these phenomena. Analysis of patterns in location and the spatial characteristics will normally be applied to all instances of such features rather than a sample. However, the population of spatially located entities may itself be thought of as a sample from the infinite number of such populations that could exist. The example applications covered in the Boxes in this chapter have used samples of entities in the interest of simplifying the presentation. One of the main reasons for analysing spatial patterns is to determine whether their distribution is random or tending towards dispersal or clustering. If the location of all features is known, why discard some instances, especially when the analytical procedures can be carried out by computer software rather than hand calculation. The spatial analysis techniques examined in this chapter have been concerned with the location of and distances between points in coordinate space, which in some cases have represented centroids of lines or polygons. Other characteristics of lines and polygons might be analysed, for example measuring complexity of polygons by means of perimeter length per unit area (see de Smith *et al.*, 2007).

References

Getis, A. and Ord, J.K. (1996) Local spatial statistics: an overview, in *Spatial Analysis: Modelling in a GIS Environment. Geoinformation International, Cambridge* (eds P. Longley and M. Batty), New York, John Wiley and Sons.

Perry, J.N., Liebhold, A.M., Rosenberg, M.S., Dungan, J., Meriti, M., Jakomulska, A. and Citron-Pousty, S. (2002) Illustrations and guidelines for selecting statistical methods for quantifying pattern in ecological data. *Ecography*, **25**, 578–600. DOI: 10.1034/j.1600-0587.2002.250507.x.

Ripley, B.D. (1979) Tests of 'randomness' for spatial point patterns. *Journal of the Royal Statistical Society, Series B*, **41**, 368–374.

de Smith, M.J., Goodchild, M.F. and Longley, P.A. (2007) *Geospatial Analysis: A Comprehensive Guide to Principles*, Leicester, Techniques and Software Tools.

Further Reading

Anselin, L. (1988) *Spatial Econometrics: Methods and Models*, Dordrecht, Kluwer Academic Publishers.

Anselin, L. (1995) Local indicators of spatial association – LISA. *Geographical Analysis*, **27**, 93–115.

Berry, B.J.L. and Marble, D.F. (1968) *Spatial Analysis – A Reader in Statistical Geography*, New Jersey, Prentice Hall.

Boots, B.N. and Getis, A. (1988) *Point Pattern Analysis*, Newbury Park, Sage.

Campbell, J. (2007) *Map Use and Analysis*, Oxford, Wm C Brown.

Clark, P.J. and Evans, F.C. (1954) Distance to nearest-neighbour as a measure of spatial relationships in populations. *Ecology*, **35**, 445–453.

Cliff, A.D. and Ord, J.K. (1973) *Spatial Autocorrelation*, London, Pion.

Cliff, A.D. and Ord, J.K. (1975) The comparison of means when samples consist of spatial autocorrelated observations. *Environment and Planning A*, **7**, 725–734.

Cressie, N.A.C. (1991) *Statistics for Spatial Data*, New York, John Wiley and Sons.

Diggle, P.J. (1983) *Statistical Analysis of Point Patterns*, London, Academic Press.

Fischer, M. and Getis, A. (2009) *Handbook of Applied Spatial Analysis: Software Tools, Methods, and Applications*, New York, Springer.

Gatrell, A.C., Bailey, T.C., Diggle, P. and Rowlingson, B.S. (1996) Spatial point pattern analysis and its application in geographical epidemiology. *Transactions of the Institute of British Geographers*, **2**, 256–274.

Geary, R.C. (1954) The contiguity ratio and statistical mapping. *The Incorporated Statistician*, **5**(3), 115–145. DOI: 10.2307/2986645.

Getis, A. and Ord, J.K. (1992) The analysis of spatial association by use of distance statistics. *Geographical Analysis*, **24**, 189–206.

Getis, A. and Franklin, J. (1987) Second-order neighbourhood analysis of mapped point patterns. *Ecology*, **68**(3), 473–477.

Griffith, D.A. (1987) *Spatial Autocorrelation: A Primer*, Washington, Association of American Geographers.

Haining, R. (1990) *Spatial Data Analysis in the Social and Environmental Sciences*, Cambridge, Cambridge University Press.

Haining, R. (2003) *Spatial Data Analysis: Theory and Practice*, Cambridge, Cambridge University Press.

Harris, R. (2016) *Quantitative Geography: The Basics*, Sage Publications Ltd.

Longley, P., Brooks, S.M., McDonnell, R. and Macmillan, B. (1998) *Geocomputation: A Primer*, Chichester, John Wiley and Sons.

Mitchell, A. (2008) *The ESRI Guide to GIS Analysis, Volume 2: Spatial Measurements and Statistics*, Redlands, ESRI Press.

Neft, D. (1968) Statistical analysis for areal distributions. Regional Science Institute, monograph series, No. 2. Regional Science Research Institute, Philadelphia, 1966. S4.75. vii and 172 pp. *Journal of the American Statistical Association*, **63**(322), 726–728.

Ord, J.K. and Getis, A. (1995) Local spatial autocorrelation statistics: distribution issues and an application. *Geographical Analysis*, **27**(4), 286–306.

11

Analysis and Modelling of Spatial Data

> *Connections between spatial patterns of geographical features and variations in the data values of their attributes and variables over space are the focus of this chapter. It extends our exploration of spatial autocorrelation and examines how spatial matrices recording the contiguity and adjacency of features and 'spatial' lagging of data values can cast new light on the implications of spatial autocorrelation for modelling relationships. We explore ways of interpolating not only surface in respect of changes in elevation as they create the sometimes undulating, sometimes flat surface of the Earth, but also surfaces created from non-physical environmental variables that can shed new light on spatial variation in geographical processes. This chapter covers some relatively advanced techniques in an introductory way to help students and researchers navigate their way through parts of the quantitative research literature.*

Learning Outcomes

This chapter will enable readers to:

- Extend their appreciation of the characteristics and implications of spatial autocorrelation;
- Calculate and apply spatial and geostatistics for investigating and coping with spatial autocorrelation in analysis;
- Consider how to include somewhat more advanced forms of spatial and geostatistical analysis of geographical datasets in an independent research investigation in Geography, Earth Science and related disciplines.

11.1 Introduction

Chapter 10 introduced spatial statistics as an area of statistical analysis techniques applied to spatial (geographical) data. It concentrated on ways of quantitatively describing spatial patterns, especially those made by points, and testing whether these patterns displayed complete spatial randomness, clustering or dispersion. Once computed, some spatial statistics measures could be converted into a related measure so that their expected value could be compared with the empirical value using 'standard' probability distributions (e.g. the Poisson or chi-square (X^2)) distributions to test Null and Alternative Hypotheses. Others relied on the Monte Carlo simulation

Practical Statistics for Geographers and Earth Scientists, Second Edition. Nigel Walford.
© 2025 John Wiley & Sons Ltd. Published 2025 by John Wiley & Sons Ltd.
Companion website: www.wiley.com/go/PracticalStatistics2e

method to create many random spatial patterns (e.g. 99,999 or 9999) based on definitive characteristics of the empirical data (e.g. number of points and study area size) and computing the spatial statistic for all of them. The average (mean) of the 99,999 or 9999 statistical quantities provides the expected random value against which the empirical one can be compared within certain confidence limits.

This chapter moves on from the previous spatial statistics techniques in two important ways: first, it is concerned with not only the spatial distribution of the geographical features themselves but also with the pattern of values for the attributes and variables relating to the features; and second, it focuses on modelling spatial data to quantify and explore the impact of spatial autocorrelation. The techniques covered are considered as a subset of spatial statistics often referred to as geostatistics. Some of the techniques, such as trend surface analysis and geographically weighted regression (GWR), may be thought of as spatial equivalents of correlation and regression. Advanced statistical techniques such as these deserve to be covered in an introductory text of this nature, if only because students may encounter them when reading the geographical, environmental and geological literature. However, a detailed account of their theoretical basis and the various permutations is more suited to specialist sources and a selection has been included as references or further reading for those wishing to explore these topics further. Spatial autocorrelation is arguably the reason why spatial and geostatistics are needed and the next section examines this concept further before we discuss the techniques themselves.

11.2   More About Spatial Autocorrelation

Spatial autocorrelation potentially 'disturbs' the ability of non-spatial correlation analysis to quantify the strength and direction of the relationship between two variables and of regression to model for such relationships because these techniques assume that each observation or entity is unrelated to any of the others in respect of the variables being analysed. They explicitly assume (require) that the individual entities are unrelated to each other in respect of their data values. The possibility that one observation possesses a certain value for the X variable because of the X values attributed to one or more other observations is not allowed. If this assumption is breached, rather than the observations being independent they might be **interdependent**. Autocorrelation occurs when some or all the observations are related to each other. There is a group of non-spatial statistics techniques including **cluster analysis**, **factor analysis** and **principal components analysis,** referred to analyses of interdependence that explore how similar observations cluster in multidimensional statistical space (as distinct from two-dimensional geographical space) or how variables overlap with each other in what they measure, and this overlapping part can be quantified as one or more new variables (factors or components). For example, a person's age, gender, employment type and level of education together contribute to their financial position, but some of these variables are likely to be correlated with each other. People with modest levels of educational achievement may be more likely to be occupied in lower paid work and be younger.

Interdependence can occur with environmental non-spatial data, for example the mean temperature on 31 December and 1 January, which are likely to be more similar to each other than either of them would be to the temperature on 30 June or 1 July. Spatial autocorrelation arises when it is locational proximity that results in observations being related, whereas in the temperature example, it was time that produced temporal autocorrelation. This reflects a first law of history like Tobler's

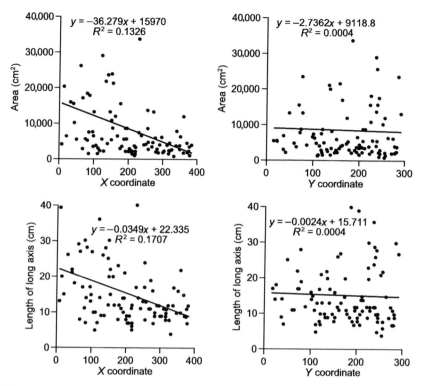

Figure 11.1 Scatter plots of full sample of moraine material by area and length of long axis against X and Y spatial coordinates.

for Geography (Tobler, 1970) that events close together in time are more likely to be similar to each other than to those further apart. Spatial and temporal autocorrelations are commonly positive in nature in the sense that the observations possess similar values for attributes and variables, whereas negative autocorrelation, when spatially or temporally close observations have dissimilar values, is rarer but by no means unknown. Many geographical phenomena display positive spatial autocorrelation; for example, people living on the same street and soil samples taken from the same field are more like each other than they are to those from locations further apart.

What would such a relationship between observations look like? Figure 11.1 helps to explain. It shows scatter plots for the complete sample of boulders, stones and pebbles on the Bossons Glacier moraine described in Box 10.1a. Rather than plotting the dependent variable (length of long axis) against the independent one (surface area), the upper and lower pairs of plots show surface area and length of the long axis plotted against the X (left side plots) and Y **coordinates** (right side plots), respectively. There are several key features to note from these scatter plots:

- the r^2 values are low, which indicates that the X and Y coordinates do not provide a strong explanation for variability in area or length;
- the relationships are all negative, although the slope of the regression line is steeper for the X coordinates;
- there are some clumps of data points where groups of observations have similar area or length values at certain X or Y coordinates.

One clear example of this is to be found just above the centre of the horizontal axis of the upper left plot where there is a group of 11 observations with low area values and X coordinates around 200. Spatial autocorrelation extends the concept of autocorrelation in two ways: first that adjacent values are strongly related and second that randomly arranged values indicates the absence of autocorrelation.

Where else are there clumps of data points in Figure 11.1? What are the combinations of variable and coordinate values at these locations?

Understanding of spatial autocorrelation owes much to earlier work concerned with **time series analysis** and the fact that geographical investigations often focus on the spatial occurrence of phenomena and the measurement of variables as they change over time. Human geographers, for example, might be interested in how disadvantaged households are distributed spatially **and** temporally between different areas. Likewise, physical geographers might be interested in the amounts of algae in lakes at different times of the year. The concept of covariance, the way in which two independent pairs of data values for variables X and Y vary together, was introduced in Chapter 8 as the starting point for understanding correlation. Dividing the covariance by the product of the square roots of the variances of X and Y produces Pearson's correlation coefficient (r). This effectively standardises the value of the coefficient to lie within the range -1.0 to $+1.0$. In Box 10.1a, we focused on the relationship between the area and long axis length of boulders, stones and pebbles on part of the moraine from Les Bossons Glacier in the French Alps, but suppose we were interested in the set of n values for one of the variables, say surface area, measured in respect of the spatially contiguous debris in a transect over the surface of the moraine. Box 11.1a illustrates the effects of spatial autocorrelation by examining the **spatial contiguity** within the subset of 19 items (boulders, pebbles and stones) lying partly or wholly within such a transect. The series of four scatter plots in Box 11.2b are known as *h*-scatter plots, where h refers to the spatial lag between data values. Temporal lags, the length of time periods or intervals in time series analysis are often constant throughout the sequence, for example daily amounts of precipitation, whereas spatial lags can be regular or irregular. In Box 11.1a, the separate items of moraine debris in the transect are not located at a regular distance apart but are lagged according to their sequential spatial contiguity.

Box 11.1a   Spatial lags.

Serial correlation coefficient for lag 1: $r_{.1} = \dfrac{\sum\limits_{h=1}^{n-1}(x_h - \overline{x}_{.1})(x_{h+1} - \overline{x}_{.2})}{\sqrt{\sum\limits_{h=1}^{n-1}(x_h - \overline{x}_{.1})^2}\sqrt{\sum\limits_{h=1}^{n-1}(x_{h+1} - \overline{x}_{.2})^2}}$

Serial correlation coefficient for k lags: $r_{.k} = \dfrac{\sum\limits_{h=1}^{n-k}(x_h - \overline{x})(x_{h+k} - \overline{x})}{\sum\limits_{h=1}^{n}(x_h - \overline{x})^2}$

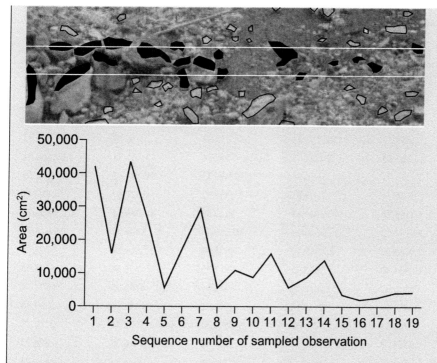

Transect through sample of debris on Les Bossons Glacier moraine.

The data values for variables measured in respect of observations located in space may be related to each other and display positive or negative spatial autocorrelation. A transect has been superimposed on to top of the image representing the part of Les Bossons Glacier's moraine and the boulders, pebbles and stones intersecting with this area have been identified and numbered 1–19 in sequence from left to right. This example deals with objects located irregularly in space, but the procedure could as easily be applied to features that are regularly spaced, for example items that are a fixed distance apart.

Autocorrelation can be examined by means of the serial correlation coefficient where there are k lags and each lag is identified a h units (e.g. h = 1, 2, 3 up to k) or (t time periods in the case of time series analysis). The 19 values for the variable measuring the surface area of these objects, denoted as x, have been tabulated in Box 11.1b and labelled as x_1 to x_{19}. In the second column of data values, they have been shifted up by one row, thus pairing the data value for one object with the next in the sequence. Lag 2 works in a similar way, but pairs one data value with the next but one in the sequence, and so for however many lags are required. Once the data values have been paired in this way, Pearson's correlation coefficients are calculated and these have been shown as h-scatter plots for spatial lags 1–4.

The r coefficients increase through lags 1, 2 and 3 (0.3341, 0.4015 and 0.7000) and the decline to 0.5156 for lag 4. This indicates that spatial autocorrelation in respect of area for these observations starts to reduce after spatial lag 3.

Box 11.1b   Linking data values by spatial lags.

Debris item	Cm^2	$x_{1...n} + h$ Lag 1 ($h = 1$)	$x_{1...n} + h$ Lag 2 ($h = 2$)	$x_{1...n} + h$ Lag 3 ($h = 3$)	$x_{1...n} + h$ Lag 4 ($h = 4$)
x_1	41,928.88	15,889.19	43,323.43	26,015.02	5536.74
x_2	15,889.19	43,323.43	26,015.02	5536.74	17,865.73
x_3	43,323.43	26,015.02	5536.74	17,865.73	29,164.62
x_4	26,015.02	5536.74	17,865.73	29,164.62	5620.59
x_5	5536.74	17865.73	29,164.62	5620.59	10,889.42
x_6	17,865.73	29164.62	5620.59	10,889.42	8805.97
x_7	29,164.62	5620.59	10,889.42	8805.97	15,814.79
x_8	5620.59	10,889.42	8805.97	15,814.79	5703.54
x_9	10,889.42	8805.97	15,814.79	5703.54	8795.05
x_{10}	8805.97	15814.79	5703.54	8795.05	13,891.54
x_{11}	15814.79	5703.54	8795.05	13,891.54	3669.44
x_{12}	5703.54	8795.05	13,891.54	3669.44	2116.38
x_{13}	8795.05	13,891.54	3669.44	2116.38	2824.25
x_{14}	13,891.54	3669.44	2116.38	2824.25	4148.04
x_{15}	3669.44	2116.38	2824.25	4148.04	4276.09
x_{16}	2116.38	2824.25	4148.04	4276.09	
x_{17}	2824.25	4148.04	4276.09		
x_{18}	4148.04	4276.09			
x_{19}	4276.09				

The serial correlation coefficients should be close to zero if they were calculated for a random set of data values and plotting a **correlogram** is a useful way of examining whether this is the case when the spacing of the observations is equal. Although the observations in our example are not spaced at a regular distance apart across the transect, they are in a unitary sequence relating to the first, second, third, fourth and so on up to $n - 1$ nearest neighbour. It is therefore not entirely inappropriate to plot the series of correlation coefficients for the lags as a correlogram (Figure 11.2). The correlograms for area and axis length show moderately strong positive correlation coefficients (in the range +0.3–+0.4) for both variables for spatial lags 1, 2 and 3 followed by a decline until lag 7. Thereafter, the area line continues to record low positive correlation coefficient values (+0.2–+0.3), whereas for axis length, they are mainly low negative ones (0.0 to −0.2).

This section has introduced some of the ways of examining spatial autocorrelation as though they could simply be migrated across from time series analysis in an unproblematic fashion. The addition of a transect to the image of the moraine is an artefact simply being used to illustrate the principles of spatial autocorrelation. There may be some underlying trend in the data values not only in respect of the portion of the moraine shown in the image but also across the entire area, which may

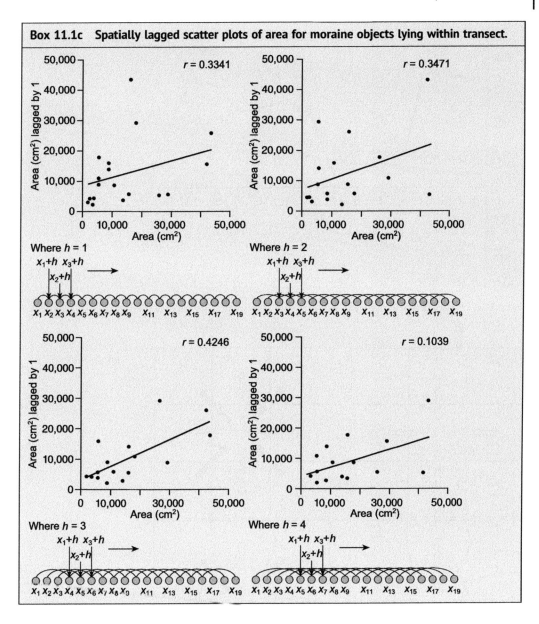

Box 11.1c   Spatially lagged scatter plots of area for moraine objects lying within transect.

relate to distance from the glacier snout, slope angle and other factors. We will return to this issue later when examining the application of trend surface analysis. A further critical issue is that a given sequence of measurements may include some rogue values or **outliers**, which distort the overall pattern. The following sections will examine a range of procedures available for examining patterns in spatial data starting with those dealing with global spatial autocorrelation and then moving to those capable of indicating local spatial autocorrelation or association.

Spatial weights are another important concept in relation to taking account of spatial autocorrelation and relate to an area's neighbourhood. There are two main types of spatial weight based

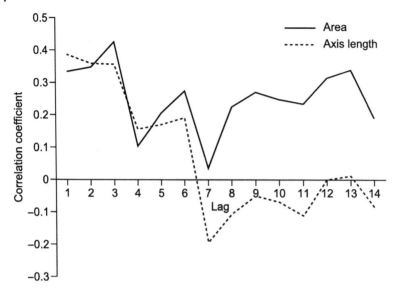

Figure 11.2 Correlograms for area and length of long axis of moraine material intersecting with transect.

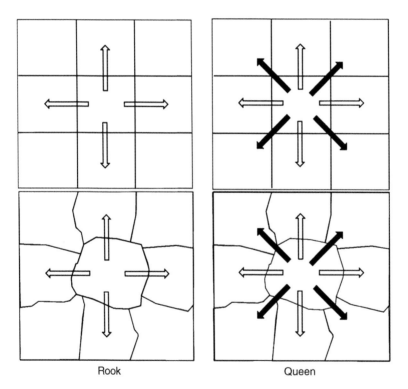

Rook   Queen

Figure 11.3 Spatial weights and contiguity.

on the contiguity or neighbourliness of an area and its neighbours. Rook and queen spatial weights make reference to the directional moves that these pieces are allowed to make in chess: a rook can move horizontally and vertically along the rows and columns of a chess board, whereas a queen can additionally move diagonally. Figure 11.3 illustrates these options in respect of regular grid of nine squares and a set of nine irregular polygons. Given that spatial autocorrelation concerns how the data value of one spatial feature is influence by surrounding features in its neighbourhood, we can see that the rook spatial weight takes account of four neighbours from the central grid square or polygon, whereas the more commonly used queen weight uses eight.

11.3 Global Spatial Autocorrelation

Geostatistical techniques for quantifying spatial autocorrelation, sometimes referred to as **Exploratory Spatial Data Analysis (ESDA)**, divide into those concerned with the global expression of this characteristic across an entire study area and others concentrating on identifying its localised outcome as expressed by variation and divergence in the dataset, which might, for example, reveal hot and cold spots. This section examines indices of global spatial autocorrelation, which summarise the extent of the characteristic across the whole of the area under study focusing on the overall pattern in the data and if there is some detectable amount of clustering. There are several measures available that are suited for use with distinct types of data. The starting point for many of the techniques is that the area or region of interest can be covered by a regular grid of squares or by a set of irregular shaped polygons. A further factor influencing the choice of technique concerns whether the values are numerical measurements or counts of nominal attributes. The data type presented by the Bossons Glacier moraine example where there are data values for 100 randomly distributed points is also covered. The essential purpose of all the techniques outlined in the following sections is to explore the correlation between the units (area or points) at various levels of spatial separation and to produce a measure that is comparable to the serial correlation coefficient used in time series analysis.

11.3.1 Join Counts Statistics

Join count statistics (JCS) focus on the patterns produced by sets of spatial units that have nominal data values by counting the number of joins or shared boundaries between areal units in different nominal categories. Most applications relate to binary data values, for example the absence or presence of a particular characteristic, although data with more than two classes can be re-grouped into a binary form. The simplest place to start with exploring JCS is the case where a regular grid of squares has been superimposed over the study area and these squares have been coded with a value of 0 and 1 to denote the binary categories (e.g. absence and presence). Chapter 10 discussed three 'standard' ways in which spatial features could be arranged, clustered together, dispersed or systematic and random Figure 11.4 illustrates the three situations with respect to a regular grid of 100 squares that belong to binary classes, here shown as either black or white. In Figure 11.4a, the 100 squares are split equally with all the white ones at the top and the black ones in the bottom. The middle grid (Figure 11.4b) has a systematic pattern of white and black squares, rather like a chess board, with no squares of the same colour sharing a boundary and only meeting at the corners. Figure 11.4c, again with half of the squares shaded black and the other half white, shows a random distribution with some same colour squares sharing edges and others meeting at corners.

  These comments have already given a clue as to how we might analyse the different patterns and to decide whether a given pattern is likely to have occurred by chance or randomly. First, consider the situation in time series analysis, where time periods are usually assumed to form a linear sequence so that one period of 24 hr (a day) is followed by another and so on and each period has one join with its predecessor and one with its successor, apart from those at the arbitrary start and end of the series. If these time periods were classified in a binary fashion (e.g. absence or

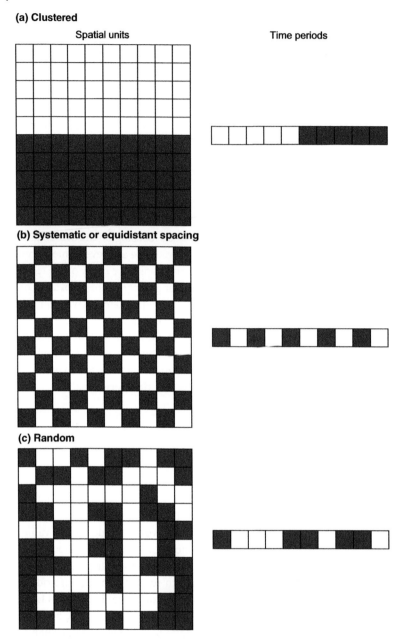

Figure 11.4 Binary join count patterns for regular grid squares and linear time series. (a) Clustered; (b) systematic or equidistant spacing; (c) random.

presence of President Biden's name on the front page of the New York Times on successive days), they could be represented as a series of black and white squares in one dimension, such as those down the right-hand side of Figure 11.4. The three linear time sequences correspond to their grid square counterparts on the left-hand side. The joins between the spatial units (squares in this case) work in two dimensions rather than the one dimension in the time series. Joins between squares in the grid occur in two ways edge to edge and corner to corner, we have already seen that these define spatial weights with the former are referred to as rook's and the latter as queen's moves. This produces up to eight possible joins for each square with those at the boundary of the study area around the edge and at the corners of the grid covering the study area having less. The edge effects can be disproportionately important if the size of the study area or the number of grid squares is relatively small, although this obviously raises the question of how small is small. Given that each square has the binary codes 0 or 1, where they join there are four possible combinations: 1-1, 0-0, 1-0 and 0-1. Counting the number of joins of these distinct types provides the basis for JCS.

How many rook's and queen's joins does each corner square have in the grids down the left-hand side of Figure 11.4? How many of both types of joins does each of the four squares at the centre of these grids have?

Each corner square has two adjacent squares, all the other squares along the boundaries or edges have three adjacencies and all the remaining squares have four. A 10×10 grid of 100 squares will therefore have 4 corner squares (8 joins), 32 side squares (96 joins) and 64 inner squares (256 joins) summing to 360, but because this has double-counted joins between adjacent squares the sum is halved to give a total of 180. The 100 squares in the grids in Figure 11.4 are equally divided between black and white; therefore, each join combination (1-1, 0-0, 1-0 and 0-1) has an equal probability of occurrence and we would expect there to be 45 joins of each type (180/4). However, we are interested in the deviation from the two extreme situations of perfect separation (Figure 11.4a) and regularity (Figure 11.4b), respectively, having 10 and 180 0-1 and 1-0 joins. The random pattern shown in Figure 11.4c has 92 0-1 and 1-0 joins, which is slightly more than the expected total of 90 but is the difference more or less than might have occurred through chance or sampling error. These comments indicate that the empirical count of each adjacency combination, with 0-1 and 1-0 being taken together since they are equally indicative of a mixed pattern, should be tested for their significance. This can be achieved by converting the difference between the observed and expected frequency into a Z score having calculated the standard deviation of the expected number of counts corresponding to each combination.

One complicating factor should be noted before examining an application of JCS, which relates to whether the data have been obtained by means of free sampling with replacement or non-free sampling without replacement. Mention of sampling might seem a little odd, since the regular grids shown in Figure 11.4 and the one in Box 11.2a are shown covering the entire study area. So, in what way can these data be thought of as a sample? In Chapter 2, we saw that the main difference between sampling with and without replacement when using non-spatial statistics is that the probability of an item being selected changes as each additional entity enters the sample from the population. Here, the issue concerns whether the probability that any square in the grid will be black or white (as we saw with the ground nesting birds in Figure 10.1a). If this probability can be determined *a priori*, for example from published figures for another location, in other words without reference to the empirical data for the study area, then free sampling applies. Sampling without replacement is much more common and its effect is to alter the expected number of joins in each combination (0-0, 1-1 and 1-0/0-1) from an equal distribution or some other hypothesised value.

The application in Box 11.2a examines the application of JCS in respect of the presence or absence of *dianthus gratianopoltanus* on the side of the Mont Cantal in the southern Auvergne region of France. The expected counts are calculated under the assumption non-free sampling, since there is no *a priori* reason to assign specific values to p and q, respectively, the probabilities of presence and absence of the species in a square. The data values in this example are nominal codes (1 and 0) relating to the presence and absence of *dianthus gratianopoltanus* and we are not interested in the number of individual plants, whereas examination of the image of the slope in Box 11.2a reveals that there is some variation in the density of occurrence. The observed numbers of B–B, B–W/W–B and W–W joins are all significantly different to what would be expected by chance, there it is reasonable to conclude that the spatial pattern is not random but indicates an underlying process in relation to the distribution of *dianthus gratianopoltanus* in this area.

Box 11.2a   Join Count Statistics (JCS)

Free sampling with replacement

Expected number of B–B joins: $E_{BB} = Jp^2$

Expected number of B–W joins: $E_{BW} = 2Jpq$

Expected number of W–W joins: $E_{WW} = Jq^2$

Standard deviation of expected B–B joins: $\sigma_{BB} = \sqrt{Jp^2 + 2Kp^3 - (J + 2K)p^4}$

Standard deviation of expected B–W joins:

$$\sigma_{BW} = \sqrt{(2J + \sum L(L + 1)pq - 4(J + \sum L(L - 1)p^2q^2}$$

Standard deviation of expected W–W joins: $\sigma_{WW} = \sqrt{Jq^2 + 2Kq^3 - (J + 2K)q^4}$

Free sample without replacement

Expected number of B–B joins: $E_{BB} = J\dfrac{n_W(n_W - 1)}{n(n - 1)}$

Expected number of B–W joins: $E_{BW} = 2J\dfrac{n_B n_W}{n(n - 1)}$

Expected number of W–W joins: $E_{WW} = J\dfrac{n_B(n_b - 1)}{n(n - 1)}$

Standard deviation of expected B–B joins:

$$\sigma_{BB} = \sqrt{E_{BB} + 2K\dfrac{n_B(n_B - 1)(n_B - 2)}{n(n - 1)(n - 2)} + [J(J - 1) - 2K]\dfrac{n_B(n_B - 1)(n_B - 2)(n_B - 3)}{n(n - 1)(n - 2)(n - 3)} - (E^2_{BB})}$$

Standard deviation of expected B–W joins:

$$\sigma_{BW} = \sqrt{\dfrac{2(J + K)n_B n_W}{n(n - 1)} + 4[J(J - 1) - 2K]\left(\dfrac{n_B(n_B - 1)n_W(n_W - 1)}{n(n - 1)(n - 2)(n - 3)}\right) - 4\left(\dfrac{Jn_B n_W}{n(n - 1)}\right)^2}$$

Standard deviation of expected W–W joins:

$$\sigma_{WW} = \sqrt{E_{WW} + 2K\frac{n_W(n_W-1)(n_W-2)}{n(n-1)(n-2)} + [J(J-1)-2K]\frac{n_W(n_W-1)(n_W-2)(n_W-3)}{n(n-1)(n-2)(n-3)} - (E_{WW}^2)}$$

Z test statistic: $z = (O_{BW} - E_{BW})/\sigma_{BW}$

W-W	B-B	W-B/B-W
2	3	4
0	7	2
0	7	2
2	6	1
3	2	4
3	5	1
5	1	3
1	7	1
0	5	4
5	0	4

W-W	5	5	2	1	0	0	0	0	1	4	39	
B-B	0	2	3	6	6	4	5	5	5	3		82	
W-B/B-W	4	2	4	2	3	5	4	4	3	2			59

Box 11.2b   Application of JCS

JCS works by examining the amount of separation between individual spatial units in respect of the nominal categories to which they have been assigned as the result of the distribution of some phenomena. The procedure involves counting the number of joins between spatial units (grid squares in this example) that fall into the different possible combinations (0-0, 1-1 and 1-0/0-1). 0 denotes the absence of the phenomenon and 1 its presence, which are here represented as white and black squares. The total number of cells in the grid (n) divides between n_W and n_B, where the subscripts denote the type of square. The total number of joins is identified as J and K is defined as $K = \sum J_i(J_i - 1)/2$, where the subscript i refers to individual squares from 1 to n. The probabilities of presence and absence of the phenomenon in any individual cells are referred to as p and q, respectively. These probabilities, which are used to calculate the expected numbers of joins in the different combinations, are determined in one of two ways depending upon whether free (with replacement) or non-free (without replacement) sampling is used. The difference between these relates to whether the probabilities are defined by *a priori* reasoning or by *a posteriori* empirical evidence. The present application is typical in so far as non-free sampling is assumed.

The present application of JCS concerns the distribution of *dianthus gratianopoltanus* on part of the slope of Mont Cantal in the Auvergne. A 10 × 10 square grid has been superimposed over

(Continued)

Box 11.2b (Continued)

the area, and the presence or absence of the species of is shown by black and white shading. There were 59 black and 41 white squares. The calculations in Box 11.2c show that the expected number of BB joins was 29.82 of BW or WB was 87.96 and 62.22 WW. The observed numbers were 82, 39, and 59, respectively (see above). The Null Hypothesis when testing JCS is that the spatial pattern of the phenomena in the grid squares is random, while the Alternative Hypothesis states it is either clustered or dispersed. Given the differences between these figures, it is not surprising that the Z tests indicate that they are significant at the 0.05 level and the spatial is not one likely to have occurred by chance. It is reasonable to conclude that there is significant spatial autocorrelation in the distribution of *dianthus gratianopoltanus* in this area.

The key stages in applying JCS are as follows:

- *Tabulate the individual squares in the grid and count the distinct types of joins for each:* this can be laborious process, since it involves inspecting each square and determining the code (0/1 or B/W) of all its neighbours;
- *Calculate the values J and K, and count the numbers of observed BB, BW/WB and WW joins:* these processes are illustrated below;
- *Calculate the counts and standard deviations of the expected number of BB, BW/WB and WW joins:* these are obtained by applying the equations appropriate to free or non-free sampling
- *Calculate the z test statistics:* the z test statistics are calculated in a comparable way to other tests and in this application are 2.54 (BB), 4.49 (BW/WB) and 6.22 (WW);
- *State Null Hypotheses and significance level:* each Null Hypothesis for the three types of join states that the difference between the observed and expected counts is not significantly greater than would occur by chance at the 0.05 level of significance.
- *Determine the probabilities of the z test statistics:* the probabilities are equal to or less than 0.01;
- *Accept or reject the Null Hypothesis:* each of the Null Hypotheses should be rejected at the 0.05 level of significance and the Alternative Hypotheses are therefore accepted leading to the conclusion that the spatial pattern tends towards being clustered.

The spatial lag in this example is 1 (i.e. adjacent grid squares), whereas further analyses could be carried out where the comparison was made between 2nd-order neighbours, this would mean that the counts of *B–B*, *B–W/W–B* and *W–W* combinations were made by 'jumping over' adjacent squares to the next but one. Similarly, queen's move adjacencies could also be included. Finally, it should be noted that grids, such as the one used in this example, are often placed over a study in an arbitrary fashion and the size and number of grid squares may be chosen for convenience rather than in a more rigorous way. There is no reason a study area should be constrained so that it is covered by a square or rectangular grid. Suppose our study area is bounded on one or more sides by coastline or river, it is highly unlikely that such natural features of the environment will be delimited by straight lines and some of the cells in the grid or lattice are likely to overlap the coast or river. These units would have a reduced chance of including or excluding the phenomenon under investigation. Examination of the image and superimposed grid in Box 11.2a shows that the size of each flower is small in relation to the size of a grid square. This results in some squares containing just one occurrence, whereas others have many, yet both are counted as a presence of the phenomenon. Smaller grid squares close to the size of each flower head would give a more realistic impression of some of the space where isolated occurrences may have distorted the situation.

Box 11.2c Calculation of JCS and significance testing.

$i = 1 \cdots n$	WW	BB	WB/BW	J_i	$J_i(J_i-1)$		$i = 1 \cdots, n$	WW	BB	WB/BW	J_i	$J_i(J_i-1)$
1	2			2	2		51	3			3	6
2	2		1	3	6		52	4			4	12
3	1		2	3	6		53	4			4	12
4		2	1	3	6		54	2		2	4	12
5		3		3	6		55		2	2	4	12
6		2	1	3	6		56		2	2	4	12
7			3	3	6		57		3	1	4	12
8		2	1	3	6		58		3	1	4	12
9		2	1	3	6		59		4		4	12
10	1		1	2	2		60		2	1	3	6
11	2		1	3	6		61	3			3	6
12		2	2	4	12		62	4			4	12
13		3	1	4	12		63	3		1	4	12
14		4		4	12		64	2		2	4	12
15		4		4	12		65		2	2	4	12
16		4		4	12		66	1		3	4	12
17		3	1	4	12		67	2		2	4	12
18		4		4	12		68	1		3	4	12
19		3	1	4	12		69		3	1	4	12
20	2		1	3	6		70		3		3	6
21	1		2	3	6		71	2		1	3	6
22		3	1	4	12		72	3		1	4	12

(Continued)

Box 11.2c (Continued)

$i = 1 \cdots n$	WW	BB	WB/BW	J_i	$J_i(J_i - 1)$
23		4		4	12
24		4		4	12
25		4		4	12
26		4		4	12
27		4		4	12
28		3	1	4	12
29		2	2	4	12
30	2		1	3	6
31		1	2	3	6
32		3	1	4	12
33		3	1	4	12
34		4		4	12
35		3	1	4	12
36		3	1	4	12
37		3	1	4	12
38	1		3	4	12
39	2		2	4	12
40	3			3	6
41	2		1	3	6
42	3		1	4	12
43	2		2	4	12
44		1	3	4	12
45	1		3	4	12

Continued

$i = 1 \cdots n$	WW	BB	WB/BW	J_i	$J_i(J_i - 1)$
73		2	2	4	12
74		3	1	4	12
75		4		4	12
76		3	1	4	12
77		3	1	4	12
78		3	1	4	12
79		3	1	4	12
80		3		3	6
81			3	3	6
82	2		2	4	12
83		2	2	4	12
84		4		4	12
85		3	1	4	12
86		3	1	4	12
87		3	1	4	12
88		3	1	4	12
89	1		3	4	12
90		1	2	3	6
91	1		1	2	2
92	3			3	6
93	1		2	3	6
94		1	2	3	6
95	1		2	3	6

46	1		3	3	4	12
47	3		3	1	4	12
48	3	1	3	1	4	12
49	2		2	2	4	12
50	1		2	2	3	6

96	2	2	1	3	6
97	1	1	2	3	6
98		1	2	3	6
99	2	2	1	3	6
100	1	1	1	2	2

$$\sum J_i/2 = 180$$
$$K = \sum J_i(J_i-1)/2 = 484$$

Expected number of B–B joins	$E_{BB} = J\dfrac{n_W(n_W-1)}{n(n-1)}$	$180\dfrac{41(40)}{100(99)} = 29.82$
Standard deviation of expected B–B joins	$\sigma_{BB} = \sqrt{\begin{array}{l} E_{BB} + 2K\dfrac{n_B(n_B-1)(n_B-2)}{n(n-1)(n-2)} + \\[2mm] [J(J-1)-2K]\dfrac{n_B(n_B-1)(n_B-2)(n_B-3)}{n(n-1)(n-2)(n-3)} - \\[2mm] (E_{BB}^2) \end{array}}$	$\sqrt{\begin{array}{l} 29.83 + 2(484)\dfrac{59(n58)(57)}{100(99)(98)} + \\[2mm] [180(179)-2(484)]\dfrac{59(58)(57)(56)}{100(99)(98)(97)} - \\[2mm] 29.82^2 \end{array}} = 3.61$
Z test statistic of B–B joins	$z = (O_{BB}-E_{BB})/\sigma_{BB}$	$(39-29.82)/3.61 = 2.54$
Probability	$p=$	$=0.0111$
Expected number of W–B/B–W joins	$E_{BW} = 2J\dfrac{n_B n_W}{n(n-1)}$	$\dfrac{2(180)(59)(41)}{100(100-1)} = 87.96$
Standard deviation of expected B–W joins	$\sigma_{BW} = \sqrt{\begin{array}{l} \dfrac{2(J+K)n_B n_W}{n(n-1)} + \\[2mm] 4[J(J-1)-2K]\left(\dfrac{n_B(n_B-1)n_W(n_W-1)}{n(n-1)(n-2)(n-3)}\right) - \\[2mm] 4\left(\dfrac{J n_B n_W}{n(n-1)}\right)^2 \end{array}}$	$\sqrt{\begin{array}{l} \dfrac{2(180+484)59(41)}{100(99)} + \\[2mm] 4[180(179)-2(484)]\left(\dfrac{59(58)41(40)}{100(99)(98)(97)}\right) - 6.45 \\[2mm] 4\left(\dfrac{180(59)(58)n_B n_W}{100(99)}\right)^2 \end{array}} = 4.49$
Z test statistic of B–W joins	$z = (O_{BW}-E_{BW})/\sigma_{BW}$	$(59-87.96)/6.45 = 4.49$
Probability	$p=$	<0.000
Expected number of W–W joins	$E_{WW} = J\dfrac{n_B(n_B-1)}{n(n-1)}$	$180\dfrac{59(58)}{100(99)} = 62.22$

(Continued)

Box 11.2c (Continued)

$i = 1\cdots n$	WW	BB	WB/BW	J_i	$J_i(J_i - 1)$

Standard deviation of expected W–W joins

$$\sigma_{ww} = \sqrt{\dfrac{E_{ww} + 2K\dfrac{n_w(n_w-1)(n_w-2)}{n(n-1)(n-2)} + \dfrac{[J(J-1)-2K]n_w(n_w-1)(n_w-2)(n_w-3)}{n(n-1)(n-2)(n-3)} - }{(E_{ww}^2)}}$$

$z = (O_{ww} - E_{ww})/\sigma_{ww}$

$p =$

Continued

$i = 1\cdots, n$	WW	BB	WB/BW	J_i	$J_i(J_i - 1)$

$$\sqrt{\dfrac{62.22 + 2(484)\dfrac{41(40)(39)}{100(99)(98)} + \dfrac{[180(179) - 2(484)]\dfrac{41(40)(39)(39)}{100(99)(98)(97)} - }{62.22^2}}} = 3.41$$

$(41 - 62.22)/3.41 = 6.22$

< 0.000

11.3.2 Moran's *I* Index with Polygon Features

Some of the issues mentioned at the end of the previous section arise from the artificial superimposition of a regular grid or lattice, or an irregular set of polygons, over a study area and that JCS applies to nominal dichotomous data values. **Moran's *I*** is a widely available technique that can be used when the study is covered by an incomplete regular grid or a set of planar (two-dimensional) polygons and the data values are real numbers rather than counts of units in nominal presence/absence categories. The difference between the presentation of the raw data for Moran's *I* compared with JCS is that, rather than tabulating count statistics, the data are organised in a three column format where the first two columns contain *X* and *Y* coordinates that relate to row and column numbers, the grid references of points features or the centroids of polygons. Figure 11.5 illustrates the procedure with respect to an incomplete lattice, irregular polygons and points. The data tables show the *X* and *Y* coordinates or row and column numbers and the third column contains the *Z*

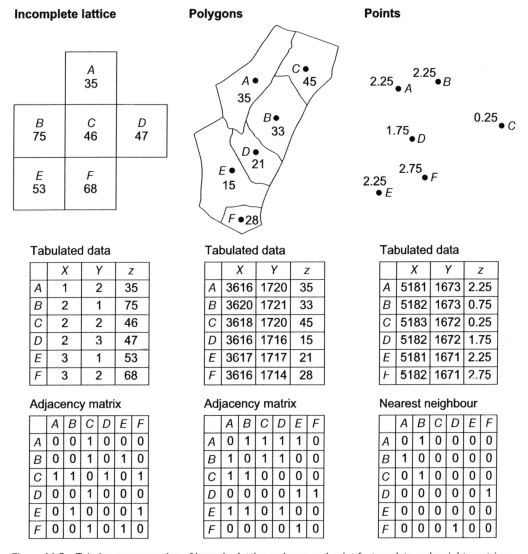

Figure 11.5 Tabular representation of irregular lattice, polygon and point feature data and weights matrices.

data values corresponding to these locations, which may be decimal values or integer counts, as in Figure 11.5. Although this process retains references to the spatial location of the features, it discards information about their topological connections, in other words whether one unit is adjacent to another. Since such information is a vital component in the analysis of spatial patterns, it is necessary to create a **weights matrix** (W) that records which units share a common boundary in the case of area data or are nearest to each other in the case of point feature data. Figure 11.5 includes the first-order weights matrices for the three types of spatial data. Normally, these weights matrices are obtained for 1st-order neighbours, but 2nd-, 3rd- or higher-order neighbours can also be used (e.g. next but one polygon or second nearest point). The weights matrix can incorporate rook's and queen's adjacencies, which can be weighted to denote their relative importance if required.

Pairs of adjacent features in a spatial pattern displaying or possessing spatial autocorrelation will both have positive **or** negative values for the variable under investigation, and if these values are large, it will indicate stronger rather than weaker spatial autocorrelation. In contrast, pairs of contiguous features where one has a positive and the other a negative value suggest the absence of spatial autocorrelation. These statements are effectively another way of describing Tobler's first law of Geography. Moran's I encapsulates the essence of these statements in a single index value. The sequence of calculations to compute Moran's I is protracted and involves the manipulation of data in matrix format. The first stage is subtraction of the overall mean of the variable Z from the data values of each spatial feature and then multiplying the result for each pair of features. The spatial weights are used to select which pairs of features are included and excluded from the final calculation of the index: those with a zero in the weights matrix (denoting non-adjacency) are omitted, whereas those with a one are included. Summing these values and dividing by the sum of the weights produces a covariance. This forms of the numerator in the equation for Moran's I, which is divided by the variance of the data to produce the index value that lies within the range −1.0 to +1.0. Values towards the extremes of this range indicate negative and positive spatial autocorrelation, respectively, whereas a value around zero usually signifies its absence. This global Moran's I statistic does not reveal where, in a geographical sense, autocorrelation occurs in the dataset, but does indicate whether it is present. This is similar to non-spatial correlation analysis, which provides information about the overall direction (positive or negative) and strength (closer to −1.0 or +1.0) of a relationship.

Adjacency and contiguity are clearly important concepts in understanding the principles underlying join counts statistics, the Global Moran's I index and other similar measures of spatial autocorrelation. Another important concept is distance between features, both points and areas, with the centroid usually marking the location in the latter case. Features that are located further apart may be expected to exert less influence on each in respect of contributing towards spatial autocorrelation compared with those that are closer together. This presumption leads to the use of inverse weighted distance as a way of considering the diminishing or decaying effect of distance as a factor on the data values of features that are further apart. Spatial weighting methods are based on contiguity (e.g. rook's and queen's adjacency for polygons) or distance using polygons' centroids or user defined X, Y coordinate pair. Other methods weight data values by focusing on each point (centroid) location and then averaging over a pre-specified number of nearest neighbours. The outcome of spatial weighting is a spatially lagged variable, which is an essential requirement for testing autocorrelation and carrying out spatial regression.

Box 11.3a illustrates the application of Moran's I in relation to the mean scores on the UK government's 2007 Index of Multiple Deprivation (IMD) for the 33 London Boroughs. The IMD is computed from a series of different variables within domains covering such topics as income, employment and social conditions. The map of the IMD for the London local authorities suggests that there were higher mean scores in many of the inner London Boroughs, whereas several of those more suburban ones had lower deprivation overall. Some degree of spatial autocorrelation seems to exist with high values clustered together near the centre and lower ones in outer areas. The application of Moran's I in Box 11.3a uses a contiguity weight (i.e. adjacent boundaries) in respect of irregular polygons. The results from the analysis provide moderate support for this claim since the Moran's I global index computes as 0.305. One explanation for this outcome is the presence of one small authority in the centre, the City of London, with an exceptionally low index value (12.84) in comparison with its seven adjacent neighbours, which apart from one have IMD values over 25.00. The Moran's I index has been re-computed leaving out the City of London Borough and the effect is to raise the Moran's I index to 0.479, which confirms the initial visual impression of strong positive spatial autocorrelation. An eastward shift in many financial and other service sector businesses from the traditional Central Business District in the City of London towards the redeveloped docklands area since the 1980s and the construction of some high rental value accommodation in the City of London help to account for its distinctive, continuing low residential population (8583 persons living at a density of 2.898 per km^2 according to the 2021 Population Census) compared with population totals and densities in the surrounding seven boroughs over 200,000 and 20 persons per km^2. These characteristics help to account for the difference between the Moran's I global index values albeit calculated from IMD data relating to 2007.

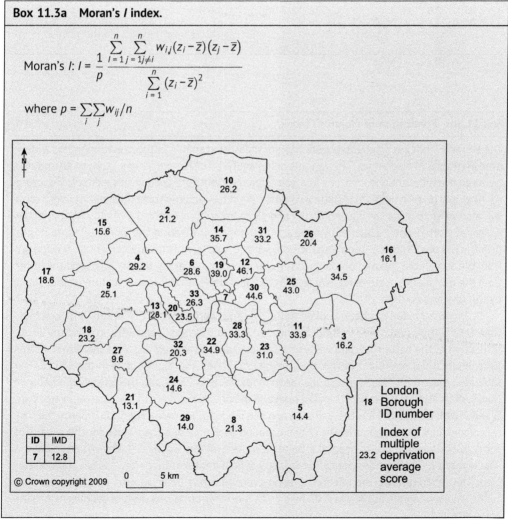

Box 11.3a   Moran's *I* index.

Moran's I: $I = \dfrac{1}{p} \cdot \dfrac{\displaystyle\sum_{i=1}^{n} \sum_{j=1, j\neq i}^{n} w_{ij}(z_i - \bar{z})(z_j - \bar{z})}{\displaystyle\sum_{i=1}^{n} (z_i - \bar{z})^2}$

where $p = \displaystyle\sum_i \sum_j w_{ij}/n$

ID	IMD
7	12.8

© Crown copyright 2009   0   5 km

18 London Borough ID number

23.2 Index of multiple deprivation average score

(Continued)

Box 11.3a   (Continued)

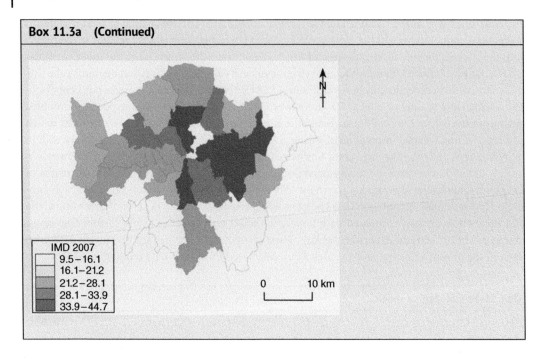

IMD 2007
- 9.5–16.1
- 16.1–21.2
- 21.2–28.1
- 28.1–33.9
- 33.9–44.7

0   10 km

Box 11.3b   Application of Moran's *I* index.

The Moran's *I* index is an adaptable statistic for measuring spatial autocorrelation that is widely used in a range of disciplines. The variable being analysed (z) has subscripts i and j to distinguish between the different observations in a pair. Binary weights are used in this example denoting whether any given pair of areas share a common boundary and the subscripts in the weight term $w_{i,j}$ also refer to a pair of observations i and j in a set with n features overall. The individual values of z are usually adjusted by subtracting the overall mean of the variable before multiplying pairs of values together. Variance/covariance 'like' quantities are computed, the former from the sum of the squared products in the C matrix that fall along the diagonal and the latter from the non-diagonal elements where there is a join between the spatial features represented by the row and column.

The Moran's *I* index has been applied to the average score variable in the 2007 Index of Multiple Deprivation (IMD) with respect to the 33 local authorities (boroughs) in London. Visual inspection of the pattern of data values for these areas on the map suggests that lower scores (less deprivation) are found in the more peripheral zones and higher ones in the centre. The tabulated data in Box 11.3c record the mean index value is 25.68 and the following adjacency matrix of 0s and 1s shows that the minimum and maximum numbers of joins between areas are 3 and 7 with a total of 164. The values along the diagonal in matrix used to compute the variance/covariance like quantities have been shaded dark grey and the values for those pairs of local authorities that are adjacent (i.e. have a 1 in the weights matrix) are shown with a border. The Moran's *I* index in this application is 0.305, which indicates moderate spatial autocorrelation. The randomisation significance testing procedure has been applied and indicates that index values is significant at the 0.05 level with 9999 permutations used to produce the reference distribution.

Box 11.3c Calculation of Moran's I.

Id No.	z		Continued Id No.	z		Continued Id No.	z
1	34.49		12	46.10		23	31.04
2	21.16		13	28.07		24	14.62
3	16.21		14	35.73		25	42.95
4	29.22		15	15.59		26	20.36
5	14.36		16	16.07		27	9.55
6	28.62		17	18.56		28	33.33
7	12.84		18	23.20		29	13.98
8	21.31		19	38.96		30	44.64
9	25.10		20	23.51		31	33.19
10	26.19		21	13.10		32	20.34
11	33.94		22	34.94		33	26.30
				Mean		$\sum x/n = 847.44/33 = 25.68$	

Adjacency matrix.

$\frac{j}{i}$	1	2	3	4	5	6	7	8	9	10	11	12	13	14	15	16	17
1	0	0	1	0	0	0	0	0	0	0	1	0	0	0	0	0	0
2	0	0	0	1	0	1	0	0	0	0	0	0	0	0	1	1	0
3	1	0	0	0	1	0	0	0	0	1	1	0	0	1	0	0	0
4	0	1	0	0	0	1	0	0	1	0	0	1	1	0	1	1	0
5	0	0	1	0	0	0	0	1	0	0	1	0	0	0	0	0	0
6	0	1	0	1	0	0	1	0	0	0	0	0	0	1	0	0	0

(Continued)

Box 11.3c (Continued)

Adjacency matrix.

$\dfrac{j}{i}$	1	2	3	4	5	6	7	8	9	10	11	12	13	14	15	16	17
7	0	0	0	0	0	1	0	0	0	0	0	1	0	0	0	0	0
8	0	0	0	0	1	0	0	0	0	0	0	0	0	0	0	0	0
9	0	0	0	1	0	0	0	0	0	0	0	0	1	1	1	0	1
10	0	1	0	0	0	0	0	0	0	0	0	0	0	1	0	0	0
11	1	0	1	0	1	0	0	0	0	0	0	0	0	0	0	0	0
12	0	0	0	0	0	0	1	0	0	0	0	0	0	1	0	0	0
13	0	0	0	1	0	0	0	0	1	1	0	1	0	0	0	0	0
14	0	1	0	0	0	1	0	0	0	0	0	0	0	0	1	0	1
15	1	1	1	1	0	0	0	0	1	0	0	1	0	0	0	0	0
16	1	0	1	0	0	0	0	0	0	0	0	0	0	0	0	0	1
17	0	0	0	0	0	0	0	0	1	0	0	0	1	1	0	0	0
18	0	0	0	0	0	1	1	0	0	0	0	1	0	1	0	0	0
19	0	0	0	1	0	0	0	0	0	0	0	0	1	0	0	0	0
20	0	0	0	0	0	0	0	1	0	0	0	1	0	0	0	0	0
21	0	0	0	0	1	0	1	0	0	0	1	0	0	0	0	0	0
22	0	0	0	0	0	0	0	1	0	0	0	0	0	0	0	0	0
23	0	0	0	0	1	0	0	0	0	0	1	0	0	0	0	0	0
24	1	0	0	0	0	0	0	1	0	0	0	0	0	0	0	0	0
25	1	0	0	0	0	0	0	0	0	0	1	1	0	0	0	0	0
26	1	0	0	0	0	0	0	0	0	0	0	1	1	0	0	1	0
27	0	0	0	0	0	0	0	0	0	0	0	0	0	0	0	0	0
28	0	0	0	0	1	0	1	0	0	0	0	0	0	0	0	0	0

i		33	32	31	30	29	28	27	26	25	24	23	22	21	20	19	18
29	0	0	0	0	0	0	0	0	0	1	0	0	0	0	0	0	0
30	0	0	0	1	1	1	1	0	0	0	1	0	0	0	0	0	0
31	0	0	0	1	1	1	0	1	0	0	0	0	0	0	0	0	0
32	0	1	0	0	1	0	0	0	0	0	0	1	0	0	0	0	0
33	0	0	0	0	0	0	0	0	0	0	1	1	0	1	0	0	0
$\sum w_i$	3	3	4	6	6	6	6	3	5	4	7	6	6	7	4	5	5

j / i	18	19	20	21	22	23	24	25	26	27	28	29	30	31	32	33
1	0	0	0	0	0	0	0	1	1	0	0	0	0	0	0	0
2	0	0	0	0	0	0	0	0	0	0	0	0	0	0	0	0
3	0	0	0	0	0	0	0	0	0	0	0	0	0	0	0	0
4	0	0	1	0	1	0	0	0	0	0	1	0	0	0	0	1
5	0	0	0	0	0	1	0	0	0	0	0	0	0	0	0	1
6	0	1	0	0	0	0	0	0	0	0	1	1	0	0	0	1
7	0	1	0	0	1	0	1	0	0	0	0	0	1	1	0	1
8	0	0	0	0	0	0	0	0	0	0	0	0	0	0	0	0
9	1	0	0	0	0	0	0	0	0	0	0	1	0	0	0	0
10	0	0	0	0	0	0	0	1	0	1	0	0	1	1	0	0
11	0	1	0	0	0	0	0	1	0	0	0	0	1	0	0	0
12	0	0	0	0	0	0	0	0	0	0	0	0	0	1	0	0
13	1	1	1	0	0	0	0	0	0	1	0	0	0	1	1	1
14	0	0	0	0	0	0	0	0	0	0	0	0	0	0	0	0
15	0	1	0	0	0	0	0	0	0	0	0	0	0	0	0	0
16	0	0	0	0	0	0	0	1	0	0	0	0	0	0	0	0
17	1	0	0	0	0	0	0	0	0	1	0	0	0	0	0	0
18	0	0	0	0	0	0	0	0	0	1	0	0	0	0	0	0

(Continued)

Box 11.3c (Continued)

$\sum w_i$	5	5	4	7	6	6	7	4	5	6	6	6	6	4	3	3	
$\dfrac{j}{i}$	18	19	20	21	22	23	24	25	26	27	28	29	30	31	32	33	
19	0	0	0	0	0	0	0	0	0	0	0	0	0	0	0	0	
20	0	0	0	0	0	0	0	0	0	0	0	0	0	0	1	1	
21	0	0	0	0	0	0	1	0	0	1	0	1	0	0	1	0	
22	0	0	0	0	0	0	1	0	0	0	1	0	0	0	0	1	
23	0	0	0	1	1	0	0	0	1	0	1	0	0	1	0	0	
24	0	0	0	0	0	0	0	0	0	0	0	1	1	1	0	0	
25	0	0	0	0	0	0	0	1	1	0	0	1	1	1	1	0	
26	0	0	0	1	0	0	0	0	0	0	0	0	0	0	0	0	
27	1	0	0	0	0	0	0	0	0	0	0	0	0	0	0	0	
28	0	0	0	1	0	1	0	1	0	0	1	0	1	0	0	0	
29	0	0	0	0	0	1	1	0	1	0	0	0	0	0	0	0	
30	0	0	0	1	0	0	0	1	0	0	0	0	0	0	0	0	
31	0	0	1	1	1	0	1	1	0	0	0	0	0	0	0	0	
32	0	0	1	0	1	0	1	0	0	1	0	0	0	0	0	1	
33	0	0	0	0	1	0	0	0	0	0	0	0	0	0	1	0	
$\sum w_i$	4	4	4	4	7	4	5	6	4	4	5	3	6	5	7	6	164

Calculation of variance/covariance-like quantities.

i	z	$z-\bar z$	j	1	2	3	4	5	6	7	8
			$z-\bar z$	8.8	−4.5	−9.5	3.5	−11.3	2.9	−12.8	−4.4
1	34.5	8.8		77.5	−39.8	−83.4	31.1	−99.7	25.9	−113.1	−38.5
2	21.2	−4.5		−39.8	20.5	42.9	−16.0	51.2	−13.3	58.1	19.8

3	16.2	−9.5	−83.4	42.9	89.8	−33.5	107.3	−27.8	121.7	41.4
4	29.2	3.5	31.1	−16.0	−33.5	12.5	−40.0	10.4	−45.4	−15.5
5	14.4	−11.3	−99.7	51.2	107.3	−40.0	128.2	−33.2	145.4	49.5
6	28.6	2.9	25.9	−13.3	−27.8	10.4	−33.2	8.6	−37.7	−12.8
7	12.8	−12.8	−113.1	58.1	121.7	−45.4	145.4	−37.7	165.0	56.2
8	21.3	−4.4	−38.5	19.8	41.4	−15.5	49.5	−12.8	56.2	19.1
9	25.1	−0.6	−5.1	2.6	5.5	−2.1	6.6	−1.7	7.5	2.6
10	26.2	0.5	4.5	−2.3	−4.8	1.8	−5.7	1.5	−6.5	−2.2
11	33.9	8.3	72.7	−37.3	−78.2	29.2	−93.5	24.2	−106.0	−36.1
12	46.1	20.4	179.8	−92.4	−193.4	72.2	−231.2	59.9	−262.2	−89.3
13	28.1	2.4	21.0	−10.8	−22.6	8.4	−27.0	7.0	−30.6	−10.4
14	35.7	10.0	88.5	−45.4	−95.2	35.5	−113.8	29.5	−129.0	−43.9
15	15.6	−10.1	−88.9	45.7	95.6	−35.7	114.3	−29.6	129.6	44.1
16	16.1	−9.6	−84.7	43.5	91.1	−34.0	108.9	−28.2	123.5	42.1
17	18.6	−7.1	−62.7	32.2	67.5	−25.2	80.7	−20.9	91.5	31.2
18	23.2	−2.5	−21.9	11.2	23.5	−8.8	28.1	−7.3	31.9	10.9
19	39.0	13.3	116.9	−60.1	−125.8	46.9	−150.3	39.0	−170.5	−58.1
20	23.5	−2.2	−19.1	9.8	20.6	−7.7	24.6	−6.4	27.9	9.5
21	13.1	−12.6	−110.8	56.9	119.2	−44.5	142.5	−36.9	161.6	55.0
22	34.9	9.3	81.5	−41.9	−87.7	32.7	−104.8	27.2	−118.9	−40.5
23	31.0	5.4	47.2	−24.2	−50.7	18.9	−60.7	15.7	−68.8	−23.4
24	14.6	−11.1	−97.4	50.1	104.8	−39.1	125.3	−32.5	142.1	48.4
25	43.0	17.3	152.0	−78.1	−163.6	61.1	−195.5	50.7	−221.8	−75.5
26	20.4	−5.3	−46.9	24.1	50.4	−18.8	60.3	−15.6	68.4	23.3
27	9.6	−16.1	−142.1	73.0	152.9	−57.1	182.7	−47.4	207.2	70.6
28	33.3	7.6	67.3	−34.6	−72.4	27.0	−86.6	22.4	−98.2	−33.4
29	14.0	−11.7	−103.1	52.9	110.9	−41.4	132.5	−34.4	150.3	51.2

(Continued)

Box 11.3c (Continued)

$\sum w_i$	4	4	4	4							164

Calculation of variance/covariance-like quantities.

			j							
			1	2	3	4	5	6	7	8
i	z	z−z̄	8.8	−4.5	−9.5	3.5	−11.3	2.9	−12.8	−4.4
30	44.6	19.0	166.9	−85.8	−179.6	67.0	−214.7	55.7	−243.5	−82.9
31	33.2	7.5	66.1	−34.0	−71.1	26.5	−85.0	22.0	−96.4	−32.8
32	20.3	−5.3	−47.1	24.2	50.6	−18.9	60.5	−15.7	68.6	23.4
33	26.3	0.6	5.4	−2.8	−5.8	2.2	−7.0	1.8	−7.9	−2.7

			j							
			9	10	11	12	13	14	15	16
i	z	z−z̄	−0.6	0.5	8.3	20.4	2.4	10.0	−10.1	−9.6
1	34.5	8.8	−5.1	4.5	72.7	179.8	21.0	88.5	−88.9	−84.7
2	21.2	−4.5	2.6	−2.3	−37.3	−92.4	−10.8	−45.4	45.7	43.5
3	16.2	−9.5	5.5	−4.8	−78.2	−193.4	−22.6	−95.2	95.6	91.1
4	29.2	3.5	−2.1	1.8	29.2	72.2	8.4	35.5	−35.7	−34.0
5	14.4	−11.3	6.6	−5.7	−93.5	−231.2	−27.0	−113.8	114.3	108.9
6	28.6	2.9	−1.7	1.5	24.2	59.9	7.0	29.5	−29.6	−28.2
7	12.8	−12.8	7.5	−6.5	−106.0	−262.2	−30.6	−129.0	129.6	123.5
8	21.3	−4.4	2.6	−2.2	−36.1	−89.3	−10.4	−43.9	44.1	42.1
9	25.1	−0.6	0.3	−0.3	−4.8	−11.9	−1.4	−5.9	5.9	5.6
10	26.2	0.5	−0.3	0.3	4.2	10.3	1.2	5.1	−5.1	−4.9
11	33.9	8.3	−4.8	4.2	68.2	168.6	19.7	82.9	−83.3	−79.4
12	46.1	20.4	−11.9	10.3	168.6	416.8	48.7	205.1	−206.1	−196.3
13	28.1	2.4	−1.4	1.2	19.7	48.7	5.7	24.0	−24.1	−22.9
14	35.7	10.0	−5.9	5.1	82.9	205.1	24.0	100.9	−101.4	−96.6

	z	z − z̄		17	18	19	20	21	22	23	24
15	15.6	−10.1		5.9	−5.1	−83.3	−206.1	−24.1	−101.4	101.9	97.0
16	16.1	−9.6		5.6	−4.9	−79.4	−196.3	−22.9	−96.6	97.0	92.4
17	18.6	−7.1		4.2	−3.6	−58.8	−145.4	−17.0	−71.6	71.9	68.5
18	23.2	−2.5		1.5	−1.3	−20.5	−50.7	−5.9	−25.0	25.1	23.9
19	39.0	13.3		−7.8	6.7	109.6	271.0	31.7	133.4	−134.0	−127.6
20	23.5	−2.2		1.3	−1.1	−17.9	−44.4	−5.2	−21.8	21.9	20.9
21	13.1	−12.6		7.3	−6.4	−103.9	−256.9	−30.0	−126.4	127.0	121.0
22	34.9	9.3		−5.4	4.7	76.4	189.0	22.1	93.0	−93.4	−89.0
23	31.0	5.4		−3.1	2.7	44.2	109.4	12.8	53.8	−54.1	−51.5
24	14.6	−11.1		6.5	−5.6	−91.3	−225.9	−26.4	−111.1	111.7	106.4
25	43.0	17.3		−10.1	8.7	142.6	352.5	41.2	173.5	−174.3	−166.0
26	20.4	−5.3		3.1	−2.7	−44.0	−108.7	−12.7	−53.5	53.7	51.2
27	9.6	−16.1		9.4	−8.2	−133.2	−329.4	−38.5	−162.1	162.9	155.1
28	33.3	7.6		−4.5	3.9	63.1	156.1	18.2	76.8	−77.2	−73.5
29	14.0	−11.7		6.8	−5.9	−96.6	−238.9	−27.9	−117.6	118.1	112.5
30	44.6	19.0		−11.1	9.6	156.5	387.0	45.2	190.4	−191.3	−182.2
31	33.2	7.5		−4.4	3.8	62.0	153.2	17.9	75.4	−75.8	−72.2
32	20.3	−5.3		3.1	−2.7	−44.1	−109.1	−12.8	−53.7	53.9	51.4
33	26.3	0.6		−0.4	0.3	5.1	12.6	1.5	6.2	−6.2	−5.9
i	z	z − z̄	*j*	17	18	19	20	21	22	23	24
			z − z̄	−7.1	−2.5	13.3	−2.2	−12.6	9.3	5.4	−11.1
1	34.5	8.8		−62.7	−21.9	116.9	−19.1	−110.8	81.5	47.2	−97.4
2	21.2	−4.5		32.2	11.2	−60.1	9.8	56.9	−41.9	−24.2	50.1
3	16.2	−9.5		67.5	23.5	−125.8	20.6	119.2	−87.7	−50.7	104.8
4	29.2	3.5		−25.2	−8.8	46.9	−7.7	−44.5	32.7	18.9	−39.1

(Continued)

Box 11.3c (Continued)

i	z	z − z̄	17	18	19	20	21	22	23	24
j			−7.1	−2.5	13.3	−2.2	−12.6	9.3	5.4	−11.1
5	14.4	−11.3	80.7	28.1	−150.3	24.6	142.5	−104.8	−60.7	125.3
6	28.6	2.9	−20.9	−7.3	39.0	−6.4	−36.9	27.2	15.7	−32.5
7	12.8	−12.8	91.5	31.9	−170.5	27.9	161.6	−118.9	−68.8	142.1
8	21.3	−4.4	31.2	10.9	−58.1	9.5	55.0	−40.5	−23.4	48.4
9	25.1	−0.6	4.2	1.5	−7.8	1.3	7.3	−5.4	−3.1	6.5
10	26.2	0.5	−3.6	−1.3	6.7	−1.1	−6.4	4.7	2.7	−5.6
11	33.9	8.3	−58.8	−20.5	109.6	−17.9	−103.9	76.4	44.2	−91.3
12	46.1	20.4	−145.4	−50.7	271.0	−44.4	−256.9	189.0	109.4	−225.9
13	28.1	2.4	−17.0	−5.9	31.7	−5.2	−30.0	22.1	12.8	−26.4
14	35.7	10.0	−71.6	−25.0	133.4	−21.8	−126.4	93.0	53.8	−111.1
15	15.6	−10.1	71.9	25.1	−134.0	21.9	127.0	−93.4	−54.1	111.7
16	16.1	−9.6	68.5	23.9	−127.6	20.9	121.0	−89.0	−51.5	106.4
17	18.6	−7.1	50.7	17.7	−94.6	15.5	89.6	−65.9	−38.2	78.8
18	23.2	−2.5	17.7	6.2	−33.0	5.4	31.3	−23.0	−13.3	27.5
19	39.0	13.3	−94.6	−33.0	176.3	−28.9	−167.1	122.9	71.1	−146.9
20	23.5	−2.2	15.5	5.4	−28.9	4.7	27.4	−20.1	−11.6	24.1
21	13.1	−12.6	89.6	31.3	−167.1	27.4	158.4	−116.5	−67.4	139.2
22	34.9	9.3	−65.9	−23.0	122.9	−20.1	−116.5	85.7	49.6	−102.4
23	31.0	5.4	−38.2	−13.3	71.1	−11.6	−67.4	49.6	28.7	−59.3
24	14.6	−11.1	78.8	27.5	−146.9	24.1	139.2	−102.4	−59.3	122.4
25	43.0	17.3	−123.0	−42.9	229.2	−37.5	−217.3	159.8	92.5	−191.0
26	20.4	−5.3	37.9	13.2	−70.7	11.6	67.0	−49.3	−28.5	58.9
27	9.6	−16.1	114.9	40.1	−214.2	35.1	203.0	−149.3	−86.4	178.5

l	z	z−z̄	25	26	27	28	29	30	31	32
28	33.3	7.6	−54.5	−19.0	101.5	−16.6	−96.2	70.8	41.0	−84.6
29	14.0	−11.7	83.4	29.1	−155.4	25.4	147.3	−108.3	−62.7	129.5
30	44.6	19.0	−135.0	−47.1	251.7	−41.2	−238.5	175.5	101.5	−209.7
31	33.2	7.5	−53.5	−18.6	99.7	−16.3	−94.5	69.5	40.2	−83.0
32	20.3	−5.3	38.1	13.3	−70.9	11.6	67.2	−49.5	−28.6	59.1
33	26.3	0.6	−4.4	−1.5	8.2	−1.3	−7.8	5.7	3.3	−6.8

		j	25	26	27	28	29	30	31	32
l	z	z−z̄	17.3	−5.3	−16.1	7.6	−11.7	19.0	7.5	−5.3
1	34.5	8.8	152.0	−46.9	−142.1	67.3	−103.1	166.9	66.1	−47.1
2	21.2	−4.5	−78.1	24.1	73.0	−34.6	52.9	−85.8	−34.0	24.2
3	16.2	−9.5	−163.6	50.4	152.9	−72.4	110.9	−179.6	−71.1	50.6
4	29.2	3.5	61.1	−18.8	−57.1	27.0	−41.4	67.0	26.5	−18.9
5	14.4	−11.3	−195.5	60.3	182.7	−86.6	132.5	−214.7	−85.0	60.5
6	28.6	2.9	50.7	−15.6	−47.4	22.4	−34.4	55.7	22.0	−15.7
7	12.8	−12.8	−221.8	68.4	207.2	−98.2	150.3	−243.5	−96.4	68.6
8	21.3	−4.4	−75.5	23.3	70.6	−33.4	51.2	−82.9	−32.8	23.4
9	25.1	−0.6	−10.1	3.1	9.4	−4.5	6.8	−11.1	−4.4	3.1
10	26.2	0.5	8.7	−2.7	−8.2	3.9	−5.9	9.6	3.8	−2.7
11	33.9	8.3	142.6	−44.0	−133.2	63.1	−96.6	156.5	62.0	−44.1
12	46.1	20.4	352.5	−108.7	−329.4	156.1	−238.9	387.0	153.2	−109.1
13	28.1	2.4	41.2	−12.7	−38.5	18.2	−27.9	45.2	17.9	−12.8
14	35.7	10.0	173.5	−53.5	−162.1	76.8	−117.6	190.4	75.4	−53.7
15	15.6	−10.1	−174.3	53.7	162.9	−77.2	118.1	−191.3	−75.8	53.9
16	16.1	−9.6	−166.0	51.2	155.1	−73.5	112.5	−182.2	−72.2	51.4
17	18.6	−7.1	−123.0	37.9	114.9	−54.5	83.4	−135.0	−53.5	38.1

(Continued)

Box 11.3c (Continued)

			j							
		$z-\bar{z}$	17.3	-5.3	-16.1	7.6	-11.7	19.0	7.5	-5.3
i	z	$z-\bar{z}$	25	26	27	28	29	30	31	32
18	23.2	-2.5	-42.9	13.2	40.1	-19.0	29.1	-47.1	-18.6	13.3
19	39.0	13.3	229.2	-70.7	-214.2	101.5	-155.4	251.7	99.7	-70.9
20	23.5	-2.2	-37.5	11.6	35.1	-16.6	25.4	-41.2	-16.3	11.6
21	13.1	-12.6	-217.3	67.0	203.0	-96.2	147.3	-238.5	-94.5	67.2
22	34.9	9.3	159.8	-49.3	-149.3	70.8	-108.3	175.5	69.5	-49.5
23	31.0	5.4	92.5	-28.5	-86.4	41.0	-62.7	101.5	40.2	-28.6
24	14.6	-11.1	-191.0	58.9	178.5	-84.6	129.5	-209.7	-83.0	59.1
25	43.0	17.3	298.1	-91.9	-278.6	132.0	-202.1	327.3	129.6	-92.3
26	20.4	-5.3	-91.9	28.3	85.9	-40.7	62.3	-100.9	-40.0	28.5
27	9.6	-16.1	-278.6	85.9	260.3	-123.4	188.8	-305.8	-121.1	86.2
28	33.3	7.6	132.0	-40.7	-123.4	58.5	-89.5	144.9	57.4	-40.9
29	14.0	-11.7	-202.1	62.3	188.8	-89.5	137.0	-221.9	-87.9	62.5
30	44.6	19.0	327.3	-100.9	-305.8	144.9	-221.9	359.3	142.3	-101.3
31	33.2	7.5	129.6	-40.0	-121.1	57.4	-87.9	142.3	56.3	-40.1
32	20.3	-5.3	-92.3	28.5	86.2	-40.9	62.5	-101.3	-40.1	28.6
33	26.3	0.6	10.6	-3.3	-9.9	4.7	-7.2	11.7	4.6	-3.3

		j			Row sum
		33			$w_{ij}(z_i-\bar{z})(z_j-\bar{z})$
	$z-\bar{z}$	0.6			
i	z	$z-\bar{z}$	33	$(z_i-\bar{z})^2$	
1	34.5	8.8	5.4	77.5	9.78
2	21.2	-4.5	-2.8	20.5	-31.35
3	16.2	-9.5	-5.8	89.8	36.72
4	29.2	3.5	2.2	12.5	-40.44

5	14.4	−11.3	−7.0	128.2	−188.73
6	28.6	2.9	1.8	8.6	29.67
7	12.8	−12.8	−7.9	165.0	−938.92
8	21.3	−4.4	−2.7	19.1	108.63
9	25.1	−0.6	−0.4	0.3	8.05
10	26.2	0.5	0.3	0.3	6.59
11	33.9	8.3	5.1	68.2	244.27
12	46.1	20.4	12.6	416.8	1106.69
13	28.1	2.4	1.5	5.7	−55.32
14	35.7	10.0	6.2	100.9	403.02
15	15.6	−10.1	−6.2	101.9	87.77
16	16.1	−9.6	−5.9	92.4	57.60
17	18.6	−7.1	−4.4	50.7	93.76
18	23.2	−2.5	−1.5	6.2	53.29
19	39.0	13.3	8.2	176.3	272.88
20	23.5	−2.2	−1.3	4.7	−2.60
21	13.1	−12.6	−7.8	158.4	556.78
22	34.9	9.3	5.7	85.7	−339.58
23	31.0	5.4	3.3	28.7	126.05
24	14.6	−11.1	−6.8	122.4	273.83
25	43.0	17.3	10.6	298.1	1012.08
26	20.4	−5.3	−3.3	28.3	−127.58
27	9.6	−16.1	−9.9	260.3	290.82
28	33.3	7.6	4.7	58.5	71.88
29	14.0	−11.7	−7.2	137.0	327.96
30	44.6	19.0	11.7	359.3	873.81

(Continued)

Box 11.3c (Continued)

i	z	j z − $\bar{z}$	33 0.6	$(z_i - \bar{z})^2$	Row sum $w_{ij}(z_i - \bar{z})(z_j - \bar{z})$
31	33.2	7.5	4.6	56.3	322.09
32	20.3	−5.3	−3.3	28.6	158.70
33	26.3	0.6	0.4	0.4	−2.85

$$\sum_1^{33}(z_i - \bar{z})^2 = 3167.58 \qquad \sum_{i=1}^{33}\sum_{j=1}^{33} w_{ij}(z_i - \bar{z})(z - \bar{z}) = 4805.35$$

Moran's I

$$I = \frac{1}{p} \frac{\sum_{i=1}^{n}\sum_{j=1,j\neq i}^{n} w_{ij}(z_i - \bar{z})(z_j - \bar{z})}{\sum_{i=1}^{n}(z_i - \bar{z})^2} \qquad \frac{33(4805.35)}{164(3167.58)} = 0.305$$

Z test statistic of Moran's I

$$Z = (i_O - \mu_e)/\sigma_e \qquad (0.305 - (-0.0297))/0.1045 = 2.63$$

Randomisation applied with 9999 permutations

Probability $\qquad p = \qquad$ 0.004

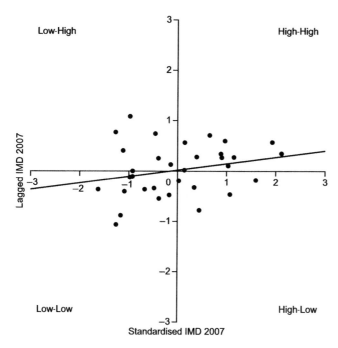

Figure 11.6 Moran's *I* scatter plot.

A useful way of visualising the extent of spatial autocorrelation is by means of a Moran's *I* scatter plot (MSP) (Anselin, 1995). Figure 11.6 plots the standardised and spatially lagged values of the 2007 IMD for the 33 London Boroughs. The MSP is divided into four segments centred on the means of the two variables that can be summarised as follows:

- *Low–Low:* spatial units where standardised and lagged values are low.
- *Low–High:* spatial units where standardised value is low and lagged value is high.
- *High–Low:* spatial units where standardised value is high and lagged value is low.
- *High–High:* spatial units where lagged and standardised values are high.

The terms High and Low in this categorisation simply refer to whether a point is higher or lower than the average (zero) at the centre of the graph. The High–High and Low–Low quadrants contain features displaying positive spatial autocorrelation, and those in the Low–High and High–Low quadrants have negative spatial autocorrelation. In the Low–Low quadrant of Figure 11.6, the point with lowest combination of values (−1.24 for standardised IMD and −1.08 for lagged IMD) is Kingston upon Thames. The map in Box 11.3a shows this borough to have an IMD value of 13.1 and its four contiguous neighbours (Richmond upon Thames (9.6), Merton (14.6), Sutton (14.0) and Wandsworth (20.3)) also have comparatively low values. Although Richmond upon Thames has the lowest IMD value of all the boroughs, two of its neighbours have comparatively high values (Hounslow at 23.2 and Hammersmith and Fulham at 28.1) and the influence of these outweighs its contiguity with Kingston upon Thames and Wandsworth. At the other end of the scale in the High–High quadrant are Newham and Tower Hamlets, respectively, with standardised and spatially lagged IMD values of −1.92 and 0.59, and 2.10 and 0.37.

> What are the 'raw' IMD values of Newham's six adjacent boroughs and what are the values for the six contiguous boroughs of Tower Hamlets? Note: the ID numbers of Newham and Tower Hamlets are 25 and 30, respectively. You will need to refer to the map in Box 11.3a.

11.3.3 Significance Testing and the Moran's *I* Index

There are two approaches to testing the significance of Moran's *I* and similar global spatial statistics such as **Geary's *C*** and the Getis and Ord's ***G* statistic** (see 11.3.5). One approach, following the standard assumption of non-spatial statistics, is to assume that the calculated statistic or index value is from a normal distribution of such quantities computed from a series of independent and identical samples. The notion of sampling with respect to spatial features has already been discussed. The second approach is rather more empirical in nature and views the set of data values that has arisen as just one of all the possible random distributions of observed data values across the set of zones or points. The number of random distributions equals the factorial of the number spatial features, for example if there are four areas the number of permutations is 24 ($4 \times 3 \times 2 \times 1$), whereas there are 8,683,317,618,811,890,000,000,000,000,000,000 (33!) ways, in which the 33 values of the 2007 IMD used in the analysis shown in Box 11.3a could be arranged across the London Boroughs.

Both methods are implemented in various software packages that carry out spatial statistics, and in the case of the randomisation approach, users are normally asked to specify the number of random permutations to be generated and the spatial index is then calculated for each of these to produce a pseudo probability distribution. The index computed from the observed data values is compared with this distribution. Both approaches to testing the significance of spatial indices involve converting the observed index value into a *Z* score in the standard way using the hypothesised or empirically derived population mean and standard deviation. For example, in the randomisation approach, the mean and standard deviation of the index in the pseudo probability distribution are used. Box 11.3a includes the results of testing the significance of the observed Moran's I value, 0.305, using one of these statistical packages. The probability of the Moran's *I* obtained from the data values can be compared with a significance level, which relates to the number of permutations used to generate the reference distribution; for example, 99 and 999 are associated with the 0.01 and 0.001 significance levels, respectively.

11.3.4 Moran's *I* Index with Point Features

The Moran's *I* Index can also be applied to point features in which case spatial neighbourliness is not signified by the adjacency of polygons but by the distance between a point and its nearest neighbour. There is another adaptation that uses distance bands in a comparable manner to Ripley's *K* function. Box 11.4a shows the Moran's *I* Index applied to the section of morainic debris on the outwash plain below Les Bossons Glacier in the French Alps. We have already discovered that the 100 items of debris are not distributed in a completely spatially random pattern but exhibit some clustering. The objective here is to discover if the lengths provide some evidence of global spatial autocorrelation. The result ($I = 0.411$) suggests this is indeed the case and the Moran's *I* scatterplot in Box 11.4d shows substantial number of the debris items classified in the High–High and the Low–Low quadrants indicating the presence of positive spatial autocorrelation.

Box 11.4a   Moran's *I* index with point data.

Moran's *I*: $I = \dfrac{1}{p} \dfrac{\sum\limits_{i=1}^{n}\sum\limits_{j=1,j\neq i}^{n} w_{i,j}(z_i - \bar{z})(z_j - \bar{z})}{\sum\limits_{i=1}^{n}(z_i - \bar{z})^2}$

where $p = \sum\limits_{i}\sum\limits_{j} w_{ij}/n$

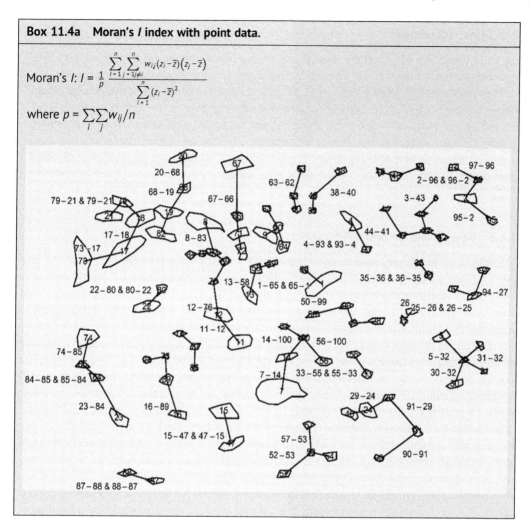

Box 11.4b   Application of Moran's *I* index.

The adaptability of the Moran's *I* index is shown by its application to a point dataset representing the centroids of the 100 items of morainic debris in the section of the outwash plain in front of Les Bossons Glacier in the French Alps. The aim is to determine whether there is spatial autocorrelation in the distribution of the length of the long axis of the debris *Z* across the area. The subscripts *i* and *j* distinguish between the different observations in a pair, which have been identified by a simple numerical identifier (1–100). Nearest neighbour weights $w_{i,j}$ have been used in this application, and the illustration in Box 11.4a shows the nearest connections between each pair. In some cases, there is a reciprocal nearest neighbour connection, such as with debris items 87 and 88 towards the lower left corner, where they are separated from the remainder. In other instances, there are small groups of three or four items where two are connected reciprocally and there is another item whose nearest neighbour is one of the former pair.

The upper part of Box 11.4c tabulates the variable *Z* (length of debris item in cm) alongside its identifier, and the lower part lists the nearest neighbour using its identifier of all 100 debris

(Continued)

Box 11.4b   (Continued)

items. Unlike in Box 11.3c, details of the procedure for computing the variance/covariance 'like' quantities are not shown in the interest of simplifying the presentation.

The mean length of debris items was 15.17 cm and most had either one or two nearest neighbours, although a few debris item 32 was the nearest neighbour of four others numbers 5, 30, 31 and 34. The Moran's *I* index in this application is 0.411, which indicates a moderately strong global spatial autocorrelation. The randomisation significance testing procedure has been applied and indicates that index values is significant at the 0.05 level ($p = 0.0029$) with 9999 permutations used to produce the reference distribution.

11.3.5   Geary *C* and Getis-Ord *G* Statistics

Two alternative techniques used to investigate global spatial autocorrelation, both of which also have localised variants, are Geary *C* (Geary, 1954) and Getis-Ord General *G* (Getis and Ord, 1992). The main difference between Moran's *I* and **Geary's *C*** is that the former uses standardised spatial covariance and the latter the sum of the squared differences between pairs of data values to measure global spatial autocorrelation. While Moran's *I* index lies between -1.0 and $+1.0$, Geary's *C* usually ranges between 0 and 2 with values above and below 1, respectively, indicating negative and positive spatial autocorrelation: a value around 1 (the expected value of *C*) indicates the data value in any one feature is unrelated those in any of the others. Statistical testing with a Null Hypothesis of complete spatial randomness can be carried out by computing a *Z* score and determining its probability, accepting or rejecting H_0 by reference to the chosen level of significance.

The **Getis-Ord General *G* statistic** is useful when the overall distribution of data values is relatively undifferentiated and you are seeking unforeseen spatial peaks of high values or dips of low values, although either outcome can be negated if both types of unexpected values occur when they can cancel each other out. Getis-Ord *G* can be tested for its significance like other inferential statistics under a Null Hypothesis of complete spatial randomness. A *Z* score is computed from the *G* statistic and its probability is determined. The interpretation of the *Z* score differs according to whether it is positive or negative: a positive score signifies that the observed *G* statistic is larger than expected and there are high values clustered in the study area; a negative *Z* score indicates the reverse that there is a clustering of low values. Importantly, the Getis-Ord *G* statistic does not contextualise the data value of any individual feature with reference to its neighbours; in some respects, it is like the difference between quantities measuring central tendency and dispersion for non-spatial data. Two sets of data values can produce identical means, but their variances and standard deviations might be quite different. A large variance and standard deviation would indicate that the values either were spread out evenly along the range or were bimodal and grouped at both ends, whereas a small variance and standard deviation denotes all the values were tightly packed around the mean. The Getis-Ord *G* statistic indicates if there is a high or low peak of values in a set that are fairly similar.

Identify some examples of geographical features that might display positive or negative global spatial autocorrelation. A useful starting point might be to think of landscapes or terrain that are generally flat or gently sloping that have occasional high and low features across their surfaces. For instance, perhaps there are some formerly glaciated terrains where moraine has been deposited as drumlins.

Box 11.4c Calculation of Moran's *I* with point data.

Debris ID No.	z		Contd … Debris ID No.	z		Contd … Debris ID No.	z		Contd … Debris ID No.	z
1	40		26	11		51	10		76	9
2	22		27	15		52	14		77	10
3	4		28	16		53	7		78	13
4	24		29	12		54	14		79	21
5	17		30	16		55	13		80	11
6	10		31	7		56	17		81	10
7	47		32	10		57	11		82	30
8	36		33	11		58	12		83	14
9	23		34	9		59	12		84	15
10	11		35	8		60	9		85	13
11	22		36	8		61	8		86	5
12	26		37	12		62	8		87	17
13	20		38	11		63	7		88	18
14	21		39	5		64	14		89	14
15	28		40	6		65	7		90	9
16	17		41	9		66	12		91	11
17	46		42	12		67	30		92	8
18	29		43	10		68	15		93	11
19	28		44	10		69	9		94	11
20	27		45	7		70	7		95	12

(Continued)

Box 11.4c (Continued)

Debris ID No.	z	Contd ... Debris ID No.	z	Contd ... Debris ID No.	z	Contd ... Debris ID No.	z
21	26	46	1	71	20	96	9
22	27	47	25	72	15	97	10
23	22	48	17	73	39	98	9
24	19	49	8	74	20	99	13
25	9	50	12	75	7	100	9
				Mean		$\sum \frac{z}{n} = 1517/100 = 15.17$	

Nearest neighbours between each i entity and its nearest neighbour j.

i	j	i	j	i	j	i	j
1	65	21	79	41	37	61	10
2	96	22	80	42	37	62	60
3	43	23	84	43	37	63	62
4	93	24	48	44	41	64	10
5	32	25	26	45	46	65	1
6	27	26	25	46	45	66	71
7	14	27	6	47	15	67	66
8	83	28	49	48	24	68	19
9	10	29	24	49	99	69	83
10	9	30	32	50	99	70	83
11	12	31	32	51	100	71	66
12	76	32	34	52	53	72	71
13	58	33	55	53	54	73	17

i	j
81	70
82	19
83	70
84	85
85	84
86	77
87	88
88	87
89	75
90	91
91	29
92	75
93	4

14	100	34	32	54	53	74	85	94	27
15	47	35	36	55	33	75	92	95	2
16	89	36	35	56	100	76	69	96	2
17	18	37	43	57	53	77	78	97	96
18	79	38	40	58	59	78	77	98	46
19	82	39	39	59	58	79	21	99	49
20	68	40	37	60	62	80	22	100	51

Sum of nearest weights for each i entity.

i	w_i	i	w_i	i	w_i	i	w_i	i	w_i
1	1	21	1	41	2	61	1	81	1
2	2	22	1	42	1	62	3	82	1
3	1	23	1	43	2	63	1	83	3
4	1	24	2	44	1	64	1	84	2
5	1	25	1	45	1	65	1	85	2
6	1	26	1	46	2	66	2	86	1
7	1	27	2	47	1	67	1	87	1
8	1	28	1	48	1	68	2	88	2
9	2	29	2	49	2	69	2	89	2
10	5	30	1	50	1	70	2	90	1
11	1	31	1	51	1	71	2	91	2
12	3	32	4	52	1	72	1	92	1
13	1	33	1	53	3	73	1	93	1
14	2	34	1	54	1	74	1	94	1
15	1	35	1	55	1	75	2	95	1
16	1	36	1	56	1	76	3	96	2

(Continued)

Box 11.4c (Continued)

Sum of nearest weights for each i entity.

i	w_i	i	w_i	i	w_i	i	w_i	i	w_i
17	2	37	3	57	1	77	2	97	1
18	2	38	1	58	2	78	1	98	1
19	2	39	1	59	1	79	2	99	2
20	1	40	2	60	1	80	1	100	3
								$\sum w_i = 151$	

Calculation of variance/covariance-like quantities (see Box 11.3c for example of computation).

$$\sum_{1}^{100} (z_i - \bar{z})'' = 4794.21 \qquad\qquad \sum_{i=1}^{100}\sum_{j=1}^{100} w_{ij}(z_i - \bar{z})(z_j - \bar{z}) = 4797.21$$

Moran's I

$$\frac{100(4797.21)}{151(7737.00)} = 0.411$$

Z test statistic of Moran's I

$Z = (i_O - \mu_e)/\sigma_e$

Randomisation applied with 9999 permutations

$$\frac{0.411 - (-0.0094)}{0.1230} = 3.418$$

Probability

$p = \qquad\qquad 0.0029$

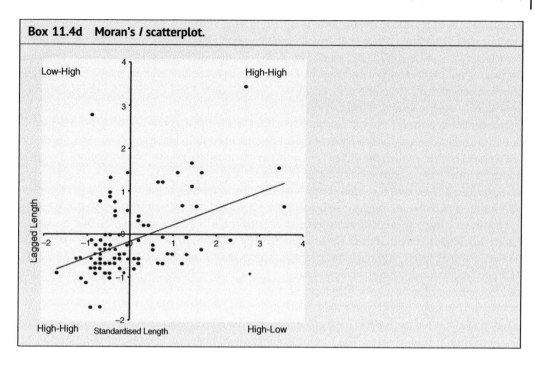

Box 11.4d   Moran's *I* scatterplot.

11.4   Local Measures of Spatial Association (LISA)

Moran's *I* index is a global measurement of the extent of spatial autocorrelation across a study area and measures the degree to which features that are spatially proximate have similar values. However, we do not yet know where, in respect of Boxes 11.3a and 11.4a, the clusters of London Boroughs or morainic debris with high and low values are located. Such local 'pockets' of positive and negative spatial autocorrelation can, at least partially, cancel each other out and lead to a deflated global index in comparison with what might have been obtained had the study area been divided into sub-areas and separate indices computed for these. Put simply, the amount of spatial autocorrelation can vary in different parts of a study area. Such variability in the distribution of spatial autocorrelation can be quantified by **Local Indicators of Spatial Association (LISA)**, and these focus on the extent to which features that are close together have similar values. The last part of the covariance/variance computation tabulation in Box 11.3c showed the row sums (*S*) obtained by summing the non-diagonal elements that were shown **bordered by a solid line** (i.e. the spatial features are joined). If each of these *S* values are standardised by dividing by the sum of the squared deviations along the diagonal elements (the variance like quantity) and the results are multiplied by the total number of spatial units (*n*), the figures obtained represent the local contribution of each row (feature) to the global spatial autocorrelation. These provide another way of computing the overall Moran's *I* index, since the sum of these local components divided by the total number of joins equals the global Moran's *I* computed from the definitional equation given in Box 11.3a:

$$I = 50.06/164 = 0.305$$

These local components are the LISA values, which can be mapped and tested for significant differences between areas. Rather than using the 'raw' LISA values, they are often standardised by dividing by the number of joins possessed by each feature (row): for example, the raw (unstandardised) LISA values of a feature with four joins is divided by 4. When this has been done for each row in the matrix, the results are in the form of a local average. Again, the row sums (*S*) can be divided by sum of the squared deviations along the diagonal and then multiplied by $n-1$ rather than n to produced standardised LISA values. The two alternative techniques of global spatial autocorrelation outlined above (Geary C and Getis-Ord General G) also have Local Indicator of Spatial Association variants. The local Getis-Ord G identifies the extent to which high and low values cluster together and is therefore often considered a form of hot spot analysis. Local Moran's I and Geary C identify statistically positive and negative spatial autocorrelation.

The calculation of these LISA values is shown in Box 11.5a with respect to the 2007 IMD mean score for the 33 London Boroughs. The effect of these adjustments is to increase global Moran's I to 0.331 (0.305 previously), and when the City of London area is excluded (see above for justification), Moran's I index becomes 0.663 compared with 0.479. The results of LISA calculation can be visualised in several ways. Box 11.5a shows the significance and cluster maps relating to the application of Moran's I LISA for the 33 London Boroughs together with the choropleth for comparison. Each LISA is tested for significance using a randomisation process to generate a reference distribution and areas are shaded according to whether their LISA is significantly different from what would be expected at the 0.05 and 0.01 levels. Although the choropleth map shows the spatial distribution of the 2007 IMD and indicates with high and low values, it has a cluttered appearance in comparison with the LISA and significance maps which focus attention on where local spatial autocorrelation is most important. The cluster maps include those areas that are significant in the four quadrants of the MSP. Taken together, these maps allow the significant combinations of positive and negative local spatial autocorrelation to be discovered. The London Boroughs used in the analyses included in Boxes 11.3a and 11.5a are in some respects large and limited in number; however, they have been used to keep the complexity of the calculations to a manageable scale. The analysis would more appropriately be carried out for smaller spatial units such as local authority wards or Middle Super Output Areas (MSOAs) across the whole of London. The scale of these units is better suited to reflecting local variations, which can easily become lost for larger areas. Nevertheless, even at the borough scale, there is some evidence of local spatial autocorrelation in certain parts of the Greater London Authority's area.

Box 11.5a   Moran's *I* Local Indicator of Spatial Association.

Local Moran's *I* Indicator of Spatial Association: $I_i = \dfrac{\sum\limits_{j=1}^{n} w_j (z_i - \bar{z})(z_j - \bar{z})}{s_z^2 \sum\limits_{j=1}^{n} w_{ij}}$

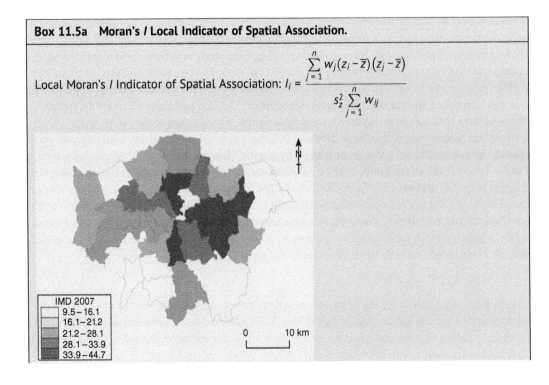

IMD 2007
- 9.5 – 16.1
- 16.1 – 21.2
- 21.2 – 28.1
- 28.1 – 33.9
- 33.9 – 44.7

0   10 km

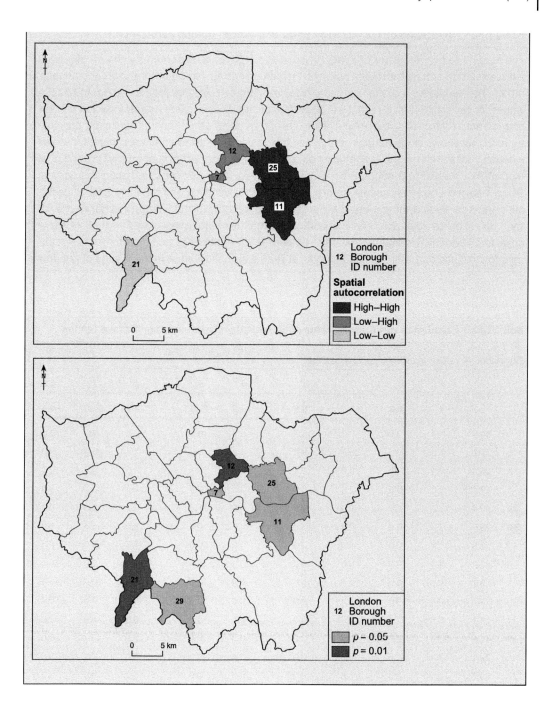

Box 11.5b   Application of the Moran's I Local Indicator of Spatial Association.

The Moran's I Local Indicator of Spatial Association focuses attention on the extent of spatial autocorrelation around individual features, including centroids as point locations for polygons, rather than providing a global summary measure. The calculations produce a Moran's I LISA values for each spatial feature that can be tested for significance using a randomisation process to generate a reference distribution. The results of the analysis are shown as two maps: the cluster map that shows combinations of high and low indices, and the significance map that shows whether a local index value is significant as certain levels. The results support the earlier global Moran's I with a pair of contiguous boroughs in south-west London (Kingston upon Thames and Sutton) having Low–Low spatial autocorrelation and Greenwich and Newham east of centre with the High–High combination. There are two boroughs, the City of London and Hackney, with the Low–High combination. The probability of the index values associated with all of these areas is ≤ 0.03, which is smaller than the 'standard' 0.05 significance, and therefore, it is reasonable to conclude that there is significant local spatial autocorrelation in these parts of London.

Box 11.5c   Calculation of Local Indicator of Spatial Association and significance testing.

Calculation of variance/covariance-like quantities.

		j	1	2	3	4	5	6	7	8
i	z	$z-\bar{z}$	8.8	−4.5	−9.5	3.5	−11.3	2.9	−12.8	−4.4
1	34.5	8.8	77.5	0.0	−16.7	0.0	0.0	0.0	0.0	0.0
2	21.2	−4.5	0.0	20.5	0.0	−3.2	0.0	−2.7	0.0	0.0
3	16.2	−9.5	−20.9	0.0	89.8	0.0	26.8	0.0	0.0	0.0
4	29.2	3.5	0.0	−2.3	0.0	12.5	0.0	1.5	0.0	0.0
5	14.4	−11.3	0.0	0.0	17.9	0.0	128.2	0.0	0.0	8.3
6	28.6	2.9	0.0	−2.2	0.0	1.7	0.0	8.6	−6.3	0.0
7	12.8	−12.8	0.0	0.0	0.0	0.0	0.0	−5.4	165.0	0.0
8	21.3	−4.4	0.0	0.0	0.0	0.0	12.4	0.0	0.0	19.1
9	25.1	−0.6	0.0	0.0	0.0	−0.4	0.0	0.0	0.0	0.0
10	26.2	0.5	0.0	−0.8	0.0	0.0	0.0	0.0	0.0	0.0
11	33.9	8.3	12.1	0.0	−13.0	0.0	−15.6	0.0	0.0	0.0
12	46.1	20.4	0.0	0.0	0.0	0.0	0.0	0.0	−43.7	0.0
13	28.1	2.4	0.0	0.0	0.0	1.4	0.0	0.0	0.0	0.0
14	35.7	10.0	0.0	−7.6	0.0	0.0	0.0	4.9	0.0	0.0
15	15.6	−10.1	0.0	11.4	0.0	−8.9	0.0	0.0	0.0	0.0
16	16.1	−9.6	−28.2	0.0	30.4	0.0	0.0	0.0	0.0	0.0
17	18.6	−7.1	0.0	0.0	0.0	0.0	0.0	0.0	0.0	0.0
18	23.2	−2.5	0.0	0.0	0.0	0.0	0.0	0.0	0.0	0.0
19	39.0	13.3	0.0	0.0	0.0	0.0	0.0	9.7	−42.6	0.0
20	23.5	−2.2	0.0	0.0	0.0	−1.9	0.0	0.0	0.0	0.0
21	13.1	−12.6	0.0	0.0	0.0	0.0	0.0	0.0	0.0	0.0

Calculation of variance/covariance-like quantities.

		j	1	2	3	4	5		6	7	8
i	*z*	$z-\bar{z}$	8.8	-4.5	-9.5	3.5	-11.3		2.9	-12.8	-4.4
22	34.9	9.3	0.0	0.0	0.0	0.0	-15.0		0.0	-17.0	-5.8
23	31.0	5.4	0.0	0.0	0.0	0.0	-15.2		0.0	0.0	0.0
24	14.6	-11.1	0.0	0.0	0.0	0.0	0.0		0.0	0.0	9.7
25	43.0	17.3	25.3	0.0	0.0	0.0	0.0		0.0	0.0	0.0
26	20.4	-5.3	-11.7	0.0	0.0	0.0	0.0		0.0	0.0	0.0
27	9.6	-16.1	0.0	0.0	0.0	0.0	0.0		0.0	0.0	0.0
28	33.3	7.6	0.0	0.0	0.0	0.0	-17.3		0.0	-19.6	0.0
29	14.0	-11.7	0.0	0.0	0.0	0.0	0.0		0.0	0.0	17.1
30	44.6	19.0	0.0	0.0	0.0	0.0	0.0		0.0	-40.6	0.0
31	33.2	7.5	0.0	0.0	0.0	0.0	0.0		0.0	0.0	0.0
32	20.3	-5.3	0.0	0.0	0.0	0.0	0.0		0.0	0.0	0.0
33	26.3	0.6	0.0	0.0	0.0	0.4	0.0		0.3	-1.3	0.0

		j	9	10	11	12	13	14	15	16
i	*z*	$z-\bar{z}$	-0.6	0.5	8.3	20.4	2.4	10.0	-10.1	-9.6
1	34.5	8.8	0.0	0.0	14.5	0.0	0.0	0.0	0.0	-16.9
2	21.2	-4.5	0.0	-0.5	0.0	0.0	0.0	-9.1	9.1	0.0
3	16.2	-9.5	0.0	0.0	-19.6	0.0	0.0	0.0	0.0	22.8
4	29.2	3.5	-0.3	0.0	0.0	0.0	1.2	0.0	-5.1	0.0
5	14.4	-11.3	0.0	0.0	-15.6	0.0	0.0	0.0	0.0	0.0
6	28.6	2.9	0.0	0.0	0.0	0.0	0.0	4.9	0.0	0.0
7	12.8	-12.8	0.0	0.0	0.0	-37.5	0.0	0.0	0.0	0.0
8	21.3	-4.4	0.0	0.0	0.0	0.0	0.0	0.0	0.0	0.0
9	25.1	-0.6	0.3	0.0	0.0	0.0	-0.3	0.0	1.2	0.0
10	26.2	0.5	0.0	0.3	0.0	0.0	0.0	1.7	0.0	0.0
11	33.9	8.3	0.0	0.0	68.2	0.0	0.0	0.0	0.0	0.0
12	46.1	20.4	0.0	0.0	0.0	416.8	0.0	34.2	0.0	0.0
13	28.1	2.4	-0.2	0.0	0.0	0.0	5.7	0.0	0.0	0.0
14	35.7	10.0	0.0	0.8	0.0	34.2	0.0	100.9	0.0	0.0
15	15.6	-10.1	1.5	0.0	0.0	0.0	0.0	0.0	101.9	0,0
16	16.1	-9.6	0.0	0.0	0.0	0.0	0.0	0.0	0.0	92.4
17	18.6	-7.1	1.4	0.0	0.0	0.0	0.0	0.0	24.0	0.0
18	23.2	-2.5	0.4	0.0	0.0	0.0	-1.5	0.0	0.0	0.0
19	39.0	13.3	0.0	0.0	0.0	67.8	0.0	33.3	0.0	0.0
20	23.5	-2.2	0.0	0.0	0.0	0.0	-1.3	0.0	0.0	0.0
21	13.1	-12.6	0.0	0.0	0.0	0.0	0.0	0.0	0.0	0.0
22	34.9	9.3	0.0	0.0	0.0	0.0	0.0	0.0	0.0	0.0
23	31.0	5.4	0.0	0.0	11.1	0.0	0.0	0.0	0.0	0.0
24	14.6	-11.1	0.0	0.0	0.0	0.0	0.0	0.0	0.0	0.0
25	43.0	17.3	0.0	0.0	23.8	58.8	0.0	0.0	0.0	0.0

(Continued)

Box 11.5c (Continued)

		j	9	10	11	12	13	14	15	16
i	z	$z-\bar{z}$	-0.6	0.5	8.3	20.4	2.4	10.0	-10.1	-9.6
26	20.4	-5.3	0.0	0.0	0.0	0.0	0.0	0.0	0.0	12.8
27	9.6	-16.1	0.0	0.0	0.0	0.0	-9.6	0.0	0.0	0.0
28	33.3	7.6	0.0	0.0	0.0	0.0	0.0	0.0	0.0	0.0
29	14.0	-11.7	0.0	0.0	0.0	0.0	0.0	0.0	0.0	0.0
30	44.6	19.0	0.0	0.0	26.1	64.5	0.0	0.0	0.0	0.0
31	33.2	7.5	0.0	0.8	0.0	30.6	0.0	15.1	0.0	0.0
32	20.3	-5.3	0.0	0.0	0.0	0.0	-1.8	0.0	0.0	0.0
33	26.3	0.6	0.0	0.0	0.0	0.0	0.0	0.0	0.0	0.0

		j	17	18	19	20	21	22	23	24
i	z	$z-\bar{z}$	-7.1	-2.5	13.3	-2.2	-12.6	9.3	5.4	-11.1
1	34.5	8.8	0.0	0.0	0.0	0.0	0.0	0.0	0.0	0.0
2	21.2	-4.5	0.0	0.0	0.0	0.0	0.0	0.0	0.0	0.0
3	16.2	-9.5	0.0	0.0	0.0	0.0	0.0	0.0	0.0	0.0
4	29.2	3.5	0.0	0.0	0.0	-1.1	0.0	0.0	0.0	0.0
5	14.4	-11.3	0.0	0.0	0.0	0.0	0.0	-17.5	-10.1	0.0
6	28.6	2.9	0.0	0.0	6.5	0.0	0.0	0.0	0.0	0.0
7	12.8	-12.8	0.0	0.0	-24.4	0.0	0.0	-17.0	0.0	0.0
8	21.3	-4.4	0.0	0.0	0.0	0.0	0.0	-10.1	0.0	12.1
9	25.1	-0.6	0.8	0.3	0.0	0.0	0.0	0.0	0.0	0.0
10	26.2	0.5	0.0	0.0	0.0	0.0	0.0	0.0	0.0	0.0
11	33.9	8.3	0.0	0.0	0.0	0.0	0.0	0.0	7.4	0.0
12	46.1	20.4	0.0	0.0	45.2	0.0	0.0	0.0	0.0	0.0
13	28.1	2.4	0.0	-1.0	0.0	-0.9	0.0	0.0	0.0	0.0
14	35.7	10.0	0.0	0.0	22.2	0.0	0.0	0.0	0.0	0.0
15	15.6	-10.1	18.0	0.0	0.0	0.0	0.0	0.0	0.0	0.0
16	16.1	-9.6	0.0	0.0	0.0	0.0	0.0	0.0	0.0	0.0
17	18.6	-7.1	50.7	4.4	0.0	0.0	0.0	0.0	0.0	0.0
18	23.2	-2.5	4.4	6.2	0.0	0.0	0.0	0.0	0.0	0.0
19	39.0	13.3	0.0	0.0	176.3	0.0	0.0	0.0	0.0	0.0
20	23.5	-2.2	0.0	0.0	0.0	4.7	0.0	0.0	0.0	0.0
21	13.1	-12.6	0.0	0.0	0.0	0.0	158.4	0.0	0.0	34.8
22	34.9	9.3	0.0	0.0	0.0	0.0	0.0	85.7	0.0	-14.6
23	31.0	5.4	0.0	0.0	0.0	0.0	0.0	0.0	28.7	0.0
24	14.6	-11.1	0.0	0.0	0.0	0.0	27.8	-20.5	0.0	122.4
25	43.0	17.3	0.0	0.0	0.0	0.0	0.0	0.0	0.0	0.0
26	20.4	-5.3	0.0	0.0	0.0	0.0	0.0	0.0	0.0	0.0
27	9.6	-16.1	0.0	10.0	0.0	0.0	50.8	0.0	0.0	0.0

j			17	18	19	20	21	22	23	24
i	z	$z-\bar{z}$	-7.1	-2.5	13.3	-2.2	-12.6	9.3	5.4	-11.1
28	33.3	7.6	0.0	0.0	0.0	0.0	0.0	14.2	8.2	0.0
29	14.0	-11.7	0.0	0.0	0.0	0.0	49.1	0.0	0.0	43.2
30	44.6	19.0	0.0	0.0	0.0	0.0	0.0	0.0	16.9	0.0
31	33.2	7.5	0.0	0.0	0.0	0.0	0.0	0.0	0.0	0.0
32	20.3	-5.3	0.0	0.0	0.0	1.7	9.6	-7.1	0.0	8.4
33	26.3	0.6	0.0	0.0	0.0	-0.2	0.0	1.0	0.0	0.0

j			25	26	27	28	29	30	31	32
i	z	$z-\bar{z}$	17.3	-5.3	-16.1	7.6	-11.7	19.0	7.5	-5.3
1	34.5	8.8	30.4	-9.4	0.0	0.0	0.0	0.0	0.0	0.0
2	21.2	-4.5	0.0	0.0	0.0	0.0	0.0	0.0	0.0	0.0
3	16.2	-9.5	0.0	0.0	0.0	0.0	0.0	0.0	0.0	0.0
4	29.2	3.5	0.0	0.0	0.0	0.0	0.0	0.0	0.0	0.0
5	14.4	-11.3	0.0	0.0	0.0	-14.4	0.0	0.0	0.0	0.0
6	28.6	2.9	0.0	0.0	0.0	0.0	0.0	0.0	0.0	0.0
7	12.8	-12.8	0.0	0.0	0.0	-14.0	0.0	-34.8	0.0	0.0
8	21.3	-4.4	0.0	0.0	0.0	0.0	12.8	0.0	0.0	0.0
9	25.1	-0.6	0.0	0.0	0.0	0.0	0.0	0.0	0.0	0.0
10	26.2	0.5	0.0	0.0	0.0	0.0	0.0	0.0	1.3	0.0
11	33.9	8.3	23.8	0.0	0.0	0.0	0.0	26.1	0.0	0.0
12	46.1	20.4	58.8	0.0	0.0	0.0	0.0	64.5	25.5	0.0
13	28.1	2.4	0.0	0.0	-6.4	0.0	0.0	0.0	0.0	-2.1
14	35.7	10.0	0.0	0.0	0.0	0.0	0.0	0.0	12.6	0.0
15	15.6	-10.1	0.0	0.0	0.0	0.0	0.0	0.0	0.0	0.0
16	16.1	-9.6	0.0	17.1	0.0	0.0	0.0	0.0	0.0	0.0
17	18.6	-7.1	0.0	0.0	0.0	0.0	0.0	0.0	0.0	0.0
18	23.2	-2.5	0.0	0.0	10.0	0.0	0.0	0.0	0.0	0.0
19	39.0	13.3	0.0	0.0	0.0	0.0	0.0	0.0	0.0	0.0
20	23.5	-2.2	0.0	0.0	0.0	0.0	0.0	0.0	0.0	2.9
21	13.1	-12.6	0.0	0.0	50.8	0.0	36.8	0.0	0.0	16.8
22	34.9	9.3	0.0	0.0	0.0	10.1	0.0	0.0	0.0	-7.1
23	31.0	5.4	0.0	0.0	0.0	10.2	0.0	25.4	0.0	0.0
24	14.6	-11.1	0.0	0.0	0.0	0.0	25.9	0.0	0.0	11.8
25	43.0	17.3	298.1	-15.3	0.0	0.0	0.0	54.5	21.6	0.0
26	20.4	-5.3	-23.0	28.3	0.0	0.0	0.0	0.0	-10.0	0.0
27	9.6	-16.1	0.0	0.0	260.3	0.0	0.0	0.0	0.0	21.6
28	33.3	7.6	0.0	0.0	0.0	58.5	0.0	29.0	0.0	0.0
29	14.0	-11.7	0.0	0.0	0.0	0.0	137.0	0.0	0.0	0.0
30	44.6	19.0	54.5	0.0	0.0	24.2	0.0	359.3	0.0	0.0
31	33.2	7.5	25.9	-8.0	0.0	0.0	0.0	0.0	56.3	0.0
32	20.3	-5.3	0.0	0.0	12.3	0.0	0.0	0.0	0.0	28.6
33	26.3	0.6	0.0	0.0	0.0	0.0	0.0	0.0	0.0	-0.5

(Continued)

Box 11.5c (Continued)

i	z	z − z̄	j 33 0.6	Row sum (S)	$\dfrac{S}{\sum(z_1-\bar{z})^2}$	$\dfrac{S}{\sum(z_1-\bar{z})^2}n-1$
1	34.5	8.8	0.0	1.96	0.001	0.02
2	21.2	−4.5	0.0	−6.27	−0.002	−0.06
3	16.2	−9.5	0.0	9.18	0.003	0.09
4	29.2	3.5	0.3	−5.78	−0.002	−0.06
5	14.4	−11.3	0.0	−31.46	−0.010	−0.32
6	28.6	2.9	0.3	4.95	0.002	0.05
7	12.8	−12.8	−1.1	−134.13	−0.042	−1.36
8	21.3	−4.4	0.0	27.16	0.009	0.27
9	25.1	−0.6	0.0	1.61	0.001	0.02
10	26.2	0.5	0.0	2.20	0.001	0.02
11	33.9	8.3	0.0	40.71	0.013	0.41
12	46.1	20.4	0.0	184.45	0.058	1.86
13	28.1	2.4	0.0	−9.22	−0.003	−0.09
14	35.7	10.0	0.0	67.17	0.021	0.68
15	15.6	−10.1	0.0	21.94	0.007	0.22
16	16.1	−9.6	0.0	19.20	0.006	0.19
17	18.6	−7.1	0.0	31.25	0.010	0.32
18	23.2	−2.5	0.0	13.32	0.004	0.13
19	39.0	13.3	0.0	68.22	0.022	0.69
20	23.5	−2.2	−0.3	−0.65	0.000	−0.01
21	13.1	−12.6	0.0	139.20	0.044	1.41
22	34.9	9.3	0.8	−48.51	−0.015	−0.49
23	31.0	5.4	0.0	31.51	0.010	0.32
24	14.6	−11.1	0.0	54.77	0.017	0.55
25	43.0	17.3	0.0	168.68	0.053	1.70
26	20.4	−5.3	0.0	−31.90	−0.010	−0.32
27	9.6	−16.1	0.0	72.71	0.023	0.73
28	33.3	7.6	0.0	14.38	0.005	0.15
29	14.0	−11.7	0.0	109.32	0.035	1.10
30	44.6	19.0	0.0	145.64	0.046	1.47
31	33.2	7.5	0.0	64.42	0.020	0.65
32	20.3	−5.3	−0.5	22.67	0.007	0.23
33	26.3	0.6	0.4	−0.48	0.000	0.00
				1048.21	0.331	10.589

11.5   Trend Surface Analysis

The most straightforward approach to introducing **trend surface analysis** is to recognise that it comprises a special form of regression. What makes it special is the inclusion of spatial location in terms of X, Y coordinates as independent variables recording the perpendicular spatial dimensions on a regular square grid. The purpose of the analysis is to define a surface that best fits these locations in space where the third dimension or the dependent variable, usually represented by the letter Z, denotes the height of the surface. The origins of the analysis lie in representing the physical surface of the Earth in which case the third dimension is elevation above a fixed datum level. However, surfaces can in principle be produced for any independent variable whose values can be theoretically conceived as varying and thus producing a surface in space. For example, land values might be expected to vary across space with peaks and troughs occurring at different places. Similarly, physical variables such as measurements of pressure and water vapour in the troposphere, the lowest layer in the Earth's atmosphere, differ over space. It is the relative concentration of different values, such as high or low amounts of water vapour that creates 'spatial features' in the atmosphere (e.g. clouds).

The starting point is a horizontal regular square grid in two dimensions (X and Y), and the end result is a surface depicting peaks and troughs, highs and lows, of the independent or Z variable: some of the Z data values will be known (i.e. have been measured in some way) and others will be produced as a result of **interpolating** additional values between them. Connecting the known and interpolated values together produces a visualisation of the continuous surface. Creating a continuous surface in this way is often referred to as 'draping', and Figure 11.7 offers an impression of the process. Starting with a simple surface extruded from ground level based on a set of 12 posts of

Figure 11.7   Graphical representation of draping a surface over a set of vertical posts with different heights.

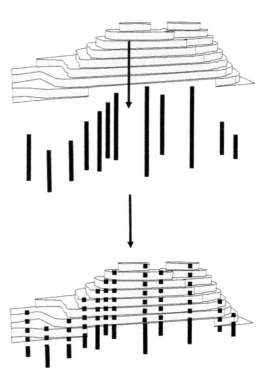

different heights. Abrupt changes in elevation between the heights of the surface occur because no attempt was made to interpolate intermediate values between those known to occur at the posts. The lower part of Figure 11.7 illustrates the surface having descended over the posts with dashed lines indicating where they would not be visible. Now imagine the posts are positions such that you, perhaps with some assistance, could throw a large piece of textile fabric over the posts so that it covered their tops and draped down to the ground. The appearance of the surface is likely to be smoother than the one shown in Figure 11.7 because the textile fabric will have found a series of curves stretching between the set of posts. There are two main approaches to deriving a trend surface known as global and local fit, and these relate to the difference between quantifying global and local spatial autocorrelation examined previously.

11.5.1 Fitting a Global Surface

Global fitting produces one mathematical function that describes the entire surface of the study area on the basis of estimating Z values for the nodes on the grid in a single operation, whereas local fitting derives multiple equations based on using a subset of points around successive individual nodes in the grid. The regression analysis techniques examined in Chapter 9 produce equations that define the line providing the best fit to the known data values, although it is acknowledged that these are unlikely to provide a 100 per cent accurate prediction of the dependent variable. The same caveat applies with trend surface analysis, and Figure 11.8 illustrates the link between prediction in linear and polynomial regression and trend surface analysis. The linear regression lines may be thought of as transects across the surface along a particular trajectory. There are known data values above and below the first, second and third degree polynomial regression lines on the left of Figure 11.8, but the lines show the overall slope.

The fitting of a global trend surface is achieved through a form of polynomial regression using the least squares method to minimise the sum of the squared deviations from the surface at the known, sampled locations (control points). The equation can be used to estimate the values of the dependent variable at any point, and this is commonly carried out for the nodes in the regular grid. The polynomial equation obtained by means of least squares fitting provides the best approximation of the surface from the available data for the control points. The process of creating a surface from the equation is commonly known as **interpolation**, since it involves determining or interpolating previously unknown Z values and connecting these together usually be means of a 'wire frame' to visualise the surface, although this term should be reserved for dealing with local variation. Figure 11.9 illustrates the outcome of this process with respect to an irregular sample of data points for elevation on the South Downs in South East England. The points from which the polynomial equation for the surface was generated are identified by white markings at the peaks.

Despite the general use of polynomial regression type equations to fit a global surface, there are some limitations to this approach. In some respects, these are further examples of the problems that were identified with respect to using regression to predict non-spatially distributed dependent variables. One of the limitations noted with respect to simple linear regression (first-order polynomial) is the potential unsuitability of defining the relationship as a straight line, especially if visualisation of the empirical data suggested some curvilinear connection. It would be just as nonsensical to claim that all surfaces are flat sloping planes. However, moving to a surface based on the second-order polynomial only introduces one maximum or minimum location on the surface and the third order only provides for one peak and one trough. Thus, global fitting does not necessarily

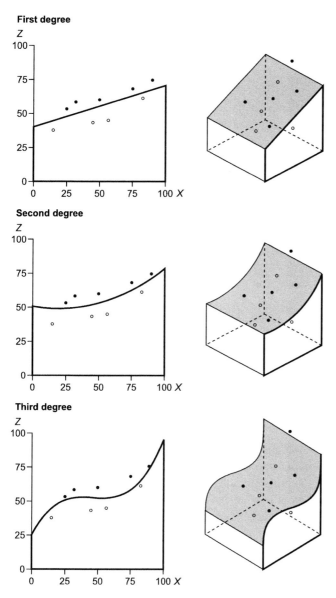

Figure 11.8 Comparison of regression lines and surfaces.

produce a very realistic surface. Another problem identified with regression in statistical analysis was that predicting the values of the dependent variable much beyond the lower and upper range of the known values of the independent variable(s) is potentially inaccurate. Similarly, the extension of a trend surface to beyond the area for which there are known data points could also be misleading. Such gaps in the observed or measured data can occur around the edges or in parts of a study area where there is a lack of control points. More realistic surfaces may be obtained by increasing the order of the polynomial equation beyond three to four, five, six or more, although the calculations involved are computationally taxing.

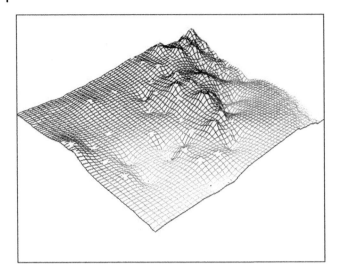

Figure 11.9 Trend surface for part of South Downs, East Sussex England.

Examine Figure 11.7 again reflecting on the issues just discussed: it shows the surface ending abruptly just beyond the limits on the ground set by the position of the 12 posts. The existing posts are fairly well distributed across the ground under the surface, but there are some gaps. How have these gaps affected the extruded surface?

Now, imagine some more posts were added, some lower and others lower than those already in place. Think about what would affect the appearance of the surface.

Box 11.6a illustrates selected aspects of the calculations involved in producing a first-order polynomial trend surface (i.e. a linear surface) in respect of the 2007 IMD for the 33 London Boroughs. The measurement of spatial autocorrelation using Moran's I has already shown this trait to be high in some boroughs just to the east of the City of London and low in others towards the south west. The equation has been computed using geostatistical software, and the predicted Z values and the residuals are tabulated in Box 11.6c. The percentile maps for the predicted figures for the linear surface and the residuals are classified into six groups and emphasise the importance of very low and very high values. The linear surface tracks north east to south west and simplifies the spatial pattern. The residuals reveal the highest positive difference was in Hackney and three of its neighbours (Newham, Lambeth and Tower Hamlets) in the central area. The largest negative residual was in Havering on the eastern edge with the next two being the City of London and Redbridge).

Box 11.6a   Trend surface analysis.

Polynomial equation for fitting a trend surface: $\hat{Z}(x,y) = \displaystyle\sum_{i=1}^{M} a_i x^{\beta_{1i}} y^{\beta_{2i}}$

Minimisation of error of estimate: $E = \displaystyle\sum_{i=1}^{L} \left(\hat{Z}(x_i,y_i) - Z_i \right)^2 \rightarrow \text{minimum}$

Linear trend (flat dipping plane): $\hat{Z}(x,y) = a_0 + a_1 x + a_2 y$

Quadratic trend (one maximum or minimum): $\hat{Z}(x,y) = a_0 + a_1 x + a_2 y + a_3 x^2 + a_4 xy + a_5 y^2$

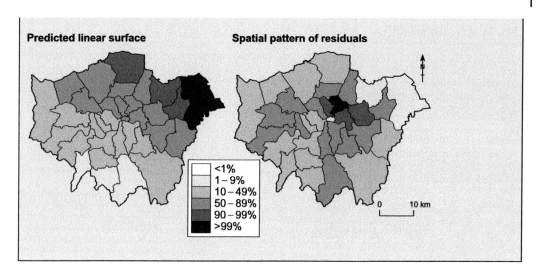

Predicted linear surface **Spatial pattern of residuals**

- <1%
- 1 – 9%
- 10 – 49%
- 50 – 89%
- 90 – 99%
- >99%

0   10 km

Box 11.6b   Application of the trend surface analysis.

The terms x and y are coordinates on the plane surface, and M is the number of degrees or orders of the polynomial equation. The series of coefficients $(a_0, a_1, \cdots a_i)$ are obtained by minimising the error of the estimation, where L is the number of sample or control points. A linear trend surface has been produced using spatial analysis software (Geoda) for the London Boroughs in respect of their average score on the 2007 IMD using rook spatial contiguity weights. The percentile maps in Box 10.6a show the predicted surface and the residuals emphasising the extreme cases. The specific 1st-order polynomial equation that best fits the empirical data is shown in Box 10.6c, and this has been used to calculate the predicted values of the IMD average score for the 500 m grid squares covering the area (Box 10.6d). This simplifies the spatial pattern and provides a clear visualisation of the north-east to south-west trending surface.

Box 11.6c   Calculation of linear trend surface.

i	X	Y	Z	$\hat{Z}$	e
1	547,980	186,083	34.5	30.50	−3.99
2	524,392	191,070	21.2	27.61	6.45
3	548,572	175,109	16.2	26.58	10.37
4	520,575	185,876	29.2	24.93	−4.29
5	542,070	166,316	14.4	22.04	7.68
6	527,756	184,528	28.6	25.87	−2.75
7	532,573	181,257	12.8	25.63	12.79
8	533,193	164,481	21.3	19.59	−1.72

(Continued)

Box 11.6c   (Continued)

i	X	Y	Z	$\hat{Z}$	e
9	516,188	181,612	25.1	22.48	−2.62
10	532,197	195,153	26.2	30.67	4.48
11	542,484	175,829	33.9	25.62	−8.32
12	533,820	185,439	46.1	27.42	−18.68
13	523,067	179,688	28.1	23.15	−4.92
14	531,141	189,534	35.7	28.39	−7.34
15	515,258	189,084	15.6	25.05	9.46
16	553,902	186,883	16.1	31.98	15.91
17	507,523	183,686	18.6	21.51	2.95
18	514,876	176,035	23.2	20.17	−3.03
19	531,241	185,097	39.0	26.78	−12.18
20	525,389	180,022	23.5	23.74	0.23
21	519,598	167,257	13.1	17.89	4.79
22	530,881	173,991	34.9	22.62	−12.32
23	537,779	174,133	31.0	24.06	−6.98
24	526,431	169,390	14.6	20.04	5.42
25	540,804	184,171	43.0	28.36	−14.59
26	543,517	189,417	20.4	30.83	10.47
27	517,275	173,187	9.6	19.60	10.05
28	533,757	176,088	33.3	23.97	−9.36
29	526,245	164,539	14.0	18.22	4.24
30	536,246	181,938	44.6	26.62	−18.02
31	538,073	189,802	33.2	29.88	−3.31
32	526,925	173,681	20.3	21.72	1.38
33	526,765	181,473	26.3	24.55	−1.75
First-order polynomial equation (linear surface)		$\hat{Z}(x,y) = a_0 + a_1 x + a_2 y$		$147.72 + 0.00020x + 0.00037y$	

Box 11.6d   Calculation of linear trend surface for grid squares.

1	527,500	197,500	32.76	34	512,500	182,500	22.77
2	532,500	202,500	36.09	35	517,500	187,500	26.10
3	537,500	207,500	39.42	36	522,500	192,500	29.43
4	512,500	192,500	27.20	37	527,500	197,500	32.76
5	517,500	197,500	30.53	38	532,500	202,500	36.09

6	522,500	202,500	33.86	39	537,500	207,500	39.42
7	527,500	207,500	37.19	40	542,500	212,500	42.75
8	532,500	212,500	40.52	41	547,500	217,500	46.08
9	537,500	217,500	43.85	42	552,500	222,500	49.41
10	542,500	222,500	47.18	43	507,500	172,500	17.23
11	547,500	227,500	50.51	44	512,500	177,500	20.56
12	507,500	187,500	23.87	45	517,500	182,500	23.89
13	512,500	192,500	27.20	46	522,500	187,500	27.22
14	517,500	197,500	30.53	47	527,500	192,500	30.55
15	522,500	202,500	33.86	48	532,500	197,500	33.88
16	527,500	207,500	37.19	49	537,500	202,500	37.21
17	532,500	212,500	40.52	50	542,500	207,500	40.54
18	537,500	217,500	43.85	51	547,500	212,500	43.87
19	542,500	222,500	47.18	52	517,500	167,500	17.24
20	547,500	227,500	50.51	53	522,500	172,500	20.57
21	552,500	232,500	53.84	54	527,500	177,500	23.90
22	557,500	237,500	57.17	55	532,500	182,500	27.23
23	507,500	182,500	21.66	56	537,500	187,500	30.56
24	512,500	187,500	24.99	57	542,500	192,500	33.89
25	517,500	192,500	28.32	58	547,500	197,500	37.22
26	522,500	197,500	31.65	59	517,500	162,500	15.02
27	527,500	202,500	34.98	60	522,500	167,500	18.35
28	532,500	207,500	38.31	61	527,500	172,500	21.68
29	537,500	212,500	41.64	62	532,500	177,500	25.01
30	542,500	217,500	44.97	63	537,500	182,500	28.34
31	547,500	222,500	48.30	64	542,500	187,500	31.67
32	552,500	227,500	51.63	65	532,500	157,500	16.15
33	507,500	177,500	19.44	66	537,500	162,500	19.48

11.5.2 Dealing with Local Variation in a Surface

The global fitted trend surface relates to the concept of regional features. Returning to the example creating surface using atmospheric pressure and water vapour to identify features in the troposphere, the cyclones or anticyclones and clouds can be viewed as regional or large-scale features. However, at a smaller scale, there will often be local variation in the variables producing highs and lows in the overall feature. Fitting the polynomial equation results in a surface, whereas the residuals, the differences between the fitted surface and known values may be thought of as local disturbances. These can be dealt with by a range of techniques that focus on groups of data points in a 'window', frame or **kernel**. These broadly divide into exact methods that produce measured values

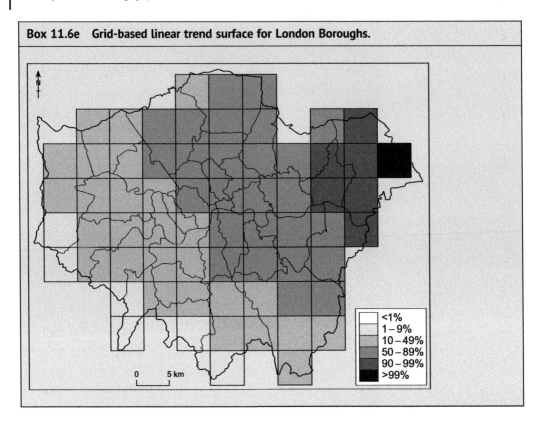

Box 11.6e   Grid-based linear trend surface for London Boroughs.

Legend:
- <1%
- 1–9%
- 10–49%
- 50–89%
- 90–99%
- >99%

0   5 km

for a series of points or areas by smoothing the original data and inexact ones that estimate a local trend and include kriging and local trend surface analysis (splines).

The simplest approach is to smooth data value means of a **moving average**, which is like the approach often adopted with time series data where a statistic for groups of days or months are calculated moving forwards or backwards in time. A spatial moving average involves partitioning the data points into groups as they fall or lie within a frame that moves across a series of data points and then interpolating by weighting the values usually according to the mid-point of the window. This is illustrated in Figure 11.10a, where the distances of the Burger King restaurants in Pittsburgh have been interpolated using a 5 km wide frame moving outwards from the city centre and the mean of their daily customer footfall (hypothetical) calculated for each frame (at 2.5, 3.5, 4.5 ... 19.5 km from the centre). The points representing the restaurants' locations have treated as though they all lie along a single X axis, but it would have been possible to use a two-dimensional frame (e.g. a square) and then calculated a weighted average within the zones formed by successive zones. The smoothing of the data achieved by a moving average is highly dependent on the size of the frame. Smoothing based on **inverse distance weighting** adjusts the value of each point in an inverse relationship to its distance from the point being estimated. The moving average in Figure 11.10a interpolated values along a one-dimensional axis from the city centre. Assuming all points to be located on the eastern side of a north south line through the centre point, it is possible to compare the previous results with inverse distance smoothing. This approach weights the values according to a predetermined number of nearest neighbours to each point being estimated. The application of inverse distance weighted interpolation to the same data values produces the

(a)

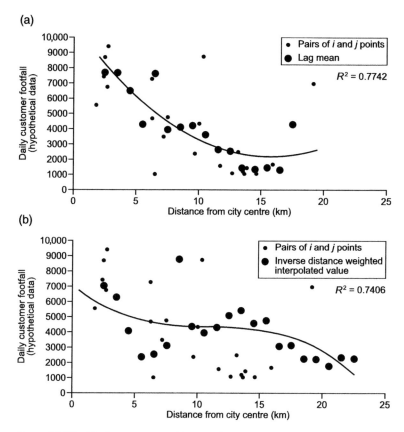

(b)

Figure 11.10 Moving average and inverse distance weighted data smoothing. (a) Moving average smoothing; (b) inverse distance weighted smoothing.

interpolated values shown in Figure 11.10b. The inverse distance used in this example is $1/d^2$, where d is the distance between the fixed points out from the city centre (2.5, 3.5, 4.5, ... 22.5 km) and their four nearest neighbours. The best-fit polynomial regression equation has been included for both sets of smoothed data: these are 2nd and 3rd degree polynomials, respectively.

Splines are another approach to dealing with local variation in a surface. They involve partitioning the surface into discrete sections each containing a group of data points and then fitting a polynomial regression equation to the data points in each group. These splines are then tied together to produce a smooth curve following the overall surface. The points where they connect are known as knots and the polynomial equation for each section is constrained to predict the same values where one section meets another. Unlike the moving average, the frames or partitions in which the splines are produced do not move across the set of data points: each has a fixed location although they can have different widths. One similarity with the moving average is that a smaller frame will reflect the local structure of the data values, whereas a wider one will produce a smoother surface.

One of the most common techniques for dealing with local variation is **kriging**. Trend surface analysis can be used to identify and then remove the overall trend in a surface before using kriging to map short range variations. Kriging is a based on inverse distance weighting and uses the local spatial structure to produce weights from which to predict the values of points. The first stage involves describing the spatial structure by means of a **semivariogram**, which involves calculating the distance between each pair of points and the square of the difference between their values for

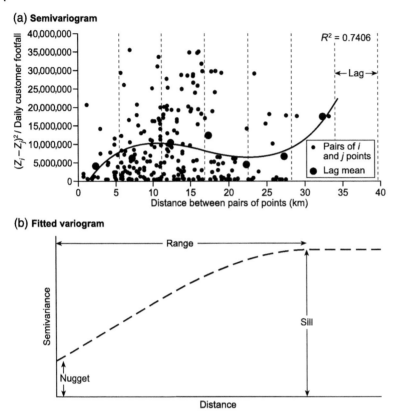

Figure 11.11 Example of semivariogram and fitted variogram used in Kriging. (a) Semivariogram with 5 km lags; (b) fitted variogram showing three main sections (nugget, range and sill).

the variable Z. These are visualised in a scatter plot that includes spatial lags known as a semivariogram: Figure 11.11 shows the semivariogram for the Burger King restaurants in Pittsburgh with 5 km lags and daily customer footfall (hypothetical values) as the variable. The mean of each group of data values within a given lag is plotted at the mid-point. The next stage involves summarising this local spatial variation by a function that best fits the means at the mid-points, which in this example is a 3rd degree polynomial. The values of neighbouring points are predicted by means of weights derived from the semivariogram. Kriging works by fitting an empirical semivariogram to a typical or model variogram. Figure 11.11b shows such a model and identifies three main sections of the fitted variogram. The nugget quantifies the uncertainty of the Z values, the range is the section over which there is correlation between distance, and the values thus represent the distance over which reliable prediction can be made and the semivariance is constant beyond the sill. Kriging is a geostatistical procedure with error estimates being produced, which can provide a useful guide as to where more detailed empirical data might be required. This contrasts with the geometrical basis of moving average and inverse distance weighting techniques.

11.6 Geographically Weighted Regression (GWR)

Recall that we have defined spatial autocorrelation as a quantitative expression of Tobler's First Law of Geography, which is that the distribution of attribute data values possessed by a set of features are likely to display some degree of positive or negative clustering. An introductory text such as this cannot attempt to explore the full extent and application of an advanced topic such as **GWR**; nevertheless, since it is now used in research not only within the geographical sciences but also across a number of other disciplines, an outline of its underlying principles and usage is appropriate (e.g. Waller *et al.*, 2007; Tu, 2011; Cellmer, Cichulska, and Bełe 2020). The term was first coined by Fotheringham, Brunsden and Charlton (1997), and readers are directed towards an early though still highly informative explanation of the statistical basis of the technique by the same authors for a more thorough coverage of the topic (Fotheringham, Brunsden and Charlton, 2002). Figure 11.12 illustrates the core of the problem, which is that the application of 'standard' regression analysis, in this instance OLS, to the X independent and Y dependent data values for the groups of points located within a circle with a fixed diameter centred on each individual point in turn will almost invariably produce different regression equations if spatial autocorrelation is present. The left side of Figure 11.12 shows 20 feature points with identical circles centred on four of them labelled, A, B, C and D, as examples. Each circle 'covers' a group of points: Point A includes five and Point D four points. Some of the individual points fall in one circle, some in two and others in three. There are some points not falling within any of the four circles shown. The right side of Figure 11.12 presents the scatter plots for the groups of points around A, B, C and D together with the OLS regression line, equation and R^2 value. The R^2 values range from 0.9099 (Point A) to 0.3027 (Point D), revealing that the independent variable X is less successful in predicting the dependent variable Y in some parts of the study area than others. The equivalent OLS regression applied to all 20 point features yields the equation $y = 1.128x + 14.38$ with $R^2 = 0.6341$, which is with the range of values supplied by the four example points.

There are various permutations of the regression model in the collection techniques known as GWR and Box 11.7a illustrates the simplest. It should also be noted that GWR parameters can

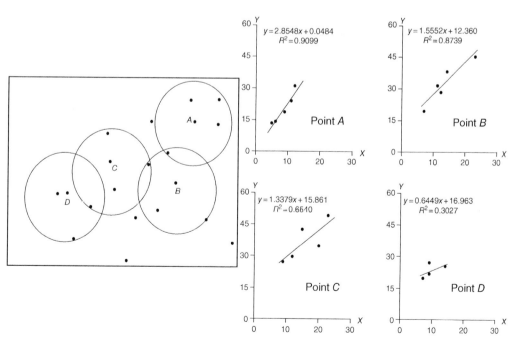

Figure 11.12   Regression analysis applied using a fixed distance 'template' illustrating effect spatial autocorrelation on OLR.

be estimated for anywhere across a study area not just for locations where there are geographical features (points or zones) with known X and Y data values for at least one independent variable and a dependent variable. The GWR example in Box 11.7a uses two continuous variables with the Gaussian normal distribution, but enhancements to the GWR technique since it was first developed have seen extension to enable count and binary (presence/absence) data to be analysed using Poisson and Logistic regression, respectively. A fixed neighbourhood or bandwidth for the circular kernel was used for the GWR application in Box 11.7a, but alternatively a chosen number of neighbours could be used for each regression equation. Some methods of searching involve an approach that seeks to minimise the value of the Akaike Information Criterion (AIC). This is a measure used to assess the output from GWR that is sometimes preferred to R^2 because it takes account of model complexity.

Box 11.7a   Geographically Weighted Regression (GWR).

Form of the GWR model: $y = X\beta(t) + \epsilon$
  Where:

  β is a group of spatially variable slope coefficients;
  t is a target point at the centre of a circular neighbourhood with radius r placed around each target point t_i;
  ϵ is an error term;

  Gaussian weight: $w_i = \left(\dfrac{-d_i}{h} \right)^2$

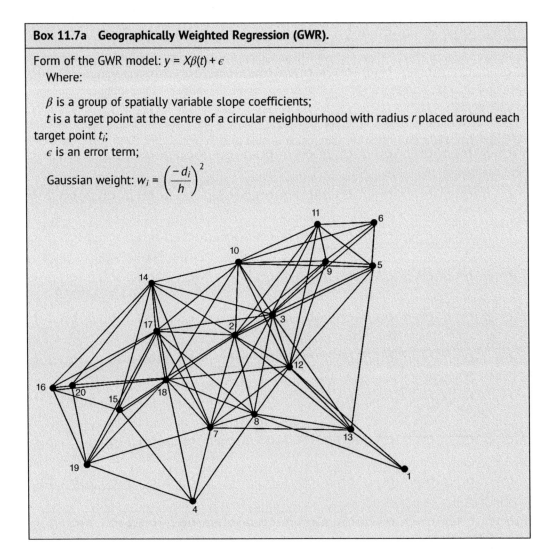

Box 11.7b   Application of the trend surface analysis.

Strictly speaking GWR should not be used with a dataset containing only 20 geographical features and is more suited to situations in excess of 100. However, for the purpose of outlining the GWR, the present example has been limited to 20 points corresponding to the centroids of a subset of the morainic debris on the outwash plain of Le Bossons Glacier in France. The complete dataset of 100 debris items as well as a transect containing a different subset have been analysed previously with modest evidence of spatial autocorrelation detected with respect to some of the measured variables, although the Global Moran's I for the subset 20 items of debris indicates a random distribution (Moran's $I = -0.006$, $Z = 0.217$ and $p = 0.826$). Application of GWR here has also been simplified by focusing on one independent variable (X), the length of the long axis of the debris in cm, and, as the dependent variable (Y), the length of its perimeter in cm. The length of the long axis is likely to determine the perimeter to some extent, although less eroded debris with a more rugged form may reflect such a connection to a lesser degree. OLS regression of the data produces the model $y = 1.128x + 14.38$ with $R^2 = 0.6341$.

The radius of the neighbourhood circle was fixed at 30 cm. The number of neighbours within this circle around the 20 items of debris ranges from 4 to 14 with the majority containing at least seven points in total (see Box 11.7c). The image above is a graphical representation of these neighbourhood connections between each target point labelled 1–20 and those surrounding points included in its GWR equation. It shows some centrally located points, such as numbers 2, 3, 12, 17 and 18, have more neighbours within 30 cm than most towards the edge of the study area (e.g. 1 and 4).

GWR has been applied using GIS software (ArcPro), and a selection of the output has been included in the lower part of Box 11.7c. For each target point, this includes the values of the intercept and slope coefficient together with their standard errors. The value of Y predicted by the individual GWR equations around each target point together with the residual and standardised residual is also given. Apart from examining the tabulated output from GWR, the modelled surface, standardised residuals, the $\beta(t)$ parameters and their estimated standard errors can be mapped as they also vary spatially. None of the standardised residuals (see Box 11.7d) lie beyond the range −2.5 to +2.5 standard deviations from the mean. Points 7 and 18 with standardised deviations in the range +1.5 to +2.5 and point 4 between −2.5 and −1.5 had local R^2 values lower than the other points (0.45, 0.61 and 0.55, respectively). This suggests these are locations where global regression model is less successful.

11.7   Closing Comments

This chapter and Chapter 10 have examined a selection of the spatial statistics techniques that are available for investigating not only the spatial patterns displayed by geographical features, but also how we can apply techniques to test whether these patterns are significantly different from complete spatial randomness. We have also seen how it is not only feasible to analyse the patterns themselves, but also that the data values of the binary, counted and continuous attributes and variables with respect to which these features have been measured may have a spatial pattern reflecting interdependence. These issues essentially derive from the implications of Tobler's (1970) First Law of

Box 11.7c Tabulated data and results of applying GWR to subset of glacial moraine debris.

t	X	Y	Identifiers of target points and their neighbours within 30-m circle.												
1	40	56	1	8	12	14	9	10	11	13	14	15	17	18	
2	23	46	2	3	5	7	8	9	10	11	12	13	14		
3	11	29	2	3	5	6	9	12	13	11					
4	20	26	4	8	9	14	15								
5	5	16	2	3	5	6	9	10	11	12	13				
6	6	17	3	5	6	9	10	11	12						
7	12	43	2	3	4	7	8	12	13	14	15	17	18	19	20
8	12	29	2	3	4	5	7	8	10	12	13	15	17	18	
9	9	23	2	3	5	6	9	10	11	12					
10	8	17	2	3	5	6	8	9	10	11	12	14	17	18	
11	8	26	2	3	5	6	9	10	11	12					
12	12	38	1	2	3	5	7	8	9	10	11	12	13	14	17	18
13	7	20	1	2	3	5	7	8	12	13					
14	12	30	2	3	7	8	10	12	14	16	17	18			
15	9	27	2	4	8	14	15	15	17	19	20				
16	9	14	14	15	16	17	18	19	20						
17	20	35	2	3	7	8	10	12	14	15	16	17	18	19	20
18	15	43	2	3	4	7	8	10	12	14	15	16	17	19	20
19	9	22	4	7	15	16	17	18	19	20					
20	14	25	7	14	15	16	17	18	19	20					

t	Intercept	SE of Intercept	Slope coefficient	SE of Slope coefficient	Predicted y	Residual	Standardised Residual	Local R^2
1	12.60	7.28	1.09	0.26	56.03	−0.03	−0.24	0.97
2	14.67	5.96	1.38	0.43	46.52	−0.52	−0.15	0.71
3	10.69	4.97	1.61	0.39	28.43	0.57	0.11	0.83

4	38.75	9.43	−0.50	0.70	28.66	−2.66	−1.69	0.55
5	3.95	9.02	2.34	1.07	15.65	0.35	0.09	0.83
6	4.35	12.17	2.28	1.63	18.06	−1.06	−0.22	0.76
7	29.27	6.69	0.42	0.43	34.34	8.66	1.68	0.45
8	22.63	6.37	0.91	0.42	33.50	−4.50	−0.86	0.53
9	7.41	4.93	1.86	0.44	24.14	−1.14	−0.21	0.87
10	9.29	5.49	1.59	0.40	22.05	−5.05	−1.04	0.86
11	7.03	5.62	1.88	0.57	22.03	3.97	0.75	0.79
12	13.47	5.18	1.52	0.39	31.73	6.27	1.17	0.74
13	17.42	5.14	0.99	0.23	24.39	−4.39	−1.12	0.91
14	11.33	7.36	1.42	0.48	28.34	1.66	0.35	0.73
15	13.13	6.88	1.26	0.48	24.51	2.49	0.51	0.69
16	5.44	8.95	1.60	0.72	19.88	−5.88	−1.29	0.65
17	14.42	6.10	1.30	0.43	40.46	−5.46	−1.21	0.67
18	18.60	5.93	1.07	0.41	34.70	8.30	1.52	0.61
19	12.60	7.28	1.09	0.26	22.63	−0.63	−0.14	0.69
20	14.67	5.96	1.38	0.43	28.93	−3.93	−0.80	0.69

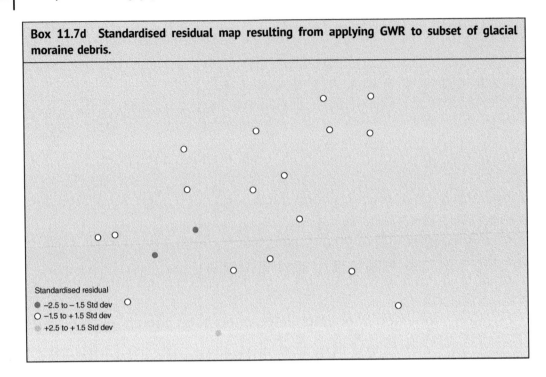

Box 11.7d Standardised residual map resulting from applying GWR to subset of glacial moraine debris.

Geography. Table 11.1 is designed to allow you to reflect on the possible interactions between clustered, random and dispersed **spatial patterns** and clustered, random and dispersed **data values**. They are referred to as grouped, random and bimodal to distinguish the characteristics of the data values of the continuous variable from the spatial pattern of the features.

There are three sets of data values all with a mean of 23 but different standard deviations, 1.5, 15.4 and 25.7 for the grouped, random and bimodal sets, respectively (see Table 11.1). Nearest neighbour analysis has been applied to the point features with Z scores and p (probability) calculated: reassuringly, the results seem to confirm the Null Hypothesis of spatial randomness can be rejected for the nominally clustered and dispersed sets of point features at the 0.05 level, whereas the ostensibly random pattern does indeed have this characteristic. The lower cells of Table 11.1 present the results of a series Global Moran's I analyses relating to each of the nine possible combinations, for example clustered spatial pattern and grouped data values, dispersed spatial pattern and random data values, random spatial pattern and bimodal data values, etc. A general rule when using Global Moran's I is that the Null Hypothesis should be retained if the p probability value is not significant. This applies to all but one of the nine analyses (dispersed and bimodal), although the result for clustered and bimodal is not much above the 0.05 level with $p = 0.054$. When you can reject the Null Hypothesis and the Z score is positive, the high and/or low data values are more spatially clustered than would be expected, whereas if Z is negative, the high and low values are more spatially dispersed indicating the operation of a competitive rather than 'cooperative' process.

Chapters 10 and 11 have introduced a selection of spatial statistics and geostatistical techniques that come under the broad headings of **ESDA** and **Spatial Data Modelling**. A focus of attention is the distribution of spatial autocorrelation and recognition that patterns of geographical features and measured variables often contravene the assumptions of classical statistics. Although trend surface and residuals analysis utilise some of the computational procedures of classical statistics,

Table 11.1 Comparison of Moran's *I* index (statistically tested) for nine datasets representing combinations of clustered, random and dispersed data values and grouped, random and bimodal point features.

	Clustered	Random	Dispersed
Data values	Mean = 23 Standard deviation = 1.5	Mean = 23 Standard deviation = 15.4	Mean = 23 Standard deviation = 25.7
Nearest neighbour	$R = 0.7235$ $Z = -2.2444$ $p = 0.0248^*$	$R = 1.0838$ $Z = 0.6806$ $p = 4.9616$	$R = 2.1399$ $Z = 11.1188$ $p = 0.0000^*$
Grouped	Moran's $I = -0.234$ Z score = -0.863 Pseudo $p = 0.388$	Moran's $I = -0.083$ Z score = -0.138 Pseudo $p = 0.900$	Moran's $I = 0.343$ Z score = 1.759 Pseudo $p = 0.078$
Random	Moran's $I = -0.243$ Z score = -0.895 Pseudo $p = 0.371$	Moran's $I = -0.178$ Z score = -0.687 Pseudo $p = 0.492$	Moran's $I = -0.203$ Z score = -0.619 Pseudo $p = 0.536$
Bimodal	Moran's $I = 0.347$ Z score = 1.925 Pseudo $p = 0.054$	Moran's $I = 0.253$ Z score = 1.756 Pseudo $p = 0.079$	Moran's $I = 0.486$ Z score = 2.287 Pseudo $p = 0.022^*$

Note: *denotes significant at 0.05 level.

such as fitting a regression equation, the failure to comply with many standard assumptions, such as data value independence and conforming to the normal probability distribution, means that calculating confidence limits for the fitted surface or applying inferential statistics is not feasible. Geographically weighted regression (GWR) (Fotheringham, Charlton and Brunsden, 1997; Fotheringham, Brunsden and Charlton, 2002) was the last statistical technique to be introduced in this chapter. It is a complex and potentially still evolving technique with various permutations. There are many examples, potentially thousands, in the published literature of researchers from Geography, Environmental and Earth Sciences using GWR as well as others from different disciplinary fields who have recognised its potential and borrowed something that geographical, spatial scientists might have wished to claim as their own. Although perhaps demanding further investigation than is a possible in an introductory text, GWR offers opportunities for students interested in modelling multivariate statistical relationships between dependent and independent variables to investigate their transition from one form to another over a surface of spatially autocorrelated geographic space.

References

Anselin, L. (1995) Local indicators of spatial association – LISA. *Geographical Analysis*, **27**, 93–115.

Cellmer, R., Cichulska, A. and Bełej, M. (2020) Spatial analysis of housing prices and market activity with the geographically weighted regression. *ISPRS International Journal of Geo-Information*, **9**, 380. DOI: 10.3390/ijgi9060380.

Fotheringham, A.S., Brunsdon, C. and Charlton, M.E. (2002) *Geographically Weighted Regression: The Analysis of Spatially Varying Relationships*, New York, John Wiley and Sons.

Fotheringham, A.S., Charlton, M. and Brunsdon, C. (1997) Measuring spatial variations in relationships with geographically weighted regression, in *Recent Developments in Spatial Analysis. Advances in*

Spatial Science (eds M. M. Fischer and A. Getis), Berlin, Heidelberg, Springer. DOI: 10.1007/978-3-662-03499-6_4.

Geary, R.C. (1954) The contiguity ratio and statistical mapping. *The Incorporated Statistician*, **5**(3), 115–145. DOI: 10.2307/2986645.

Getis, A. and Ord, J.K. (1992) The analysis of spatial association by use of distance statistics. *Geographical Analysis*, **24**, 189–206.

Tobler, W. (1970) A computer movie simulating urban growth in the Detroit region. *Economic Geography*, **46**(2), 234–240.

Tu, J. (2011) Spatially varying relationships between land use and water quality across an urbanization gradient explored by geographically weighted regression. *Applied Geography*, **31**(1), 376–392. DOI: 10.1016/j.apgeog.2010.08.001.

Waller, L.A., Zhu, L., Gotway, C.A., Gorman, D.M. and Gruenewald, P.J. (2007) Quantifying geographic variations in associations between alcohol distribution and violence: a comparison of geographically weighted regression and spatially varying coefficient models. *Stochastic Environmental Research and Risk Assessment*, **21**, 573–588. DOI: 10.1007/s00477-007-0139-9.

Further Reading

Berry, B.J.L. and Marble, D.F. (1968) *Spatial Analysis – A Reader in Statistical Geography*, New Jersey, Prentice Hall.

Besag, J. and Newell, J. (1991) The detection of clusters in rare diseases. *Journal of the Royal Statistical Society Series A*, **154**, 143–155.

Brunsdon, C., Fotheringham, A.S. and Charlton, M.E. (1999) Some notes on parametric significance tests for geographically weighted regression. *Journal of Regional Science*, **39**, 497–524.

Cliff, A.D. and Ord, J.K. (1973) *Spatial Autocorrelation*, London, Pion.

Cliff, A.D. and Ord, J.K. (1981) *Spatial Processes: Models and Applications*, London, Pion.

Clifford, P. and Richardson, S. (1985) Testing the association between two spatial processes. *Statistics and Decisions*, **2**, 155–160.

Cressie, N.A.C. and Chan, N.H. (1989) Spatial modelling of regional variables. *Journal of American Statistical Association*, **84**, 393–401.

Fischer, M. and Getis, A. (2009) *Handbook of Applied Spatial Analysis: Software Tools, Methods, and Applications*, New York, Springer.

Fotheringham, A.S. and Rogerson, P.A. (2008) *Handbook of Spatial Analysis*, London, Sage.

Fotheringham, A.S., Brunsdon, C. and Charlton, M.E. (2000) *Quantitative Geography: Perspectives on Spatial Data Analysis*, London, Sage.

Fotheringham, A.S., Charlton, M.E. and Brunsdon, C. (1998) Geographically weighted regression: a natural evolution of the expansion method for spatial data analysis. *Environment and Planning A*, **30**, 1905–1927.

Gao, X., Asami, Y. and Ching, C.-J.F. (2006) An empirical evaluation of spatial regression models. *Computers and Geosciences*, **32**, 1040–1051.

Griffith, D.A. (1987) *Spatial Autocorrelation: A Primer*, Washington, Association of American Geographers.

Griffith, D.A. (1996) Computational simplifications for space-time forecasting within GIS: the neighbourhood spatial forecasting model, in *Spatial Analysis: Modelling in a GIS Environment. Cambridge: Geoinformation International* (eds P. A. Longley and M. Batty), New York, John Wiley and Sons.

Hardisty, J., Taylor, D.M. and Metcalfe, S.E. (1995) *Computerised Environmental Modelling*, Chichester, John Wiley and Sons.

Harris, R. (2016) *Quantitative Geography: The Basics*, Sage Publications Ltd.

Longley, P., Brooks, S.M., McDonnell, R. and Macmillan, B. (1998) *Geocomputation: A Primer*, Chichester, John Wiley and Sons.

Mitchell, A. (2008) *The ESRI Guide to GIS Analysis, Volume 2: Spatial Measurements and Statistics*, Redlands, ESRI Press.

Sawada, M. (1999) ROOKCASE: an excel 97/2000 visual basic (VB) add-in for exploring global and local spatial autocorrelation. *Bulletin of the Ecological Society of America*, **80**, 231–234.

Smith, C.H. (1986) A general approach to the study of spatial systems. Two examples of Application. *International Journal of General Systems*, **12**(4), 385–400. doi:10.1080/03081078608934944.

de Smith, M.J., Goodchild, M.F. and Longley, P.A. (2007) *Geopatial Analysis: A Comprehensive Guide to Principles*, Leicester, Techniques and Software Tools.

Tukey, J.W. (1972) Some graphic and semigraphic displays, in *Statistical Papers in Honour of George W Snedecor* (ed. T. A. Bancroft), Ames, Iowa State University Press.

Tukey, J.W. (1977) *Explanatory Data Analysis*, Reading, Addison-Wesley.

Section VI

Practical Application

12

Practicalities of Applying, Interpreting and Visualising Quantitative Analysis in Geographical Projects

Students of Geography and Environmental and Geological Sciences, as we approach the end of the first 25 years of the 21st century, have a range of information technology available for carrying out quantitative analysis and visualisation of geographical (geospatial) datasets in a way that students of 30, 40, 50 or more years ago could only have dreamt of. However, the challenges of how to apply, interpret and visualise in charts or as maps the output or results of the analysis remain less straightforward. This chapter aims to provide a 'roadmap' to these issues in four different but related ways: first, by reviewing the results of the statistical techniques that have been analysed in previous chapters; second, by highlighting how researchers present and discuss the results of their research in journal articles; third, by exploring the practical aspects of using information technology; and lastly, by outlining some typical projects of a scale that might be carried by students for their independent research investigation in Geography, Earth and Environmental Science and related disciplines.

Learning Outcomes

This chapter will enable readers to:

- Review the results of the quantitative analysis outlined in previous chapters;
- Describe the ways in which academic researchers present their research results in published journal articles;
- Prepare for analysing their own project dataset using appropriate information technology;
- Reflect on the design and methods suitable for an independent research investigation in Geography, Earth and Environmental Sciences.

12.1   Introduction

This chapter aims to focus on the practicalities of carrying out quantitative analysis, reporting and visualising the outcomes of research using geographical (geospatial) datasets. It explicitly seeks to help students of Geography, Environmental and Earth Sciences with performing these tasks (analysis, reporting and visualising) for datasets in their own independent but guided research projects. The chapter concentrates on four aspects to address these issues. First, we revisit the results of using the quantitative techniques and statistical tests that have been covered in Chapters 6–11. Second, a small selection of articles published in open-access journals are discussed to examine how results are described and presented in graphs, tables, maps, etc. Third, the capture and transfer of data, potentially collected or obtained by different means, into a format whereby they can be analysed and visualised using information technology is examined. Finally, three undergraduate-style projects are described, and some analysis is carried out on their datasets and the results are outlined and visualised. The four approaches to the overall practical focus of this chapter seek to make the challenge of planning and undertaking a modest-scale research project less daunting and doing the analysis and discussing the results more rewarding.

12.2 Summary of Results from Quantitative Analysis of Previously Used Datasets

Eleven datasets have been analysed in several ways in Chapters 6–11 with the aim of introducing and explaining a range of spatial and non-spatial statistical (quantitative) techniques. The quantity of data (number of entities and variables) was deliberately restricted in size to limit (or make more manageable) the computational details contained in the boxes. The questions, often expressed in the form of Null and Alternative Hypotheses, that these datasets were used to answer were not presented as part of a coherent, well-specified research project. The questions for each individual dataset were isolated and unconnected with each other and were simply there to show how a quantitative technique worked. The questions asked and answered for each dataset were unstructured and not thematic, because they were to demonstrate the techniques. The following sub-sections with their individual tables review and recapitulate the results obtained from the analysis of the different datasets. The self-assessment questions in each section are intended to prompt ideas about how the analyses were carried out and how the summarised results might have been included in a more structured research question. The sections are organised simply in the order in which the datasets were introduced in Chapters 6–11, and the tables include the box numbers enabling readers to refer to where the technique was applied.

12.2.1 Isle of Wight Residents' Survey Dataset

This dataset was based on a sample of residents on the Isle of Wight of the south coast of England with the aim of exploring which other counties in England and Wales they might consider living. The overall sample was divided randomly into two sub-samples: one group was asked this question and shown a map on which the English and Welsh counties were named, and the second sub-sample was shown a similar map without the county names. In both cases, survey participants were asked to rank the top five counties where they would consider living. These rank scores were aggregated across all respondents, which resulted in some counties not being mentioned by anyone and other popular locations ranked by several people. The results from applying a series of parametric and non-parametric tests to these aggregate ranked scores are summarised in Table 12.1. It did not seem to make any difference whether people were shown a map with or without the names of the counties; the observed difference in ranking could easily have arisen by chance. The hypothesised mean rank score (0.283) that would have occurred if people had randomly listed counties lies outside the 95 per cent confidence limits for the Z and t tests, whereas the observed mean rank of 0.404 falls within these confidence intervals. The non-parametric Wilcoxon signed ranks test supports the parametric Z and t test results. The tests that were applied have answered the specific questions they were asked, but inevitably having completed them we are left wanting more information. Reflect on the following self-assessment questions, and propose some additional topics that could have been obtained in the questionnaire survey.

Table 12.1 Results obtained from applying statistical tests to Isle of Wight residents' survey dataset.

	Isle of Wight residents' survey	
Quantitative Technique	Result	Box number
Z test	$Z = 2.163$ and $p = 0.015$: the Alternative Hypothesis was accepted that residents preferentially chose certain counties in the rest of England and Wales as places they might live. If respondents had chosen counties at random, the mean score per county would have been 0.283 rather than 0.404. The top five counties in descending order were Devon, Cornwall, Dorset, Surrey and Greater London.	6.1a
t test	$t = 2.942$ and $p = 0.006$ with 29 degrees of freedom: this result also indicates acceptance of the Alternative Hypothesis that certain counties in England and Wales were selected as places to live and the same set of counties had the highest differences between the sample mean and the hypothesised population mean (zero) rank score.	6.2a
Two sample t test	$t = 0.613$ and $p = 0.470$ with 64 degrees of freedom: the Null Hypothesis was accepted indicating that presenting one sample with a map of counties in England and Wales showing their names and another sample with a similar but unlabelled map produced a difference in preferred counties that could easily have been the result of chance.	6.4a
Two sample F test	$F = 1.434$ and $p = 0.307$ with 29 and 35 degrees of freedom, respectively, the sample with the larger and smaller variance: the variances of the populations from which the two samples of Isle of Wight residents were selected are the same, and the difference in sample variances has arisen through chance.	6.5a
Z and t confidence limits	95% confidence limits for Z are +2.94--+0.513; 95% confidence limits for t are +0.231--+0.576: the mean rank score for residents who were shown a map with county names in England and Wales was 0.404. The Z confidence interval range is narrower around this value than the t confidence interval, but both exclude the hypothesised mean score if respondents had chosen randomly was 0.283.	6.7a
Wilcoxon signed ranks	$W(Z) = 0.150$ and $p = 0.881$: the Null Hypothesis was accepted that mean rank scores for the two samples of Isle of Wight residents who looked at the named and unnamed county maps are not significantly different from each other at the 0.05 level.	7.1a

> The statistical results raise further questions. Why should Surrey and Greater London be included alongside three generally rural and coastal counties? What reasons might people have for wanting to move to another county in England and Wales? Had respondents previously lived on the mainland of England and Wales, or had they lived all their lives on the Isle of Wight?

12.2.2 Les Bossons Glacier Meltwater Stream Dataset

The dataset was created following a single day of fieldwork in late September on the outwash (sandur) plain below the snout of Les Bossons Glacier in the French Alps below Mont Blanc. The valley containing the glacier and outwash plain has a northerly aspect, and on a sunny day, the shadow cast by the mountains on the southern side moves off the area over which the meltwater stream flows by mid-morning as the sun rises and returns by late afternoon as the sun starts to set at a time related to the season of year. Temperature measurements were recorded at identical

Table 12.2   Results obtained from applying statistical tests to Les Bossons Glacier meltwater stream dataset.

	Les Bossons Glacier meltwater stream	
Quantitative Technique	Result	Box number
Paired t test	$t = 7.425$ and $p = 0.000000992$ with 17 degrees of freedom: the Alternative Hypothesis was accepted given the extremely low probability of having obtained the test statistic by chance. This clearly indicates that the difference between the paired temperature measurements of the meltwater stream at the glacier snout and the exit of the outwash plain increases during the day. The greatest difference occurred at 12.30 hrs.	6.3a

half-hourly intervals by two groups of researchers one based at a safe location near the snout of the glacier and the other at the point where the stream descends from the gently sloping outwash plain down a steeper gradient to join the L'Arve River in the main valley. These pairs of water temperature measurements were tested using the paired sample t test, which revealed that there was an extremely low probability that the difference between the samples was the result of chance or sampling error (see Table 12.2). Looking at the recorded temperatures and considering the topography of the area, it is apparent that the sun has a warming influence on the water as it flows across a series of bifurcated channels. This result confirms what might be expected, but it does raise other issues, some of which are outlined in the following self-assessment questions.

> It is not surprising that the warmer ambient temperature during the middle of the day would increase the temperature of water in the stream and that this would be more noticeable at the furthest distance from the glacier snout. Is there a difference in meltwater temperature in a transverse direction across the outwash plain? Did the temperature in shallow, slow-moving channels become higher than in faster-flowing, deeper channels? How might the fieldwork be changed so that this new perspective on the topic could be addressed?

12.2.3   Mid-Wales Village Residents' Shopping Survey

The dataset was created from the answers given to a questionnaire survey of residents in four villages in rural Mid-Wales in the late 1990s. The sampled residents in each village were selected at random. The settlements were small enough for the researcher to walk along the roads and public footpaths in each so that a list of all addresses could be created, and these were allocated an identification number. Sets of random numbers were generated, 20 per cent more than were required to allow for replacements should a household refuse to participate in the survey, and these identifier numbers were used to select households at random. Each of these households was visited and invited to take part in the survey and signing a consent form, with one person completing the questionnaire. The preliminary F test confirmed there was a significant difference in the mean distance travelled for their main shopping by households in the four settlements, and a subsequent Kruskal–Wallis test confirmed also a significant difference in the median distance (see Table 12.3). Paradoxically, when the distance was classified into bands the Kolmogorov–Smirnov test produced a non-significant result. When exploring the mode of transport used to carry out the weekly shop, the test results indicated that the normal mode was significantly different from what would have been expected either if they had conformed to the proportionate distribution found in a previous national survey or if they were distributed equally across the modes of transport. However, there was no significant difference in the choice of transport mode between the four villages, which might

Table 12.3 Results obtained from applying statistical tests to Mid-Wales village residents' shopping survey dataset.

Mid-Wales village residents' shopping survey		
Quantitative Technique	**Result**	**Box number**
One-way analysis of variance (ANOVA)	$F = 3.36$ and $p = 0.025$ with 60 and 3 degrees of freedom, respectively, for the within-group and between-group variance. The test uses the sample variances to determine if there is a difference between their means. The test probability is less than the chosen level of significance (0.05), and therefore the Alternative Hypothesis was accepted that there is a significant difference in the mean distance travelled for shopping between residents in the four Mid-Wales villages.	6.6a
Pearson's chi-square (χ^2) with one variable (univariate)	$\chi^2 = 9.12$ and $p = 0.023$ using equal proportions to calculate expected frequencies, and $\chi^2 = 14.80$ and $p = 0.002$ using proportions in a previous national survey to calculate expected frequencies. The Alternative Hypothesis was accepted in both cases, and households carried out their weekly shopping using a mode of transport that was different from equal preference and from the results of a national survey.	7.2a
Kolmogorov–Smirnov D test	$D = 0.77$ and $p = 0.17$ with 64 observations: the Null Hypothesis was accepted that the distance travelled for shopping by residents in four Mid-Wales villages when grouped into distance bands is not significantly different from what would be expected to occur at random.	7.3a
Pearson's chi-square (χ^2) with two variables (bivariate)	$\chi^2 = 10.44$ and $p = 0.316$ with 9 degrees of freedom: the Null Hypothesis was accepted, and the conclusion was reached that there was no significant difference between the choice of mode of transport used for weekly shopping by residents in the four villages.	7.5a
Kruskal-Wallis H test	$H = 31.43$ and $p < 0.001$ with 3 degrees of freedom: the Alternative Hypothesis was accepted that there was a significant difference in the median distance travelled by households for their weekly shop in the four Mid-Wales villages.	7.7a
Phi (ϕ) correlation coefficient (tested by χ^2)	$\phi = 0.089$; $\chi^2 = 1.497$; and $p = 0.221$: the Null Hypothesis was accepted, leading to the conclusion that the weak relationship between the presence of a laptop and gender of the head of household was no different from what would have occurred by chance.	8.4a

suggest that the experience of living in rural Mid-Wales is similar for people in different settlements. The survey questionnaire also asked about the availability of a laptop in the household and the self reported gender of the person who considered himself or herself to be the head of the household. Correlating these two variables using the Phi coefficient produced a result close to 0.0 (no relationship), which was also not significant and could have arisen by chance. It is often the case, when a survey has been completed and analysed, that further topics that could have been explored come to the researcher's attention. The following self-assessment feature raises some topics that could have been explored in this project, which might produce a more robust piece of research. Reflect on how you would frame questions for an expanded questionnaire that could seek information about these topics.

The survey of how residents in four villages in Mid-Wales carry out their weekly shopping missed the opportunity to explore some different issues. The time taken to carry out a journey, especially in rural areas, may be more important than the physical distance travelled. It is possible that when people made their weekly shopping journey, they also did something else, such as visiting friends, eating lunch in a pub or walking in the hills. Asking about the presence of a laptop suggested the researcher was aware of the increasing importance of information technology, but perhaps there should also have been a question about internet availability, which can be an important issue in some of the more remote rural areas.

How could these additional topics make a redesigned project more robust?

12.2.4 Parked Cars and Other Vehicles along One Side of a Road

This dataset was produced by the researcher selecting a stretch of road in a residential area and printing out a map showing 5-m road segments. This map was taken in the field between 12.00 and 14.00 hours on a weekday so that each segment could be marked as with or without a parked vehicle. The road had a mixture of houses with and without driveways, and these occurred in groups; for example, there were six groups of three neighbouring houses without driveways, 16 semi-detached properties and four detached houses with large gardens. This subset of 20 properties each had a drive. Two statistical tests were used to discover if the spatial patterns of occupied and unoccupied segments were randomly distributed. The outcome indicated that this was indeed the case with the Null Hypotheses being accepted for both tests (see Table 12.4). This is a comparatively simple dataset, and even if significant test results had been obtained, they might not have provided any clear indication of why vehicles were parked in certain segments. It would be problematic to argue that vehicles parked in segments outside houses without a driveway belonged to the occupants of those houses. Some could belong to friends who were visiting the residents or to a carer who was assisting someone with a disability. Similarly, houses with a driveway might include one, two, three or more adults who could all drive and owned vehicles, some of which were parked in the road because there was not enough space for all of them on the drive. Analysing a spatial pattern on its own is often insufficient to answer pertinent research questions. Examine the following self-assessment points and decide if additional data collection could have been undertaken that would provide clearer answers about why some segments of the road had vehicles parked in them, and others did not.

> The original fieldwork was carried out in the middle of the day and recorded if a vehicle was parked in each segment. How could a longer period of observation in the field provide more information about whether the vehicles belonged to drivers living in the adjacent house? How could you discover if there were any houses where there was not enough space in the driveway for all the vehicles driven by their residents? What time of day would be most suitable for gaining this additional information? Would the researcher need to recruit a co-worker to help with the data collection?

Table 12.4 Results obtained from applying statistical tests to parked car and other vehicles dataset.

Parked cars and other vehicles along one side of a road		
Quantitative Technique	**Result**	**Box number**
Wald–Wolfowitz runs test	$W–W(Z) = 0.134$ and $p = 0.897$: the Null Hypothesis was accepted. The sequence of occupied and unoccupied segments of the road with kerbside parked vehicles was not significantly different from what might be expected to have occurred if drivers had parked at random.	7.4a
Mann–Whitney *U* test	$U(Z) = 0.290$ and $p = 0.772$: the Null Hypothesis was accepted at the 0.05 level of significance. The presence or absence of vehicles in the 25-m road segments occurred randomly.	7.6a

12.2.5 Pittsburgh Fast Food Restaurants

The location of business premises is inevitably subject to change over time, and the addresses of the 64 fast food restaurants for two companies (Burger King and McDonald's) in Pittsburgh were obtained by searching on the internet in the early 2000s. These were plotted on a base map, and the outline of the city's built-up area was added to provide a context. Difficulties exist with obtaining commercially sensitive information such as the daily customer footfall for each of the outlets, which meant that hypothetical data were estimated. The dataset distinguished between the locations of the two companies, and in this respect, the statistical analyses present genuine differences in their spatial distribution across the city and its suburbs. The three types of correlation analysis (Pearson's product moment, Spearman's rank and Kendall's τ) concentrated on Burger King outlets and revealed a statistically significant (at 0.05 level) moderate negative relationship between distance from the city centre and daily customer footfall (see Table 12.5). This result was a starting point for modelling the relationship between these variables to determine if the distance from the city centre (the independent variable) might act as a predictor of the dependent variable, daily customer footfall. Such information might be useful for determining locations for opening new outlets and for closing some of the existing ones. When tested, the whole regression model and the intercept and slope coefficient were statistically significant at the 0.05 significance level. The coefficient of determination (R^2) value was less than 45 per cent (<0.45), which confirms that distance is a factor in explaining customer footfall, but that other variables contribute to the popularity of fast food outlets.

The fast food restaurants dataset also provided the opportunity to explore the spatial distribution of outlets across the city asking the question: were they distributed in a completely random spatial pattern? The group of spatial statistics used to answer this question produced some slightly contradictory results. Nearest neighbour analysis indicated that taken together the restaurants tended towards a dispersed pattern, which was statistically significant. Ripley's K function indicated some clustering in the second distance band out from the city centre. The quadrat-based techniques applied to just the Burger King restaurants produced contrasting results: the variance to mean ratio suggested they were randomly distributed across the quadrats, whereas using a χ^2 test indicated their frequency distribution across the quadrats was significantly different from what would be expected to occur randomly according to the Poisson distribution. Repeating these analyses with smaller quadrats could produce different outcomes. The results from the spatial and non-spatial statistics suggest some further reflection on the project design is required as indicated in the self-assessment questions below.

> The customer footfall data values were estimated in good faith, but how could a researcher have obtained reliable empirical data about the daily footfall of customers? What other factors could be added into the regression model as independent variables? How would you collect data about them for each of the restaurants? Is the customer profile for fast food restaurants near the centre of the city likely to be different from those in the suburbs? Is the mode of transport for getting to the restaurants likely to be different for these two types of locations? What effect might any difference have on the density of restaurants between the city centre and the suburbs?

12.2.6 Sample of Farms in South East England During the Early Years of World War II

The dataset was based on information in three surveys of farms in parts of south east England carried out in 1941, 1978 and 1998. It is a subset of the farms that were included in each of the surveys and illustrates a cross-sectional study of three periods in the development and modernisation of British agriculture. Curvilinear (polynomial) regression was used to model the relationship between farm size (area) and total horsepower of wheeled (non-tracklaying) tractors on the farms

Table 12.5 Results obtained from applying statistical tests to Pittsburgh fast food restaurants dataset.

	Pittsburgh fast food restaurants	
Quantitative Technique	**Result**	**Box number**
Pearson's product moment correlation r (tested using t test)	$r = -0.577$, $t = 0.511$ and $p = 0.00000537$: the moderately strong negative correlation between the distance of Burger King restaurants in Pittsburgh and the number of daily customers (hypothetical data values) is statistically significant with customer numbers decreasing with distance from the city centre.	8.1a
Spearmans rank correlation r_s (tested using t test)	$r_s = -0.582$, $t = 15.900$ and $p < 0.05$: the rank scores of the distance of Burger King restaurants in Pittsburgh and the number of daily customers (hypothetical data values) produce a slightly stronger negative correlation coefficient, which is significant at the 0.05 level when tested using the t test.	8.2a
Kendall's τ correlation (tested using Z test)	$\tau = -0.429$; $Z = 2.792$ and $p = <0.005$: the correlation coefficient indicated a slightly weaker negative relationship between the rank scores of the distance of Burger King restaurants in Pittsburgh and the number of daily customers (hypothetical data values).	8.3a
Ordinary least squares (OLS) regression	Burger King restaurants $\hat{y} = 7326.74 - 336.61x$ $R^2 = 0.34$; McDonald's restaurants $\hat{y} = 7802.1 - 251.42x$ $R^2 = 0.42$: OLS regression suggests that distance from the city centre in Pittsburgh was a slightly stronger predictor of customer footfall (hypothetical data values) in the case of McDonald's restaurants compared with Burger King outlets.	9.1a
ANOVA test of OLS regression models; t test of OLS regression intercept (a) and slope coefficient (b)	Burger King restaurants $F = 10.26$ and $p = 0.04$; McDonald's restaurants $F = 27.84$ and $p = <0.000$: both regression models were significant at the 0.05 level. Burger King restaurants $t = 3.20$ and $p = 0.04$; McDonald's restaurants $t = 5.28$ and $p = <0.000$: the Alternative Hypotheses were accepted in respect of the intercept (a) and slope coefficient (b) for both OLS regression equations using the t test.	9.2a
Nearest Neighbour Index (tested using Z test)	$R = 1.69$, $Z = 10.277$ and $p = <0.001$: the R statistic indicated a dispersed pattern of fast food restaurants, and the Alternative Hypothesis was accepted that this is unlikely to have occurred through chance when a Z test was applied to the R statistic.	10.2a
Ripley's K function (point features)	Ripley's K function chart showed there was a clustering of fast food restaurants in distance band 2 (21.0–41.9 m from the city centre), but in other distance bands the value of K was within the 95% confidence limits.	10.3a
Variance to mean ratio (tested using t test and χ^2 test)	VMR $= 0.66$; $t = -1.33$, $p = 0.220$; $\chi^2 = 5.336$, $p = 0.721$: the Null Hypothesis was accepted that the Burger King restaurants were located in a random pattern within a grid overlain on the map when both statistical tests were applied.	10.4a
χ^2 test applied to spatially gridded data	$\chi^2 = 69.74$ and $p = <0.000$ with 5 degrees of freedom: the Alternative Hypothesis was accepted that the location of the Burger King restaurants in Pittsburgh when overlain by a grid of nine quadrats (squares) is significantly different from what would be expected to occur at random according to the Poisson probability distribution.	10.5a

Table 12.6   Results obtained from applying statistical tests to south east England farm sample survey dataset.

Sample of farms in south east England surveyed by questionnaire in 1941, 1978, and 1998		
Quantitative Technique	Result	Box number
Curvilinear (polynomial) regression	First-order (OLS) regression: $\hat{y} = 0.033x + 29.77, R^2 = 0.11$;	9.3a and 9.4a
	Second-order polynomial $\hat{y} = 7.55 + 0.17x - 0.000095x^2$, $R^2 = 0.33$;	
	Third-order polynomial $\hat{y} = 0.76 + 0.25x - 0.0003x^2 + 0.000000008X^3$, $R^2 = 0.34$;	
	Fourth-order polynomial $\hat{y} = 37.66 - 0.36x + 0.0022x^2 - 0.0006x^3 + 0.99x^4$, $R^2 = 0.42$: curvilinear (polynomial) regression of the relationship between farm area (independent variable) and total horsepower of wheeled tractors (dependent variable) indicated that the fourth-order polynomial gave a slightly closer fit between the empirical data and the regression equation. The first-order, OLS, regression gave a poorer fit than any of the curvilinear models according to the coefficient of determination R^2.	
Multivariate regression	$\hat{y} = -1.098 - 0.034x_1 + 1.228x_2$, $R^2 = 36.5$: the multivariate regression model provides a moderate explanation for farm size in 1998 using the area in 1941 and 1978 as independent variables. The squared multiple correlation coefficient indicates 0.365 or 36.5% of the total variance is accounted for by the model.	9.5a
ANOVA test of multivariate regression model; t test of multivariate regression intercept (a) and slope coefficients (b)	$F = 4.89$ and $p = 0.21$ with degrees of freedom 2 and 17, respectively, for the regression and residual variances: the Null Hypothesis was accepted that the multivariate regression model with two independent variables (farm size in 1941 and in 1978) was not significant when used to predict farm area in 1998.	9.6a
	Slope coefficient b_1, $t = 0.183$ and $p = 0.857$; slope coefficient b_2, $t = 3.079$ and $p = 0.007$; intercept (a), $t = 0.005$ and $p = 0.996$: the t tests applied to the slope coefficients for the two independent variables (farm size in 1941 and in 1978) produced different results with the former not significant and the latter significant, and the t test for the intercept was also not significant. Overall, the results indicate that farm size at both of the earlier dates was not a good predictor of farm area in 1998.	

in 1941 (see Table 12.6). Tractors are an example of a fixed resource or input: put simply on a small farm, there may not be enough land to one tractor to be used to its full capacity, whereas on a larger farm it might be used up to its maximum. If the farmer of the larger farm acquired more land, it would be necessary to buy another 'whole' tractor of the same capacity or trade in the first one and buy a more powerful one with greater capacity. This illustrates the principle of a fixed, indivisible resource (input), although in practice there might be other options, such as hiring a contractor to work the extra land. The fourth-order polynomial regression model gave slightly closer fit to the empirical data than the first, second and third orders, which may be related to the point about tractors being fixed resources.

The three surveys were carried out over a period of nearly 60 years, and this allowed the analysis to use multivariate regression to examine if the total farm area in 1998 could be predicted from the farm's size in 1941 and 1978. The regression model and testing its significance required careful interpretation, but in summary previous farm area appears to offer a partial predictor of subsequent farm size, but it is evident there were other factors that should be included as independent variables. The following self-assessment offers a brief historical narrative about how the British agricultural industry changed between 1941 and 1998 and asks what additional variables you would like to find in the archive survey data for 1941 and 1978 that might allow a more complex, robust multivariate regression analysis.

Farms are complex businesses, and traditionally ownership or tenancy has often passed between generations in a family. It is not unusual to find farms that have been in the same family for over 100 years. However, during the period covered by the surveys a process of agricultural modernisation occurred and some features of this process, such as investment in farms run by managers appointed by financial institutions (e.g. pension funds and insurance companies), the introduction of policy measures to protect biodiversity and the countryside, and the diversification of agricultural businesses, are likely to have impacted on the simple model of economic growth, leading to an increase in farm area.

In 1998, recognising the impossibility of going back in time to re-survey the farms in 1941 and 1978, what additional variables would you look for in the complete archived 1941 and 1978 survey data files that might be added to the multivariate regression model that could improve its predictive power in respect of farm size in 1998?

12.2.7 Moraine Debris on Les Bossons Glacier Outwash (Sandur) Plain

The fieldwork carried out on the outwash (sandur) plain of Les Bossons outlined in Section 12.2.2 also allowed the researchers to photograph areas where there was debris from the moraine. It was assumed that spatial sorting had occurred because of fluvial processes across the area, which would have resulted in smaller debris being transported preferentially in comparison with larger items. The debris was also likely to have become aligned with the predominant direction of flow even if individual items had fallen from the lateral moraines rather than emerging with the meltwater stream from the glacier snout. The size, orientation and shape of the items varied across the area, and one of these photographs was used to generate the moraine debris dataset. A sample of 100 items in the photograph was selected at random, which were digitised with the length of their perimeter and long axis as well as its orientation computed in the digitising software. Statistical techniques were applied to determine whether these variables provided evidence of spatial autocorrelation, which might be expected if spatial sorting of debris was a prominent process. Ordinary least squares regression between surface area (independent variable) and axis length (dependent variable) indicated that the size of the residuals along the regression line rose as the size of debris items increased: in other words, the regression model was less successful in predicting debris axis length for larger items (see Table 12.7). Analysing the regression between spatially lagged debris items (the relationship between one debris item and its first, second, third, fourth, etc., nearest neighbour) provided evidence of spatial autocorrelation up to lag 3. This result suggested that further investigation was needed and global Moran's *I* was applied, which confirmed that a moderate level of statistically significant spatial autocorrelation was present. Geographically weighted regression was applied in recognition that the overall ordinary least squares regression was likely to be less successful in its predictive power across some parts of the area covered by the photograph.

Table 12.7 Results obtained from applying statistical tests to Les Bossons Glacier moraine debris dataset.

Moraine debris on Les Bossons Glacier outwash (sandur) plain		
Quantitative Technique	**Result**	**Box number**
OLS regression and spatial autocorrelation	$\hat{y} = 7.87 + 0.001x; R^2 = 0.905$: fitting an ordinary least squares model to a sub-sample of moraine debris with size (area) as the independent variable and long axis length as the dependent variable indicated the significant explanatory power of surface area as a predictor of axis length. However, examination of the residuals suggested these grew larger as debris surface area increased, which pointed to the presence of autocorrelation.	10.1a
Spatial lags and autocorrelation	Ordinary least squares regression r coefficients increased through lags 1, 2 and 3 (0.3341, 0.4015 and 0.7000) and declined for lag 4–0.515,56. The effect of spatial autocorrelation started to decrease after lag 3.	11.1a
Global Moran's I with point features (tested using Z test)	Moran's $I = 0.411$, $Z = 3.418$ and $p = 0.0003$: the Alternative Hypothesis was accepted on the basis that the statistic was significant at the 0.05 level and its value (0.411) indicated the presence of moderate global spatial autocorrelation in the overall distribution of morainic debris on Les Bossons Glacier outwash plain.	11.4a
Geographically weighted regression	None of the standardised residuals were below -2.5 or higher than $+2.0$ standard deviations from the mean, although two points were in the range $+1.5$–$+2.5$ and one point was between -2.5 and -1.5; these three points had R^2 values lower than the other points (0.45, 0.61 and 0.55). At these locations, the independent variable X (length of the long axis) was less successful as a predictor of the dependent variable Y (length of its perimeter).	11.7a

The self-assessment questions for this dataset focus on other questions that might be asked in relation to the geomorphology of the fieldwork area as a whole and might involve looking at the overview image of the study area in Chapter 6 again and the description of the meltwater stream dataset in Section 12.2.2 as well as searching for a contemporary aerial image of the area.

> The research team have the opportunity to visit the outwash (sandur) over a series of three days. What other aspects of its physical geography, including geomorphology, hydrology and vegetation, could be investigated? What data collection procedures could be developed to address these additional questions? What statistical techniques could be used to analyse other datasets that might be collected?

12.2.8 Slit Trenches Dug for Defensive Purposes on the South Downs, England, During World War II

This dataset was created by digitising from aerial photographs the slit trenches on the South Downs, England, which were dug as part of the defensive works undertaken during WWII to deter enemy invasion. They were intended to prevent enemy gliders from landing silently on this upland area of south east England, which in places reaches an elevation of approximately 300 m. The South Downs has a rolling landscape with dry valleys and close-cropped grass. They are part of the Weald-Artois

Table 12.8 Results obtained from applying statistical tests to World War II slit trench on the South Downs, England, dataset.

Slit trenches dug for defensive purposes on the South Downs, England, during World War II		
Quantitative Technique	**Result**	**Box number**
Ripley's K function (linear features)	Ripley's K function chart showed there was a clustering of slit trenches in all distance bands apart from the first (0.0–309.9 m) from their centre with the observed value of K outside the 95% confidence limits.	10.6a

Anticline formed during the Alpine orogeny that arched across south east England and northern France. Visual inspection of the trenches suggests that they were not randomly distributed across the study area but might have been clustered together or dispersed (see Table 12.8). Applying Ripley's K function to the trenches using the centre point of each as its location revealed that at all distance bands apart from 0.0 to 309.9 m, they were more clustered than would have been expected to occur if they had been dug at random. The following self-assessment question explores the military strategy that could have justified such an arrangement.

> Given the result of analysing the spatial pattern of the slit trenches, discuss the strategic thinking that might have underlain their clustering in distance bands. How could such an arrangement of tranches have impeded the landing of gliders on the Downs?
> How could the orientation of the trenches be analysed?

12.2.9 Fields Identified for Ploughing Up on Farms on the South Downs, England, During World War II

The British government was concerned about increasing food supplies during WWII and instigated a campaign to encourage farmers to plough up relatively unproductive land, which was sometimes weed-infested, poorly managed permanent grass, and to re-sow it to crops that could be used to increase home-grown food production. A survey carried out during the early years of WWII (the National Farm Survey, 1941–1943) was the means whereby these plough-up fields were identified. Applying Ripley's K function to the spatial distribution of the plough-up fields in a case study area on the South Downs, England, revealed strong, statistically significant evidence to conclude that these fields were clustered together rather than being randomly distributed or dispersed across the area (see Table 12.9). Given that many farms in the mid-20th century in this part of England were compact with a limited number having parcels of land detached from the main holding, this statistical evidence of plough-up clustering hints at certain farms being more likely than others to have unproductive land and that it was not a randomly distributed phenomenon. The following self-assessment questions prompt suggestions about aspects of these plough-up fields that might be worthy of statistical analysis.

> Were fields identified by surveyors as suitable for plough up more likely to occur on lower or higher land on the South Downs? Did they occur preferentially on land with a northerly, easterly, southerly or westerly aspect? How much land was designated for re-sowing with different types of crops (wheat, barley, oats, potatoes, etc.)? What was the mean area of plough-up fields? Were they significantly larger or smaller than other fields in the study area?

Table 12.9 Results obtained from applying statistical tests to plough-up fields on the South Downs, England, dataset.

Fields identified for ploughing up on farms on the South Downs, England, during World War II		
Quantitative Technique	**Result**	**Box number**
Ripley's K function (polygon features)	Ripley's K function chart revealed there was a clustering of field designated for ploughing up in distance bands 2–8 (1960.0–3919.9 m to 13 720.0–15 679.9 m) from their centre with the observed value of K outside the 95% confidence limits.	10.7a

12.2.10 Distribution of *Dianthus gratianopolitanus* on Mont Cantal in the Auvergne, France

The dataset was created by overlaying a photograph of part of the slope of the extinct Mont Cantal volcano in the Massif Central, France, with a grid of squares and then examining each square to determine if it contained at least one *D. gratianopolitanus* (common name Cheddar Pink) plant. The presence and absence of this species in the squares enabled them to be shaded black or white respectively and the join count statistical technique was applied to discover if the spatial pattern was random or otherwise (dispersed/clustered). Each type of join (B–B, W–W and B–W/W–B) between adjacent squares was found to be statistically significant at the 0.05 level and provides evidence of spatial autocorrelation being present (see Table 12.10). The self-assessment question for this dataset prompts further research to discover the ecological context in which this plant species occurred on the slopes of Mont Cantal.

A single plant species often occurs in conjunction with other species as they fulfil their niche roles in an ecosystem. Further research using the grid squares in the study area could involve counting the number of distinct species found in each square. Some species might be competitive with each other, and it could be the case that some squares without any *D. gratianopolitanus* might be dominated by another plant species. What environmental conditions promote or inhibit the growth of *D. gratianopolitanus*? Is there any evidence of conditions unsuitable for this species in those grid squares where it is absent?

Table 12.10 Results obtained from applying statistical tests to *Dianthus gratianopolitanus* on Mont Cantal in the Auvergne, France, dataset.

Distribution of *D. gratianopolitanus* on Mont Cantal in the Auvergne, France		
Quantitative Technique	**Result**	**Box number**
Join count statistics (tested using Z test)	Expected B–B joins $= 29.82$, $Z = 2.54$ and $p = 0.011$; expected number of W–B/W–B joins $= 87.96$, $Z = 4.49$ and $p = <0.000$; expected W–W joins $= 62.22$, $Z = 6.22$ and $p = <0.000$: the Alternative Hypotheses were accepted for each type of join, which supports the argument that global spatial autocorrelation was present as evidenced by a clustering of *D. gratianopolitanus* in the quadrats on Mont Cantal in the Auvergne.	11.2a

12.2.11 Index of Multiple Deprivation in the 33 London Boroughs, 2007

The dataset focused on the 2007 Index of Multiple Deprivation (IMD) for the 33 London boroughs. It was extracted from the national IMD for 2007, one of a series of such secondary sources of data analysed by Geographers and others to explore the different forms of deprivation and disadvantage in the population. The 2007 IMD included a series of domains (income, employment, health and disability, education, crime, barriers to housing and services, and living environment), which could be explored on their own and were combined into a unified overall deprivation score. The global Moran's *I* statistic indicated the presence of statistically significant spatial autocorrelation across the boroughs (see Table 12.11). The local indicator of spatial association revealed where clusters of positive and negative spatial autocorrelation occurred locally: there were two boroughs in each of High–High and Low–Low categories (positive spatial autocorrelation) and two in the Low–High one (negative). Trend surface analysis added to these results by suggesting there was less deprivation in boroughs towards south-west London compared with those in the north-east. The following self-assessment questions invite you to consider how this analysis could be enhanced.

> The population of London in 2007 was 7.7 million, and the IMD analysis focused on 33 spatially large areas, which may have hidden local pockets of deprivation within boroughs that were relatively advantaged overall. Search the internet to discover if the 2007 IMD data are available for smaller areas, which could be analysed using the techniques presented here. What do we discover about concepts such as deprivation by investigating them at different geographical scales (for varied sizes of geographical units)? What measurement of size is important in such analyses? Is it population total, physical area or some other measure?
>
> Did areas with high overall IMD scores also have high ranks for all of the domains, or did some areas have a mixture of high and low values for the separate domains?

Table 12.11 Results obtained from applying statistical tests to the 2007 Index of Multiple Deprivation in London Boroughs dataset.

Index of Multiple Deprivation in the 33 London boroughs, 2007		
Quantitative Technique	**Result**	**Box number**
Global Moran's *I* with polygon features (tested using *Z* test)	Moran's $I = 0.305$, $Z = 2.63$ and $p = 0.004$: the Alternative Hypothesis was accepted that there was evidence of moderate global spatial autocorrelation in the distribution of deprivation across the 33 London boroughs as a whole.	11.3a
Local indicator of spatial association Moran's *I* with polygon features (tested using *Z* test)	Low–Low LISA category – two boroughs in south-west London (Kingston upon Thames and Sutton); High–High LISA category – two boroughs east of the centre (Greenwich and Newham); and Low–High LISA category – two central London boroughs (City of London and Hackney): these were significant at the 0.05 level and indicate positive local spatial autocorrelation for the HH and LL boroughs and negative for the LH ones.	11.5a
Trend surface analysis	$Z_{x,y} = 147.72 + 0.00020x + 0.00037y$: the predicted values ($Z$) of the IMD score for the 500-m grid squares simplified the overall spatial pattern and suggested deprivation followed a north-east to south-west downward trending surface.	11.6a

12.3 Describing and Presenting Quantitative Results in Geographical Journal Articles

Students writing up their independent research projects are doing something not unlike researchers crafting articles for publication in academic journals. This section examines a small selection of five open-access journals from across the spectrum of geographical and environmental research to identify generic ways in which datasets, quantitative statistical analysis and results are presented. We focus on the ways and means by which researchers have described, tabulated and illustrated their research in journal publications and aim to offer a practical demonstration of best practice. Open-access journals have become increasingly popular during the early decades of the 21st century, although archived publications that are frequently available to students and researchers through institutional subscription will allow you to explore a far greater range of output than is feasible here. The articles have been selected not only for reasons of easy access and thematic coverage but also because the authors have used a range of statistical techniques described in previous chapters. Therefore, they offer practical examples of how to write up your independent project research in a style considered acceptable for academic writing. The selected journal articles are fully referenced in the Selected Reading and References at the end of this chapter, but for convenience, Table 12.12 summarises their authorship and thematic focus. Each article has been identified by a letter to simplify the following discussion rather than giving its full citation.

The thematic substance and argument of the research reported in the articles is not the focus of attention here; instead, the approach taken examines them in a way like medics triaging patients or casualties. However, rather than determining the urgency of their need for and the nature of any treatment required, the aim is to highlight a set of key 'signs and symptoms' in the article in relation to the data, analysis and reporting of the research. The five articles reveal a mixture of primary and secondary sources of data being used either individually or in combination with each other (see Table 12.13). Articles B and C, falling broadly under the headings of environmental science and geomorphology, had involved researchers going into the field and collecting physical samples of material from the environment. Articles A and D used secondary sources of data: in the first instance, a survey of farming households; in the second, official population census and population register data. Article E involved a single questionnaire survey, which when distributed via email produced a low response rate, which was subsequently boosted by mailing out a hard copy to '20 Geography departments on the South Island and Upper North Island, randomly sampled from the original list' (Taylor, 2011: 191). Three of the five articles stated the software used for analysis: A and B used the R statistics package, which has become increasingly widely used on account of its

Table 12.12 Selected open-access journal articles' themes and authorship.

Identifier	Authors	Title
A	Dallimer *et al.*, 2019	Who uses sustainable land management practices and what are the costs and benefits? Insights from Kenya.
B	Hinshaw *et al.*, 2022	Development of a geomorphic monitoring strategy for stage 0 restoration in the South Fork McKenzie River, Oregon, USA.
C	Krishnaswamy *et al.*, 2001	Spatial patterns of suspended sediment yield in a humid tropical watershed in Costa Rica.
D	Krisjane *et al.*, 2023	Uneven geographies: ageing and population dynamics in Latvia.
E	Taylor, 2011	Year 11 Geography teachers' response to the Darfield earthquake.

Table 12.13 Selected open-access journal articles: themes and authorship.

			Article		
	A	B	C	D	E
Study area(s)	Three counties in Kenya.	South Fork McKenzie River, Oregon, USA.	Terraba River basin, Costa Rica.	Latvia.	New Zealand.
Data source(s)	Secondary source: non-random surveys of 320 farming households in 2014.	Primary source: field survey data collection in 40 hexagonal plots each with four 1 m radius circular subplots selected randomly from 4000 hexagons.	Primary source: water sampling to measure sediment concentration, monthly (more frequent during storm events). Secondary source: rainfall and fluvial data. Data collected for eight stations. Landsat Multispectral Scanner imagery.	Secondary source: population census and population register; anonymised, geocoded individual-level data. Spatial analysis at sub-municipal level and hexagonal grid units.	Primary source: email and postal questionnaire survey of randomly selected schools. Low response rate (28 teachers), but widely spread.
Software used	*R package.*	*R package.*	Not stated	ESRI ArcGIS	Not stated
Analysis techniques	Regression with 21 independent (explanatory) variables.	*t* test, two sample *t* test, regression and Wilcoxon signed ranks.	Regression and spatial analysis.	Spatial autocorrelation (global Moran's *I*) and spatial clustering analysis (Getis-Ord Gi).	Pearson's chi square.
Descriptive statistics	Described in text and tabulated.	Described in text and tabulated.	Described in text and tabulated.	Described in text, tabulated and mapped.	Described in text and tabulated.
Statistical results	Regression models described in text and tabulated.	Test results described in text and tabulated.	Regression models described in text.	Global Moran's *I* tabulated.	Test results described in text.
Graph(s)/ chart(s)	Graphs of significant regression models.	Graphs of significant results.	Regression scatterplots.	Gender age structure chart.	Graphs of frequency statistics.
Map(s)	None.	Aerial image of the study area.	Maps of erosion indices.	Getis-Ord Gi mapped.	Map of sample distribution.

flexibility; Article D, which applied spatial statistics, used one of the widely available GIS software packages (ESRI ArcGIS). Statistical techniques covered in Chapters 6–11 were used in all articles, although some extended these to more complex versions.

It is not surprising that all articles provided a summary of the variables as descriptive statistics before covering the application and results from the statistical analysis. There was some discussion of descriptive statistics in the text of all articles, and in most cases, this was linked to tabulated and/or graphical illustrations of key information. Articles A, B and D included at least one table that summarised the results produced from the statistical analysis. This type of summary can be a very concise and informative way of presenting complex information, for example allowing readers to discover those variables with significant and non-significant test results succinctly. This approach may be compared with the other two articles: in C regression coefficients and the intercepts together with their statistical testing results are described in the text; Article E outlines Pearson's chi-square test results in an equivalent way. All articles include some form of graphical illustration to support text-based description of statistical results and where appropriate maps have also been included. The focus on spatial statistics in article D has led to illustrating the hot and cold spots (locations of spatial autocorrelation) identified by the Getis-Ord Gi analysis becoming a central part of the research output.

Search one of the websites that lists published academic journal articles, such as Google Scholar or the Web of Science, to find an article that interests you. Explore the ways in which the author(s) discussed the data analysed, the statistical techniques used and the ways in which results and findings were presented. The aim is to find an article written by researchers in Geography, Environmental or Earth Sciences that includes some form of statistical or spatial statistics analysis. Rather than searching for articles on a specific topic, such as coastal erosion or football stadia, you are likely to have more success by using combinations of search terms such as the following:

regression, ecology and probability;
questionnaire, non-parametric and health;
geostatistics, geomorphology and glaciers;
correlation, survey and urban.

You should also specify that the discipline of the journals to be searched is Geography, Environmental or Earth Sciences.

12.4 Software for Quantitative Analysis

Mather (1991: 52) suggested 'the most probable reason' for geography students dislike statistics 'is the perceived association between statistics and tedious calculation'. More than 30 years since his assertion and the virtual eradication of a need for 'tedious calculation' through the spread of statistical software, some reluctance remains for students of Geography, Environmental and Earth Sciences to engage with statistics without a certain amount of hesitancy. Information technology had already made 'tedious calculation' unnecessary in 1991, although learning how to use software was not a simple task. There were many things you needed to know, or more correctly not to forget, when trying to persuade computer software in the early 1990s to give you what you wanted. This is not the place for a detailed account of the development of information technology, whether for statistical or other purposes. However, it is now feasible to claim that the widespread availability and the comparatively straightforward user interfaces present on inexpensive devices as well as the licensing arrangements in place between software providers and educational institutions allowing free access makes applying statistical techniques in student projects a less challenging task than it was previously. Therefore, it is perhaps surprising that students from the 'Millennials' generation (born 1981–1996) and 'generation Z' (born 1997–2012), who have been exposed to information technology throughout their adult lives and in the latter case from their early years, should still display some aversion to statistics.

Undoubtedly information technology has removed 'tedious calculation' from teaching and learning about statistics, but recent problems have arisen, and some old ones remain. There may be an assumption that computers will give you the correct answer to a question, but the adage 'garbage-in-garbage-out' (GIGO) applies. You can ask the correct question of incorrect data. Recall that calculating the mean of a set of numbers involves adding them together and dividing by how many numbers there are (e.g. $5 + 1 + 3 = 9/3 = 3$). Suppose the second data value had accidentally been entered into the computer as 11 instead of 1, the statistical software would have reported the mean as more than double to true mean ($5 + 11 + 3 = 19/3 = 6.6$). The computer gave the correct answer to the question it was asked, but one of the numbers was 'garbage'. Once we have entered our project data into statistical software, it is very easy to ask it to produce all the results we need as well as a lot that we do not. Mather (1991: 28) also suggested that there would come a time when expert systems would exist capable of storing 'the wide-ranging knowledge covering the nature of the [statistical] techniques and the assumptions on which they are based'. The growth of artificial intelligence (AI) in recent years attests to Mather's foresight. Such expert systems will eventually free researchers and students from the need to decide what statistical techniques should be applied and what the results indicate, but unless we are to relinquish our quest for discovery and seeking answers to questions the need to devise new, original research seems likely to continue. The following subsections introduce selected key aspects of using information technology for quantitative analysis and visualisation of results.

12.4.1 Data Input and Storage

The chapters in this text introducing the nature of data and the spatial and non-spatial statistical techniques that can be used for their analysis include an illustration of the calculations involved, not so that readers would be able to carry out 'tedious calculation', but in order to help with explaining where the numbers produced by statistical software come from and what they mean. The means of collecting data change over time. Chapter 3 included ways such as social media, crowd-sourced data and online surveys that were not common when the first edition was published. Students are using these methods and established procedures in their independent projects to discover answers to their research questions. Each project is unique, but we propose some scenarios that characterise feasible options for data collection and assembly in student projects:

1) Carry out an online questionnaire survey of individuals, charities, businesses, etc., and follow this with interviews lasting up to 30 minutes with a subset of questionnaire survey respondents to explore key topics in greater depth.
2) Download a dataset from the internet that includes all entities of the required type in the study area, such as burglaries or point pollution incidents, that relate to a certain time period.
3) Record stream flow measurements, and take water samples for water chemistry analysis for sample points on 10 streams in a drainage basin.
4) Capture digital photographic images of historical census documents in a national archive, and download digital historical large-scale topographic maps showing land parcels from the same era.
5) Extract the tracklogs from GPS devices used by a sample of participants who had completed a simple background questionnaire and a map their routes followed on a topographic map in conjunction with IMD area data to determine the types of area visited.

Some online sources of data also offer the opportunity to produce descriptive statistics and graphs and to carry out simple spatial interrogations such as measuring distance or area, although these are often restricted in the complexity of facilities available. Online questionnaire services usually include tools to summarise the responses to questions, but these rarely go beyond simple descriptive statistics and graphics such as pie and bar charts. The range of analysis available through such interactive web-based services should not be regarded as defining the range of analysis required to answer your research questions.

These scenarios reveal there is a range of ways in which researchers can obtain and combine their data and that some of these involve using data already stored in a digital and analysable format and others require transfer from one medium to another. For example, digitally recorded interviews may be put through speech recognition software to create data suitable for text analysis, land parcels shown on historical topographic maps may need to be digitised for use in a GIS or chemical concentrations in water samples may need to be entered via a keyboard into software on a computer. The data referred to in some of these scenarios will have already undergone a certain amount of 'processing' before reaching the researcher, for example a dataset downloaded from the internet or a tracklog on a GPS device. Some scenarios incorporate raw data that have not yet been processed in any way and need to be entered or input onto a computer, thus turning them into a digital rather than analogue format. The streamflow, water chemistry and images of historical documentary data are likely to require conversion before starting their statistical analysis. Direct entry involves typing at a keyboard, digitising, scanning or using another device capable of capturing data and will usually be carried out in conjunction with the software that will be used for analysis.

The transfer of data from one format to another or from one computer to another has become a simpler task over recent decades as interoperability, the ability to exchange and use of data and information on computer systems and in software manufactured by different companies, has become normalised. One persistent feature remains, which offers a conceptual and practical framework for understanding data input and storage. This is the simple rectangular data matrix of rows and columns, in which a single row stores all the data for one entity (observation, feature, etc.) and each column holds one variable (or attribute) for all entities. This simple structure is a way of organising all the variables and all the entities, in other words a dataset, such that it is possible to issue instructions to software that will allow the retrieval of a specific piece of information. One important drawback of the simple rectangular matrix as a data storage structure is that it requires the data rows for all entities to have the same variables (attributes) even if the principal entities themselves may be subdivided into 'sub-entities' that can occur a different number of times for each entity in the dataset. Households provide a classic example of where this situation arises. Households are made up of individual persons, but the number of persons in a household can change over time and a set of households at a given point in time are likely to have different numbers of persons. Each person in a household may also be subdivided, perhaps in relation to their activity or lifestyle. For example, the research might be interested in variables associated with their separate visits to a gym, cinema or restaurant over the previous month. Some people might go to a gym three times a week, others once a fortnight and yet others not at all, and a similar or greater frequency of visits to a cinema or restaurant is also possible. The whole entity (household) and its different 'sub-entities' (persons and visits to gym, cinema and restaurant) can be thought of as tiers or levels in a nested hierarchy, which has given rise to a second important data structure as an alternative to the simple rectangular matrix.

Table 12.14 uses the example of survey of households, the persons in these households and the social media platforms they use; the upper part illustrates this simple data structure and the lower

Table 12.14 Simple rectangular data matrix and nested hierarchical data structures.

Simple rectangular data matrix

	3 household variables common to all persons			Individual person variables, repeating sets of 4 variables for each person												Individual person social media variables, repeating sets of 2 variables for each social media for each person in each household											
	Household			Person 1				Person 2				Person 3				Person 1 Social Media 1		Person 1 Social Media 2		Person 2 Social Media 1		Person 2 Social Media 2		Person 3 Social Media 1		Person 3 Social Media 1	
Entity	1	2	3	1	2	3	4	1	2	3	4	1	2	3	4	1	2	1	2	1	2	1	2	1	2	1	2
1																											
2																											
3																											

Household 1 has 3 persons, persons 1 and 3 both use 1 social media platform and person 2 uses 2 social media platforms.
Household 2 has 3 persons, person 1 does not use any social media platform, person 2 uses 1 social media platform and person 3 uses 2 social media platforms.
Household 3 has 1 person who uses 2 social media platforms.
Total data entries = 27 x 3 = 91

Nested Hierarchical Data

Entity	Person ID	Social media ID	Combined row (record) type	Variables			
1	0	0	100 (household)	Household variable 1	Household variable 2	Household variable 3	
1	1	0	110 (person)	Person variable 1	Person variable 2	Person variable 3	Person variable 4
1	1	1	111 (smedia)	Social media variable 1	Social media variable 2		
1	2	0	120 (person)	Person variable 1	Person variable 2	Person variable 3	Person variable 4
1	2	1	121 (smedia)	Social media variable 1	Social media variable 2		
1	2	2	122 (smedia)	Social media variable 1	Social media variable 2		
1	3	0	130 (person)	Person variable 1	Person variable 2	Person variable 3	Person variable 4
1	3	1	131 (smedia)	Social media variable 1	Social media variable 2		
2	0	0	200 (Household)	Household variable 1	Household variable 2	Household variable 3	

				Person variable 1	Person variable 2	Person variable 3	Person variable 4
2	1	0	210 (person)	Person variable 1	Person variable 2	Person variable 3	Person variable 4
2	2	0	220 (person)	Person variable 1	Person variable 2	Person variable 3	Person variable 4
2	2	1	221 (smedia)	Social media variable 1	Social media variable 2		
2	3	0	230 (person)	Person variable 1	Person variable 2	Person variable 3	Person variable 4
2	3	1	231 (smedia)	Social media variable 1	Social media variable 2		
2	3	2	232 (smedia)	Social media variable 1	Social media variable 2		
3	0	0	300 (household)	Household variable 1	Household variable 2	Household variable 3	
3	1	0	310 (person)	Person variable 1	Person variable 2	Person variable 3	Person variable 4
3	1	1	311 (smedia)	Social media variable 1	Social media variable 2		
3	1	2	312 (smedia)	Social media variable 1	Social media variable 2		

Total data entries = 3×3 household variables + 7×4 person variables + 9×2 social media variables = 55 section a more complex structure. Three households with the following characteristics are used to explore these data structures:

- Household 1 has 3 persons; persons 1 and 3 both use 1 social media platform, and person 2 uses 2 social media.
- Household 2 has 3 persons; person 1 does not use any social media platform, person 2 uses 1 social media and person 3 uses 2 social media.
- Household 3 has 1 person who uses 2 social media platforms.

The rectangular data matrix requires 91 pieces of data (variables) to be entered and stored for each household, irrespective of the number of persons present and their usage of social media. The maximum household size in the three example households is three, but if a complete survey included larger households and some people used more social media platforms the total number of variables per household would increase. Some of the data items are redundant in the rectangular matrix because the three households do not have the same number of persons and each person does not use the same number of social media platforms. An alternative data structure, nested hierarchical data, takes advantage of this characteristic by having separate rows for the principal entity, household in this instance, and for each type of 'sub-entity'. Here tier two 'sub-entities' are persons and tier three 'sub-entities' are social media platforms. The data for each principal entity and 'sub-entity' are stored in separate rows with the first variable or field used to instruct the computer software about what type of row and variables are stored. Creating this type of hierarchically nested data structure may require more planning and increases the number of rows entered, but in this example a reduction of nearly 50 per cent to 55 data items is achieved.

Identify two other examples of entities that can be subdivided into 'sub-entities', and list some of the variables that could be recorded for the entity as a whole, for its first and second tiers of subdivision.

Hint: might such a hierarchical data structure be relevant to fast food restaurant companies?

12.4.2 Data Processing and Statistical Software

Software or programs may be viewed as sets of instructions issued to a computer enabling a user to achieve a certain outcome. Researchers and students in the 1960s and 1970s seeking to use computers for statistical analysis were often faced with the challenge of learning how to write programs for a specific piece of analysis. Most Geography, Environmental and Earth Science students will now use packaged software to carry out their statistical analysis. These packages have bundled together all the instructions you will need to produce your analytical results, although ongoing refinement of such software results in new versions becoming available on a regular, annual basis. Once you are familiar with the basics of such a package, the arrival of an updated version does not usually mean 'unlearning' what you have already learnt but involves adapting to a different interface or making use of additional statistical techniques. Some researchers have veered away from packaged software over the past 25 years or so, preferring to use a programming language for statistical computing, such as Python or R. However, for those readers seeking to 'dip their toes' into the murky waters of quantitative statistical analysis, packaged software represents a safer entry point.

Using packaged statistical software involves applying three main types of operation on your data for entities and their variables: planning and organising; selecting, calculating or classifying; and carrying out statistical techniques using definitional equations. Careful planning at the data collection stage of a project can help to avoid problems when capturing data in a digital format. We have seen that the simplest way of organising your data is as a rectangular matrix. The term metadata refers to 'data about data' and provides a way of instructing the software about what it can expect to find in your dataset. The scales of measurement (nominal, ordinal, interval and ratio) introduced in Chapter 4 is one piece of metadata, but we may also want to store other types of information such as dates, times and text characters (words) as well as specifying if numeric variables are integers (whole numbers) or decimals, including the number of decimal places. The text strings associated with nominal attributes often contain a large number of characters, such as the descriptions of dwelling types and of a person's economic status, and instead of entering each in full every time it occurs, they are usually given unique code numbers, entering these once and informing the software that these are labels corresponding to the data values (codes) that have been entered.

Table 12.15 illustrates performing some of these operations on a dataset using the household, person and social media entities and variables outlined in Table 12.14. The first column in Table 12.15 lists the short variable identifiers used in Table 12.14 to indicate the three household-level variables (HVar1 to HVar3), the four person-level variables (PVar1 to PVar4) and the two social media variables (SVar1 to SVar2). Columns two and three describe these variables and how they were recorded on the questionnaire. Some were recorded as real numbers, year arrived at the present address, a person's age, hours employed in the previous week and minutes per day spent on a social media platform; other variables were captured as numbers that represented codes for groups of words, dwelling type, gender, economic status and name of social media platform. The household's postcode was stored as an alphanumeric string (letters and numbers). Two of the original variables were used to calculate new variables (see column four in Table 12.15). HVar2, used to record the year the household moved to their present address, was turned into a new variable (NVar1) by subtracting the year of arrival from 2024 (current year) to obtain the number of years the household had lived there. There are 1440 minutes in a day, and the variable SVar2, the number of minutes spent per day on first named social media platform, was converted into the percentage of the minutes in a day spent on that platform. There would be a similar computation for a person's second social media platform if required, and both would also need to be repeated for each person in the household (see Table 12.14).

The second way of creating new variables is by recoding or classifying the data recorded on the questionnaire, which has been shown for two of the original variables (PVar1 and PVar4) and for the two newly computed variables (NVar1 and NVar2) in Table 12.15. The precise way of specifying such recoding or classification varies between software, and the example value ranges (e.g. lowest through 9, 10 through 19, 20 through 29 and 30 through highest for NVar1) shown in column five are illustrative. The new variable in each contains numeric codes (1, 2, 3, 4, etc.) that correspond to the ranges. The last column shows the data labels that could be specified for some of the original and new variables, apart from the household's postcode. The postcode was recorded because the researcher intended to join this data to an external data source of georeferenced postcodes that would enable to approximate location to be mapped, thus illustrating the geographical spread of respondents to the survey.

Earlier in this section, we mentioned a third type of operation that users apply to their data using packaged statistical software, namely, to carry out certain analytical techniques. The precise details will vary between software, but the essential principles involve selecting the type of analysis required (e.g. paired t test, Pearson's chi-square test, Spearman's rank correlation or simple linear regression), specifying the variables and choosing options including the format of output to be produced (tables, graphs, etc.). The nature of the choices made for each of these will depend on the analysis to be undertaken. Regression techniques will require at least one dependent and

Table 12.15 Variable naming and labelling, and creation of new variables by computation, recoding or classification.

Variable identifier	Variable description	Data recorded	New variable created by computation	New variable created by recoding, classification or other processing		Data labels	
				Recode NVar1	Value	Code	Label
HVar1	Postcode	Standard format (e.g. BN19RH, CO43SQ and KT12EE)	Not applicable	Join recorded postcode with data from an external source of georeferenced postcodes		Not applicable	
HVar2	Year arrived at address	Recorded as year (e.g. 1977, 1982 and 1987) on questionnaire	Years at present address: NVar1 = 2024 – HVar2	Recode NVar1			
				Lowest through 9	1	1	Less than 10 years
				10 through 19	2	2	10–19 years
				20 through 29	3	3	20–29 years
				30 through highest	4	4	30 years and over
HVar3	Dwelling type	Recorded as 1, 2, 3, 4, 5, etc., corresponding to pre-coded list of dwelling types on questionnaire	Not applicable	Not applicable		Code	Label
						1	Detached
						2	Semi-detached
						3	Terraced
						4	Flat or apartment
						5	Other
PVar1	Age	Recorded as age in years (26, 47, etc.) on questionnaire	Not applicable	Recode PVar1	Value	Code	Label
				18 through 24	1	1	18–24 years
				25 through 34	2	2	25–34 years
				35 through 54	3	3	35–54 years
				55 through 74	4	4	55–74 years
				75 through 84	5	5	75–84 years
				85 through highest	6	6	85 years and over

				Code	Label
PVar2	Gender	Recorded as 1, 2 or 3 corresponding to pre-coded self-reported gender on questionnaire	Not applicable	1	Male
				2	Female
				3	Other
PVar3	Economic status	Recorded as 1, 2, 3, 4, 5, 6, 7, 8 or 9 corresponding to pre-coded list of economic status categories on questionnaire	Not applicable	Code	Label
				1	Working as employee
				2	Working self-employed/ freelance
				3	Working paid or unpaid in own/family's business
				4	Away from work maternity/ paternity or sick leave
				5	Student or training scheme
				6	Retired
				7	Long-term sick or disabled
				8	Looking after home or family
				9	Other

				Recode PVar4	Value
PVar4	Hours paid work	Recorded as hours worked in the previous week for a person in any form of paid employment at PVar3 on questionnaire (only applies to persons with codes 1–8 at PVar3)	Not applicable	1 through 15	1
				16 through 30	2
				31 through 48	3
				49 through highest	4

				Code	Label
				1	15 hours or less
				2	16–30 hours
				3	31–48 hours
				4	49 hours or more

				Code	Label
SVar1	Name of social media platform	Recorded as 1, 2, 3, 4, 5, 6, 7, 8, 9, 10 or 11 corresponding to pre-coded list of economic status categories on questionnaire	Not applicable	1	Facebook
				2	YouTube
				3	Instagram
				4	WhatsApp
				5	WeChat
				6	TikTok
				7	Messenger
				8	LinkedIn

(Continued)

Table 12.15 (Continued)

Variable identifier	Variable description	Data recorded	New variable created by computation	New variable created by recoding, classification or other processing	Value	Code	Label
						9	Telegram
						10	Snapchat
						11	Other
SVar2	Minutes per day on social media platform	Recorded as minutes per day on questionnaire	Percentage of 1440 minutes/day on social media: NVar2 = (SVar2 $*$ 100) / 1440	Recode NVar2			
				Lowest through 4.9	1	1	Less than 5%
				5.0 through 9.9	2	2	5-9.9%
				10.0 through 14.9	3	3	10-14.9%
				15.0 through 19.9	4	4	15-19.9%
				20 through 24.9	5	5	20-24.9%
				25 through highest	6	6	More than 25%

independent variable, although more of each type could be included if a series of models are to be examined. If confidence intervals are required, the 95 or 99 per cent versions could be selected. Finally, some statistical software allows for sophisticated treatment of missing data.

Missing data can occur for several reasons, but it means that a dataset is not as complete as it could be. The survey of households, the persons in them and each person's use of two main social media platforms outlined in Tables 12.14 and 12.15 will be used to illustrate some types of missing data. First, some persons in a household may be unwilling to give their age in years and others using two social media platforms might use one much more than the other and be unable to answer with accurate figures for the number of minutes per day spent on each. Any missing ages would leave gaps in the dataset, and estimates of time spent on a social media platform would either be accepted as possibly unreliable or removed by the researcher. The questionnaire has been designed so that one person answers the household-level questions and the individuals complete the person- and social media-level questions. Delivering such a questionnaire online would raise challenges as it would either require everyone in a household to complete it together or one after the other, or for one or more persons to answer the person and social media questions for everyone even if they are not present at the time to confirm the information. Such a strategy is likely to lead to an indeterminate amount of inaccuracy being present. The researcher could decide to visit households in the survey at a suitable time of day with repeat visits as necessary to obtain the details of persons not present on the first occasion. However, after four visits and still with no answers from the last person in a household, the researcher might have to treat this as missing data.

The presence of missing data, or more correctly the absence of all expected data, is not only associated with datasets produced by surveys through people refusing to answer certain questions, providing unreliable answers or being uncontactable, but also occurs with other sources of data. Researchers have become increasingly attracted to converting historical documents (maps, reports, statistical accounts and narrative descriptions) into digital data with the intention of applying modern quantitative and qualitative analytical techniques to them. Researchers using such archived sources may come across missing or damaged records, resulting in an incomplete dataset. One notable example in the UK concerns the 1931 census records. British historical census documents are released under the 100-year disclosure rule for research and personal study; the most recent set becoming available are those from the 1921 population census. However, the 1931 census records were unfortunately destroyed by an explosion and fire on 20 December 1942 (unconnected with wartime enemy action) in their storage location in Hayes, Middlesex (General Register Office, 1942). Missing data can also arise because of operational or technical difficulties with equipment, such as when digitally recording interviews the speaker's voice may lack clarity, making automatic or manual transcription inaccurate. Satellite imagery and aerial photography, especially during the early years of their development, were prone to the vagaries of weather conditions with clouds obscuring areas of images collected with the result that some parts of the earth's surface had pixels with missing data in the images (NASA (National Aeronautics and Space Administration), n.d.). Techniques have been developed to remediate such data losses (Kuemmerle, Damm and Hostert, 2008). Similarly, equipment for recording data in the field can develop faults or simply batteries might discharge, which can also lead to an incomplete dataset.

This discussion of missing data has been included at this point because some statistical software provides users with options as to how gaps in their dataset should be treated, which depends on how the gaps in the dataset have occurred. If the gaps in a variable are completely random and unrelated to any other variable, for example if our survey outlined above has people with ages ranging from low to high values across a wide distribution, we can conclude any persons missing an age data value are not from any particular age group: younger nor older people would not seem to have excluded themselves. If inspection of the distribution of ages provided by respondents suggests there are less people aged 75 years and older, but there is also a wide spread of ages above this threshold, it indicates the gaps have occurred for reason unrelated to other variables, perhaps because some older people held a more traditional view about not revealing their age to a stranger. Finally, if, when examining the distribution of ages, there is almost complete absence of persons from one age range (e.g. 18–24 years), it might indicate this group is not represented in the survey and was for some reason a 'hard to reach' group (e.g. because of their busy, hectic lifestyle).

The default solution to missing data, gaps in the data values of the variable(s) included in a specific piece of analysis in most statistical software, is to exclude those entities (records) from that analysis if a variable with missing data is involved. An alternative option is to exclude entities (records) for all parts of the analysis. The difference between these options is examined in Table 12.16 with respect to computing Pearson's product moment correlation coefficients between three variables (A, B and C) with an expected 10 data values in each. The left side of the upper part of Table 12.16 is the complete dataset with 10 values for three variables; the righthand side shows two values missing for different entities in variables A and C (numbers 5 and 9, and 3 and 8, respectively). The three sections in the lower part of Table 12.16 give the correlation coefficients: first for all data values; second, for when entities with both values are present for each pair of variables; third, for only those entities with a complete set of data for all variables (entities 3, 5, 8 and 9 omitted). The latter means that the data values for only six of the entities are correlated, whereas in the

Table 12.16 The effect of differences in treatment of missing data in correlation analysis.

Entity	No missing data			Missing data in A and C		
	A	B	C	A	B	C
1	23.0	17.0	43.0	23.0	17.0	43.0
2	24.0	33.0	23.0	24.0	33.0	23.0
3	33.0	34.0	48.0	33.0	34.0	Missing
4	47.0	56.0	58.0	47.0	56.0	58.0
5	32.0	52.0	37.0	Missing	52.0	37.0
6	14.0	26.0	28.0	14.0	26.0	28.0
7	65.0	46.0	15.0	65.0	46.0	15.0
8	42.0	20.0	55.0	42.0	20.0	Missing
9	22.0	19.0	31.0	Missing	19.0	31.0
10	37.0	38.0	22.0	37.0	38.0	22.0

	Pearson's product moment correlation coefficients								
	No missing data			Pairwise exclusion of missing data			Listwise exclusion of missing data		
	A	B	C	A	B	C	A	B	C
A	1.000	0.571	0.002	1.000	0.626	−0.111	1.000	0.777	−0.111
B		1.000	−0.21		1.000	0.202		1.000	0.168
C			1.000			1.000			1.000
N		10	10		8	6		6	6

second instance only two are excluded from each pair of variables, thus maximising the analysis of the available data.

12.5 Introduction to Human and Physical Geography Projects

The three projects outlined here are a little larger in size because their data are not being used to explain the purpose and calculations involved with the techniques as in Chapters 6–11. Instead, the purpose of the projects and the data analysed is to work through the stages of a modest-sized investigation of the type undertaken by Geography, Earth and Environmental Science students as part of their undergraduate and postgraduate studies. The projects use 'genuine' data, although some preliminary pre-processing may have been undertaken as detailed below to avoid any anomalies, such as grid references lying outside the geographical area to which the dataset relates: details of any adjustments are given in the relevant section below. Grid references recorded incorrectly but still lying with the area will not be adjusted. Typical word limits for undergraduate and postgraduate projects are within the range of 10,000–15,000 words, but the following descriptions are shorter and effectively summarise what might be crafted by individual students. The following subsections for the projects include a short background context to the topic, a summary of the data and methods used, commentary and presentation of the results obtained and a brief discussion of the conclusions and possible further research. These are intended to reflect the typical content of a student project report or dissertation without providing a comprehensive literature review.

12.5.1 Crime and Deprivation

12.5.1.1 Research Context

Most crimes are an innately geographical phenomenon: someone goes from a location, the origin, to a crime site and then departs either back to the same location as the origin or somewhere else. Most crimes are therefore inherently spatial, and even some 'modern' criminal activity carried out in cyberspace involves one or more persons using information technology from a fixed or mobile location and for example transferring money from one financial account to another held in banks that have their own location. It is only necessary to recall televised crime dramas to understand the importance of Geography whether this is a map with pins stuck in it depicting where suspects live or photographs of the crime scene. The use of stop and search or frisk and search procedures dates back to at least the 1960s (Lautenschlager, 2020) and represents an attempt by law enforcement agencies and policymakers to prevent or detect certain types of crime as they occur. There has been concern that police have used this measure in a way that has targeted certain ethnic groups and genders (Bowling and Phillips, 2007; Meng, 2017; Shiner *et al.*, 2018) as well as more disadvantaged members of society (Coid *et al.*, 2021; Suss and Oliveira, 2023). This project explores this topic in a county in south east England that borders London but extends into more rural countryside localities where the influence of metropolitan lifestyle might be expected to be less. The project aims to address the following objectives or questions:

- Is there evidence of targeting of stop and search in more deprived areas?
- Is stop and search used disproportionately towards certain groups in society based on race and gender?
- If either or both occur, does their prevalence diminish further away from London?

The research objectives indicate that both non-spatial and spatial statistical techniques would be used in the analysis as well as summary of descriptive statistics for the datasets.

12.5.1.2 Data and Methods

The focus of this project is on the county of Essex including the unitary authorities (Southend-on-Sea and Thurrock) that became equivalent independent upper-tier local authorities providing all services in 1998, thus reconstituting the former whole county of Essex. Three datasets have been used together with supplementary statistics from the 2021 Population Census:

- the stop and search records in the county for all months in 2019 from the police data archive (https://data.police.uk/data/archive/);
- the 2019 English Index of Deprivation (EIofD) for the Lower Super Output Area (LSOA) spatial units from (https://www.gov.uk/government/statistics/english-indices-of-deprivation-2019)
- digitised boundaries of the LSOAs obtained from the UKBorders, part of the UKDataservice (https://borders.ukdataservice.ac.uk/bds.html)

The EIofD data used in the analysis were the ranks and percentile categories for the 1077 LSOAs in Essex. Nationally the ranks ranged from most deprived (1st) to least deprived (32,844th). LSOAs in Essex spanned virtually the full range from the most deprived one located in Tendring district through to one near the upper end of the spectrum at 32,817th in the neighbouring local authority of Colchester. The percentile categories were produced by sorting all LSOAs according to their rank and then dividing this listing into 10 per cent bands. The first stage in the data processing involved joining the EIofD ranks and percentile categories for the 1077 LSOAs to the digitised boundaries of these areas. The 2019 EIofD data were created for the 2011 LSOAs, and 28 of these were affected by boundary changes in preparation for the release of 2021 census statistics. Inspection of the 2011 and 2021 LSOA boundaries enabled the 2021 census data to be reaggregated to fit exactly to the 2011 boundary set.

The original stop and search data included the following variables:

- type (3 nominal categories)
- date and time
- part of policing operation (2 nominal categories)
- latitude (364 blanks)
- longitude (364 blanks)
- gender (2 nominal categories and 732 blanks)
- age (1108 blanks)
- self-defined ethnicity (17 nominal categories and 723 blanks)
- officer-defined ethnicity (4 nominal categories and 877 blanks)
- legislation (7 nominal categories)
- object of search (8 nominal categories)
- outcome (7 nominal categories and 225 blanks)
- outcome linked to the object of search (2 nominal categories)
- removal of more than just outer clothing (2 nominal categories and 723 blanks)

Once read into Excel, the occurrence of missing data (blank entries) for some variables was noted, as indicated, and any records with blanks were deleted. Four new variables were created:

- data and time were recoded to month;
- age was recoded into five groups (1: <10 years; 2: 10–17 years; 3: 18–24 years; 4: 25–34 years; 5: over 34 years);
- self-defined ethnicity was recoded into five groups (Asian, Asian British or any other Asian background; Black, Black British/Welsh, Caribbean or African; mixed or multiple ethnicity; white and other) corresponding to those used in the 2021 Population Census (produced by collapsing the same set of self-defined ethnicities);

- the number of stop and search actions was recoded into percentiles (similar to the EIofD percentiles).

The reduced set of data records were loaded into GIS software (ArcPro) and created as point features using latitude and longitude grid references. These coordinates were converted to the British National Grid to assist with mapping onto the LSOA polygons. This reveals that some of the points fell outside of the LSOA polygons and these records were also excluded from further analysis. Overall, starting from 17,146 stop and search entities, a total of 15,164 were used in the analysis with 1982 (11.5%) excluded.

12.5.1.3 Results

The police carried out stop and search actions in all Essex local authorities in 2019: there were under 500 in Brentwood, Maldon, Rochford and Uttlesford (8.3% of the total), and over 1500 in Chelmsford, Colchester and Southend-on-Sea (42.5%). Police stop and searches were carried out throughout Essex, although none took place in 108 (10.0%) of the LSOAs. There is visual evidence from Figure 12.1 that searches were clustered in the main cities and towns, such as Basildon, Chelmsford, Colchester and Southend-on-Sea, and nearest neighbour analysis for all points and for each of the five self-defined ethnic groups supports this argument with the Z statistics for all having extremely low probabilities (<0.000001). Some 90 per cent were carried out on males and 74.4 per cent were applied to persons only (i.e. not with a vehicle that was also searched) with the balance also including a search of vehicles. There was little seasonal variation through the year with between 5.7 and 9.6 per cent in all months apart from August (11.3%). The large majority of

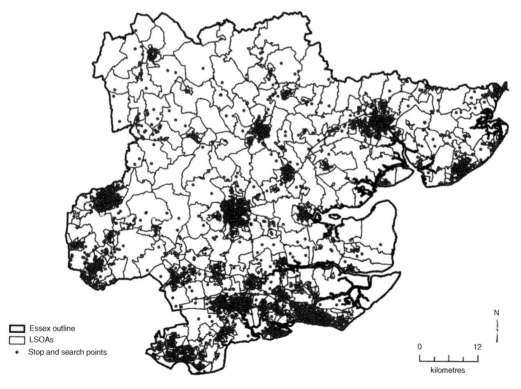

Figure 12.1 Spatial distribution of stop and search actions carried out in Essex Lower Super Output Areas in 2019.

stop and search actions were carried under two pieces of legislation, the Misuse of Drugs Act 1971 (section 23) (66.6%) and the Police and Criminal Evidence Act 1984 (section 1) (30.2%). Controlled drugs were the object of the search in 66.9 per cent of cases with offensive weapons or firearms a distant second at 13.3 per cent. The outcome of the search in 72.0 per cent of cases was 'No further action disposal', although 1709 (11.3%) resulted in an arrest being made. Two-thirds of the searches were made of people aged 10–17 or 18–24 years (28.4 and 35.8%, respectively); these were followed by 25–34 year olds (19.7%), and all but four of the remainder were of people aged 35 and over.

The first research objective focuses on whether stop and search actions were disproportionately concentrated in the area with higher levels of deprivation, which was investigated statistically in two ways with ordinal variables: first, by applying Kendall's tau to the recoded EIofD ranks and grouped count of stop and search actions per LSOA; second by using Spearman's rank correlation to assess the strength of any relation between ungrouped EIofD ranks and count of searches. Kendall's tau tested the Null Hypothesis that any difference between the LSOAs in respect of these variables was the result of chance or sampling error. The test statistic was -11.748 with a probability <0.001, which indicated that the Null Hypothesis should be rejected. The recoding of the variables has been taken further in Table 12.17 to provide a simpler illustration of the association between these ordinal variables: the ten percentile ranges have been collapsed into those for groups 1 to 3, 4 to 6 and 8 to 10 in each case. The table shows the observed and expected number of LSOAs with lower and higher index ranks and the count of searches. The expected number of LSOAs in the top left cell in the table (higher deprivation and lower searches) was three times more than occurred, whereas in those areas with more searches and lower deprivation (top right cell) the number of searches carried out was double what would be expected. The lower deprivation row in the table indicated the opposite occurred. Applying Spearman's rank correlation to the same pair of variables produces a coefficient of -0.329, which is significant at the 0.01 level thus adding to the statistical evidence of a moderate negative relationship between the variables.

The second research objective investigates whether certain social groups were more likely than others to experience a stop and search action by the police. Our descriptive analysis of the data revealed that most searches were made of men, and although all age ranges were included (including 4 children under 10 years), the age ranges 10–17 and 18–24 were predominant. Our main focus for this objective is on ethnicity, which was quantified by two variables: the person's self-defined and the officer-defined ethnic group. Unfortunately, the two assessments of ethnicity have a different number of groups. There was a high level of agreement (in excess of 90.0%) between the self-defined and the officer-defined ethnicity for Asian, Black and White persons who had experienced a stop and search, but, although 90.3 per cent of the self-defined mixed or multiple ethnic persons were in the officer-defined other ethnic group, when adding in those who self-defined as other there is much less agreement (32.0%). Self-defined ethnicity was used in the analysis because this

Table 12.17 Ranked English Index of Deprivation and count of stop and searches made in Essex Lower Super Output Areas in 2019 (expected counts in brackets).

	Lower stop and search (1–3)	Medium stop and search (4–7)	Higher stop and search (8–10)	Total
Higher deprivation (1–3)	28 (62)	71 (82)	107 (62)	206
Medium deprivation (4–7)	132 (143)	196 (196)	149 (143)	477
Lower deprivation (8–10)	164 (119)	163 (157)	67 (118)	394
Total	206	477	394	1077

represents how people identified themselves and because it corresponds with the groups in the 2021 census statistics, which also required people to self-define their ethnic group into the same categories. Rather than analysing the count data of each ethnic group in the LSOAs for the search data and census statistics, the variables were converted into percentages. For example, the percentage of usual Asian residents and the percentage of all searches experienced by Asian persons in each LSOA. The two-year difference between the datasets has been disregarded on the basis that the distribution of the five ethnic groups was unlikely to have changed between 2019 and March 2021 when the census took place. This produced five pairs of variables that were investigated using Wilcoxon's signed ranks test. Table 12.18 summarises the results. The Wilcoxon signed ranks test takes

Table 12.18 Application of Wilcoxon's signed ranks test to five ethnic groups in relation to stop and searches made in Essex Lower Super Output Areas in 2019.

		N of LSOAs	Mean rank	Wilcoxon test statistic	Probability
Percentage Asian in 2021 – percentage Asian stop and searches 2019	Negative Ranks (As21 < AsSS)	159	727.68		
	Positive Ranks (As21 > AsSS)	809	436.71		
	Ties	0			
	Total	968		−13.654	<0.001
Percentage Black in 2021 – percentage Black stop and searches 2019	Negative Ranks (Bl21 < BlSS)	380	716.87		
	Positive Ranks (Bl21 > BlSS)	585	331.09		
	Ties (Bl21 = BlSS)	3			
	Total	968		−4.545	<0.001
Percentage mixed/multiple in 2021 – percentage mixed/multiple stop and searches 2019	Negative Ranks (MM21 < MMSS)	0	0.00		
	Positive Ranks (MM21 > MMSS)	969	485.00		
	Ties	0[i]			
	Total	969		−26.965	<0.001
Percentage other in 2021 – percentage other stop and searches 2019	Negative Ranks (Ot21 < OtSS)	1	19.00		
	Positive Ranks (Ot21 > OtSS)	945	473.98		
	Ties	23			
	Total	969		−26.641	<0.001
Percentage White in 2021 – percentage White stop and searches 2019	Negative Ranks (Wh21 < WhSS)	416	324.65		
	Positive Ranks (Wh21 > WhSS)	552	604.97		
	Ties	0			
	Total	968		−11.429	<0.001

account of both the sign and magnitude of the differences between the ranks. All test results were significant at the 0.001 level, but only for the Black ethnic group the result was based on the mean of the positive ranks, which indicates that the number of searches in LSOAs involving people from this ethnic group was higher than expected, whereas for other groups it was lower.

The third research objective explicitly concerns a spatial aspect of stop and search actions by investigating if their distribution varies in relation to deprivation across Essex and if any connection appears to diminish for LSOAs further away from the London metropolis. Getis-Ord hot spot analysis was applied to the percentages of usual residents according to the 2021 census and searches of persons in each of the five self-defined ethnic groups. The results are shown as pairs of maps in Figure 12.2 and reveal the extent to which high and low values cluster together in a statistically significant fashion with 90, 95 and 99 per cent confidence. Maps on the left of Figure 12.2 indicate if there were high or low clusters of these ethnic groups according to the 2021 census statistics. Asian, Black, mixed/multiple and other groups were clustered together in high numbers in some LSOAs close to Greater London Authority towards the south-west, although there were also similar clusters in urban centres in different parts of the county. The 2021 population in the White ethnic group was widely distributed, but there were cold spots in some of the LSOAs where other groups were concentrated. There is also a contrast in the geography of stop and search actions between incidents involving members of the White ethnic group and the other four groups. There were cold spots for the former in many of the larger, more rural LSOAs towards the north and centre of the county. The hot spots for searches involving members of the other ethnic groups were discrete and focused in their location, but often not coinciding with where substantial clusters of usual residents in these groups occurred according to the 2021 census statistics.

Finally, the possible connection between stop and search and deprivation has been explored using geographically weighted regression with the number of stop and search occurrences in LSOAs as the dependent variable and their rank score on the 2019 EIofD as the independent variable. Figure 12.3 shows the spatial distribution of standardised residuals: large and positive values (>+1.5 std. dev.) indicate high rates (hot spots) of stop and search actions associated with deprivation, and large and negative values (<−1.5 std. dev.) represent low stop and search rates (cold spots) related to deprivation. There were few hot spots detected, just one on the northern fringe of Colchester, but there were several cold spots in Basildon, Colchester, Southend-on-Sea, Thurrock and Waltham Cross. Rather than diminishing with distance from London, these results support the argument that settlement size and population density are likely to be more important factors.

12.5.1.4 Discussion and Limitations

It is usually possible to undertake more analysis of your project data as you become familiar with the variables and realise there are other directions in which to take the investigation. The scope here has necessarily been limited to what is relevant for the research objectives. Some limitations can be identified in most research, and in this instance, the decision to focus on LSOAs in one county could be challenged but was considered realistic for a student-style project and on account of the proximity to London especially given the topical issue of 'county lines' (Coomber and Moyle, 2018). Returning to the research objectives, the analysis has shown a strong connection between deprivation and stop and search occurrences together with a disproportionate tendency for members of some ethnic groups to be searched; especially if they are away from the areas where this section of the population live in significant numbers. Relatively few searches were carried out in the more sparsely populated, rural areas, and most LSOAs without any stop and search occurrence were in such parts of the county. There was less evidence of proximity to London being

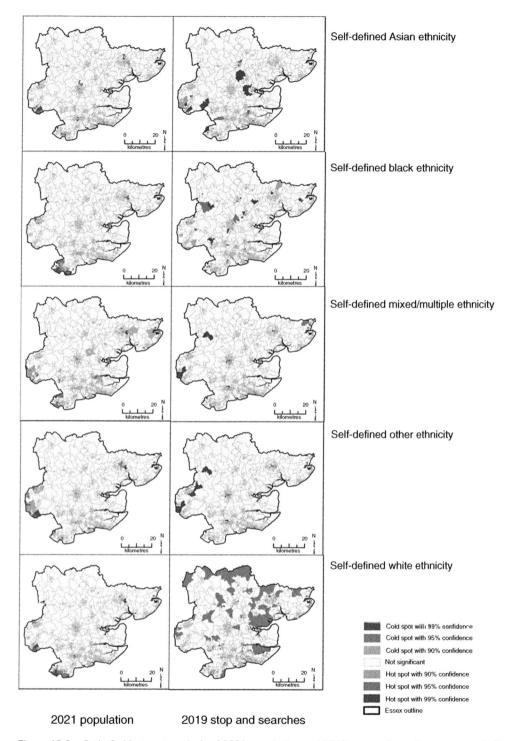

Self-defined Asian ethnicity

Self-defined black ethnicity

Self-defined mixed/multiple ethnicity

Self-defined other ethnicity

Self-defined white ethnicity

Cold spot with 99% confidence
Cold spot with 95% confidence
Cold spot with 90% confidence
Not significant
Hot spot with 90% confidence
Hot spot with 95% confidence
Hot spot with 99% confidence
Essex outline

2021 population   2019 stop and searches

Figure 12.2 Getis-Ord hot spot analysis of 2021 population and 2019 stop and search occurrences in Essex Lower Super Output Areas for five ethnic groups (self-defined).

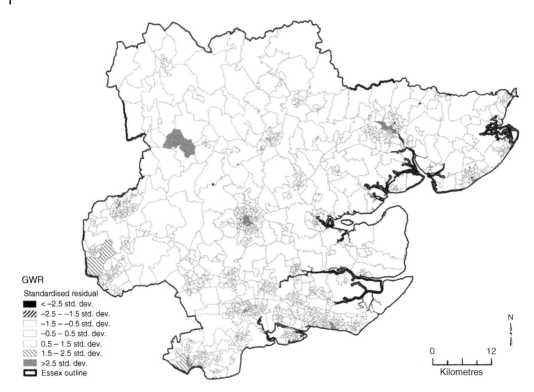

Figure 12.3 Geographically weighted regression of stop and search occurrences (dependent) and Index of Deprivation rank (independent) in Essex Lower Super Output Areas in 2019.

important, and further research could introduce population density and total population per LSOA as explanatory variables.

> The archive of crime data used as the source for the stop and search data also includes other types of crime data reported monthly going back to 2010, although there was a change in the way crime was recorded in January 2014. Review this data source and reflect on ways the current analysis could be extended, for example, in spatial coverage or time period.

12.5.2 Pick Your Own Farm Enterprises

12.5.2.1 Research Context
Agriculture and farming have been established as a focus of geographical research for many decades, probably centuries. The centre of attention has pivoted from the farm itself, as expressed in spatial distribution of agricultural production (Coppock, 1964, 1971) and farmer's role in rural society towards the environmental impact of farming practice, the diversification of farm business and wider concern for food security and the supply chain in recent decades (Marsden, Banks and Bristow, 2000; Morris, 2005; Watts, Ilbery and Maye, 2008). One topic featuring strongly the early years of this transition concerned direct marketing of farm products to consumers. One of the first forms of direct marketing that emerged was Pick Your Own (PYO) (Weimer, 1978; Bowler, 1981a, 1981b), whereby farmers invited consumers onto their farms to harvest mostly horticultural type crops, although farm shops, roadside sales and more recently produce boxes also feature. The prevalence

of signs advertising PYO on trips through the countryside and in local or regional newspapers has reduced since the 1990s. Farms continuing their direct marketing enterprises had often upgraded their amenities to include all-year-round shops, bought-in produce, play areas for children, catering facilities and special events (e.g. a Santa's Grotto at Christmas) and to advertise their produce and prices on the World Wide Web (WWW). This project is a preliminary exploration of whether direct marketing and especially PYO remains a viable option for farms. The investigation begins with previous published research as a starting point for developing a new project, which represents a sound approach for undergraduate students. The project aims to address the following questions:

- Is there evidence of change in PYO on farms between the 1970s (Bowler, 1981b) and early 2000s?
- What factors potentially influence the inclusion of direct marketing enterprises on farms?
- What were the dynamics of farmers starting or stopping direct marketing enterprises during the period they had occupied their farms?

These research questions will be explored by means of descriptive statistics as well as applying non-spatial statistical techniques to a survey dataset.

12.5.2.2 Data and Methods

Bowler's (1981a) research suggested that English counties could be divided in growth phase groups (before 1972, 1972–1976 and after 1976) in respect of their farms' engagement with PYO in certain counties in England. It was decided that a postal questionnaire would be the most suitable means of collecting sufficient data to address these questions given the geographical spread of these counties from Cornwall in the south-west, Lancashire in the north and Norfolk in East Anglia. This raised the issue of how to assemble a list of farms with direct marking enterprises for selecting a sample and distributing a questionnaire. A list of suitable farms from which to select a sample was compiled from published business directories such as *Harvest Times*, *Pick of the Crop Guide*, *Pick Your Own Farms* and *Yellow Pages* using PYO, Farm Shop Fruit and Vegetable Growers, and Market Garden categories. Postal distribution was the most suitable means of delivering the questionnaire in 2002 at a time before email and social media were widespread. A total of 407 questionnaires were sent out, and 70 usable replies were returned after posting a reminder second copy of the questionnaire giving a response rate of 17.1 per cent. Most returns came from farms in the south-west region (34.4%), the east region (27.1%) and south east (20.8%). Figure 12.4 shows the farms responding to the survey and non-response farms in relation to Bowler's (1981a) PYO growth phases. Calculating the nearest neighbour index for both sets of point features shows them to have a statistically significant clustered spatial distribution, which suggests that despite the low response rate achieved the data realistically provide valid insights into the direct marketing sector of English farming in the early 2000s.

The questionnaire (see Figure 12.5) was designed with some questions requiring the respondent to fill in answers with factual information and others with pre-defined coded responses, mainly Yes or No. Background information was limited to seeking the age of the farmer and, if different, of the person managing the direct marketing enterprise. Data obtained from the questionnaires were held in a simple rectangular data matrix (see above), but separate data files were also created for the following:

- farm-level variables relating to total area, area of different crops and livestock numbers
- the areas of each crop grown for PYO or other direct marketing
- current and former direct marketing enterprise-level variables for PYO, farm shop, roadside sales and other types

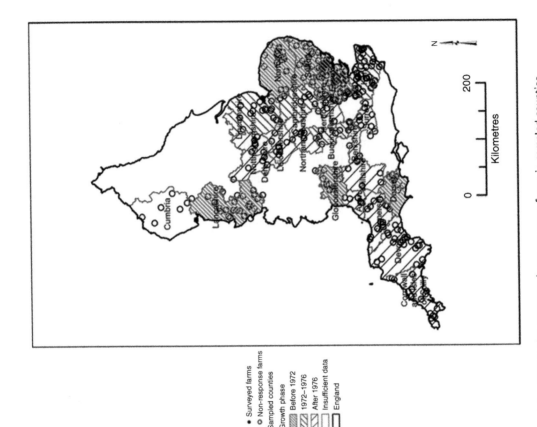

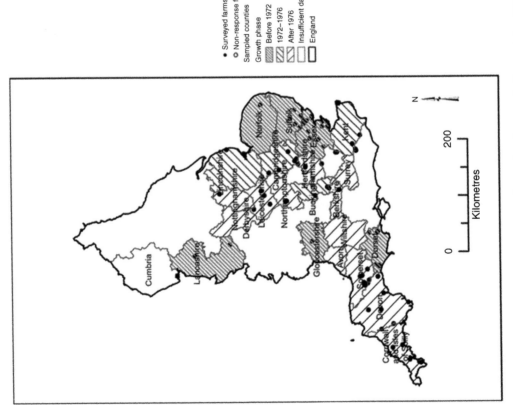

Figure 12.4 Distribution of farms responding to the direct marketing postal questionnaire survey and non-response farms in sampled counties.

Direct Marketing on Farms Questionnaire

Please enter your farm's postcode

Please leave BLANK any boxes or columns that do NOT apply to you or your farm.
The questionnaire has 4 Sections.
Please circle the relevant answer OR write in the information.

Date of completion   __ / __ /02

Section 1: Farm and Farmer Characteristics

1.1 What is the TOTAL area of land on this farm?

Please indicate whether areas are in acres OR hectares

	Acs / ha	If any production is sold directly to consumers please give % of total
Please enter the amount of the TOTAL farm land that is: Owned		
Rented		
Please enter the amount of the TOTAL farm land that is:		
Woodland (productive and non-productive)		
Temporary grassland and fodder crops		
Permanent grassland and rough grazing		
Arable crops (e.g. wheat, barley, oats, oilseed rape, field beans, etc.)		
Field-scale horticultural crops (e.g. potatoes, cabbages, strawberries, carrots, flowers, etc.)		
Crops grown under cover (polytunnels, glass, etc)		

1.2 What livestock, if any, are kept on this farm this year? Please indicate TOTAL numbers (head) as requested.

Dairy cows (milking herd)	
Beef cattle (animals being fattened)	
Breeding ewes	
Lambs	
Broiler chickens	
Laying chickens	
Ducks	
Turkeys	
Other livestock, please specify	

1.3 Management of the farm

How many years has your family been on this farm?
How many years have you managed this farm?
Do you run this farm as: A sole trader?   A partnership?   A limited company?
Other, please specify

1.4 Farm location

Is the entrance to your farm located on a main road (A or B class)?   YES / NO
Approximately how many miles by road is your farm from the nearest settlement with a population of at least 10,000?

1.5 Workforce

How many members of your family work:
On the farm for an average of at least 15 hours per week?
On the farm for an average of less than 15 hours per week?
Away from the farm in non-farm employment?

How many non-family workers are employed:
To do agricultural work for an average of at least 15 hours per week?
To do agricultural work for an average of less than 15 hours per week?
To do non-agricultural work for an average of at least 15 hours per week?
To do non-agricultural work for an average of less than 15 hours per week?

Section 2: CURRENT Direct Marketing Enterprises

2.1 Are you or any members of your family running a **Pick Your Own (PYO)** enterprise from this farm?   YES / NO

2.2 Are you or any members of your family running a **Farm Shop** enterprise from this farm?   YES / NO

2.3 Are you or any members of your family running **Roadside or Casual Sales** from this farm? (e.g. eggs, flowers, honey, other produce, etc.)   YES / NO

2.4 Are you or any members of your family running any other **Direct Marketing** enterprise from this farm? (e.g. delivery round or mail order)   YES / NO

If NO to ALL of these, please go to Section 3 on Page 3

For EACH of the above answered YES, please answer the questions in the table below entering the information in the appropriate column. Ignore any columns that do not apply.

Table 2	CURRENT Pick Your Own (PYO)	CURRENT Farm Shop		CURRENT Roadside/ Casual Sales		CURRENT Other Direct Marketing	
When did you or other members of your family start running the enterprise(s)?							
Why did you or other members of your family start running the enterprise(s)?							
Please list below the crops that are grown for PYO (up to 10 crops) and their area this year in acres or ha (please state which).	Area (acres/ha)						
1							
2							
3							
4							
5							
6							
7							
8							
9							
10							
Please tick the types of goods currently sold in the Farm Shop, at the Roadside or by Other Direct Marketing AND indicate their percentage of sales in a typical month.		✓ if sold	% sales	✓ if sold	% sales	✓ if sold	% sales
Fresh food produced on the farm							
Fresh food bought in or sold on behalf of another farm							
Processed food goods made on your farm (e.g. yoghurt, jams, prepared meals, drinks, etc.)							
Processed food goods made elsewhere (e.g. yoghurt, jams, prepared meals, drinks, etc.)							
Flowers and plants							
Craft goods							
Non-food goods							
Who manages the enterprise? (e.g. wife, daughter, son, father, friend, etc.)							
Are any of the goods on sale certified organic?	YES / NO	YES / NO		YES / NO		YES / NO	
Does your advertising state that you sell organic produce?	YES / NO	YES / NO		YES / NO		YES / NO	

Figure 12.5 Postal questionnaire used in survey of farms with direct marketing enterprises. Source: author's survey.

Table 2 Continued

	CURRENT Pick Your Own (PYO)	CURRENT Farm Shop	CURRENT Roadside/ Casual Sales	CURRENT Other Direct Marketing
How often is the range of crops or goods for sale changed?				
Weekly	1	1	1	1
Monthly	2	2	2	2
Seasonally	3	3	3	3
Annually or less often	4	4	4	4
What facilities are provided for customers visiting the farm?				
Off road car parking	YES / NO	YES / NO	YES / NO	
Children's play area	YES / NO	YES / NO	YES / NO	
Refreshments	YES / NO	YES / NO	YES / NO	
Toilets	YES / NO	YES / NO	YES / NO	
Leaflets about crops or goods available	YES / NO	YES / NO	YES / NO	
General tourist advertising leaflets/posters	YES / NO	YES / NO	YES / NO	
Other				
Where do you advertise the enterprise(s)?				
Local press	YES / NO	YES / NO	YES / NO	YES / NO
Local radio	YES / NO	YES / NO	YES / NO	YES / NO
Roadside notices	YES / NO	YES / NO	YES / NO	YES / NO
Leaflet distribution	YES / NO	YES / NO	YES / NO	YES / NO
Yellow Pages	YES / NO	YES / NO	YES / NO	YES / NO
Internet	YES / NO	YES / NO	YES / NO	YES / NO
Other, please specify	YES / NO	YES / NO	YES / NO	YES / NO
What aspects of the goods on sale are emphasised when advertising the enterprise(s)?				
During which months are the direct marketing enterprises open for business? (e.g. Mid-May to end September)	to	to	to	to
What are your standard opening hours during this period?	to	to	to	to
Are the opening hours varied according to the availability of crops or goods for sale?	YES / NO	YES / NO	YES / NO	YES / NO

Section 3: PREVIOUS Direct Marketing Enterprises on this farm

3.1 Have you or any members of your family ever run a **Pick Your Own (PYO)** enterprise from this farm? YES / NO

3.2 Have you or any members of your family ever run a **Farm Shop** enterprise from this farm? YES / NO — **If NO to ALL of these, please go to Section 4 on Page 5**

3.3 Have you or any members of your family ever run **Roadside or Casual Sales** from this farm? (e.g. eggs, flowers, honey, other produce, etc.) YES / NO

3.4 Have you or any members of your family ever run any other **Direct Marketing** enterprise from this farm (e.g. delivery round or mail order)? YES / NO

For EACH of the above answered YES, please answer the questions in the table below entering the information in the appropriate column. Ignore any columns that do not apply.

Table 3	PREVIOUS Pick Your Own (PYO)	PREVIOUS Farm Shop	PREVIOUS Roadside/ Casual Sales	PREVIOUS Other Direct Marketing
When did you or other members of your family stop running the enterprise(s)?				

Table 3 Continued

	PREVIOUS Pick Your Own (PYO)	PREVIOUS Farm Shop		PREVIOUS Roadside/ Casual Sales		PREVIOUS Other Direct Marketing	
Why did you or other members of your family stop running the enterprise(s)?							

Please list below the crops that were grown for PYO (up to 10 crops) and their area in the final year in acres or ha (please state which).

Area (acres/ha)
1.
2.
3.
4.
5.
6.
7.
8.
9.
10.

	PREVIOUS Farm Shop		PREVIOUS Roadside/ Casual Sales		PREVIOUS Other Direct Marketing	
Please tick the types of goods that were sold in the Farm Shop, at the Roadside or by Other Direct Marketing AND indicate their percentage of sales in a typical month (final year).	✓ if sold	% sales	✓ if sold	% sales	✓ if sold	% sales
Fresh food produced on the farm						
Fresh food bought in or sold on behalf of another farm						
Processed food goods made on your farm (e.g. yoghurt, jams, prepared meals, drinks, etc.)						
Processed food goods made elsewhere (e.g. yoghurt, jams, prepared meals, drinks, etc.)						
Flowers and plants						
Craft goods						
Non-food goods						

	PREVIOUS Pick Your Own (PYO)	PREVIOUS Farm Shop	PREVIOUS Roadside/ Casual Sales	PREVIOUS Other Direct Marketing
Who managed the enterprise? (e.g. wife, daughter, son, father, friend, etc.)				
Were any of the goods on sale certified organic? Does your advertising state that you sell organic produce?	YES / NO	YES / NO	YES / NO	YES / NO
How often was the range of crops or goods for sale changed?				
Weekly	1	1	1	1
Monthly	2	2	2	2
Seasonally	3	3	3	3
Annually or less often	4	4	4	4
What facilities were provided for customers visiting the farm?				
Off road car parking	YES / NO	YES / NO	YES / NO	YES / NO
Children's play area	YES / NO YES / NO	YES / NO YES / NO	YES / NO YES / NO	YES / NO
Refreshments	YES / NO	YES / NO	YES / NO	YES / NO
Toilets	YES / NO	YES / NO	YES / NO	YES / NO
Leaflets about crops available	YES / NO	YES / NO	YES / NO	YES / NO
General tourist advertising leaflets/posters	YES / NO	YES / NO	YES / NO	YES / NO
Other				

Figure 12.5 (Continued)

Table 3 Continued	PREVIOUS Pick Your Own (PYO)	PREVIOUS Farm Shop	PREVIOUS Roadside/ Casual Sales	PREVIOUS Other Direct Marketing
Where did you advertise the enterprise(s)?				
Local press	YES / NO	YES / NO	YES / NO	YES / NO
Local radio	YES / NO	YES / NO	YES / NO	YES / NO
Roadside notices	YES / NO	YES / NO	YES / NO	YES / NO
Leaflet distribution	YES / NO	YES / NO	YES / NO	YES / NO
Yellow Pages	YES / NO	YES / NO	YES / NO	YES / NO
Internet	YES / NO	YES / NO	YES / NO	YES / NO
Other, please specify........................	YES / NO	YES / NO	YES / NO	YES / NO
What aspects of the goods on sale were emphasised when advertising the enterprise(s)?				
During which months was/were the direct marketing enterprise(s) open for business? (e.g. Mid-May to end September, or All year)	to	to	to	to
What were your standard opening hours during this period?	to	to	to	to
Were the opening hours varied according to the availability of crops or goods for sale?	YES / NO	YES / NO	YES / NO	YES / NO
Was this shorter a period than the average in previous years?	YES / NO	YES / NO	YES / NO	YES / NO

Section 4: Background Characteristics

6.1 How old are you, preferably in years, or one of the following age groups?	6.2 If there is any Direct Marketing on the farm, how old is the person managing that enterprise, preferably in years, or one of the following age groups?
.... years	.... years
OR	**OR**
Under 25	Under 25
25-34	25-34
35-44	35-44
45-54	45-54
55-64	55-64
65 and over	65 and over

Finally, thank you for completing this questionnaire. I assure that all information will be treated in the strictest confidence and no details about identifiable individual or farms will be released.

Please **SEND / DO NOT SEND** a copy of the summary results from the survey. (Please delete as appropriate).

Figure 12.5 (Continued)

Separation of the data in this way allowed analysis to be carried out on diverse groups of variables in a simple fashion. Some recoding of variables captured from data recorded on the questionnaires was also undertaken, for example the text responses to the type of facilities available were classified into categories.

12.5.2.3 Results

The mean total area of farms in the survey was 75.3 ha compared with 54.4 ha for all English farms at the time, whereas the ratio of owned to rented land overall was very similar to the national figure at 69:31. Approximately two-thirds of the mean area of productive land on the farms was under arable cropping, temporary or permanent grass resulting in a larger than usual area devoted to horticultural cropping including those grown specifically for PYO and other forms of direct marketing. The mean area of land used for horticulture was 10.0 ha (England average 0.9 ha), and 5.6 ha (57.5%) was for PYO crops. The size of the agricultural workforce reduced during the second half of the 20th century, leading to greater dependency on family and seasonal, casual labour especially for planting and harvesting work with horticultural crops. One feature of PYO enterprises was that farmers could have their horticultural crops harvested by members of the public and sell them to this 'unpaid labour' at a lower price than through normal retail outlets. Farmers in the survey had a high level of family labour engaged in farm work (20% had more than 3 family members working on the farm), but most farms (56.0%) still employed non-family workers. Importantly, a third of farms employed non-family employees for work not directly concerned with agricultural production.

The first research question seeks to establish if there was statistically significant evidence of change in PYO on English farms between the 1970s (Bowler, 1981a, 1981b) and the early 2000s. There are insufficient farms in the 2002 survey for a county-by-county analysis, but Bowler (1981b: 149) discovered that 'over 20 per cent of [PYO] farms' sold more than nine crops: the survey data used in this project also found a similar percentage (25.7 per cent of farms with PYO) also sold this number of crops. He also found that 32 per cent of farms north of a line between Cheshire and Lincolnshire sold only soft fruit: in the 2002 survey, all 11 farms north and 69 per cent of farms south of this line only sold this type of horticultural crop. Soft fruits and vegetables were the most common combination of crop types (14.5% of farms) for farms with more than one type, which again corresponds with Bowler's (1981b) findings. However, the small group of farms with top fruits (apples, cherries, pears and plums) were all in Essex and were not as widespread as reported by Bowler (1981b: 147) in the traditional fruit-growing areas of 'South East England (the counties of Essex, Kent, Sussex) and the West Midlands (Vale of Evesham)', although this may have been influenced by the low number of survey responses. He also suggested about 40 per cent of farms showed a degree of 'commercialisation' (Bowler, 1981b: 149) by having a farm shop as well as PYO direct marketing. The later survey included two further types of direct marketing (roadside sales and others (e.g. delivery round or mail order)): there were 24 (34.8%) farms with PYO and at least one of the three other types.

The second research objective aimed to discover the importance of several factors potentially accounting for variation in engagement with direct marketing enterprises on their farms. Six factors were considered as contributing to this decision (farm area, balance between owned and rented land, age of farmer, years managed the farm, location on A or B class road and the type of business). Only a limited number of categories could feasibly be included in any statistical analysis of these variables given the number of survey responses. The Kruskal–Wallis H test was carried out to test for a significant difference in respect of the six factors between three types of farms: PYO only; PYO with any number of the other three types of direct marking; other three types but no PYO. The test results summarised in Table 12.19 clearly show that the survey had not elicited the reasons why farms differentially engaged with direct marketing enterprises, despite its presence showing similarities to earlier research on the topic. None of the variables tested had probabilities below the 0.05 threshold for being significant. Although such a negative outcome may be disappointing, it nonetheless indicates that other variables account for the differences that were discovered by the survey.

Table 12.19 Application of Kruskal–Wallis test to three categories of farms in England in respect of factors potentially contributing to engagement with direct marketing in 2002.

		N of farms	Mean Rank	Kruskal-Wallis test statistic	Probability
Total farm area (<25.0 ha; 25.0–49.9 ha; >49.9 ha)	PYO only	12	36.83		
	PYO with other types of direct marketing	24	28.50		
	Only other types of direct marketing	30	36.17		
	Total	66		2.958	0.228
Balance between owned and rented land (75% or more tenanted; not less than 25% tenanted and not less than 25% owned; 75% or more owned)	PYO only	12	32.54		
	PYO with other types of direct marketing	22	30.33		
	Only other types of direct marketing	30	36.42		
	Total	66		2.042	0.360
Farmer's age (<40 years; 40–59 years; >59 years)	PYO only	12	28.58		
	PYO with other types of direct marketing	22	34.55		
	Only other types of direct marketing	30	32.57		
	Total	64		1.074	0.584
Years managed the farm (<15 years; 15–24 years; >24 years)	PYO only	12	32.75		
	PYO with other direct types of marketing	23	33.11		
	Only other types of direct marketing	29	31.91		
	Total	64		0.063	0.969
Location on A or B class road (yes; no)	PYO only	12	30.33		
	PYO with other types of direct marketing	24	30.33		
	Only other types of direct marketing	29	36.31		
	Total	65		2.203	0.332
Type of farm business (sole trader; partnership; limited company)	PYO only	12	26.58		
	PYO with other types of direct marketing	23	32.04		
	Only other types of direct marketing	29	35.31		
	Total	65		2.313	0.315

Note: Four farms that had stopped direct marketing or had other missing data have been excluded.
Source: author's survey.

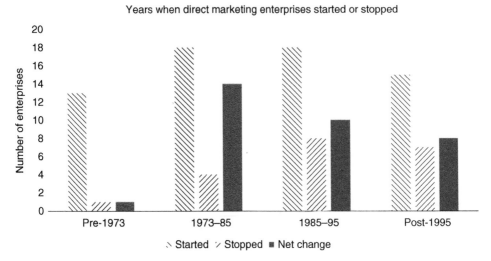

Figure 12.6 Starting and stopping of direct marketing enterprises on survey farms.

The third research objective concerned the dynamics of farmers starting or stopping direct marketing enterprises during the period they had occupied their farms. Across all farms, there were 112 such enterprises and 64 farmers were able to recollect the year direct marketing started and 20 when it stopped in at least one of the four categories with the balance unable to specify a start-up year. The mean start-up year was 1984 and the mean stop year was 1990. Figure 12.6 shows that the growth anticipated by Bowler (1981b) occurred, although the cessation of enterprises 1985–1995 and after 1995 suggests a change in the market either because consumers preferences had changed or farmers sought other outlets for their production. Recognising the limited number of farms in the survey, Kruskal–Wallis tests were carried out to investigate if the mean year when direct marketing enterprises started or stopped was significantly different for the four types: the results are summarised in Table 12.20. The test statistic ($H = 9.103$) for year started is significant at the 0.05 level ($p = 0.028$).

Table 12.20 Application of Kruskal–Wallis test for a significant difference between the four types of direct marketing enterprise in respect of when they were started or stopped.

		N of enterprises	Mean year (integer)	Mean Rank	Kruskal–Wallis test statistic	Probability
Year started	PYO	19	1981	26.71		
	Farm shop	27	1982	29.44		
	Roadside sales	8	1989	39.44		
	Other	10	1992	46.20		
	Total	64	1984		9.103	0.028
Year stopped	PYO	7	1993	11.50		
	Farm shop	4	1992	10.88		
	Roadside sales	7	1990	9.29		
	Other	2	1976	10.50		
	Total	20	1990		0.913	0.516

Examination of the mean year each type reveals that roadside sales and other types started later than the PYO and farm shops, which corresponds with Bowler's (1981b) comments about farms linking these two types of activity together. The limited number of direct marketing enterprises recorded as having stopped may account for the insignificant test result. However, the evidence that the 'stop years' for three of the types were between 1990 and 1993 may indicate the onset of new market conditions to which some farmers were responding earlier than others.

Farmers were asked their reasons for starting or stopping the different types of direct marketing enterprises. There were 82 and 21 comments about reasons for starting and stopping these enterprises, which have not been coded for analysis here, but the following examples provide a flavour of the reasons why some farmers decided to start or stop these activities:

- Only way to make fruit pay (started PYO in 1968)
- Small acreage, adjacent town and need better returns (started PYO in 1975)
- Established business took over – fitted with objectives (started PYO in 1999)
- Towns with large supermarket stores (stopped PYO in 1985)
- To add value to visits (started farm shop in 1980)
- To grow Christmas trees (started roadside sales in 1975)
- Selling through farmers' market reaches the wider market (started other in 1999)
- Father-in-law when he retired (stopped PYO in 1962)
- Crop failures, public lost interest and public's shopping habits changed with arrival in nearby towns with large supermarket stores (stopped PYO 1995)
- Loss of customers due to supermarkets (stopped farm shop in 1992)
- Developed into a farm shop (stopped roadside sales in 1982)
- Produce kept getting stolen (stopped roadside sales in 1994)
- Main farming activities increased (stopped other in 1950)

These verbatim comments indicate a mixture of internal and external factors that contributed to farmers' decision-making in relation to engaging with the direct marketing sector.

12.5.2.4 Discussion and Limitations

The results from the survey and analysis presented illustrate how a student-style project can extend or develop previously published research. This approach may be helpful to students facing the challenge of devising an independent research project that might need to sustain their interest for a period of up to 12 months in some universities. The previous research provides a platform from which new ideas can be developed and change can be quantified. The survey carried out for the project presented here had a response rate and number of respondents that are typical of what might be achieved when resources, including especially time, are limited. Statistical analysis of survey data commonly involves applying non-parametric procedures, and, in this instance, the Kruskal–Wallis H test was used given that there was an interest in discovering differences between two types of entity: first, farms and farmers in respect of their characteristics; second distinct types of direct marketing enterprise. Spatial statistics have not been central to the analysis carried out here, but applying nearest neighbour analysis to the point features locating the sample respondents and the farms in the sampling frames enabled the two patterns to be compared. Examination of the postal questionnaire in Figure 12.5 shows that several questions asked have not be included in the analyses presented here. Use the following self-assessment suggestions to reflect on how the analysis could be extended, perhaps by including some of the qualitative text responses.

The first posting out of the questionnaire was followed up by a second approach after three weeks. How else might the researcher have attempted to boost the number of responses given that publications listing names, addresses and telephone numbers of farms with the different types of direct marketing enterprises had been used as the basis for sending out over 400 questionnaires?

In total, 70 fully completed questionnaires represent quite a successful outcome for a survey of this type where the researcher does not have face-to-face contact with the intended participants. Reflect on how the analysis presented here could be extended using the quantitative data collected from the questionnaires. Are there any new variables that could be created by recoding or collapsing of categories?

An indication of the verbatim text comments, using these examples as well as your reading of the literature on the topic of direct marketing, proposes coding schemes for the responses.

12.5.3   Trees and Greenspaces in Urban Environments

12.5.3.1   Research Context

The United Nations announced in 2018 that 55 per cent of the world's population lived in urban areas, a figure which was expected to rise to 68 per cent by 2050 (United Nations, 2018). The increased proportion of people sharing these spaces has been accompanied by growing interest in the interactions between people and the environments in which they play out their lives. Research has focused on a number of issues, but significant among these are the impact on people's physical and mental health (Salmond *et al.*, 2014; Turner and Cavender, 2019; Wolf *et al.*, 2020), alleviation of air pollution (Baris, Sahin and Yazgan, 2009; Selmia *et al.*, 2016) and ecological services through connectivity (Von Thaden *et al.* 2021). Much of this research has assumed that members of the public hold positive attitudes towards the presence of trees and greenspaces in the urban landscape, although some authors have recently offered a more nuanced perspective (Moffat *et al.*, 2024). They noted 'younger age groups identifying less enthusiasm' (Moffat *et al.*, 2024 : 4) towards trees especially for those persons with fewer years spent in formal education. This project delves into some of these issues by exploring the distribution of population, trees and greenspaces in one of London's 33 boroughs, Greenwich. This inner London borough lies south of the River Thames, although is treated as an outer borough for statistical purposes. The project aims to address the following objectives or questions:

- Is there a difference between the greenspace land parcels with and without local authority-maintained trees?
- Is there evidence of population and greenspace showing similar or different spatial distributions?
- How accessible in terms of distance are greenspaces with different land uses for the population of the study area?

The research objectives indicate that both non-spatial and spatial statistical techniques would be used in the analysis as well as the summary of descriptive statistics for the datasets.

12.5.3.2   Data and Methods

The project focuses on Greenwich, an inner London borough lying south of the River Thames, although the area is treated as an outer borough for statistical purposes. Three datasets obtained from secondary sources have been used for this project: the greenspace land cover data from Ordnance Survey's Mastermap accessed through the Digimap service (https://digimap.edina.ac.uk/);

2021 usual resident and household counts from the Office for National Statistics available through NOMIS (https://www.nomisweb.co.uk/sources/census_2021); and local authority-maintained trees data from the London Data Store part of the Greater London Authority (https://www.data.gov.uk/dataset/df223a81-879f-46a0-88db-4417069ea8c4/local-authority-maintained-trees).

Selected variables in these datasets have been chosen to address the research questions or objectives. Pre-processing of these datasets was done to select those data records relating to the study area and limit variables to those necessary for the analysis:

Mastermap greenspaces subset:

- Geolocation (polygon)
- Primary land use function (18 categories)
- Parcel area (m^2)

2021 Population Census:

- Geolocation – centroid (point)
- Usual residents count (in unit postcode)
- Households count (in unit postcode)

Local authority-maintained trees:

- Geolocation (point)
- Common name

The data records for Greenwich were loaded into GIS software (ArcPro), and the non-spatial variables for the greenspace parcels were read into SPSS™ for statistical analysis. There were 4031 unit postcode centroids in the study area, and the Euclidean distance to the nearest greenspace polygon over 100 m^2 in area and in certain categories (see below) was determined in the GIS. The new distance variable together with the primary land use function were then added as additional variables to the population dataset. This subset of larger polygons in the following categories were used when investigating the third research question:

- Allotments or community-growing spaces
- Amenity – residential or business
- Amenity – transport
- Bowling green
- Camping or caravan park
- Cemetery
- Golf course
- Institutional grounds
- Natural (mostly inland water or beach areas on the River Thames)
- Other sports facility
- Play space
- Playing fields
- Public park or garden
- Religious grounds
- School grounds

These land use categories were selected on the basis that they are either accessible to the public or are visible because of their size and thus more likely to contribute positively to physical and mental health and well-being. Private gardens were excluded, despite being the most numerous type of greenspace (see below) because the benefits potentially deriving from the large majority of these were more likely to be for the residents living in the property associated with the private garden.

12.5.3.3 Results

There were a total 142,684 greenspace land parcels distributed across 18 primary function categories and by far the most common were private gardens (83.6%), followed by amenity land associated with residential or business premises (7.1%). The least numerous were parcels categorised as allotments or community-growing spaces, bowling greens, camping or caravan parks and tennis courts together with those where the land use was changing (all less than 100 occurrences). The primary functions of greenspaces with local authority-maintained trees ($N = 776$) showed a rather different pattern with amenity spaces connected with transport or with residential or business premises the most common (73.2 and 17.3%, respectively). The greenspaces with local authority-maintained trees were in eight of the functional categories, and these were the focus of the first research objective or question. Differences in the area of these different categories of greenspace were investigated by applying the Mann–Whitney test and converting the test statistics to Z. The results, tabulated in Table 12.21, show statistically significant test results at the 0.05 level for both categories of amenity land, institutional grounds and private gardens. The difference in area for other sports facility land was very close to being statistically significant (0.051). There were

Table 12.21 Application of Mann–Whitney test for a significant difference between the area of greenspace land parcels in Greenwich categorised according to the presence or absence of local authority-maintained trees.

	Greenspace without/ with LA-maintained trees (1)/(2)	N	Mean area (m^2)	Mean Rank	Z test statistic	Probability
Amenity – residential or business	1	10,028	655.59	5055.93	−7.601	<0.001
	2	134	1031.27	6995.18		
Amenity – transport	1	2659	199.02	1509.84	−13.740	<0.001
	2	568	372.87	2101.60		
Golf course	1	253	4581.99	127.25	−0.880	0.379
	2	1	1392.70	192.00		
Institutional grounds	1	2660	405.69	1332.44	−2.522	0.012
	2	8	594.27	2020.38		
Land use changing	1	75	3191.81	38.04	−2.306	0.021
	2	2	29 614.84	773.50		
Other sports facility	1	571	1895.11	286.20	−1.953	0.051
	2	3	19 487.64	758.00		
Private garden	1	119,186	91.63	59 610.70	−5.365	<0.001
	2	58	182.85	301.71		
Public park or garden	1	1976	2002.89	989.20	−0.746	0.46
	2	2	724.05	639.00		

Sources: Local authority maintained trees data used under Open Government Licence (https://www.nationalarchives. gov.uk/doc/open-government-licence/version/3/). Copyright © Crown copyright and database rights 2024 Ordnance Survey (AC0000851941)

very small numbers of land parcels with local authority-maintained trees in the other sports facility, golf course, land undergoing a change of use and public park or garden categories, which raises questions about the reliability of their test results.

The second research question or objective, concerning any difference in the spatial distribution of greenspace land parcels and the usual resident population in 2021, has been investigated using spatial statistics. Getis-Ord Gi hot spot analysis has been applied to enable comparison between the spatial distribution of significant high and low spatial clusters of these variables. Figure 12.7 compares the results of the pair of hot spot analyses. No significant clusters at either the 99 or 95 per cent levels of confidence were revealed for the greenspace polygons across the study area. However, some larger polygons were spatially clustered in different parts of the borough, with substantial areas towards the north-east, centre, south and south east. Arguably the 2021 usually resident population showed spatial divergence with significant hot and cold spots being identified in different parts of the study area. Some cold spots occurred towards the west and central south of the borough with hot spots mainly towards the north-east, although in general not overlapping with the greenspaces' clusters. Outliers have also been identified using local Moran's *I* with respect to the same pair of variables (Figure 12.8). There was a substantial cluster of Low–Low greenspace polygons stretching east–west across the study area, which were gardens, whereas there were several spatial clusters of High–High polygons in various parts of the borough. Low–High outliers were not apparent, but there were some scattered High–Low outliers in a range of locations.

The third research question or objective focuses attention on the straight line distance between the usually resident unit postcode population geolocated to the centroids of these units and the nearest greenspace land parcels in 15 selected functional land use categories (excluding private gardens) potentially of benefit in respect of human physical and mental health and well-being. The first analysis applies the Kruskal–Wallis test, the non-parametric equivalent of ANOVA, to investigate if there was a significant difference between the mean distance, land parcel area and population total across the 15 categories. Table 12.22 summarises the test results and shows the mean ranks and mean values of the three variables. Recall that the test simply reports on whether there is a significant difference overall, although it is realistic to examine the mean values of the variables to gauge the extent of the differences. Postcode centroids with their nearest greenspace over 30 m were closest to areas classified as allotments and community-growing spaces, amenity transport, institutional, religious and school grounds. In contrast, mean distances under 15 m were for caravan and camping parks, cemeteries and golf courses. The mean area of the nearest greenspace parcels to the postcode centroids was unsurprisingly largest for land classified as golf course, playing field and cemetery, whereas the smallest were bowling greens and play spaces. The population totals showed the least variation in their mean across the 15 categories, although the difference was significant at the 0.05 level ($P = 0.028$), and this reflects the way in which addresses are allocated to postcodes and the design of census geography.

The second approach to examining the third research objective applied spatial statistics in the form of Getis-Ord Gi optimised hot spot analysis in two ways: first, to the distance between postcode centroids and their nearest greenspace; second to the population total of the postcodes. There were statistically significant hot and cold spots for both variables, and there is some suggestion in Figure 12.9 that there was a degree of overlap between them. Again, optimised outlier analyses have been carried out and there is a suggestion in Figure 12.10 that High–Low and Low–High outliers for the nearest distance variable were on the fringes of the hot and cold spots. The outliers for usually resident 2021 population were rather more varied in the location, especially in respect of the Low–Low spatial cluster.

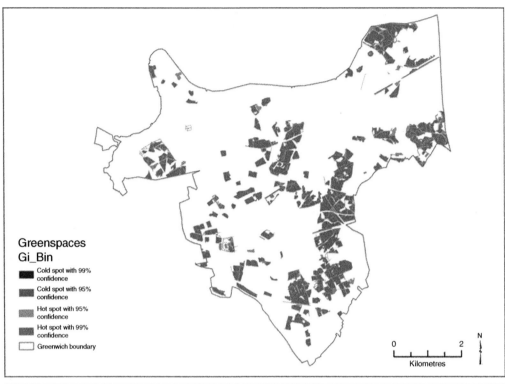

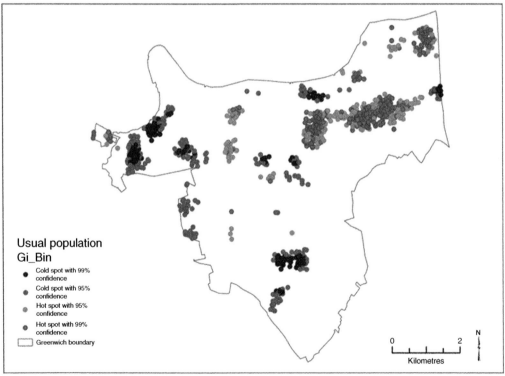

Figure 12.7 Optimised hot spot analysis for area (m^2) of greenspaces and postcode unit total usual population in 2021 in Greenwich.

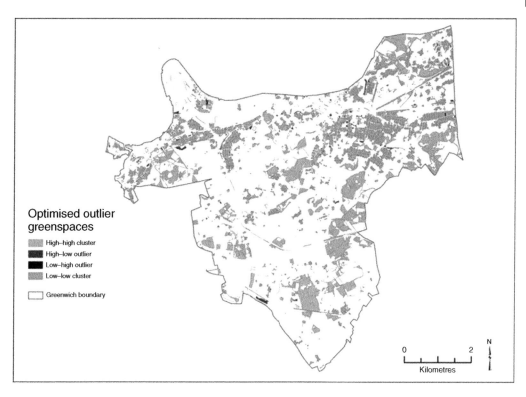

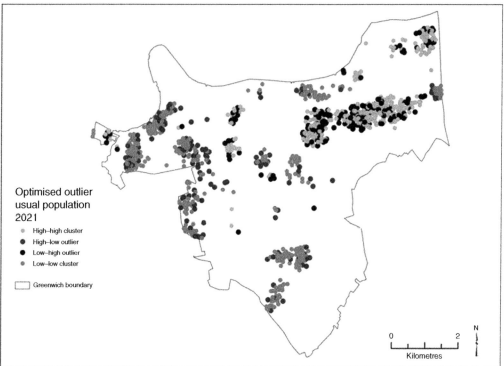

Figure 12.8 Optimised outlier analysis for area (m²) of greenspaces and postcode unit total usual population in 2021 in Greenwich.

Table 12.22 Application of Kruskal–Wallis test for a significant difference between usually resident population, its distance from and area of greenspace land parcels in Greenwich categorised according to different functional types.

	N of postcodes	Mean	Mean Rank	Kruskal–Wallis test statistic	Probability
Distance		Distance (m)		207.094	<0.001
	42	33.84	2264.02		
Amenity – residential or business	2162	29.17	1990.64		
Amenity – transport	749	31.63	2178.55		
Bowling green	3	42.61	2785.67		
Caravan or camping park	4	0.49	403.25		
Cemetery	28	14.10	1121.61		
Golf course	23	4.83	597.57		
Institutional grounds	185	38.02	2328.11		
Natural	18	18.15	1595.50		
Other sports facility	55	23.88	1489.16		
Play space	86	24.60	1457.86		
Playing fields	32	20.25	1951.88		
Public park or garden	210	22.20	1497.37		
Religious grounds	142	48.27	2611.06		
School grounds	292	35.53	2145.21		
Area		Area (m^2)		763.653	<0.001
Allotments and community growing spaces	42	4656.20	3163.98		
Amenity – residential or business	2162	2243.52	1951.81		
Amenity – transport	749	515.01	1507.92		
Bowling green	3	312.34	1677.00		
Caravan or camping park	4	2657.15	2814.50		
Cemetery	28	15,233.57	3648.18		
Golf course	23	163,887.79	3688.00		
Institutional grounds	185	1617.66	1904.62		
Natural	18	11,907.16	2888.89		
Other sports facility	55	13,871.58	3652.13		
Play space	86	364.94	3065.07		
Playing fields	32	22,183.75	1599.53		
Public park or garden	210	9047.03	3164.40		
Religious grounds	142	476.51	1307.23		
School grounds	292	2227.63	2297.76		

Table 12.22 (Continued)

	N of postcodes	Mean	Mean Rank	Kruskal–Wallis test statistic	Probability
Population		Persons (integer)		25.750	0.028
Allotments and community growing spaces	42	67	2222.76		
Amenity – residential or business	2162	63	2015.04		
Amenity – transport	749	63	2054.18		
Bowling green	3	51	1749.33		
Caravan or camping park	4	31	1045.50		
Cemetery	28	66	2163.23		
Golf course	23	46	1552.91		
Institutional grounds	185	59	1850.70		
Natural	18	50	1588.50		
Other sports facility	55	54	1988.72		
Play space	86	77	1758.37		
Playing fields	32	58	2405.47		
Public park or garden	210	66	2114.45		
Religious grounds	142	61	1944.81		
School grounds	292	64	2067.07		

Source: Copyright © Crown copyright and database rights 2024 Ordnance Survey (AC0000851941).

12.5.3.4 Discussion and Limitations

The analysis and results presented here show how students can adapt previously published research to provide a new perspective and add to existing knowledge. The project has also highlighted that non-spatial and spatial statistical techniques can be combined to provide answers to complementary research questions. The decision to focus on a single London borough leads to the project being of a manageable size for students who may have limited time and resources in which to complete their independent study. Most student projects, indeed, most geographically focused projects, involve selecting an area or location for study. The selection of a bounded study area, such as Greenwich, inevitably raises questions about the artificial nature of many types of boundaries. Some of the people living in Greenwich in 2021 may well have lived closer to greenspaces lying outside the borough boundary, but such polygons were not considered in the analysis. Given that the datasets used were readily available and, in the case of students or lecturers based in UK educational institutions are generally freely accessible, it would be straightforward to undertake the same types of analyses in other London boroughs for comparative purposes. Some of the datasets are freely available under open licence arrangements and could therefore be used outside the UK, OpenStreetMap (OSM) data could provide suitable greenspace data, or equally other countries may have similar data sources available to those used here. The following self-assessment is designed to encourage readers to reflect on ways in which this project could be extended and developed in new directions.

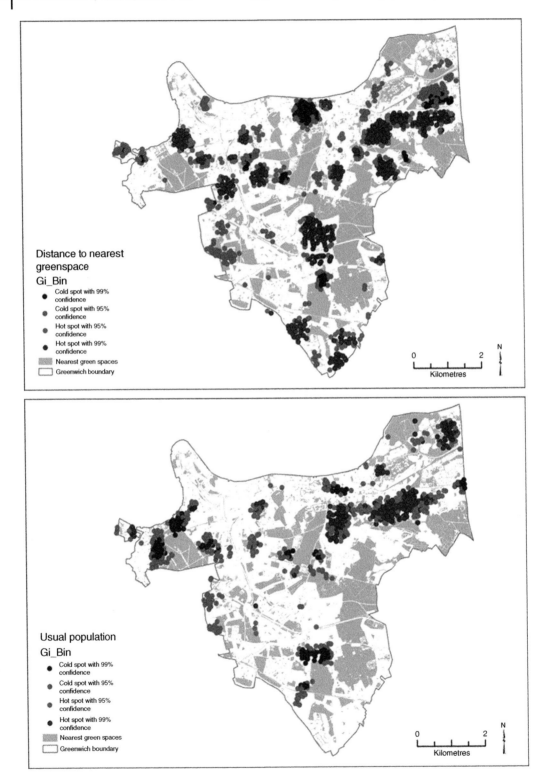

Figure 12.9 Optimised hot spot analysis for distance to nearest greenspace in 15 selected functional land use categories and postcode unit total usual population in 2021 in Greenwich.

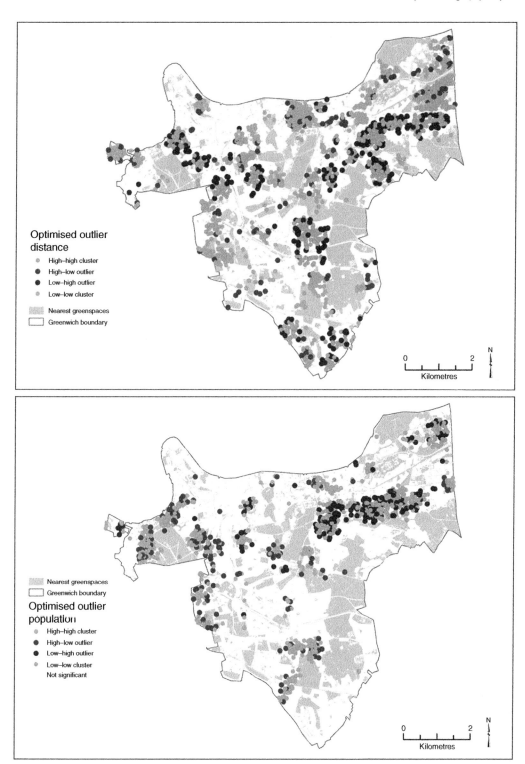

Figure 12.10 Optimised outlier analysis for distance to nearest greenspace in 15 selected functional land use categories and postcode unit total usual population in 2021 in Greenwich.

There are a number of ways in which this project could be extended. Reflect on the following suggestions, and outline any additional datasets that might be needed and/or any further processing of those used here.

- Instead of investigating the total population, consider differentiating between various groups, for example those with low and high levels of deprivation and disadvantage.
- The 15 functional land use categories include those that are less accessible to some groups in the population or may be inaccessible altogether because they are on private land. Perhaps examine further details of the Mastermap greenspaces dataset and decide whether a different selection of land use might provide more robust answers to the research questions.
- The analysis of trees in the borough focused on those that were local authority maintained. Obviously, there are likely to be many more trees in the borough that contribute to its physical environment. Explore the internet to see if there are other datasets available that fill this gap.
- The project has concentrated on analysing secondary sources of data and made some implicit assumptions about people's attitudes towards greenspaces and trees in urban environments. Examine ways in which primary data could be obtained that would add strength to the research.

12.6 Closing Comments

The emphasis in this chapter has been on the practicalities of moving on from the formality of explaining how different non-spatial and spatial statistics work to applying them in research and describing the results produced. Students in many different degree disciplines produce something as part of their studies: architecture students design a new structure, engineers develop a prototype device, computer scientists write software and drama students create an innovative performance. The common feature of these examples is that students use the knowledge, skills and understanding they have gained from reading and practical study to produce something that is their own work. Geography, Earth and Environmental Science students also produce something new when they carry out their own investigation into a topic fitting within the context of these disciplines to produce their own interpretations and conclusions. The dissertation or report written for assessment purposes will normally outline the wider implications of the research findings within the context of the background literature and offer recommendations for addressing the topic in the future. All these examples across the disciplines aim to encourage the development of problem-solving skills that students carry forward into the workplace. Producing these 'somethings' in the safety of your studies seeks to make doing similar things in the workplace, where the consequences may be more significant (in a non-statistical sense), more successful.

The application of quantitative techniques as covered in this text, or qualitative approaches as examined elsewhere (Cope and Elwood, 2009; DeLyser *et al.*, 2009; Flowerdew and Martin, 2005) to investigate data in a consistent and rigorous way, is just one of the skills needed to produce a research dissertation or report. This chapter has focused on four practical ways of helping you to achieve this: first, reviewing and identifying unanswered questions arising from the analyses carried out in the previous chapters; second, examining how researchers have presented and discussed their findings in published journal articles; third, outlining how to carry out your analysis using information technology and software; and fourth, by describing three mini projects of the scope typically undertaken by students and showing how these can develop on previously published research. Lastly, researchers are sometimes challenged and question their results when application of quantitative techniques fails to support their initial ideas and the alternative hypotheses about what they are aiming to discover are rejected. Accepting Null Hypotheses somehow seems like defeat, and hiding such negative findings is not a good academic practice. These negative outcomes leave open the opportunity for you or someone else to try again, and so our knowledge and understanding grow inexorably.

References

Baris, M.E., Sahin, S. and Yazgan, M.E. (2009) The contribution of trees and green spaces to the urban climate: the case of Ankara. *African Journal of Agricultural Research*, **4**(9), 791–800. http://www.academicjournals.org/AJAR.

Bowler, I.R. (1981a) Some characteristics of an innovative form of agricultural marketing. *Area*, **13**(4), 307–314. https://www.jstor.org/stable/20001750.

Bowler, I.R. (1981b) Self-service down on the farm. *Geography*, **66**(2), 147–150. https://www.jstor.org/stable/40570347.

Bowling, B. and Phillips, C. (2007) Disproportionate and discriminatory: reviewing the evidence on police stop and search. *Modern Law Review*, **70**(6), 936–961. DOI: 10.1111/j.1468-2230.2007.00671.x.

Coid, J., Zhang, Y., Zhang, Y., Hu, J., Thomson, L., Bebbington, P. and Bhui, K. (2021) Epidemiology of knife carrying among young British men. *Social Psychiatry and Psychiatric Epidemiology*, **56**, 1555–1563. DOI: 10.1007/s00127-021-02031-x.

Coomber, R. and Moyle, L. (2018) The changing shape of street-level heroin and crack supply in England: commuting, holidaying and cuckooing drug dealers across 'county lines'. *British Journal of Criminology*, **58**, 1323–1342. DOI: 10.1093/bjc/azx068.

Cope, M. and Elwood, S. (2009) *Qualitative GIS: A Mixed Methods Approach*, Sage Publications Ltd., p. 192.

Coppock, J.T. (1964) Crop, livestock and enterprise combinations in England and Wales. *Economic Geography*, **40**(10), 65–81. DOI: 10.2307/142174.

Coppock, J.T. (1971) *An Agricultural Geography of Great Britain*, Cambridge, CABI.

Dallimer, M., Stringer, L.C., Orchard, S.E., Osano, P., Njoroge, G., Wen, C. and Gicheru, P. (2019) Who uses sustainable land management practices and what are the costs and benefits? Insights from Kenya. *Land Degradation and Development*, **29**, 2822–2835. DOI: 10.1002/ldr.3001.

DeLyser, D., Herbert, S., Aitken, S., Crang, M. and McDowell, L. (2009) *The Sage Handbook of Qualitative Geography*, Sage Publications, Ltd., p. 448.

Flowerdew, R. and Martin, D. (2005) *Methods in Human Geography: A Guide for Students Doing a Research Project*, Routledge, p. 392.

General Register Office (1942) Destruction of 1931 Census Records. The National Archives, reference RG20/109/E.66/02.

Hinshaw, S., Wohl, E., Burnett, J.D. and Wondzell, S. (2022) Development of a geomorphic monitoring strategy for stage 0 restoration in the South Fork McKenzie River, Oregon, USA. *Earth Surface Processes and Lanforms*, **47**, 1937–1951. DOI: 10.1002/esp.5356.

Krishnaswamy, J., Richter, D.D., Halpin, P.N. and Hofmockel, M.S. (2001) Spatial patterns of suspended sediment yields in a humid tropical watershed in Costa Rica. *Hydrological Processes*, **15**, 2237–2257. DOI: 10.1002/hyp.230.

Krisjane, Z., Berzins, M., Krumins, J., Apsite-Berina, E. and Balode, S. (2023) Uneven geographies: ageing and population dynamics in Latvia. *Regional Science Policy and Practice*, **15**(4), 893–908. DOI: 10.1111/rsp3.12648.

Kuemmerle, T., Damm, A. and Hostert, P. (2008) A method to detect and correct single-ban missing pixels in Landsat TM and ETM+ data. *Computers and Geosciences*, **34**(5), 445–455. DOI: 10.1016/j.cageo.2007.05.016.

Lautenschlager, R. (2020) Surveillance Hot Spots: The Geography of Stop and Frisk in Nine U.S. Cities. PhD thesis, The University of Miami. https://scholarship.miami.edu/esploro/outputs/doctoral/Surveillance-Hot-Spots-The-Geography-of/991031453886402976/filesAndLinks?index=0

Marsden, T., Banks, J. and Bristow, G. (2000) Food supply chain approaches: exploring their role in rural development. *Sociologia Ruralis*, **40**(4), 424–438. DOI: 10.1111/1467-9523.00158.

Mather, P.M. (1991) *Computer Applications in Geography*, Chichester, Wiley.

Meng, Y. (2017) Profiling minorities: police stop and search practices in Toronto, Canada. *Journal of Studies and Research in Human Geography*, **11**(1), 6–23. DOI: 10.5719/hgeo.2017.111.1.

Moffat, A.J., Ambrose-Oji, B., Clarke, T.-K., O'Brien, L. and Doick, K.J. (2024) Public attitudes to urban trees in Great Britain in the early 2020s. *Urban Forestry and Urban Greening*, **19**, 1–15. DOI: 10.1016/j.ufug.2023.128177.

Morris, C. (2005) Negotiating the boundary between state-led and farmer approaches to knowing nature: an analysis of UK agri-environment schemes. *Geoforum*, **37**(1), 113–127. DOI: 10/1016/j.geoforum.2005.01.003.

NASA (n.d.) Landsat missions: data loss. https://www.usgs.gov/landsat-missions/data-loss.

Salmond, J.A., Tadaki, M., Vardoulakis, S., Arbuthnott, K., Coutts, A. Demuzere, M., Dirks, K.N., Heaviside, C., Lim, S., Macintyre, H., McInnes, R.N. and Wheeler, B.W. (2014) Health and climate related ecosystem services provided by street trees in the urban environment. 11th International Conference on Urban Health, Manchester, UK. 6 March 2014.

Selmia, W., Webera, C., Rivièreb, E., Blonda, N., Mehdia, L. and Nowak, D. (2016) Air pollution removal by trees in public green spaces in Strasbourg city, France. *Urban Forestry and Greening*, **17**, 192–201. DOI: 10.1016/j.ufug.2016.04.010.

Shiner, M., Carre, Z., Delsol, R. and Eastwood, N. (2018) The Colour of Injustice: 'Race', Drugs and Law Enforcement in England and Wales – A Briefing Paper. StopWatch, Release, London School of Economics. https://www.release.org.uk/sites/default/files/pdf/publications/The%20Colour%20of%20Injustice.pdf.

Suss, J.H. and Oliveira, T.R. (2023) Economic inequality and the spatial distribution of stop and search: evidence from London. *The British Journal of Criminology*, **63**, 828–847. DOI: 10.1093/bjc/azac069.

Taylor, M. (2011) Year 11 geography teachers' response to the Darfield earthquake. *New Zealand Geographer*, **67**, 190–198. DOI: 10.1111/j.1745-7939.2011.01211.x.

Turner, J.B. and Cavender, N. (2019) The benefits of trees for liveable and sustainable communities. *Plants, People, Planet*, **1**, 323–335. DOI: 10.1002/ppp3.39.

United Nations 2018 68% of the world population projected to live in urban areas by 2050 Accessed: 23 April 2024 https://www.un.org/development/desa/en/news/population/2018-revision-of-world-urbanization-prospects.html.

Von Thaden, J., Badillo-Montaño, R., Lira-Noriega, A., García-Ramírez, A., Benítez, G., Equihua, M., Looker, N. and Pérez-Maqueo, O. (2021) Contributions of green spaces and isolated trees to landscape connectivity in an urban landscape. *Urban Forestry and Greening*, **64**, 127277. DOI: 10.1016/j.ufug.2021.127277.

Watts, D.C.H., Ilbery, B. and Maye, D. (2008) Making reconnections in agro-food geography: alternative systems of food provision, in *The Rural* (ed. R. J. C. Munton), London, Routledge.

Weimer, J. (1978) Pick your own and roadside stands. Who's buying and why? *Food Review/National Food Review*, **5**(1), 23–25.

Wolf, K.L., Lam, S.T., McKeen, J.K., Richardson, G.R.A., van den Bosch, M. and Bardekjian, A.C. (2020) Urban trees and human health: a scoping review. *International Journal of Environmental Research and Public Health*, **17**, 4371. DOI: 10.3390/ijerph17124371.

Glossary

A Posteriori **Probability** A probability that takes into account additional 'after the event' information.

A Priori **Probability** Probability determined from past evidence or theory.

Acceptance Region The section of the range of probability associated with acceptance of the null hypothesis in hypothesis testing (e.g. $1.00->0.05$).

Alternative Hypothesis The opposite of the Null Hypothesis.

Analysis of Variance (ANOVA) A statistical procedure used to examine whether the difference between three or more sample means indicates whether they have come from the same or different populations.

Association Variability of one variable can to some degree be accounted for by the other variable.

Attribute A unit of categorical data relating to an observation. Also refers to information about features or elements of a spatial database.

Autocorrelation Correlation between observations that are separated by a fixed time interval or unit of distance.

Average Term commonly, but inaccurately, used to refer to the arithmetic mean, median or mode.

Bias Systematic deviation from a true value.

Bimodal Describes a variable with two modes.

Binomial Distribution A probability distribution relating to a series of random events or trials each of which has two outcomes with known probabilities.

Bivariate Analysis where there are two variables of interest.

Bivariate Normal Distribution A joint probability distribution between two variables which follow the normal distribution and are completely uncorrelated.

Canonical Correlation Analysis A method of correlation analysis that summarises the relationship between two groups of variables.

Categorical Data Data relating to the classification of observations into categories or classes.

Categorical Data Analysis A collection of statistical techniques used to analyse categorical data.

Census The collection of data about all members of a (statistical) population, which should not contain random or systematic errors.

Central Limit Theorem Theorem underpinning certain statistical tests (e.g. Z and t tests) that state the sampling distribution of the mean approaches normality as the sample size increases, irrespective of the probability distribution of the sampled population.

Central Tendency (Measures of) A summary measure describing the typical value of a variable or attribute.

Centroid (with spatial data) The centre of an area, region or polygon and is effectively the 'centre of gravity' in the case of irregular polygons.

Practical Statistics for Geographers and Earth Scientists, Second Edition. Nigel Walford.
© 2025 John Wiley & Sons Ltd. Published 2025 by John Wiley & Sons Ltd.
Companion website: www.wiley.com/go/PracticalStatistics2e

Chi-square Distribution Probability relating to continuous variable that is used in statistical testing with frequency count data.

Chi-square Curve A series of curves which vary according to the degrees of freedom and approximate the probability histogram of the Chi-square statistic.

Chi-square Statistic Statistic measuring the difference between observed and expected frequency counts.

Chi-square Test Statistical test used with discrete frequency count data used to test goodness-of-fit.

Class Interval A set of non-overlapping ranges or bins for grouping entities when producing a histogram or similar descriptive frequency distribution.

Cluster(ed) Sampling The entire population is divided into groups (clusters) before randomly selecting observations from the groups.

Coefficient of Variation Defined as the standard deviation divided by the mean of a variable.

Conditional Probability The probability of one event occurring is conditional upon whether or not a second event occurs. For example, if two outcomes (events) are possible, but they are mutually exclusive (cannot both occur), and event A occurs, the probability of B is necessarily zero (non-occurrence).

Confidence Intervals (Limits) A pair of values or two sets of values that define a zone around an estimated statistic obtained from a sample within which the corresponding population parameter can be expected to lie with a specified level of probability.

Contributed Geographic Information One of the crowdsourced data collection methods in which the people creating the data are not aware of the content being shared.

Convenience sampling Non-probability method of sampling when some members of a population are difficult to find or select and limited numbers are available.

Co-ordinate System A set of coordinate axes with a known metric.

Co-ordinates One, two or three ordinates defined as numeric values that are mutually independent and equal the number of dimensions in the coordinate space.

Correlation Measurement of linear association between variables, which does not imply a causal relationship.

Correlation Coefficient A measure of the strength and direction of the relationship between two attributes or variables that lies within the range -1.0 to $+1.0$.

Correlation Matrix Correlation coefficients between a group of variables that is symmetrical on either side of the diagonal where the correlation of the variables with themselves equals 1.0.

Correlogram A plot of spatially or temporally lagged correlation coefficients.

Covariance The expected value of the result of multiplying the deviations (differences) of two variables from their respective means which indicates the amount of their joint variance.

Critical Value Value of a test statistic at which the Alternative Hypothesis would start to be accepted at a given level of significance.

Cross-sectional Data Data relating to observations at a given point in time.

Crosstabulation Joint frequency distribution of two or more discrete variables.

Crowdsourced Data Crowdsourcing involves gathering data or information from a large number of people usually through the Internet or smartphone.

Curvilinear regression Regression models attempt to fit a curved rather than a straight line between dependent and independent variables.

Data (datum) Recorded counts, measurements or quantities made on observations (entities).

Data Mining Discovery of hidden patterns in large datasets.

Datum The origin, orientation and scale of a coordinate system tying it to the Earth.

Degrees of Freedom The smallest number of data values in a particular situation needed to determine all the data values (e.g. n−1 for a mean).

Dependent Variable A set of one or more variables that functionally depend on another set of one or more independent variables.

Descriptive Statistics A group of statistical techniques used to summarise or describe a data set.

Deviation Difference between a certain data value and a reference datum, commonly the mean, including when the mean is the set of values along a regression line.

Dichotomous An outcome or variable with only two possible values.

Discrete Distribution A probability distribution for data values that are discrete or integer as opposed to continuous.

Discrete Random Variable A random variable with a finite (fixed) number of possible values.

Dispersion (Measures of) A measure that quantitatively describes the spread of a set of data values.

Distance Matrix A matrix of measurements that quantifies the dissimilarity or physical distance between all pairs of observations in a population or sample.

Entity, Element, Observation or Phenomenon One of several terms used to identify the 'things' that are the subject of quantitative analysis. Geographical and environmental entities are usually associated with location above, on or under the Earth's surface.

Error Difference between the estimated and true value.

Estimation Use of information from a sample to guess the value of a population parameter.

Expected Value The value of a statistical measure (e.g. mean or count) expected according to the appropriate random probability distribution.

Explanatory Variable An alternative name for an independent variable.

F Distribution A group of probability distributions with two parameters relating to degrees of freedom of the numerator and the denominator that is used to test for the significance of differences in means and variances between samples.

Feature An abstract representation of a real-world phenomenon or entity.

Geocoding Allocation of coordinates or alphanumeric codes to reference data to geographical locations.

Geographically Weighted Regression A form of regression analysis that fits different regressions at different points across a study area thus weighting by spatial location.

Georeferencing Assignment of coordinates to spatial features tying them to an Earth-based coordinate system.

Geospatial Data Data relating to any real-world feature or phenomenon concerning its location of itself or in relation to other features.

Histogram A chart or graph with a set of class intervals that summarises the frequency distribution of data values. Visually the bins into which the observations are grouped sit along the horizontal axis.

Hypothesis An assertion about one or more attributes or variables.

Hypothesis Testing Statistical procedure for deciding between null and alternative hypotheses in significance tests.

Independent Events (Values) The situation in which the occurrence or non-occurrence of one event, or the measurement of a specific data value for one observation is totally unaffected (independent of) the outcome for any other event or observation.

Independent Variable A set of one or more variables that functionally control another set of one or more independent variables.

Inferential Statistics A group of statistical techniques including confidence intervals and hypothesis tests that seek to discover the reliability of an estimated value or conclusion produced from sample data.

Interdependence The data values for a variable measured in respect of a group of observations are not independent of each other. The data values for some of the entities affect the values of other entities.

Interquartile Range A measure of dispersion that is the difference between 1st and 3rd quartile values.

Interval Scale A measurement scale where 0 does not indicate the variable being measured is absent (e.g. temperature in degrees Celsius), which contrasts with the ratio scale.

Inverse Distance Weighting A form of spatial data smoothing that adjusts the value of each point in an inverse relationship to its distance from the point being estimated.

Join Count Statistics Counts of the number of joins or shared boundaries between area features that have been assigned nominal categorical values.

Judgmental (Purposive) Sampling A non-random method for selecting observations to be included in a sample based on the investigators' judgement that generates data less amenable than random methods to statistical analysis and hypothesis testing.

Kolmogorov–Smirnov Test (one sample) A statistical test that examines whether the cumulative frequency distribution obtained from sample data is significantly different from that expected according to the theoretical probability distribution.

Kolmogorov–Smirnov Test (two sample) A statistical test is used to determine whether two independent samples are probably drawn from the same or different populations by reference to the maximal difference between their cumulative frequency distributions.

Kriging Kriging uses inverse distance weighting and the local spatial structure to predict the values and points and to map short-range variations.

Kruskal-Wallis Test The non-parametric equivalent of ANOVA that tests if three or more independent samples could have come from the same population.

Kurtosis A measure of the height of the tails of a distribution: positive and negative kurtosis values respectively indicate relatively more and less observations in comparison with the normal distribution.

Lag (Spatial or Temporal) A unit of space or time between observations or objects that is used in the analysis of autocorrelation.

Level of Significance Threshold probability that is used to help decide whether to accept or reject the null hypothesis (e.g. 0.05 or 5%).

Likert Scale A style of a survey question that presents respondents with a numerical scale, typically 5, 7 or 9 options, enabling the collection of data on people's behaviour, opinions or views, for example ranging from complete disagreement to complete agreement with a set of statements.

Linear Regression Form of statistical analysis that seeks to find the best fit linear relationship between the dependent variable and independent variable(s).

Local Indicator of Spatial Association A quantity or indicator, such as Local Moran's I, measuring local pockets of positive and negative spatial autocorrelation.

Mann Whitney *U* Test A non-parametric statistical test that examines the difference between medians of an ordinal variable to determine whether two samples or two groups of observations come from one population.

Mean The arithmetic average of a variable for population or sample is a measure of central tendency.

Mean Centre The central location of a set of point features that is at the intersection of their mean X and Y coordinates.

Mean Deviation Average of the absolute deviations from a mean or median.

Measurement Error Difference between observed and true value in the measurement of data is usually divided into random and systematic errors.

Median Middle value that divides an ordered set of data values into two equal halves and is the 50th percentile.

Median Centre A central location in a set of point features that occurs at the intersection of the median values of their X and Y coordinates or which divides the points into four groups of equal size.

Mode Most frequently occurring value in a population or sample.

Moran's *I* A quantitative measure of global spatial autocorrelation.

Multimodal Describes a variable with more than two modes.

Multiple Regression Form of regression analysis where there are two or more independent variables used to explain the dependent variable.

Multivariate Analysis where there are more than two variables of interest.

Nearest Neighbour Index An index is used as a descriptive measure to compare the patterns of different categories of phenomena within the same study area.

Nominal Scale Categories are used to distinguish between observations where there is no difference in magnitude between one category and another.

Non-parametric Tests Form inferential statistics that make only limited assumptions about the population parameters.

Non-probability Sample A sample selected without using a random procedure.

Normal Curve and Distribution Bell-shaped, symmetrical and single peaked probability density curve with tails extending to plus and minus infinity.

Normality Property of a random variable that conforms to the normal distribution.

Null Hypothesis Deliberately cautious hypothesis that in general terms asserts that a difference between a sample and population in respect of a particular statistic (e.g. mean) has arisen through chance.

Observed Value The value of a statistical measure (e.g. mean or count) obtained from sample data.

One-tailed (Sided) Test Statistical test in which there is good reason to believe that the difference between the sample and population will be in a specific direction (i.e. more or less) in which case the probability is halved.

Ordinal Scale Data values that occur in an ordered sequence where the difference in magnitude between one observation and another relates to its rank position in the sequence.

Ordinary Least Squares Regression Most common form of ordinary linear regression that uses least squares to determine the regression line. (see also Simple Linear Regression)

Outlier An observation with an unusually high or low data value.

P Value P value is the probability that random variation could have produced a difference from the population parameter as obtained from the observed sample.

Paired-sample Data Measurement of observations on two occasions in respect of the same variable or single observations that can be split into two parts.

Parameter A numerical value that describes a characteristic of a probability distribution or population.

Parametric Tests Type of inferential statistics that make stringent assumptions about the population parameters.

Pearson's Correlation Coefficient A form of correlation analysis where parametric assumptions apply.

Percentile The percentage of data values that are less than or equal to a particular point in the percentage scale from 1 to 100 (e.g. the data value at which 40% of all values are less this one is 40th percentile).

Poisson Distribution A probability distribution where there is a discrete number of outcomes for each event or measurement.

Polygon A representation of area features.

Polynomial A function, for example a regression equation containing N terms for the independent variable raised to the power of N.

Population A complete set of objects or entities of the same nature (e.g. rivers, human beings, etc.).

Precision The degree of exactness in the measurement of data.

Probability A number between zero (0.0) and unity (1.0) (or 0.0 and 100.0 per cent) that indicates an event (outcome) will occur. An event or outcome is usually specified as a numerical quantity.

Probability Sample Observations for inclusion in a sample are selected using a random procedure giving all members of the population a known chance of being chosen without bias.

Purposive Sampling See judgmental sampling.

Quadrat A framework of usually regular, contiguous, square units superimposed on a study and used as a means of counting and testing the randomness of spatial patterns.

Qualitative Variable (Attribute) A variable whose values are collected as descriptive adjectives (e.g. employed/unemployed, fluvial/pluvial flooding, owner occupied/rented) that can be used as nominal categories.

Quartile The 1st, 2nd and 3rd quartiles correspond with the 25th, 50th and 75th percentiles.

Quota Sampling Non-probability approach similar to convenience sampling, but additionally seeks to select potential participants in different subgroups of the population. The researcher chooses individuals or entities to satisfy a predetermined quota in each subgroup (stratum).

Random Error The part of the overall error that varies randomly from the measurement of one observation to another.

Random Sampling Sometimes called simple random sampling, this involves selecting objects or entities from a population such that each has an equal chance of being chosen.

Range A measure of dispersion that is the difference between the maximum and minimum data values.

Rank Correlation Coefficient A method of correlation analysis that is less demanding than Pearson's and involves ranking the two variables of interest and then calculating the coefficient from the rank scores.

Raster Representation of spatial data as values in a matrix of numbered rows and columns that can be related to a co-ordinate system in the case of geospatial data.

Ratio Scale A scale of measurement with an absolute zero where any two pairs of values that are a certain distance apart are separated by the same degree of magnitude.

Raw Data Qualitative or quantitative data (images, numbers, text, words, etc.) that have been captured or collected and stored digitally, but which have not undergone any processing to turn them into information or statistics.

Regression (Analysis) A type of statistical analysis that seeks the best fit mathematical equation or model between dependent and independent variable(s).

Regression Line Line that best fits a scatter of data points drawn using the intercept and slope parameters in the regression equation.

Relative Frequency Distribution Summary frequency distribution showing relative percentage or proportion of observations in discrete categories.

Residuals The differences between observed and predicted values commonly used in regression analysis.

Root Mean Square (RMS) The square root of the mean of the squares of a set of data values.

R-squared The coefficient of determination acts as a measure of the goodness of fit in regression analysis and is thought of as the proportion or percentage of the total variance accounted for by the independent variable(s).

Sample A subset of objects or entities selected by some means from a population and intended to encapsulate the latter's characteristics.

Sample Size The number of observations in a sample.

Sample Space The set of all possible outcomes from an experiment.

Sample Survey A survey of a sample of objects or entities from a population.

Sampling The process of selecting a sample of objects or entities from a population.

Sampling Distribution Frequency distribution of a series of summary statistics (e.g. means) calculated from samples selected from one population.

Sampling Frame The set of objects or entities in a population that are sampled.

Scatter Plots (Graphs) A visual representation of the joint distribution of observed data values of two (occasionally three) variables for a population or sample that uses symbols to show the location of the data points on X, Y and possibly Z planes.

Simple Linear Regression Simplest form of least squares regression with two parameters (intercept and slope).

Skewness A measure of the symmetry (or lack of it) of a probability distribution.

Smoothing A way of reducing noise in spatial or temporal datasets.

Snowball sampling Non-probability sampling method that starts with an initial group of conveniently selected individuals and then proceeds to extend outwards from these early participants by obtaining contact details to enable the sample size to be grown.

Spline The polynomial regression lines obtained for discrete groups of points that are tied together to produce a smooth curve following the overall surface in an alternative method of fitting a surface.

Standard Deviation A measure of dispersion that is the square root of the variance and used with interval and ratio scale measurements.

Standard Distance A measure of the dispersion of data points around their mean centre in two-dimensional space.

Standard Error A measure of the variability of a sample statistic as it varies from one sample to another.

Standardisation (Standardise) Transformation of a variable or set of data values such that its mean becomes zero and its standard deviation is 1.0 by subtracting the mean from each data value and dividing the results by the standard deviation of the complete set of values.

Standard Normal Distribution A version of the normal distribution where the mean is zero and the standard deviation is 1.0.

Standard Score The number of units of the standard deviation that an observation is below or above the mean.

Statistic Either a number used as a measurement or a quantity calculated from sample data.

Statistical Test A procedure used in hypothesis testing to evaluate the null and alternative hypotheses.

Stratified Random Sampling A method of selecting objects or entities for inclusion in a sample that involves separating the population into homogenous strata on the basis of some criterion (e.g. mode of travel to work) before randomly sample from each stratum either proportionately or disproportionately to the proportion of the total entities in each one.

(Student's) t-distribution A bell-shaped, symmetrical and single-peaked continuous probability distribution that approaches the normal distribution as the degrees of freedom increase.

Subset A group or portion of the original collection of observations, which may be subject its own analysis or visualisation.

Systematic Error A regular repeated error that occurs when measuring data about observations.

Systematic Sampling A method of selecting objects or entities from a population for inclusion in some regular sequence (e.g. every 5th person).

Test Statistic A numerical quantity computed from a set of data values according to the specification of a statistical test. The test statistic is used to test or challenge the Null and Alternative Hypotheses in relation to the level of significance, which is the chance that the Null Hypothesis is true.

Time Series Data that includes measurements of the same variable at regular intervals over time.

Time Series Analysis Group of statistical techniques used to analyse time series data.

Trend Surface Analysis The 'best fitting' of an equation to a set of data points in order to produce a surface, for example by means of a polynomial regression equation.

t-statistic Test statistic whose probabilities are given by the *t*-distribution.

t-test A group of statistical tests that use the *t*-statistic and *t*-distribution which are in practice rather less demanding in their assumptions than those using the normal distribution.

Two-tailed (sided) Test Most practical applications of statistical testing are two-tailed, since there is no good reason to argue that the difference being tested should be in one direction or the other (i.e. positive or negative) and the probability is divided equally between both tails of the distribution.

Type I Error A Type I error is made when rejecting a null hypothesis that is true and so concluding that an outcome is statistically significant when it is not.

Type II Error A Type II error is committed when a null hypothesis is accepted that is false and so inadvertently failing to conclude that a difference is statistically significant.

Univariate Analysis where there is one variable of interest.

Variance A measure of dispersion that is the average squared difference between the mean and the value of each observation in a population or sample with respect to a particular variable measured on the interval or ratio scale.

Variance/Mean Ratio A statistic used to describe the distribution of events in time or space that can be examined using the Poisson distribution.

Variate An alternative term for variable or attribute, although sometime reserved for things that are numerical measurements (i.e. not attribute categories).

Variogram (Semi-variogram) Quantification of spatial correlation by means of a function.

Vector Representation of the extent of geographic features in a coordinate system by means of geometric primitives (e.g. point, curve and surface).

Volunteered Geographic Information One of the crowdsourced data collection methods where the data are deliberately created and shared.

Weighted Mean Centre The mean centre of a set of spatial points that is weighted according to the value or amount of a characteristic at each location.

Weights Matrix A matrix of numbers between all spatial features in a set where the numbers, e.g. 0 and 1, indicate the contiguity, adjacency or neighbourliness of each pair of features.

Wilcoxon Signed Ranks Test A non-parametric equivalent of the paired sample *t*-test that tests a null hypothesis that difference between the signed ranks is due to chance.

Wilk's Lambda A multivariate test of difference in mean for three or more samples (groups).

Z distribution Probability distribution that is equivalent to the standardised normal distribution providing the probabilities used in the *Z* test.

Z Statistic Test statistic whose probabilities are given by the *Z* distribution.

Z Test A parametric statistical test that uses the *Z* statistic and *Z* distribution which makes relatively demanding assumptions about the normality of the variable of interest.

Z-score A Z-score standardises the value of a variable for an observation in terms of the number of standard deviations that it is either above or below the sample or population mean.

Index

Practical Statistics for Geographers and Earth Scientists, Second Edition. Nigel Walford.
© 2025 John Wiley & Sons Ltd. Published 2025 by John Wiley & Sons Ltd.
Companion website: www.wiley.com/go/PracticalStatistics2e

Printed and bound by CPI Group (UK) Ltd, Croydon, CR0 4YY

02/01/2025

14620080-0001